Monographs in Theoretical Computer Science
An EATCS Series

Editors: J. Hromkovič G. Rozenberg A. Salomaa
Founding Editors: W. Brauer G. Rozenberg A. Salomaa

On behalf of the European Association
for Theoretical Computer Science (EATCS)

T0191596

For further volumes:
http://www.springer.com/series/776

Sergey Kitaev

Patterns in Permutations and Words

 Springer

Dr. Sergey Kitaev
Department of Computer and Information Sciences
University of Strathclyde
Livingstone Tower
26 Richmond Street
Glasgow, G1 1XH
United Kingdom
sergey.kitaev@cis.strath.ac.uk

Series Editors
Prof. Dr. Juraj Hromkovič
ETH Zentrum
Department of Computer Science
Swiss Federal Institute of Technology
8092 Zürich, Switzerland
juraj.hromkovic@inf.ethz.ch

Prof. Dr. Grzegorz Rozenberg
Leiden Institute of Advanced
Computer Science
University of Leiden
Niels Bohrweg 1
2333 CA Leiden, The Netherlands
rozenber@liacs.nl

Prof. Dr. Arto Salomaa
Turku Centre of Computer Science
Lemminkäisenkatu 14 A
20520 Turku, Finland
asalomaa@utu.fi

ISSN 1431-2654
ISBN 978-3-642-26987-5   ISBN 978-3-642-17333-2 (eBook)
DOI 10.1007/978-3-642-17333-2
Springer Heidelberg Dordrecht London New York

ACM Computing Classification (1998): G.2, F.2

Printed on acid-free paper

Springer is part of Springer Science+Business Media (www.springer.com)

Foreword

The study of patterns in permutations and words has a long history. For example, starting in the 1880s, MacMahon began his study of combinatorial objects. In particular, he gave generating functions for the distribution of inversions in permutations and words which in modern day terminology correspond to occurrences of the pattern 21. The study of descents or consecutive occurrences of the pattern 21 goes back even further in that Leonhard Euler in 1749 introduced polynomials of the form $\sum_{k=0}^{n}(k+1)^{n}t^{k} = \frac{A_n(t)}{(1-t)^{n+1}}$ to help in the evaluation of the Dirichlet η-function $\eta(s) = \sum_{n=1}^{\infty}\frac{(-1)^{n-1}}{n^s}$. The polynomial $A_n(t)$, which is now known as the Eulerian polynomial, is the generating function of the number of descents over the symmetric group S_n, i.e. $A_n(t) = \sum_{\sigma \in S_n} x^{\text{des}(\sigma)}$. The Eulerian polynomials were studied from a combinatorial point of view by Foata and Schützenberger in 1970. The study of permutations and words that have regular patterns also has a long history. For example, André in 1879 showed that the exponential generating function of up-down permutations is $\sec(x) + \tan(x)$ and Carlitz in 1973 found the generating functions for up-down words. These and other works led to the study of permutation statistics which became an active area of research starting in the 1970s and remains an active area of research up to this day.

The origin of the modern day study of patterns in words can be traced back to papers by Rotem, Rogers, and Knuth in the 1970s and early 1980s. The first systematic study of permutation patterns was not undertaken until the paper by Simion and Schmidt which appeared in 1985. Today the study of patterns in permutations and words is a very active field as is evidenced by the exceedingly long bibliography for this volume. At this point, there is a rich collection of tools that has been developed to study a variety of problems such as how to count the number of permutations and words that avoid a given pattern or collection of patterns or how to find the generating function for the number of occurrences of a pattern or collection of patterns. The notion of patterns in permutations and words has proved to be a useful language in a variety of seemingly unrelated problems including the theory of Kazhdan-Lusztig polynomials, singularities of Schubert varieties, Chebyshev polynomials, rook polynomials for Ferrers boards, and various sorting algorithms including sorting stacks and sortable permutations. In addition, the study of patterns in permutations and words also arises in computational biology and theoretical physics.

The present book is a welcome addition to the literature on permutation patterns. In this book, Sergey Kitaev not only systematizes and organizes a vast number of results on patterns in permutations and words that have appeared in the literature, but he also describes the connections of the study of patterns in permutations and words with the various fields mentioned in the previous paragraph. The study of patterns in permutations and words also leads to a large number of interesting questions in bijective combinatorics. For example, we say that two permutations σ and τ are Wilf-equivalent if for all n, the number of permutations of S_n which avoid σ, i.e. which have no occurrences of the pattern σ, is the same as the number of permutations of S_n which avoid τ. There is a vast literature on classifying which pairs of permutations are Wilf-equivalent and such a classification naturally leads to questions of how to find bijective proofs of such equivalences. Similarly, there is a large literature which shows that various other combinatorial objects such as partially ordered sets, lattice paths, planar maps, and labeled trees are equinumerous with certain classes of permutations which avoid a given pattern. This book provides a valuable introduction to the study of bijective questions that arise in the study of patterns in permutations and words. Another novel feature of this book is that Kitaev puts particular emphasis on various generalizations of classical patterns in permutations and words such as partially ordered patterns, barred patterns, vincular patterns, bivincular patterns, and place-difference-value patterns. This is the first book that provides a comprehensive look at such generalizations of classical patterns. Thus, this book provides a welcome and valuable addition to the study of patterns in permutations and words. I believe that it will provide a valuable introduction for graduate students and researchers who want to pursue research in the study of patterns in permutations and words as well as a valuable reference for experts in the field.

Jeffrey B. Remmel

San Diego, February 20, 2011

Preface

This book deals with occurrences of patterns in permutations and, more generally, in words with repeated letters. An occurrence of a pattern in a word is a subsequence of the word whose letters appear in the same order of size as the letters in the pattern. As a simple example, an occurrence of the pattern 321 is simply a decreasing subsequence of length 3, such as the one formed by the letters 632 in the permutation 4631725.

The patterns we deal with have been studied sporadically, often implicitly, for over a century, but in the last 20 years or so, the area has grown dramatically, resulting in many hundreds of published papers and in the annual conference "Permutation Patterns" organized for the first time at the University of Otago in Dunedin, New Zealand, in 2003. The introduction of the area of (permutation) patterns is traditionally attributed to Donald Knuth and in particular to exercises on pages 242–243 in his first volume of "The Art of Computer Programming" [540] in 1968, while the first systematic study of pattern avoidance was done by Rodica Simion and Frank W. Schmidt [721] in 1985.

There are several survey papers on this subject [136, 183, 515, 512, 352, 749, 751, 805] and the entire Volume 9 (2) of the *Electronic Journal of Combinatorics* is devoted to it. Moreover, there are proceedings of the Permutation Patterns 2007 conference, edited by Steve Linton, Nik Ruškuc and Vincent Vatter, and published in the London Mathematical Society Lecture Note Series, Cambridge University Press (vol. 376). The book "A walk through combinatorics" [141] by Miklós Bóna contains material on permutation patterns, while the book "Combinatorics of permutations" [137] by the same author provides a comprehensive and accessible introduction to so-called *classical permutation patterns*. Finally, there is the book "Combinatorics of compositions and words" [461] by Silvia Heubach and Toufik Mansour directly related to the subject.

However, the area has grown far beyond the content of the books mentioned above. The notion of a "pattern" has been extended in many different ways, often bringing new connections to other disciplines. Even such an important and relatively well-studied class of patterns as "generalized patterns" (called "vincular patterns" in this book) introduced by Eric Babson and Einar Steingrímsson in 2000, received almost no attention in the books by Bóna and was not considered for permutations in the book by Heubach and Mansour. One of the goals of this book is to introduce a new notation for vincular patterns, because it has been somewhat confusing so far.

The two main objectives of this book are the following:

1. **To provide a motivation** to study patterns by demonstrating as many links to other areas of research as possible — Chapters 2 and 3 are entirely dedicated to this, and many other parts of the book contain such motivating material. These links provide connections to different combinatorial structures appearing in the literature, but also references to computer science (*sorting, generating,* and *complexity issues*), statistical mechanics (*Partially Asymmetric Simple Exclusion Process*), and computational biology (*whole genome duplication-random loss model*).

2. **To be comprehensive** in mentioning existing publications, and in sketching new research directions and trends in the field. In particular, we hope to have gathered, in our bibliography, almost all published papers related to the area (the book contains more than 800 references). Of course, there is a price to pay for being comprehensive, while keeping the size of the book within reasonable limits — we give very few proofs and there are no exercises. However, references are given to all results mentioned, so that the interested reader should have no problems finding the details.

While the book mentions several original results from papers in preparation, a couple of important topics were not covered the way they deserve to be. These topics include, but are not limited to, the intensively studied theory of *pattern classes* (see Remark 6.1.65 for a collection of references on it) and *enumeration schemes* (see [662, 663, 781, 816]). Moreover, we do not discuss at all a couple of research directions, for example, the results on the Möbius function on posets defined by different notions of pattern containment (see [107, 122, 202, 513, 708, 752]). Except for that, the book is a comprehensive collection of up-to-date results on patterns, most of which will be accessible to a broad audience, from undergraduate students to active researchers in the area of patterns in words and permutations, or adjacent fields.

The book is organized as follows.

- In Chapter 1 we introduce the main classes of patterns of interest in this book (classical patterns, barred patterns, vincular patterns, bivincular patterns, and partially ordered patterns) and also provide examples of typical problems on these patterns. Moreover, we list a bibliography related to each of the pattern classes.

- In Chapters 2 and 3 we provide motivation points to study patterns in words and permutations. They include links to theoretical computer science through several sorting devices, planar maps and relevant objects, Schubert varieties and Kazhdan-Lusztig polynomials, computational biology through the tandem duplication-random loss model, statistical mechanics through the Partially Asymmetric Simple Exclusion Process, the theory of partially ordered sets, classification of Mahonian statistics, bijective combinatorics through encoding combinatorial objects by pattern-restricted permutations, and more.

- In Chapter 4 we present the more than thirty year history of bijections between permutations avoiding any classical pattern trivially equivalent to 321 and any pattern trivially equivalent to 132. We discuss a recent classification of these bijections and a philosophical question on what is a "good" bijection from the point of view of bijective combinatorics. Additionally, this chapter provides a collection of approaches to deal with classical patterns, along with general theorems about Wilf-equivalence of certain classical patterns.

- Chapter 5 contains an overview of results on consecutive patterns, and various approaches to studying them. For example, in this chapter we discuss the symbolic method, the symmetric functions approach, and the cluster and chain methods.

- In Chapters 6 and 7 we collect known (mostly enumerative) results on various patterns other than consecutive ones.

- In Chapter 8 we discuss several topics without a common thread. These topics include simple permutations, pattern matching problems, Gray codes, packing patterns, a link to combinatorics on words, universal cycles, simsun permutations, and games on patterns in permutations.

- In Chapter 9 we present several extensions and generalizations of the study of patterns discussed in the previous chapters.

- Appendix A serves as an easy access to the definitions for most of the permutation/word statistics and number sequences appearing in this book, while in Appendix B we provide a couple of basic algebraic facts, including the notion of Chebyshev polynomials of the second kind.

Acknowledgments

A significant part of the book was written during the author's six month visit to the Polytechnic University of Catalonia, Barcelona, Spain, in Summer-Fall 2010. The visit was supported by a grant from Iceland, Liechtenstein and Norway through the EEA Financial Mechanism, and supported and coordinated by Universidad Complutense de Madrid. I am grateful to Marc Noy and Anna de Mier for their hospitality during my stay in Barcelona.

I am deeply indebted to the following mathematicians who kindly agreed to carefully proofread different parts of the book, providing substantial feedback: Sara Billey, Alexander Burstein, Mark Dukes, Jeff Liese, Aaron Robertson, Christopher Severs, Rebecca Smith, Einar Steingrímsson, Vince Vatter and Lauren Williams.

Special thanks go to Lara Pudwell and Manda Riehl, who each proofread two chapters providing valuable comments, to Phil Watson who went through the entire book giving many useful suggestions, and to Henning Úlfarsson, who not only proofread a section of this book, but also provided technical assistance related to LaTeX issues. Moreover, I would like to thank the following individuals for helpful discussions related to the book: Mike Atkinson, Anders Claesson, Vít Jelínek, Eva Jelínková, Marc Noy and Don Rawlings.

I would like to thank my parents, Polina Kitaeva and Alexander Filin, my grandmother Fedocia Semenova (who is about to turn 93!) and my parents-in-law, Sergey Shuiskii and Nina Shuiskaya, for their constant support. I also want to thank all my friends and colleagues for making my life so interesting.

The final acknowledgements go to my wonderful wife Daria Kitaeva and my son Daniel Kitaev for their love and inspiration. It is to them I dedicate this book.

Sergey Kitaev
Reykjavík, March 11, 2011

Notation

$\lvert A \rvert$	the number of elements in a set A
$\mathrm{asc}(\pi)$	the number of ascents in π
$\mathrm{Av}(P)$	the set of all permutations avoiding each pattern in P
$\mathrm{Av}_\ell(P)$	the set $\cup_{n\geq 0} W_{n,\ell}(P)$
b.g.f.	bivariate generating function
BP	barred pattern
BVP	bivincular pattern
C_n	the n-th Catalan number
$C(x)$	the g.f. $\frac{1-\sqrt{1-4x}}{2x}$ for the Catalan numbers
$c(\pi)$	the complement of a permutation π
$\mathcal{D}_n$	the set of all Dyck paths on $2n$ steps
$\mathrm{des}(\pi)$	the number of descents in π
$\mathbb{E}$	the set of even numbers $\{0, 2, 4, \ldots\}$
e.g.f.	exponential generating function
F_n	the n-th Fibonacci number
$f.g(\pi)$	the composition $f(g(\pi))$
g.f.	generating function
$i(\pi)$	the inverse of a permutation π
$\mathrm{inv}(\pi)$	the number of inversions in a permutation π
$[\ell]$	the set $\{1, 2, \ldots, \ell\}$
$[\ell]^n$	the set of all words of length n over $[\ell]$
$[\ell]^*$	the set $\cup_{n\geq 0}[\ell]^n$
M_n	the n-th Motzkin number
$\mathbb{N}$	the set of natural numbers $\{0, 1, 2, \ldots\}$
n-permutation	permutation of length n
$[n]_q$	$1 + q + q^2 + \cdots + q^{n-1} = \frac{1-q^n}{1-q}$
$[n]_{p,q}$	$\frac{p^n-q^n}{p-q}$
$[n]_q! = [n]!$	$[n]_q[n-1]_q \cdots [1]_q$
$[n]_{p,q}!$	$[n]_{p,q}[n-1]_{p,q} \cdots [1]_{p,q}$
$\begin{bmatrix} n \\ k \end{bmatrix}_q$	$\frac{[n]_q!}{[k]_q![n-k]_q!}$

$\mathbb{O}$	the set of odd numbers $\{1, 3, 5, \ldots\}$
$\mathbb{P}$	the set of positive integers
POP	partially ordered pattern
$\mathbb{Q}$	the set of rational numbers
$\mathbb{R}$	the set of real numbers
$r(\pi)$	the reverse of a permutation/word π
$\mathrm{red}(\pi)$	the reduced form of a permutation π
$\mathcal{S}_n$	the set of all permutations of length n
S_n	the n-th (large) Schröder number
$\mathcal{S}_n(P)$	the set of all n-permutations avoiding each pattern in P
$\mathcal{S}_n^k(P)$	the set of all n-permutations containing k occurrences of patterns in P
$\mathcal{S}_\infty$	the set $\cup_{n \geq 0} \mathcal{S}_n$
$s_n(P)$	$|\mathcal{S}_n(P)|$
$s_n^k(P)$	$|\mathcal{S}_n^k(P)|$
$S(n, k)$	Stirling number of the second kind
$U_n(x)$	the n-th Chebyshev polynomial of the second kind
VP	vincular pattern
$W_{n,\ell}(P)$	the set of all words in $[\ell]^n$ that avoid each pattern in P
$W_{n,\ell}^k(P)$	the set of all words in $[\ell]^n$ that contain exactly k occurrences of patterns in P
$w_{n,\ell}(P)$	$|W_{n,\ell}(P)|$
$w_{n,\ell}^k(P)$	$|W_{n,\ell}^k(P)|$
$\mathbb{Z}$	the set of integers
$\alpha_1 \oplus \alpha_2$	$12[\alpha_1, \alpha_2]$
$\alpha_1 \ominus \alpha_2$	$21[\alpha_1, \alpha_2]$
ε	the empty word/permutation
$p_1 \sim p_2$	p_1 and p_2 are Wilf-equivalent
$p_1 \sim_k p_2$	p_1 and p_2 are k-Wilf-equivalent
$p_1 \sim_* p_2$	p_1 and p_2 are strongly Wilf-equivalent
$\sigma[\alpha_1, \alpha_2, \ldots, \alpha_m]$	the inflation of σ by $\alpha_1, \alpha_2, \ldots, \alpha_m$

Contents

Chapter 1

What is a pattern in a permutation or a word?

We begin with some basic definitions.

Definition 1.0.1. A *word* is a sequence whose symbols (or *letters*) come from a set called an *alphabet*. Alphabets in this book are finite, and the most typical alphabet here is of the form $[\ell] = \{1, 2, \ldots, \ell\}$. Most of the words we deal with in this book are finite as well. We let $[\ell]^n$ denote the set of all words of length n over $[\ell]$ and $[\ell]^* = \cup_{n \geq 0}[\ell]^n$. A word from an ℓ-letter alphabet is often referred to as an "ℓ-*ary word*".

Example 1.0.2. 2411121 and 25554 are words over the alphabet $[5] = \{1, 2, 3, 4, 5\}$, while *abbacca* is a word over the alphabet $\{a, b, c\}$.

$$[3]^2 = \{11, 12, 13, 21, 22, 23, 31, 32, 33\}.$$

It is easy to show that there are k^n different words of length n over a k-letter alphabet.

Definition 1.0.3. A *permutation* of length n (also referred to as an n-permutation) is a one-to-one function from an n-element set to itself. We write permutations as words $\sigma = \sigma_1\sigma_2\cdots\sigma_n$, whose letters are distinct and usually consist of the integers $1, 2, \ldots, n$ (where σ_i is the image of i under σ). This way of writing permutations is often referred to as *one-line notation*. We let $\mathcal{S}_n$ denote the set of all permutations of length n, and $\mathcal{S}_\infty = \cup_{n \geq 0}\mathcal{S}_n$.

Example 1.0.4. 413265 is a 6-permutation, whereas 265, *bca*, and 132 are permutations of length 3. $\mathcal{S}_1 = \{1\}$, $\mathcal{S}_2 = \{12, 21\}$, $\mathcal{S}_3 = \{123, 132, 213, 231, 312, 321\}$, etc.

It is easy to prove that there are $n!$ distinct n-permutations.

Remark 1.0.5. In this book, ε is used to denote the empty word/permutation.

Definition 1.0.6. A *subsequence* of a word or permutation w is the word obtained by striking out some (possibly none) of the letters in w. A *factor* of w is a subsequence consisting of contiguous letters in w; that is, u is a factor of w if there exist (possibly empty) words w_1 and w_2 such that $w = w_1 u w_2$.

Example 1.0.7. If $w = 134214451$, then examples of subsequences of w are 321441, 4141, 3, 1151, and 342, whereas examples of factors are 1342, 34, 21445, and 5.

There are countless notions of (often prohibited or forbidden) patterns in different combinatorial objects, e.g. graphs, matrices, words, permutations, compositions, etc. Even if one considers only permutations and words, one meets many notions of a pattern in these objects in the literature. Roughly speaking, for our purposes, patterns in permutations are permutations with additional restrictions, and patterns in words are certain restricted words (possibly permutations). More precisely, the main focus of this book is a particular notion of a pattern and some related modifications of this notion studied in algebraic combinatorics. These are

- *classical patterns*, introduced for permutations by Knuth [540, pp. 242-243] in 1968 but studied intensively for the first time by Simion and Schmidt [721] in 1985; this notion was extended to words by Burstein [195] in 1998, and generalized for words (by allowing repetitions in patterns) by Burstein and Mansour [208] in 2002; two books [137, 141] by Bóna appeared in 2004 and 2006 which discuss classical patterns for permutations, and the book [461] by Heubach and Mansour appeared in 2010 and discusses these patterns for words;

- *barred patterns*, or *BPs* introduced for permutations by West [799] in 1990;

- *vincular patterns*, or *VPs* (known in the literature as *generalized patterns* or *GPs*) introduced for permutations by Babson and Steingrímsson [64] in 2000; this notion was extended to words by Burstein and Mansour [210] in 2003; the book [461] by Heubach and Mansour discusses vincular patterns for words;

- *bivincular patterns*, or *BVPs* introduced for permutations by Bousquet-Mélou, Claesson, Dukes, and Kitaev [161] in 2010; this notion cannot be extended directly to words;

- *partially ordered patterns*, or *POPs* (also known as *partially ordered generalized patterns* or *POGPs*) introduced for permutations by Kitaev [504] in 2003; this notion was extended to words by Kitaev and Mansour [514] in 2003, and it is discussed for words in the book [461] by Heubach and Mansour.

The goal of this chapter is to introduce the above-mentioned patterns, and the rest of the book will be mostly concerned with results on these patterns, in particular showing relations of these patterns to other combinatorial objects and other disciplines, thus motivating further study. However, Chapter 9 introduces other notions of patterns in permutations and words, as well as on some other relevant objects, and discusses known results on them.

Our approach to introducing the patterns is to provide careful definitions of patterns in permutations together with examples in Sections 1.1–1.5, and then to extend slightly these definitions, whenever it makes sense, to the case of words in Section 1.6.

Remark 1.0.8. One of the most important missions of this book is to promote the expression "vincular patterns" instead of "generalized patterns," and to promote the new notation for vincular patterns instead of the notation existing in the literature and discussed in Remark 1.3.4 below. As a matter of fact, many people have found the original notation for vincular patterns (GPs) inconvenient, and there were constantly talks on changing it. Different options appeared in various discussions, out of which the one presented in Section 1.3 was chosen. Being a compact one, the new notation for vincular patterns is a consistent generalization of that for classical patterns and it allows a consistent extension to the notation for bivincular patterns introduced in Section 1.4. The new notation for (bi)vincular patterns was created in collaboration with Anders Claesson, Vít Jelínek, Eva Jelínková, and Henning Úlfarsson, and it has appeared in a recent preprint [771] by Úlfarsson.

Before we continue, let us introduce the following notions.

Definition 1.0.9. The *reduced form* of a permutation σ on a set $\{j_1, j_2, \ldots, j_r\}$, where $j_1 < j_2 < \cdots < j_r$, is the permutation σ_1 obtained by renaming the letters of the permutation σ so that j_i is renamed i for all $i \in \{1, \ldots, r\}$. In other words, to find the reduced form of a permutation σ on r elements, we replace the i-th smallest letter of σ by i, for $i = 1, 2, \ldots, r$. We let $\mathrm{red}(\sigma)$ denote the reduced form of σ.

Example 1.0.10. $\mathrm{red}(453) = 231$, while $\mathrm{red}(3174) = 2143$.

Definition 1.0.11. In this book, for two functions f and g, the composition $f(g(x))$ is denoted $f.g(x)$ or $f \circ g(x)$.

Definition 1.0.12. The *reverse* of a permutation $\sigma = \sigma_1 \sigma_2 \cdots \sigma_n$ is the permutation $r(\sigma) = \sigma_n \sigma_{n-1} \cdots \sigma_1$. The *complement* $c(\sigma)$ of σ is the permutation $\sigma_1' \sigma_2' \cdots \sigma_n'$ where $\sigma_i' = n + 1 - \sigma_i$. That is, the complement substitutes the largest element of a permutation by the smallest one, the next largest letter by the next smallest one, etc. The *inverse* is the regular group theoretical inverse on permutations; that is, the σ_i-th position of the inverse $i(\sigma)$ is occupied by i. The reverse, complement, and inverse are called the *trivial bijections* from $\mathcal{S}_n$ to itself. One can also discuss compositions of trivial bijections, e.g. $r.c(\sigma)$ is the composition of the complement and reverse applied to σ (the complement is applied first), and $i.r(\sigma)$ is the composition of the reverse and inverse.

Example 1.0.13. One can see that $r(14253) = 35241$, $c(14253) = 52413$, $i(14253) = 13524$, $r.c(14253) = c.r(14253) = 31425$, $i.r(14253) = 53142$, and $i.c.r(14253) = 24135$.

There are (at least) two common way to represent a permutation by a matrix, which is very helpful in many different contexts. Unless otherwise specified, we will normally be using the following definition.

Definition 1.0.14. For a permutation $\pi = \pi_1\pi_2\ldots\pi_n$, its *permutation matrix* is obtained by placing a dot in column i of an $n \times n$ array and row π_i from below.

Example 1.0.15. The permutation matrix corresponding to 63712584 is given in Figure 1.1.

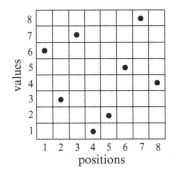

Figure 1.1: The permutation matrix corresponding to the permutation 63712584.

Definition 1.0.16. An *ascent* in a permutation is a letter followed by a larger letter. A *descent* in a permutation is a letter followed by a smaller letter.

Example 1.0.17. In the permutation 342156, the letters 3, 1, and 5 are ascents, whereas the letters 4 and 2 are descents. In Figure 1.1, the ascents are 1, 2, 3, and 5. It is easy to see that in an n-permutation the number of ascents plus the number of descents is always $n - 1$.

Definition 1.0.18. A permutation is *alternating* if it starts with an ascent and then descents and ascents come one after the other. Formally speaking, an alternating permutation $\pi = \pi_1 \pi_2 \cdots \pi_n$ has the property that $\pi_{2i-1} < \pi_{2i} > \pi_{2i+1}$ for $1 \leq i \leq \lfloor n/2 \rfloor$. Alternating permutations are also called *zigzag permutations* and *up-down permutations* in the literature.

Example 1.0.19. 4723165 is an alternating permutation, while 3156724 is not because of the factor 56 and the descent at the first position. The permutation in Figure 1.1 is not alternating.

Definition 1.0.20. A permutation π is *reverse alternating* if $c(\pi)$ is alternating. Reverse alternating permutations are also called *down-up permutations* in the literature.

Example 1.0.21. Examples of reverse alternating permutations are as follows: 31524, 213, and 326451. On the other hand, 3214 and 2413 are not reverse alternating.

Remark 1.0.22. There is sometimes confusion in the literature between alternating and reverse alternating permutations: some authors prefer to switch the definitions, others refer to *both* alternating and reverse alternating permutations as alternating permutations. One should pay attention to the context when dealing with this problem. However, up-down and down-up permutations are always well-defined. It is well known that (reverse) alternating permutations are counted by *Euler (zigzag) numbers* whose e.g.f. is $\tan x + \sec x$ (see [271]).

Definition 1.0.23. An *interval* in a permutation is a factor that contains a set of contiguous values. In particular, every permutation is an interval, and every letter is also an interval.

Example 1.0.24. 423 is an interval in 1642375, but 642 and 123 are not intervals as in the former case the values are not contiguous whereas in the second case we do not have a factor.

Definition 1.0.25. We say that a permutation is *irreducible* (also called *indecomposable*) if no proper initial factor of it is an interval. The *number of components, comp*, is the statistic defined as the number of ways to factor a given permutation $\pi = \sigma\tau$ so that each letter in non-empty σ is smaller than any letter in τ; the empty τ gives 1 such factorization. (See Definition 1.0.32 for the notion of statistic and Table A.1 for a list of permutation statistics appearing in this book.)

Example 1.0.26. comp(3125476) $= 3$ because of the three factorizations: 312|5476, 31254|76, and 3125476|. The permutation 63712584 in Figure 1.1 is irreducible, while the permutation 43125786 in Figure 1.2 has three irreducible components.

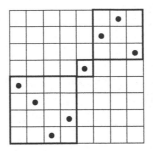

Figure 1.2: The permutation 43125786 has three irreducible components.

Definition 1.0.27. Suppose P is a set of patterns (of types defined in Sections 1.1–1.5 below or in any other relevant sense). We let $\mathcal{S}_n(P)$ denote the set of all n-permutations avoiding (in any specified sense) *each* pattern in P. We also let $s_n(P)$ be the cardinality of $\mathcal{S}_n(P)$, that is, $s_n(P) = |\mathcal{S}_n(P)|$, and $\mathrm{Av}(P) = \mathcal{S}(P) = \cup_{n \geq 0}\mathcal{S}_n(P)$. $\mathrm{Av}(P)$ is usually referred to as a *pattern class*. For the sake of convenience, we omit P's braces thus having, e.g. $\mathcal{S}_5(132, 4312)$ instead of $\mathcal{S}_5(\{132, 4312\})$, $\mathcal{S}_3(312)$ instead of $\mathcal{S}_3(\{312\})$, etc. More generally, we let $\mathcal{S}_n^k(P)$ denote the set of n-permutations containing k occurrences of patterns from P and $\mathcal{S}^k(P) = \cup_{n \geq 0}\mathcal{S}_n^k(P)$. In particular, $\mathcal{S}_n(P) = \mathcal{S}_n^0(P)$. Usually, the cardinality of P is 1 when dealing with $\mathcal{S}_n^k(P)$. We let $s_n^k(P) = |\mathcal{S}_n^k(P)|$.

Remark 1.0.28. We note that in the literature, the notation $\mathrm{Av}(P)$ is used for so-called *permutation classes* defined below (see Definition 2.2.38) rather than for sets of permutations avoiding each pattern from P, which does not have to be a permutation class (see Example 2.2.39). However, no confusion should arise about this, as we will be stating explicitly when we mean classes rather than just sets.

Definition 1.0.29. Similarly to Definition 1.0.27, we define $W_{n,\ell}(P)$ to be the set of all words in $[\ell]^n$ that avoid *each* pattern in P and $\mathrm{Av}_\ell(P) = \cup_{n \geq 0}W_{n,\ell}(P)$. Also, $W_{n,\ell}^k(P)$ is the set of all words in $[\ell]^n$ that contain exactly k occurrences of patterns in P. Finally, we let $w_{n,\ell}(P) = |W_{n,\ell}(P)|$ and $w_{n,\ell}^k(P) = |W_{n,\ell}^k(P)|$.

Definition 1.0.30. In the case of permutations, two patterns p_1 and p_2 are called *Wilf-equivalent*, which is denoted $p_1 \sim p_2$, if $s_n(p_1) = s_n(p_2)$ for all $n \geq 0$. More generally, patterns p_1 and p_2 are called *k-Wilf-equivalent*, which is denoted $p_1 \sim_k p_2$, if $s_n^k(p_1) = s_n^k(p_2)$ for all $n \geq 0$. If $s_n^k(p_1) = s_n^k(p_2)$ for all $n, k \geq 0$, then p_1 and p_2 are *strongly Wilf-equivalent* which is denoted $p_1 \sim_* p_2$. If $p_1 \sim_* p_2$, we also say that p_1 and p_2 have the same *distribution*. Similarly, in the case of words, patterns p_1 and p_2 are called *Wilf-equivalent*, in symbols $p_1 \sim p_2$, if $w_{n,\ell}(p_1) = w_{n,\ell}(p_2)$ for all $n, \ell \geq 0$. Also, patterns p_1 and p_2 are called *k-Wilf-equivalent*, in symbols $p_1 \sim_k p_2$, if $w_{n,\ell}^k(p_1) = w_{n,\ell}^k(p_2)$ for all $n, \ell \geq 0$. Finally, if $w_{n,\ell}^k(p_1) = w_{n,\ell}^k(p_2)$ for all $n, \ell, k \geq 0$, then p_1 and p_2 are *strongly Wilf-equivalent*, in symbols $p_1 \sim_* p_2$.

Definition 1.0.31. Everything in Definition 1.0.30 can be generalized to considering two sets of patterns rather than two patterns. In particular, two sets of patterns, P_1 and P_2, are called *Wilf-equivalent*, which is denoted $P_1 \sim P_2$, if $s_n(P_1) = s_n(P_2)$ for all $n \geq 0$.

Frequently in this book we will be discussing *statistics*. Below we give some relevant definitions. We also refer to Section A.1 for a list, together with definitions and examples, of permutation/word statistics we will utilize in this book.

Definition 1.0.32. A *statistic* on a set of objects A is just a function from A to $\{0, 1, 2, \ldots\}$.

Example 1.0.33. *The number of ascents* (see Definition 1.0.16) in a permutation π, denoted $\mathrm{asc}(\pi)$, is an example of a statistic. Another statistic is "*the number of descents,*" denoted $\mathrm{des}(\pi)$.

Definition 1.0.34. We say that two vectors $(s_1, s_2, \ldots, s_k)$ and $(s_1', s_2', \ldots, s_k')$ of statistics on sets A and B, respectively, have *the same distribution*, or are *equidistributed*, if

$$\sum_{a \in A} x_1^{s_1(a)} x_2^{s_2(a)} \cdots x_k^{s_k(a)} = \sum_{b \in B} x_1^{s_1'(b)} x_2^{s_2'(b)} \cdots x_k^{s_k'(b)}.$$

To show equidistribution, we often find a bijective map that *translates*, or *sends*, one tuple of statistics to the other.

Example 1.0.35. Suppose in Definition 1.0.34 we have $A = B = \mathcal{S}_n$, $s_1 = s_1' = \mathrm{asc}$, and $s_2 = s_1' = \mathrm{des}$. Then $(\mathrm{asc}, \mathrm{des})$ and $(\mathrm{des}, \mathrm{asc})$ are equidistributed on $\mathcal{S}_n$ as can be seen by applying the reverse operation r to each permutation in $\mathcal{S}_n$. Indeed, under the reverse operation a permutation having k ascents and ℓ descents is mapped to a permutation having ℓ ascents and k descents, so the following polynomials are equal:

$$\sum_{\pi \in \mathcal{S}_n} x_1^{\mathrm{asc}(\pi)} x_2^{\mathrm{des}(\pi)} = \sum_{\pi' \in \mathcal{S}_n} x_1^{\mathrm{des}(\pi')} x_2^{\mathrm{asc}(\pi')}.$$

Remark 1.0.36. Any pattern introduced below can be seen as a statistic on permutations/words, since we can count occurrences of these patterns. In this book, we use $p(\pi)$ to denote the number of occurrences of a pattern p in a word or permutation π. More generally, the expression $(p_1 + p_2 + \cdots + p_k)(\pi)$ will denote the total number of occurrences of the patterns from the set $\{p_1, p_2, \ldots, p_k\}$ in π.

1.1   Classical patterns in permutations

Definition 1.1.1. An occurrence of a pattern τ in a permutation σ is "classically" defined as a subsequence in σ (of the same length as τ) whose letters are in the same relative order as those in τ. Formally speaking, for $r \leq n$, we say that a permutation $\sigma = \sigma_1\sigma_2\cdots\sigma_n$ in the symmetric group $\mathcal{S}_n$ has an occurrence of the pattern $\tau \in \mathcal{S}_r$ if there exist $1 \leq i_1 < i_2 < \cdots < i_r \leq n$ such that $\tau = \mathrm{red}(\sigma_{i_1}\sigma_{i_2}\cdots\sigma_{i_r})$.

Example 1.1.2. The permutation 241563 has four occurrences of the pattern 231, which are the subsequences 241, 453, 463, and 563. On the other hand, the permutation $\pi = 31254$ *avoids* the pattern 231; that is, π has no occurrences of 231.

Some people prefer to think of occurrences of patterns in terms of permutation matrices. For example, to see the pattern built by the letters 4, 6, and 3 in the permutation $\sigma = 241563$ above, we first draw the permutation matrix corresponding to σ by placing a dot in column i and row σ_i from below. Next, we strike out all rows but rows 3, 4, and 6, and all columns but the three columns corresponding to the dots in rows 3, 4, and 6. We rename rows and columns of the smaller matrix accordingly to obtain the permutation matrix corresponding to the pattern 231 as shown in Figure 1.3.

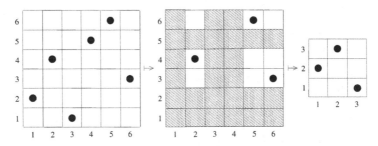

Figure 1.3: An example of the permutation matrices approach to define occurrences of patterns in permutations.

We note that it is often beneficial to use the matrix representation of the permutations in question as it better reveals the structure of permutations, gives a clue to enumeration, etc. However, there is yet another, though not as common, way to think of an occurrence of a pattern in a permutation. Imagine a permutation written in the standard *two-line notation*. For each i, join i in the top row with i in the bottom row by a line. Suppose now that as above we are given the permutation 241563 and we are asked what pattern the letters 4, 6, and 3 form. We erase all letters and all lines corresponding to them except for the letters 3, 4, and 6. We then rename 3, 4, and 6 by 1, 2, and 3 respectively to see that the pattern in question is 231. See Figure 1.4 illustrating this approach.

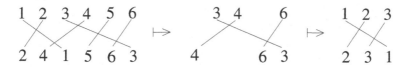

Figure 1.4: An example of the two-line notation approach to define occurrences of patterns in permutations.

Remark 1.1.3. We note that in many places in the literature one sees dashes between the letters of a classical pattern, in order to be consistent with the terminology discussed in Remark 1.3.4 dealing with a generalization of the notion of a classical pattern. For example, 1-3-2 would be used instead of 132, 2-3-4-1 instead of 2341, etc. However, assuming the notation introduced in Section 1.3 will be well-received in the pattern literature, no terminology ambiguity will appear in upcoming papers.

Remark 1.1.4. In the literature one meets other ways to say that "a pattern τ occurs in a permutation π," for example, "τ is involved in π" or "π contains τ".

1.2 Barred patterns (BPs) in permutations

Definition 1.2.1. A barred pattern of length r is a permutation in $\mathcal{S}_r$ some of whose letters are barred. We let $\overline{\mathcal{S}}_r$ denote the set of all barred patterns of length r.

Example 1.2.2. $2\overline{4}153$ and $\overline{31}5\overline{2}4$ are barred patterns from $\overline{\mathcal{S}}_5$. $\overline{\mathcal{S}}_1 = \{1, \overline{1}\}$ and

$$\overline{\mathcal{S}}_2 = \{12, \overline{1}2, 1\overline{2}, \overline{12}, 21, \overline{2}1, 2\overline{1}, \overline{21}\}.$$

Definition 1.2.3. For $\overline{\tau} \in \overline{\mathcal{S}}_r$, let τ be the pattern obtained by removing all the bars in $\overline{\tau}$ (if any), and τ' be the pattern obtained from $\overline{\tau}$ by removing all the barred letters (if any) and bars and taking the reduced form of the rest of $\overline{\tau}$. For example, if $\overline{\tau} = 5\overline{3}2\overline{1}4$ then $\tau = 53214$ and $\tau' = 312$. We say that a permutation π *avoids* $\overline{\tau}$ if *each* occurrence of τ' in π (if any) is part of an occurrence of τ in π. Finally, the number of occurrences of $\overline{\tau}$ in π is the number of occurrences of τ' in π that are not parts of occurrences of τ in π.

Example 1.2.4. The permutation $\pi = 1625374$ avoids the pattern $\overline{\tau} = 4\overline{1}32$ since the only occurrences of $\tau' = 321$ in π are the subsequences 653 and 654, and both of them are parts of the corresponding occurrences, 6253 and 6254, of the pattern $\tau = 4132$ in π.

Pattern	Occurrence in 241563
231	241, 453, 463, 563
2̲3̲1	241, 563
23̲1̲	241, 463, 563
⌊231	241
231⌋	563
231̲⌋	453, 463, 563

Table 1.1: Examples of occurrences of vincular patterns in the permutation 241563.

Example 1.2.5. The permutation 45213 contains four occurrences of the pattern $\overline{\tau} = 2\overline{3}1$ which are the subsequences 52, 51, 53, and 21.

1.3   Vincular patterns (VPs) in permutations

In an occurrence $\sigma_{i_1}\sigma_{i_2}\cdots\sigma_{i_k}$ of a classical pattern $\tau = \tau_1\tau_2\cdots\tau_k$ in a permutation $\sigma = \sigma_1\sigma_2\cdots\sigma_n$, there are no requirements on the number of letters in σ between σ_{i_j} and $\sigma_{i_{j+1}}$. However, a natural restriction would be to require $i_{j+1} = i_j + 1$ for some js; that is, to require that σ_{i_j} and $\sigma_{i_{j+1}}$ are adjacent in σ for the chosen js. Other restrictions the classical patterns cannot accommodate are requirements for an occurrence of τ in σ to begin with σ_1 and/or end with σ_n. A "vincular pattern" is a generalization of the notion of a classical pattern, introduced to allow for these requirements.

Definition 1.3.1. A *vincular pattern* (abbreviated *VP*) τ of length k is a permutation in $\mathcal{S}_k$ some of whose consecutive letters may be underlined. If τ contains $\underline{\tau_i\tau_{i+1}\cdots\tau_j}$ then the letters corresponding to τ_i, τ_{i+1}, ..., τ_j in an occurrence of τ in a permutation must be adjacent, whereas there is no adjacency condition for non-underlined consecutive letters. Moreover, if τ begins with ⌊τ_1 then any occurrence of τ in a permutation σ must begin with the leftmost letter of σ, and if τ ends with τ_k⌋ then any occurrence of τ in σ must end with the rightmost letter of σ.

Table 1.1 provides examples of occurrences of several vincular patterns with *underlying permutation* 231 in the permutation 241563.

For another example, the pattern $\tau = 12\underline{34}5$ occurs in the permutation $\sigma = 243568179$ four times, as the subsequences 24568, 24579, 24679, and 35679. Note that, for instance, the subsequence 24589 of σ is not an occurrence of τ since 8 and 9 are not adjacent in σ.

For matrix representation of permutations the requirement "to be in adjacent positions" is clearly translated to the requirement "to be in adjacent columns".

The following subclass of vincular patterns is of special interest to us and it is considered in Chapter 5.

Definition 1.3.2. Patterns *all* of whose letters are underlined are called *consecutive*. Thus an occurrence of a consecutive pattern in a permutation corresponds to a (contiguous) factor of the permutation. Consecutive patterns are also known in the literature as *"segmented patterns," "segmental patterns,"* and *"subword patterns,"* however these expressions are less common.

Example 1.3.3. There are two occurrences of the consecutive pattern 1234 in the permutation 245783169, namely, the factors 2457 and 4578, while this permutation avoids the pattern 4321.

Remark 1.3.4. As mentioned above, "vincular patterns" are known in the literature as "generalized patterns" or "GPs". Since there are many papers written on GPs, we find it convenient to state here the original notation for these patterns, even though it will not be used in this book. As already mentioned in Remark 1.1.3 above, a classical pattern is a permutation where every pair of adjacent letters is separated by a dash. In order to impose the requirement that two adjacent letters in a GP must be adjacent in an occurrence of it in a permutation, we simply remove the dash between the letters. Thus, comparing the GP notation with the VP notation, we have, for example, the following identities: 2-3-1 = 231, 3214 = 3214, 12-43 = 1243, etc. Finally, the requirement for an occurrence of a pattern to begin or to end with the first or the last letter of a permutation is denoted in the GP notation by placing [or] in front or behind the pattern, respectively. Thus, for instance, [124-3 and 2-3-1-4] in the GP notation correspond to [1243 and 2314] in the VP notation, respectively.

It is remarkable that in some cases the language of VPs can be used instead of the language of classical patterns or BPs, and vice versa. We illustrate this observation with the following three simple propositions. We provide two of them with proofs to demonstrate a way to deal with patterns.

Proposition 1.3.5. *([251]) For any $n \geq 0$, $\mathcal{S}_n(2\underline{13}) = \mathcal{S}_n(213)$. Thus, $\mathrm{Av}(2\underline{13}) = \mathrm{Av}(213)$.*

Proof. If an n-permutation π avoids the pattern 213, then it clearly avoids the VP $2\underline{13}$ as well. On the other hand, if π contains an occurrence of 213, among all such occurrences we pick a subsequence bac of π having the minimum number, say k, of letters between a and c. See the schematic view of the permutation matrix of π in the left picture in Figure 1.5. If $k = 0$, then we get an occurrence of 2$\underline{13}$ and we are done. Assuming $k > 0$, consider the letter, say x, immediately to the right of a. If $x > b$ then bax is an occurrence of 2$\underline{13}$; if $x < b$ then bxc is an occurrence of the pattern 213 with $k-1$ letters between x and c. In both cases, we get a contradiction with the assumption on minimality of k with the given properties. Thus, π contains 2$\underline{13}$ and we are done.   □

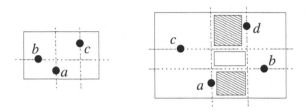

Figure 1.5: Two permutation diagrams related to Propositions 1.3.5 and 1.3.6.

Proposition 1.3.6. *([262]) For any $n \geq 0$, $\mathcal{S}_n(3\underline{14}2) = \mathcal{S}_n(41\overline{3}52)$. Thus, $\mathrm{Av}(3\underline{14}2) = \mathrm{Av}(41\overline{3}52)$.*

Proof. If an n-permutation π avoids the BP $41\overline{3}52$, then it also avoids the VP $3\underline{14}2$ as any occurrence of the 3142 pattern in π is necessarily involved in an occurrence of the 41352 pattern. Conversely, suppose π does not avoid $41\overline{3}52$. That means that there exists a subsequence of π, say $cadb$, that forms the pattern 3142 and that is not a part of an occurrence of the pattern 41352. Among all such subsequences, we can choose one having the minimum number of letters, say k, between a and d. See the sketch of the permutation matrix of π in the right picture in Figure 1.5: the white rectangle in the middle of the matrix indicates that no letter between a and d has value between b and c. Now, if $k = 0$ then $cadb$ is actually an occurrence of 3$\underline{14}$2 and we are done. However, if $k > 0$ then we can consider the letter x next to the right of a: if $x > c$ then $caxb$ is an occurrence of 3$\underline{14}$2; if $x < c$ then $cxdb$ is an occurrence of the pattern 3142 having $k - 1$ letters between x and d. In any case, we get a contradiction with the assumption on minimality of k satisfying our conditions. Thus, π contains 3$\underline{14}$2 and we are done.   □

Another result to be be used in the book is the following proposition that can be proved similarly to Proposition 1.3.6.

Proposition 1.3.7. *For any $n \geq 0$, $\mathcal{S}_n(2\underline{14}3) = \mathcal{S}_n(21\overline{3}54)$. Thus, $\mathrm{Av}(2\underline{14}3) = \mathrm{Av}(21\overline{3}54)$.*

Remark 1.3.8. We note that not all barred patterns can be expressed in terms of vincular patterns. For example, the pattern $3\bar{5}241$, appearing in the context of so-called *2-stack sortable permutations* (see Subsection 2.1.1), cannot be expressed in terms of vincular patterns.

1.4 Bivincular patterns (BVPs) in permutations

Vincular patterns allow us to have some control over positions in which we want our pattern to occur or not to occur. However, we do not have any control over the values of occurrences of such patterns — all we know is that these values must be in the relative order specified by a given pattern. The main idea of *bivincular patterns* is not only to have control over adjacent positions, but also over adjacent values. A basic motivation for introducing bivincular patterns in [161] was finding the minimal superset of vincular patterns that is not only closed under the reverse and complement operations, but also under the inverse operation (see Remark 1.4.4 below), in order to copy behavior of classical patterns in this respect (the inverse of a vincular pattern is not well-defined). In either case, bivincular patterns are useful for descriptive purposes and are interesting due to their connections to other objects to be discussed in Section 3.2.

Definition 1.4.1. A *bivincular pattern* (abbreviated *BVP*) τ of length k is a permutation in $\mathcal{S}_k$ written in two-line notation (that is, the top row of τ is $12 \cdots k$ and the bottom row of τ is a permutation $\tau_1\tau_2 \cdots \tau_k$) where the bottom row is a usual vincular pattern defined in Definition 1.3.1, and some consecutive elements in the top row can be overlined. We have the following conditions on the bottom and top lines of τ:

- If the bottom line of τ contains $\underline{\tau_i\tau_{i+1} \cdots \tau_j}$ then the letters corresponding to τ_i, $\tau_{i+1}, \ldots$, τ_j in an occurrence of τ in a permutation must be adjacent, whereas there is no adjacency condition for non-underlined consecutive letters. Moreover, if the bottom row of τ begins with $\lfloor\tau_1$ then any occurrence of τ in a permutation σ must begin with the leftmost letter of σ, and if the bottom row of τ ends with $\tau_k\rfloor$ then any occurrence of τ in σ must end with the rightmost letter of σ.

- If the top line of τ contains $\overline{i(i+1) \cdots j}$ then the letters corresponding to τ_i, $\tau_{i+1}, \ldots$, τ_j in an occurrence of τ in a permutation must be adjacent in value, whereas there is no value adjacency restriction for non-overlined letters. Moreover, if the top row of τ begins with $\overline{1}$ then any occurrence of τ in a permutation σ must begin with the smallest letter of σ, and if the top row of

τ ends with $\overline{n}$ then any occurrence of τ in σ must end with the largest letter of σ.

Taking into account that Definition 1.4.1 is rather complex, we provide several examples illustrating the concept. Notice that all of the examples contain at least one restriction on values (the top row) as otherwise we would be dealing with vincular or even classical patterns.

Example 1.4.2. Consider occurrences of the following bivincular patterns in a permutation $\sigma = \sigma_1 \sigma_2 \cdots \sigma_n$:

- $\frac{\overline{12}}{\overline{12}}$. If $\sigma_i \sigma_j$ is an occurrence of the pattern, then $j = i + 1$ (σ_i and σ_j are adjacent), $\sigma_i = n - 1$, and $\sigma_j = n$ (σ_i and σ_j are the next largest and the largest letters in σ, respectively). For example, the permutation 2341 contains one occurrence of the pattern (the factor 34), whereas the permutation 25143 avoids the pattern. Clearly, any permutation contains at most one occurrence of $\frac{\overline{12}}{\overline{12}}$.

- $\frac{\lceil 12}{21 \rfloor}$. If $\sigma_i \sigma_j$ is an occurrence of the pattern, then $\sigma_j = 1$ (σ_j is the smallest letter in σ) and $j = n$ (σ_j is the rightmost letter in σ). For example, 35241 contains four occurrences of the pattern (the subsequences 31, 51, 21, and 41), whereas 2134 avoids the pattern. In general, it is easy to see that if the last (rightmost) letter of an n-permutation σ is 1, then the permutation contains $n - 1$ occurrences of $\frac{\lceil 12}{21 \rfloor}$, while if the last letter of σ is not 1, then σ avoids the pattern.

- $\frac{\overline{123}}{\lfloor 213 \rfloor}$. If $\sigma_i \sigma_j \sigma_\ell$ is an occurrence of the pattern, then $\sigma_i \sigma_j \sigma_\ell = (a + 1)a(a + 2)$ for some $1 \le a \le n - 2$, $i = 1$, and $\ell = n$. For example, the permutation 316524 contains one occurrence of the pattern (the subsequence 324), whereas 42351 avoids it. It is straightforward to see that an n-permutation σ avoids $\frac{\overline{123}}{\lfloor 213 \rfloor}$ unless $n \ge 3$, $\sigma_1 = a > 1$, and $\sigma_n = a + 1$ (in which case there is one occurrence of the pattern).

- $\frac{\overline{123}}{321}$. If $\sigma_i \sigma_j \sigma_\ell$ is an occurrence of the pattern, then $\sigma_i \sigma_j \sigma_\ell = n(n - 1)a$ for $1 \le a \le n - 2$, and $j = i + 1$. For example, 246513 contains two occurrences of the pattern (the subsequences 651 and 653), whereas 213465 avoids $\frac{\overline{123}}{321}$. In general, if an n-permutation σ does not contain the factor $n(n - 1)$, or it contains this factor at the very end (as the two rightmost letters of σ) then σ

avoids the pattern; otherwise σ contains as many occurrences of $\overline{\underline{1\overline{23}}}_{3\underline{21}}$ as there are letters to the right of $n(n-1)$ in σ.

- $\overline{\underline{1234}}_{3412}$. If $\sigma_i\sigma_j\sigma_\ell\sigma_m$ is an occurrence of the pattern, then $\sigma_\ell < \sigma_m < \sigma_i < \sigma_j$, $\sigma_m = \sigma_\ell + 1$, and $j\ell m = a(a+1)(a+2)$ for some $2 \le a \le n-2$. For example, the permutation 4167235 contains two occurrences of the pattern (the subsequences 4723 and 6723), whereas 416325 avoids it.

- $\overline{\underline{1234}}_{4213}$. If $\sigma_i\sigma_j\sigma_\ell\sigma_m$ is an occurrence of the pattern, then $\sigma_\ell < \sigma_j < \sigma_m < \sigma_i$, $\sigma_m = \sigma_j + 1$, $j = i+1$, and $m = \ell + 1$. For example, the permutation 563124 contains one occurrence of the pattern (the subsequence 6324), whereas 23154 avoids it.

- $\overline{\underline{12\cdots k}}_{12\cdots k}$. An occurrence of the pattern is a subsequence $a, a+1, \ldots, a+k-1$ of σ where $1 \le a \le n-k+1$. For example, for $k = 4$, the permutation 672134859 contains two occurrences of the pattern (the subsequences 6789 and 2345), whereas 32187645 avoids it. Notice that the inverse of the pattern (see Remark 1.4.4 for definition of inverse on bivincular patterns) is the regular vincular pattern $\underline{12\cdots k}$.

- $\overline{\underline{12\cdots k}}_{12\cdots k}$. An occurrence of the pattern is a factor $a(a+1)\cdots(a+k-1)$ of σ where $1 \le a \le n-k+1$. For example, for $k = 3$, the permutation 72345618 contains three occurrences of the pattern (the factors 234, 345, and 456), whereas 452136 avoids it. Notice that the pattern $\overline{\underline{12\cdots k}}_{12\cdots k}$ is a fixed point under the inverse operation on patterns (see Remark 1.4.4 for definition of inverse on bivincular patterns).

Remark 1.4.3. Originally, bivincular patterns were introduced in [161] using permutation matrices. For example, the pattern $\overline{\underline{123}}_{231}$ studied in [161] was denoted by

where the thick vertical line indicates that in an occurrence of the pattern the first and the second letters must be adjacent, whereas the first and the last letters must be adjacent in value (the first letter must be one larger than the third one). The requirement to begin (resp., end) with the leftmost (resp., rightmost) letter in a permutation is achieved by making the leftmost (resp., rightmost) vertical line thick, whereas the requirement to involve in an occurrence of the pattern the smallest (resp., largest) letter in a permutation is achieved by making the bottom (resp., top) horizontal line thick.

Remark 1.4.4. The inverse operation can be defined naturally on bivincular patterns by taking the regular group theoretical inverse of the bottom permutation and then interchanging the roles of overlines/upper hooks with underlines/lower hooks. For example, the inverse of $\overline{1234}\atop\underline{2413}$ is $\overline{1234}\atop\underline{3142}$. The best way to convince oneself that the definition makes sense is to use the permutation matrix definition of bivincular patterns discussed in Remark 1.4.3. Indeed, taking the inverse of the underlying permutation corresponds to reflecting the dots with respect to the second main diagonal. Simultaneously, the thick horizontal and vertical lines are reflected with respect to the same line making them vertical and horizontal, respectively. Figure 1.6 is an example of such a reflection.

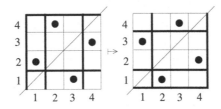

Figure 1.6: An example of taking the inverse of a bivincular pattern.

1.5   Partially ordered patterns

Note that a common feature of the patterns introduced so far is that in an occurrence $\sigma_{i_1}\sigma_{i_2}\cdots\sigma_{i_k}$ of a pattern we can compare any two elements; that is, given σ_{i_m} and σ_{i_ℓ}, $m \neq \ell$, we can say whether $\sigma_{i_m} < \sigma_{i_\ell}$ or $\sigma_{i_m} > \sigma_{i_\ell}$. The idea of partially ordered patterns is to allow the possibility for some letters to be incomparable in the sense that we do not know the relative order of these letters. Partially ordered patterns generalize the concept of classical and vincular patterns (which are based on totally ordered alphabets). However there is no such direct relation to bivincular patterns since there is no presentation of partially ordered patterns in terms of permutation matrices.

Definition 1.5.1. Let $\mathcal{A}$ be a partially ordered set (poset) having k elements. A *partially ordered pattern* (*POP*) $\tau = \tau_1\tau_2\cdots\tau_k$ of length k is essentially a vincular pattern (introduced in Definition 1.3.1) except that instead of a permutation from $\mathcal{S}_k$ one has a permutation of the letters in $\mathcal{A}$. An occurrence of τ in a permutation $\sigma = \sigma_1\sigma_2\cdots\sigma_n$ is a subsequence $\sigma_{i_1}\sigma_{i_2}\cdots\sigma_{i_k}$ satisfying the vincular conditions (respecting underlines and left/right hooks, if any), and additionally to that, if $\tau_m < \tau_\ell$ then $\sigma_{i_m} < \sigma_{i_\ell}$ (the converse is not true!).

As in the case of defining bivincular patterns, we provide below several examples illustrating the concept. Sometimes in defining posets we use *Hasse diagrams* (pictures showing *covering relations* in posets); in other cases, we list explicitly all relations in a given poset. However, in the examples below we provide both such descriptions for the reader to be comfortable with either of them.

Example 1.5.2. Consider occurrences of the following POPs in a permutation $\sigma = \sigma_1\sigma_2\cdots\sigma_n$:

- Let $\mathcal{A} = \{a, c, x\}$ where $a < c$ and $x < c$. The Hasse diagram corresponding to $\mathcal{A}$ is

  If $\sigma_{i_1}\sigma_{i_2}\sigma_{i_3}$ is an occurrence of, say, the pattern xac in σ, then the only condition on σ_{i_j}s is that σ_{i_3} is the largest letter out of the three letters (in fact, the pattern axc is exactly the same as the pattern xac as the roles played by the minimal elements a and x are indistinguishable). For example, the permutation 31452 contains four occurrences of the pattern xac (the subsequences 314, 315, 345, and 145), while the permutation 45231 avoids it.

- Let $\mathcal{A} = \{a, c, x\}$ where $a < x$ (c is incomparable to both a and x). The Hasse diagram corresponding to $\mathcal{A}$ is

  If $\sigma_{i_1}\sigma_{i_2}\sigma_{i_3}$ is an occurrence of, say, the pattern $x\underline{ac}$ in σ, then $\sigma_{i_1} > \sigma_{i_2}$ and σ_{i_2} and σ_{i_3} must be adjacent. For example, the permutation 241536 has four occurrences of $x\underline{ac}$ (the subsequences 215, 415, 436, and 536), while the permutation 13452 avoids it. Notice that unlike the previous example, permutations of a, x, and c produce different patterns as the corresponding poset, after removing labels, does not have *indistinguishable elements*.

- Let $\mathcal{A} = \{a, c, v, x\}$ where $c > v$, $c > x$, and $v > a$. The Hasse diagram

corresponding to $\mathcal{A}$ is

If $\sigma_{i_1}\sigma_{i_2}\sigma_{i_3}\sigma_{i_4}$ is an occurrence of, say, the pattern $\underline{\text{caxv}}$ then $i_1 = 1$, $i_3 = i_2 + 1$, σ_{i_1} is the largest letter out of the four letters, and $\sigma_{i_4} > \sigma_{i_2}$. For example, the permutation 6723541 contains three occurrences of the pattern (the subsequences 6235, 6234, and 6354).

- Let $\mathcal{A} = \{a, c, v, x\}$ where $v > a$, $v > x$, and $c > x$. The Hasse diagram corresponding to $\mathcal{A}$ is

If $\sigma_{i_1}\sigma_{i_2}\sigma_{i_3}\sigma_{i_4}$ is an occurrence of, say, the pattern $\underline{xv}ac$ then $i_2 = i_1 + 1$, $i_4 = i_3 + 1$, $\sigma_{i_2} > \sigma_{i_1}$, $\sigma_{i_2} > \sigma_{i_3}$, and $\sigma_{i_4} > \sigma_{i_1}$. For example, the permutation 2514736 has five occurrences of the pattern (the subsequences 2514, 2547, 2536, 1436, and 4736).

- Let $\mathcal{A} = \{a, c, e, v, x\}$ where $e > v$, $e > c$, $v > a$, $v > x$, and $c > x$. The Hasse diagram corresponding to $\mathcal{A}$ is

If $\sigma_{i_1}\sigma_{i_2}\sigma_{i_3}\sigma_{i_4}\sigma_{i_5}$ is an occurrence of, say, the pattern $v\underline{ae}x\underline{c}$ then $i_3 = i_2 + 1$, $i_4 = i_3 + 1$, $i_5 = n$, σ_{i_3} is the largest letter out of the five letters, $\sigma_{i_1} > \sigma_{i_2}$, $\sigma_{i_1} > \sigma_{i_4}$, and $\sigma_{i_5} > \sigma_{i_4}$. For example, the permutation 15372846 contains three occurrences of the pattern (the subsequences 53726, 52846, and 72846).

Remark 1.5.3. Partially ordered patterns can be thought of as a way to encode certain sets of vincular patterns, which is convenient in many situations. For example, to avoid the pattern $\underline{axc}$ in permutations based on the poset

is the same as to simultaneously avoid the (usual) vincular patterns $\underline{123}$, $\underline{132}$, and $\underline{231}$.

Remark 1.5.4. As discussed in [504] and in a couple of other sources, for some posets it is convenient to use labels based on natural numbers with natural ordering where non-intersecting chains receive numbers marked in different ways (labels from a chain are marked in a uniform way, e.g. using the same number of primes). For example, if our partially ordered set a pattern is built on is $\mathcal{A} = \{1', 2', 1'', 2'', 3''\}$, then it is understood that the only relations on $\mathcal{A}$ are $1' < 2'$ and $1'' < 2'' < 3''$ (the letters with one prime are incomparable with the letters with double prime). Additionally, if letters with no primes are used, they are understood to be comparable with all other letters and the order is determined by comparing elements after erasing primes. For example, if $\mathcal{A} = \{1, 2', 3', 2'', 3'', 4\}$ then 4 is understood as the largest letter, 1 is understood as the smallest letter, and the other relations are $2' < 3'$ and $2'' < 3''$. However, in the rest of the book, whenever we deal with POPs, we explicitly describe relations on the elements, so there is no need to remember this remark.

1.6 Patterns in words

In Sections 1.1–1.5 we provided definitions of classical patterns, barred patterns, (bi)vincular patterns, and partially ordered patterns in permutations. Bivincular patterns have no direct analogue in the case of words, and we omit them from consideration in this section. Moreover, "barred patterns in words" do not seem to appear in the literature, and we do not consider them in this section, even though this notion is well defined. The only difference in the case of occurrences of the other types of patterns (defined above for permutations) in words over ordered alphabets is that we allow repetitions of letters in patterns and words. Thus, essentially the same definitions work, and we do not state them again. Instead, we provide examples of occurrences of patterns of the types of interest in words. Note that in these examples we mark certain letters in permutations to indicate occurrences of patterns in question.

Example 1.6.1. Below we present examples of occurrences of classical patterns in words.

- The pattern 312 occurs six times in the word 4425143: twice as subsequence 423 (namely, **4**42**5**1**4**3 and **4**425**14**3), twice as subsequence 413 (namely, **4**425**1**4**3** and 44**2**5**14**3), once as subsequence 514 (namely, 442**5**1**4**3), and once as subsequence 513 (namely, 442**5**1**4**3).

- The pattern 1213 occurs five times in the word 232425: twice as subsequence 2325 (namely, **232**4**2**5 and **2**3**242**5), twice as subsequence 2425 (namely, **2**3**242**5 and 23**242**5), and once as subsequence 2324 (namely, **2324**25).

- The pattern 1221 occurs twice in the word 143253154: as subsequence 1331 (namely, **1**4**3**2**5****3**15**4**) and as subsequence 4554 (namely, 1**4**32**5****3**1**54**).

Example 1.6.2. Below we present examples of occurrences of vincular patterns in words.

- The pattern 3̲1̲2 occurs three times in the word 4425143 (namely, **44**25143, 4**42****5**143, and 44**25****1**43) as the first and the second letters in an occurrence of the pattern must be adjacent.

- The pattern ⌊3̲1̲2⌋ occurs twice in the word 4425143 (namely, **44**251**43** and **4****42****5**1**43**) as any occurrence of the pattern must begin with the leftmost letter and end with the rightmost letter of the word.

- The pattern 12̲2̲3̲ occurs twice in the word 2411445 (namely, **2**41**14****45** and **2**411**44****5**) as in an occurrence of the pattern in a word, the first two letters must be adjacent, and the last two letters must be adjacent.

- The pattern⌊2̲12̲1̲ occurs three times in the word 31223231 (namely, **3**1**2**2**32****3**1, **3**12**2****3****23**1, and **3**1223**23****1**) as any occurrence of the pattern must start with the leftmost letter in the word, and it must end with two adjacent letters.

Example 1.6.3. Below we present examples of occurrences of partially ordered patterns in words.

- The pattern *ax̲ac̲*⌋ based on the poset

  does not occur in the word 23214, occurs twice in the word 121213 (namely, **12**1**21****3** and **12**1**213**), and it occurs once in 122114 (namely, **122114**).

- The pattern *cca̲x* based on the poset

  occurs four times in the word 3512134512: **3**5**1**2**1**34512, **3**512**134**512, **351****2**1**3**4512, and **3**512134**512**.

- The pattern $avcxav|$ based on the poset

  occurs four times in the word 14412354413: **144**12354413, 144123**54413**, 14412**354**413, and 14412**354413**. Notice that, say, 14412**354413** is *not* an occurrence of the pattern as $c = x = 4$ in this case and $x \not< c$.

- The pattern $v\underline{cv}axc$ based on the poset

  occurs twice in the word 314321241: **314**32**1**241 and 3143**21241**.

- The pattern $\underline{av}xcc$ based on the poset

  occurs five times in the word 242315161526: **242**315**1**61526, **242**31516**1**526, **242315**161526, 242315**161526**, and 242315**161526**.

1.7 Some typical problems related to patterns

One of the most fundamental questions we can ask in the theory of patterns in words and permutations is as follows.

Problem 1.7.1. For a given pattern p (of any specified type) find its *distribution* over all permutations (resp., words) of given length; that is, find $s_n^k(p)$ (resp., $w_{n,\ell}^k(p)$) for all $n, k, \ell \geq 0$. More generally, we can ask the same question for a set of patterns P instead of a single pattern.

Unfortunately, the problem above is very difficult, and there are very few patterns for which we can solve it. However, it is not a hopeless problem, at least for some classes of patterns. Almost certainly the most popular case of Problem 1.7.1 in the literature is the following:

Problem 1.7.2. For a given pattern p find the number of permutations (resp., words) of given length that avoid p, that is, find $s_n(p)$ (resp., $w_{n,\ell}(p)$) for all $n, \ell \geq 0$. Also, we can state the same problem for a set of patterns P instead of a single pattern.

It is intriguing that the patterns of a given length cannot necessarily be ordered according to avoidance: Stankova and West [737] observed that $s_n(53241) < s_n(43251)$ for all $n \leq 12$ while $s_{13}(53241) > s_{13}(43251)$. However, Stankova and West conjectured (for classical patterns) that such ordering can be done asymptotically (a pattern τ_1 is *asymptotically more avoidable* than a pattern τ_2 if $s_n(\tau_1) > s_n(\tau_2)$ for sufficiently large n).

One particular case of Problem 1.7.2 is the following problem, which is attracting significant attention from many researchers, and is probably the best known unsolved question in the area.

Problem 1.7.3. Find $s_n(1324)$. For $n \geq 1$, the sequence $s_n(1324)$ begins as follows: $1, 2, 6, 23, 103, 513, 2762, \ldots$. Marinov and Radoičić [624] provided the first 20 terms of the sequence using a generating tree approach, while Albert et al. [29] found a few extra terms.

This book will familiarize the reader with all known cases of patterns for which Problems 1.7.1 and 1.7.2 are solved. Ideally, as an answer to such a problem one would like to have a formula in *closed form*, like n^2, or $k\binom{n}{3}$, or $\frac{k\ell(n+n^2)}{2}\binom{2n+1}{4}$, etc. However, if instead of a closed formula one is able to get the *generating function* (*g.f.*), or the *exponential generating function* (*e.g.f.*) for, say, the $s_n(p)$ in question, that is, the functions

$$G_p(x) = \sum_{n \geq 0} s_n(p)x^n \quad \text{or} \quad E_p(x) = \sum_{n \geq 0} s_n(p)\frac{x^n}{n!},$$

respectively, one is normally equally satisfied with such an answer. After all, obtaining generating functions often leads to closed formulas and/or aids in finding *asymptotics* for the numbers involved. The last observation brings us to another typical problem that is often possible to solve without finding exact formulas for the numbers involved.

Problem 1.7.4. For a given pattern p, find the asymptotic growth of $s_n(p)$. The same problem can be stated more generally to study $s_n^k(p)$ or $w_{n,\ell}^k(p)$ for given k and ℓ.

Yet another way to answer Problems 1.7.1 and 1.7.2 is to provide an expression involving summation/product symbols, or a recurrence relation, possibly involving auxiliary sequences. Often, we can turn such expressions into (exponential) generating functions. However, even if one is unsuccessful in getting any expressions for the numbers involved, a problem related to the following definition may arise.

Definition 1.7.5. A sequence a_n is said to be *P-recursive*, which is short for *polynomially recursive*, if there are polynomials $p_0(n)$, $p_1(n)$, $\ldots$, $p_k(n)$ so that

$$p_k(n)a_{n+k} + p_{k-1}(n)a_{n+k-1} + \cdots + p_0(n)a_n = 0.$$

For instance, the sequence $a_n = n!$ is *P*-recursive since $a_{n+1} - (n+1)a_n = 0$.

It is known that a sequence is *P*-recursive if and only if its generating function is *D-finite*, which means that its derivatives span a finite dimensional vector space over $\mathcal{C}(x)$ (see [741]). In either case, a question that may arise is as follows.

Problem 1.7.6. For a given set of patterns P, is the sequence $s_n(P)$ *P*-recursive? More generally, assuming that $s_n^{k_1,\ldots,k_r}(p_1,\ldots,p_r)$ is the number of permutations that have k_i occurrences of a pattern p_i, for each $i \in [r]$, is the sequence $s_n^{k_1,\ldots,k_r}(p_1,\ldots,p_r)$ *P*-recursive?

Noonan and Zeilberger [644] conjectured that for classical patterns p_is, one has that $s_n^{k_1,\ldots,k_r}(p_1,\ldots,p_r)$ is always *P*-recursive. However, Zeilberger [352] believes that this conjecture is false, and, in particular, the sequence $s_n(1324)$ is not *P*-recursive. Stanley [805] also believes that the conjecture of Noonan and Zeilberger is false.

Another fundamental problem on the patterns is related to the (k-)Wilf-equivalence defined in Definition 1.0.30.

Problem 1.7.7. For given patterns p_1 and p_2 determine if $p_1 \sim p_2$, or, more generally, if $p_1 \sim_k p_2$ for all/some k. An adjacent problem is as follows: Find necessary and sufficient conditions for two patterns to be (k-)Wilf-equivalent.

Note that to solve Problem 1.7.7 one does not necessarily need to know formulas for the numbers involved (knowing the formulas would normally automatically solve the problem on k-Wilf equivalence): such formulas may be unknown, or known just for one pattern. A traditional way to provide an affirmative answer to Problem 1.7.7 in case of unknown numbers is to construct a bijection between the classes of permutations involved. In certain cases, to be discussed in this book, it is possible to provide a *Wilf classification* for (k-)Wilf-equivalences, namely, to state explicitly classes containing (k-)Wilf-equivalent patterns. For example, as we will see, there is only one Wilf-equivalence class for classical patterns of length 3, while there are three such classes for classical patterns of length 4.

The following problem is of special importance as it provides a motivation to study patterns in words and permutations — we will see material related to it in this book, in particular, in Chapter 2.

Problem 1.7.8. For a given set of patterns P, find other (combinatorial) objects related to $\mathcal{S}_n(P)$, $n \geq 0$, or $\mathcal{S}_n^k(P)$ for $n \geq 0$ and some k, or $\mathrm{Av}(P)$. The same problem can be stated for $W_{n,\ell}(P)$, $W_{n,\ell}^k(P)$, and $\mathrm{Av}_\ell(P)$.

Problem 1.7.8 is probably not transparent at first glance, though there is a well-established procedure for how to look for connections to other objects while dealing with patterns. For example, let us consider the pattern 132 and ask whether the class of permutations avoiding 132, $\mathrm{Av}(132)$, is related to any other combinatorial objects.

Either one is able to find a formula for $s_n(132)$ right away, or, as happens more often, one, using his/her favorite software, or even by hand, generates $\mathcal{S}_0(132)$, $\mathcal{S}_1(132)$, $\mathcal{S}_2(132)$, and so on. Then one obtains the initial values of the sequence $s_0(132)$, $s_1(132)$, $s_2(132)$, etc., which are, in this case, 1, 1, 2, 5, 14, 42, ... One could recognize the *Catalan numbers* immediately, but such knowledge is not necessary thanks to the *Online Encyclopedia of Integer Sequences* [726], *OEIS*, by Sloane which is published electronically at `http://www.research.att.com/~njas/sequences/`. We type in the initial values of our sequence and find that it matches the initial values of the sequence A000108 (see the corresponding screenshot in Figure 1.7). Of course, the match we got does not prove that the number of permutations avoiding 132 is given by the Catalan numbers, but it gives an indication that this may be the case (the more initial terms are calculated to look for a match, the stronger such indication is). So, the match suggests a solution to Problem 1.7.2 for $p = 132$, and beyond that, it suggests other objects equinumerous with $\mathcal{S}_n(132)$ and thus related to our pattern-avoiding class. Thus, to solve Problem 1.7.8 in this case we should try to build bijections between $\mathcal{S}_n(132)$ and other objects mentioned, for example the set of *ordered rooted trees* with n nodes, not including the root. Any such bijection would normally give automatically the number of permutations avoiding the pattern 132, the Catalan numbers (there are some exceptions when sequences in OEIS are published based on calculations of initial terms rather than on exact enumeration — one should look carefully at the references provided by OEIS). And vice versa, if we were to find the cardinality of $\mathcal{S}_n(132)$, we would establish a connection of our pattern class to the objects listed under A000108; finding explicit bijections is then still desirable and often is a non-trivial thing to do. Note that one is, more and more often, not interested in just finding any bijection, but a bijection preserving as many *statistics* as possible. This idea will be explained in detail in Chapter 4 on examples of bijections between 321- and 132-avoiding permutations.

If one does not find a match in OEIS for the pattern class in question, one can try another feature provided by OEIS, namely "Superseeker," that will try to relate your sequence of numbers to existing sequences through various transformations. Refer to `http://www.research.att.com/~njas/sequences/ol.html` for more de-

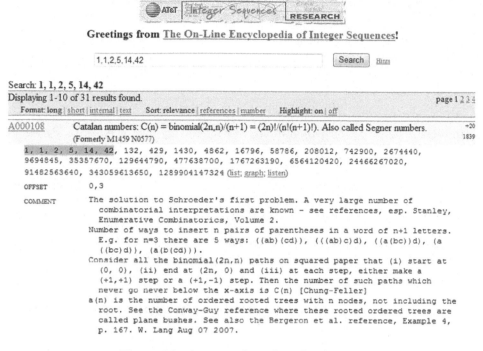

Figure 1.7: A screenshot of an OEIS page.

tails.

It should be mentioned that finding bijections between objects is not just an enjoyable and often complicated process revealing the structural similarities involved, but also may have various applications. For example, in order to prove an equidistribution result on permutations avoiding both the patterns 3142 and 2413 (to be discussed in more detail in Chapter 2), Claesson et al. [262] used a bijection between the permutations and so-called $\beta(1,0)$-*trees* and then an involution on the trees. It still remains an open question to prove directly the equidistribution result.

There are many other directions of possible research on patterns to be discussed in this book. Below we state just a few of them.

Problem 1.7.9. Given a pattern p, what time do we need to find whether p occurs in an n-permutation π? Find an algorithm of minimum complexity detecting if p occurs in π.

Problem 1.7.10. For a given pattern τ what is the maximum number of occurrences of τ in an n-permutation? Which permutation has the maximum? It is a related problem to find the *packing density* of τ, to be discussed later in the book.

Problem 1.7.11. For a given *permutation sorting device* (such devices are to be discussed in Chapter 2), find a set of restrictions in terms of patterns, whenever such a set exists, giving the set of sortable permutations. Then enumerate the set of sortable permutations.

Problem 1.7.12. (To be discussed in detail later in the book.) Given a set of pattern-restricted words or permutations, provide a *Gray code* or *universal cycle* for it.

Finally, it is a very popular direction of research to study the problems listed above on restricted sets of words or permutations (for example, on *involutions*, *even permutations*, 132-avoiding permutations, etc.).

Problem 1.7.13. Given a somehow restricted set of words or permutations, solve Problems 1.7.1—1.7.12, or any other relevant problem, on this set.

1.8   Bibliography related to the introduced patterns

Not that much is published so far related to bivincular patterns ([255, 263, 318, 321, 143, 145, 161, 527, 528, 557, 636, 771, 772, 811]) and partially ordered patterns ([204, 448, 457, 509, 512, 524, 510, 511, 514]) so these remain attractive research directions yet to be explored. Similarly, barred patterns do not appear that often in the literature ([84, 79, 82, 160, 161, 214, 215, 243, 242, 322, 324, 356, 323, 325, 549, 662]). However, classical and vincular patterns are relatively well studied, yet there are more questions we can ask than answers we can give. Several PhD theses involving classical and vincular patterns have been published ([102, 126, 195, 252, 353, 402, 419, 434, 481, 504, 562, 626, 634, 642, 647, 661, 662, 692, 728, 778, 797, 799]). The following papers in the literature will be discussed, among other things, in this book in varying levels of detail (from merely mentioned to thorough proofs or sketches of proofs):

- On classical patterns: [9, 10, 14, 24, 31, 15, 28, 23, 12, 20, 19, 29, 143, 30, 22, 18, 159, 17, 25, 26, 27, 16, 14, 21, 37, 41, 43, 48, 55, 47, 58, 54, 53, 51, 50, 46, 44, 45, 52, 65, 66, 72, 73, 74, 78, 77, 84, 82, 83, 81, 79, 87, 101, 115, 120, 114, 498, 124, 730, 424, 142, 140, 139, 138, 136, 134, 133, 135, 150, 132, 130, 131, 127, 129, 128, 153, 155, 154, 217, 160, 162, 158, 165, 164, 170, 184, 186, 187,

188, 181, 198, 203, 196, 201, 208, 200, 220, 243, 247, 451, 259, 270, 292, 304, 302, 317, 320, 316, 319, 323, 337, 336, 335, 342, 341, 339, 340, 334, 338, 351, 350, 358, 354, 361, 366, 368, 308, 384, 407, 416, 427, 437, 440, 439, 436, 435, 441, 459, 466, 467, 468, 469, 477, 478, 479, 480, 483, 484, 491, 706, 492, 522, 536, 535, 534, 533, 540, 539, 543, 547, 546, 545, 549, 551, 565, 575, 606, 613, 43, 600, 602, 622, 601, 592, 594, 582, 599, 596, 590, 611, 608, 621, 589, 619, 588, 616, 618, 586, 617, 614, 585, 615, 623, 624, 631, 635, 640, 638, 637, 644, 643, 663, 556, 673, 674, 682, 681, 680, 678, 679, 691, 696, 697, 695, 698, 694, 713, 719, 720, 721, 729, 727, 738, 737, 735, 734, 765, 764, 781, 807, 779, 782, 780, 783, 777, 710, 785, 795, 794, 798, 801, 802, 800, 805, 813, 815, 816];

- On vincular patterns: [34, 38, 42, 61, 64, 92, 105, 143, 103, 219, 310, 109, 346, 106, 197, 210, 433, 209, 262, 500, 265, 253, 264, 251, 279, 277, 282, 357, 39, 356, 355, 359, 360, 383, 382, 401, 438, 465, 520, 510, 563, 509, 516, 517, 519, 514, 505, 518, 503, 620, 610, 597, 598, 593, 591, 587, 629, 651, 650, 672, 671, 677, 751, 753, 793, 792].

Notice that even though we have cited more than 320 publications directly related to the patterns discussed so far, we left aside most of the references related to these patterns in *compositions* (which are essentially patterns in words with restricted sums of elements) that can be found in the book [461]. Moreover, we left aside most of the various generalizations/modifications of the notion of a "pattern in a word or permutation" to be discussed later in the book (e.g. [290, 445, 530, 526, 525, 521, 528, 561]). Finally, it is worth mentioning that much in the *theory of word/permutation statistics* (studying functions from words/permutations to natural numbers) is directly related to our theory of patterns. Indeed, for example, the classical statistic "*descent*" defined in words/permutations as positions having elements followed by smaller elements is nothing but occurrences of the vincular pattern $\underline{21}$, whereas the number of *inversions* in a permutation $\pi = \pi_1\pi_2\cdots\pi_n$ (that is, the number of pairs $i < j$ such that $\pi_i > \pi_j$) is nothing but the number of occurrences of the classical pattern 21; well-studied *alternating permutations* are nothing but permutations simultaneously avoiding three vincular patterns $\underline{123}$, $\underline{321}$, and $\underline{12}$, etc. Such (mostly permutation) statistics are considered in the book [422] by Goulden and Jackson, and in the book [743] by Stanley. Even though we do not intend to discuss word/permutation statistics much in this book (except for a few cases relevant to our considerations), taking into account their intimate connection to the theory of patterns in words and permutations, we have decided to provide an exhaustive list of references related to the statistics studied in the literature, thus generating a comprehensive bibliography of publications related in one way or another to word/permutation patterns: [2, 1, 3, 7, 6, 9, 5, 8, 33, 68, 70, 69, 67, 75, 76, 86, 85, 91, 90, 163, 89, 103, 111, 110, 113, 119, 123, 125, 149, 147, 146, 329, 144, 147, 160, 156, 442, 169, 176, 177, 178, 175, 173,

174, 179, 192, 197, 211, 212, 220, 222, 223, 225, 291, 226, 224, 231, 230, 229, 228, 227, 234, 233, 232, 236, 238, 240, 241, 245, 266, 269, 268, 267, 279, 277, 276, 282, 278, 285, 299, 293, 303, 307, 306, 313, 289, 315, 327, 331, 343, 347, 348, 365, 368, 369, 367, 374, 378, 377, 376, 395, 396, 394, 393, 397, 388, 400, 392, 391, 390, 401, 389, 405, 404, 411, 412, 416, 418, 427, 429, 430, 432, 438, 443, 445, 450, 455, 456, 463, 470, 471, 474, 473, 494, 493, 495, 502, 501, 537, 538, 548, 550, 553, 569, 568, 567, 574, 576, 575, 607, 612, 610, 605, 600, 653, 630, 628, 633, 699, 632, 635, 638, 639, 646, 648, 657, 664, 668, 669, 667, 666, 665, 670, 675, 677, 676, 674, 681, 684, 683, 685, 688, 690, 686, 701, 704, 718, 719, 720, 722, 725, 739, 724, 723, 804, 746, 750, 755, 756, 796, 760, 759, 761, 767, 790].

Chapter 2

Why such patterns? A few motivation points

While getting familiar with results on patterns in later chapters, one not only has a chance to appreciate the beauty of combinatorics of patterns on words and permutations (which is interesting in its own right), but also to learn about connections of the field to other branches of combinatorics, mathematics, and theoretical computer science. The main application of patterns so far is that in many situations they provide a convenient language for describing various (combinatorial) objects. Such descriptions can be used in establishing properties (e.g., equidistribution results) of objects related to words or permutations restricted somehow by patterns (e.g., avoiding certain patterns). However, in most of the cases considered in the literature, the prime interest in linking pattern-restricted permutations/words to other objects is finding a bijection between the structures involved rather than also trying to find immediate applications.

In either case, the current chapter contains several of the most striking connections between patterns and other objects. Many more such connections will be given in later chapters.

2.1 Sorting permutations with stacks and other devices

There is a long line of papers in the computer science literature dedicated to problems of *sorting permutations* (arranging permutations in increasing order) with different devices, e.g., *stacks*, *queues*, and *deques* (see, for example, [18, 21, 33, 54, 56, 44, 45, 373, 52, 49, 63, 35, 325, 351, 148, 256, 59, 136, 134, 133, 135, 157, 156, 206, 205, 339,

423, 435, 441, 539, 658, 703, 705, 729, 716, 727, 762, 799, 801, 800, 813, 815, 774, 775]). We refer to [136] and [137, Chapter 8] by Bóna for introductions to the area of sorting we are interested in, and to [539] by Knuth for general sorting algorithms. The original appearance of the theory of patterns was due to its connections to sorting with stacks as discovered by Knuth [540, pp. 242–243] in 1968. Even these days, one of the main applications of our patterns is providing a language for describing sets of permutations sortable by given devices.

In this section we will illustrate how patterns can be used in the sorting industry. Notice that even though the results presented in this section on relations between patterns and sorting are rather comprehensive, we still leave aside some of them. For example, we do not discuss in any details the fact that the matrices corresponding to $Av(3142, 2413)$ are exactly those which do not fill up under *"bootstrap percolation"* [716] (as we will see in Subsection 2.2.5, $Av(3142, 2413)$ is enumerated by the *large Schröder numbers*; see Subsection A.2.1 for definitions).

2.1.1   Sorting with k stacks in series

A *stack* is a last-in first-out linear sorting device with *push* and *pop operations* (also known as *insert* and *remove operations*). In other words, a stack is a container for a linear sequence (in our case, for a permutation) that one is allowed to change by inserting new items (one at a time) at its tail and by removing tail items (again, one at a time). Initially the stack is empty and then a sequence of insertions interleaved with removals is made. Thus an input permutation is transformed thereby into an output permutation.

The *greedy* algorithm we are interested in for stack sorting a permutation $\pi = \pi_1\pi_2 \cdots \pi_n$ works as follows. We start with pushing π_1 onto the stack. Next, if $\pi_2 < \pi_1$ then we push π_2 onto the stack to be on top of π_1; on the other hand, if $\pi_2 > \pi_1$, we pop π_1 off and let π_2 enter the stack. More generally, suppose, at some point, the letters $\pi_1, \pi_2, \ldots, \pi_i$ have all been added to the stack (some of them could be still in the stack, others have been popped off), so we are reading π_{i+1}. We push π_{i+1} onto the stack if and only if π_{i+1} is less than the top element of the stack (which is easily seen to be π_i). Otherwise, we pop elements off the stack, one by one, until π_{i+1} is less than the top remaining stack element and then we push π_{i+1} onto the stack. When no more elements remain to be pushed onto the stack, we pop off all elements of the stack until it is empty. This produces a permutation $S(\pi)$ as output.

Definition 2.1.1. If $S(\pi)$ is the identity permutation $12 \cdots n$, we say that π is *stack-sortable*. More generally, if $S^k(\pi)$ is the identity permutation (S^k is the result of application of S k times), we say that π is *k-stack sortable* (π is sorted with k

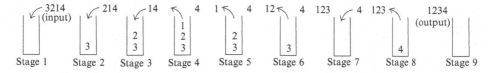

Figure 2.1: Stack sorting the permutation 3214.

stacks in series). We let $W_{n,k}$ denote the set of all k-stack sortable n-permutations.

Remark 2.1.2. There are other notions of a "k-stack sortable permutation" in the literature, which are different from that introduced in Definition 2.1.1 (see [54, 634]). To distinguish the permutations in Definition 2.1.1, they are sometimes called *West-k-stack sortable permutations* (West considered them in [799]). However, since we do not discuss the other notions of k-stack sortable permutations in much detail in this book, we omit "West-" in Definition 2.1.1, which should not cause any confusion.

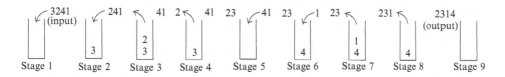

Figure 2.2: Stack sorting the permutation 3241.

Figure 2.1 shows an example of a stack-sortable permutation (3214), while Figure 2.2 shows an example of a non-stack-sortable permutation (3241). As a matter of fact, the permutation 3241 is 3-stack sortable but not 2-stack sortable which is illustrated in Figure 2.3 (a single adjacent transposition makes the permutation 3214 much harder to sort).

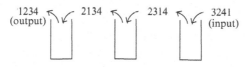

Figure 2.3: Sorting the permutation 3241 with 3 stacks in series.

It is clear that the set of stack-sortable permutations is closed under pattern containment (meaning that any subsequence of a stack-sortable permutation in reduced form is stack-sortable), since removing a letter from the input and ignoring the insertion and removal operations on it gives a proper computation that applies to the shorter input. However, before stating results on patterns related to stack sorting, we would like to define the stack sorting procedure in an alternative (equivalent) way (introducing relevant notions/notations), that is not only normally more convenient to deal with, but also allows several modifications of the procedure, not to be discussed here.

We define the *stack sorting operator* S recursively on an n-permutation as follows. For the empty permutation ε, $S(\varepsilon) = \varepsilon$. If $\pi \neq \varepsilon$ is an n-permutation, decompose π as $\pi = LnR$, where L and R are, possibly empty, factors to the left and to the right of n, respectively. Then

$$S(\pi) = S(L)S(R)n.$$

Going again through the examples above we see that using the new definition, $S(3241) = 2314$ and $S(3214) = S^3(3241) = 1234$ which matches our previous computations.

It is easy to see that $W_{n,n-1} = \mathcal{S}_n = \mathcal{S}_n(\emptyset)$, that is, all n-permutations can be sorted by $n - 1$ applications of the S operator, and this can be seen as avoidance of the empty set of patterns. The first part of the following proposition, that Knuth [540, pp. 242-243] left as an exercise to the reader, began the theory of patterns in permutations, and it provides the first explicit application of patterns in computer science. To give an idea of approaches to use, we will provide a proof of Proposition 2.1.3. However, almost all of the upcoming propositions/theorems in this section are stated without proofs (explicit references to the results are given though).

Proposition 2.1.3. $W_{n,1} = \mathcal{S}_n(231)$. *Moreover,* $|W_{n,1}| = s_n(231) = \frac{1}{n+1}\binom{2n}{n}$, *the n-th Catalan number.*

Proof. Let us prove that $W_{n,1} = \mathcal{S}_n(231)$. Suppose $\pi_i \pi_j \pi_k$ is an occurrence of the pattern 231 in an n-permutation π, that is, $\pi_k < \pi_i < \pi_j$. At some point π_i will enter the stack, and before π_j can do so, π_i must leave the stack so π_k cannot be to the left of it in $S(\pi)$, so π is not stack-sortable.

Conversely, suppose π is not stack-sortable. Then $S(\pi)$ contains a 2-letter subsequence $\pi_i \pi_k$ such that $\pi_i > \pi_k$ ($\pi_i \pi_k$ is an *inversion* in $S(\pi)$). Thus, π_i left the stack before π_k arrived there. This can only happen if there is a letter π_j such that $\pi_i \pi_j \pi_k$ is a subsequence in π and $\pi_j > \pi_i$. But then, $\pi_i \pi_j \pi_k$ is an occurrence of the pattern 231 in π.

To prove the second part of the proposition, decompose a 231-avoiding n-permutation π as $\pi = LnR$, where L and R are the possibly empty factors of π to the left and to the right of the largest letter n, respectively. To avoid 231 each letter of L must be smaller than any letter of R, and L and R in reduced form must be 231-avoiding permutations. This brings us to the following recursion in which i can be viewed as the length of L:

$$s_n(231) = \sum_{i=0}^{n-1} s_i(231)s_{n-i-1}(231).$$

This recursion is a well-known recursion for the Catalan numbers and we are done.

$\square$

In his PhD thesis West [799] proved the following theorem (notice the appearance of a barred pattern).

Theorem 2.1.4. *([799])* $W_{n,2} = \mathcal{S}_n(2341, \bar{3}5241)$.

West [799] conjectured the following formula for the number of 2-stack sortable permutations

$$(2.1) \qquad s_n(2341, 35241) \;=\; \frac{2(3n)!}{(2n+1)!(n+1)!}$$

which was first proved by Zeilberger [815], who found the functional equation

$$x^2 G^3(x) + x(2+3x)G^2(x) + (1 - 14x + 3x^2)G(x) + x^2 + 11x - 1 = 0$$

for the generating function $G(x) = \sum_{n\geq 0} s_n(2341, \bar{3}5241)x^n$ and then used Lagrange's inversion formula to solve it. Two bijective proofs [323, 423] of the conjecture by West appeared later and they connected together different combinatorial objects involving several classes of pattern-restricted permutations. Both of the bijective proofs rely on the known result on the number of *rooted non-separable planar maps* [190, 191]. Some further details on the proofs and related things are to be discussed in Section 2.11.

It should be mentioned that refinements for the number of stack-sortable and 2-stack sortable permutations are know when *descents* (occurrences of the pattern $\underline{21}$) are taken into account. In the first case, one gets the *Narayana numbers* (see [745]). More precisely, the number of stack-sortable n-permutations with m descents is shown to be equal to

$$(2.2) \qquad \frac{1}{n}\binom{n}{m}\binom{n}{m+1}.$$

For the number of 2-stack sortable n-permutations with m descents one gets the following formula (see [135] by Bóna and references therein):

$$(2.3) \qquad \frac{(n+m)!(2n-m-1)!}{(m+1)!(n-m)!(2m+1)!(2n-2m-1)!}.$$

Another refinement for counting 2-stack sortable permutations is the following theorem by Dulucq et al [323].

Theorem 2.1.5. *([323]) The number of 2-stack sortable n-permutations having k right-to-left maxima (see Definition A.1.1) is*

$$\frac{k+1}{(2n-k+1)!} \sum_{j=k+1}^{\min\{n+1,2k+2\}} \frac{(3k-2j+2)(2j-k-1)(j-2)!(3n-j-k+1)!}{(n-j+1)!(j-k-1)!(j-k)!(2k-j+2)!}.$$

To complete the relevant enumeration story, one should mention a result of Bousquet-Mélou [156] related to study of 2-stack sortable permutations subject to five statistics (including permutation length and the number of descents). It was shown that the five-variable generating function in question is *algebraic* of degree 20.

Coming back to using our patterns in describing sortable sets, West [799] showed that $W_{n,n-2}$ are precisely those permutations that do not have suffix $n1$. We state this as the following proposition that involves a bivincular pattern.

Proposition 2.1.6. *([799]) $W_{n,n-2} = \mathcal{S}_n(\begin{smallmatrix} \lceil 1 2 \rceil \\ 2 1 \rfloor \end{smallmatrix})$.*

It is straightforward to see from Proposition 2.1.6, that the number of n-permutations sortable by applying the operator S $n-2$ times is $s_n(\begin{smallmatrix} \lceil 1 2 \rceil \\ 2 1 \rfloor \end{smallmatrix}) = n! - (n-2)!$.

To our best knowledge, there is no known "nice" pattern description of the set $W_{n,n-3}$, although the cardinality of this set is found by Claesson et al. [256]:

$$(2.4) \qquad |W_{n,n-3}| = \frac{(n-3)!}{2}(2n^3 - 6n^2 - 5n + 16)$$

which holds for $n \geq 4$. Permutations from $W_{n,n-4}$ that are sortable by $n-4$ passes through a stack are also studied in [256].

Regarding other ways to define the notion of a k-stack sortable permutation (different from West-k-stack sortable permutations), Atkinson et al. [54] considered permutations that can be sorted by two stacks in series with each stack *remaining sorted from top to bottom*. This set of permutations, M, cannot be characterized by a finite number of classical patterns, but it is given by avoiding the following infinite set of patterns:

$$\{2(2m-1)416385\cdots(2m)(2m-3) \mid m = 2,3,\ldots\}.$$

Further, Atkinson et al. [54] showed that M is equinumerous with Av(1342), which was counted by Bóna (see Table 6.2).

Finally, Murphy [634] considered sorting with k stacks in series in more generality (many more operations are allowed) and proved that for $k \geq 2$ stacks, the set of sortable permutations cannot be characterized by a finite set of forbidden classical patterns.

2.1.2   Sorting with k stacks in parallel

Figure 2.4 shows an example of sorting the permutation 2341 with 2 stacks in parallel.

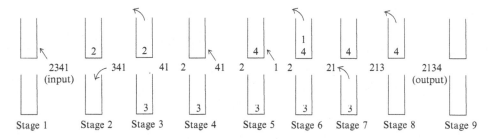

Figure 2.4: Sorting the permutation 2341 with 2 stacks in parallel.

As one sees, the permutation 2341 is not sortable this way, even though, of course, the class of such sortable permutations is larger than the class of 1-stack sortable permutations. We consider parallel sorting to give yet another example of a negative result: it is shown in [373, 762] that for $k \geq 2$, *no* finite set of forbidden *classical* patterns can characterize the set of permutations sortable with k stacks in parallel. It is no surprise that the enumeration problem related to the sortable permutations is unsolved for $k \geq 2$. What we do know [774, 775] is that for $k \leq 3$, the permutations sortable with k stacks in parallel can be recognized in time $O(n \log n)$ while for larger k, that recognition is NP-complete.

2.1.3   Input-restricted and output-restricted deques

An *input-restricted deque*, introduced by Knuth [540] is similar to a stack in that it has a push operation, but the pop operation can remove an element from either end of the deque. A successful sorting of a permutation requires the existence of a sequence involving the allowed operations that leads to the increasing permutation. Of course, we now have more possibilities to sort a permutation. For example, the reader may check that the permutation 2341 requires three stacks in series to be sorted while it can be sorted with a single input-restricted deque as shown in Figure 2.5.

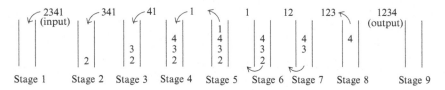

Figure 2.5: Sorting the permutation 2341 with an input-restricted deque.

On the other hand, not all permutations can be sorted with an input-restricted deque as shown in Figure 2.6.

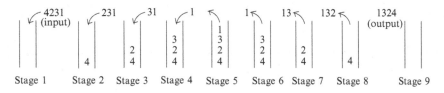

Figure 2.6: Sorting the permutation 4231 with an input-restricted deque.

Knuth [540] proved the following theorem where he used the so-called *kernel method* for proving the second part.

Theorem 2.1.7. *([540]) The set of n-permutations that can be sorted by an input-restricted deque is given by $\mathcal{S}_n(4231, 3241)$. The number $s_n(4231, 3241)$ of the sortable n-permutations is given by the $(n-1)$-th Schröder number S_{n-1}.*

We refer to [540] and to Subsection A.2.1 for more information on Schröder numbers. These numbers, no surprise, appear in the context of *output-restricted deques* as well when we are allowed to push letters at either end, but to pop them only from the top end. Knuth [540] show that the number of such permutations is given by the Schröder numbers, while West [801] shows the relation of these permutations to pattern-avoidance:

Theorem 2.1.8. *([801]) Av(2431, 4231) is the set of all permutations that can be sorted using the output-restricted deque. Consequently, by the corresponding result of Knuth [540], $s_n(2431, 4231) = S_{n-1}$, the $(n-1)$-th Schröder number.*

We note the the second part of Theorem 2.1.8 can be obtained from Theorem 2.1.7 using the trivial bijections as discussed in the following remark.

Remark 2.1.9. Since $r.c.i(4231) = 4231$ and $r.c.i(3241) = 2431$, we have that a permutation π avoids the patterns 4231 and 3241 if and only if the permutation $r.c.i(\pi)$ avoids the patterns 4231 and 2431, that is,

$$\mathrm{Av}(2431, 4231) = \{r.c.i(\pi) | \pi \in \mathrm{Av}(4231, 3241)\}$$

and, in particular, $s_n(4231, 3241) = s_n(2431, 4231) = S_{n-1}$.

Definition 2.1.10. We call the sets $\mathrm{Av}(4231, 3241)$ and $\mathrm{Av}(2431, 4231)$ *input-restricted deque permutations* and *output-restricted deque permutations*, respectively.

Nothing is known on enumeration of (general) deque-sortable permutations (in such deques, we can push and pop letters at either end; Knuth posed the problem to study such permutations), although Pratt [658] proved that deque-sortable permutations are characterized by avoiding a certain *infinite* set of patterns.

2.1.4 Sorting with pop-stacks

Avis and Newborn [63] defined the following (less powerful) modification of the stack sorting procedure which they call "*sorting with pop-stacks*". A *pop-stack* is similar to a stack except that the pop operation unloads the entire stack (in the last-in, first-out manner). Figure 2.7 gives an example of a pop-sortable permutation (32154), while Figure 2.8 provides an example of a non-pop-sortable permutation (53412).

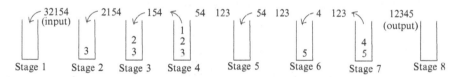

Figure 2.7: Pop-sorting the permutation 32154.

Definition 2.1.11. A permutation is called *layered* if it consists of a disjoint union of factors (the *layers*) so that the letters decrease within each layer, and increase between the layers. For example, 2136547 is a layered permutation with layers 21, 3, 654, and 7. It is an easy exercise to show that $\mathcal{S}_n(231, 312)$ is exactly the set of layered permutations of length n.

Proposition 2.1.12. ([63]) *The set of pop-sortable permutations of length n is $\mathcal{S}_n(231, 312)$. Thus, the number of such permutations is $s_n(231, 312) = 2^{n-1}$.*

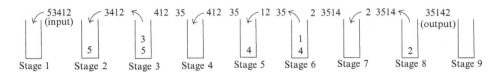

Figure 2.8: Pop-sorting the permutation 53412.

Proof. It is straightforward to see that any layered permutation is sortable with a pop-stack (the first layer of the form $i(i-1)\ldots 1$, $i \geq 1$, will be turned into $12\ldots i$ and the rest will be done by induction on length with the trivial base case — the permutation 1).

Conversely, assuming a permutation π is not pop-stack sortable, the output permutation must contain an inversion $\pi_i\pi_j$ ($\pi_i > \pi_j$). Thinking on what could make π_i precede π_j in the output permutation, we can see that either π contains a subsequence $\pi_i\pi_k\pi_j$ with $\pi_k < \pi_j$, or π contains a subsequence $\pi_i\pi_k\pi_j$ with $\pi_k > \pi_i$. Thus, π either contains the pattern 312, or the pattern 231, or both, and thus $\pi \notin \mathcal{S}_n(231, 312)$.

To enumerate $\mathcal{S}_n(231, 312)$, think of creating a layered permutation of length n by inserting the letters $1, 2, \ldots, n$, one by one, starting with placing 1. Assuming $i-1$ letters have already been placed, $2 \leq i \leq n$, to avoid the patterns 231 and 312, we have two choices for placing i: either immediately to the left of $i-1$ or at the rightmost end of the $(i-1)$-permutation. Thus, $s_n(231, 312) = 2^{n-1}$. $\qquad\square$

Proposition 2.1.12 justifies the fact that the permutation 32154 is sortable this way while 53412 is not.

Avis and Newborn [63] generalized Proposition 2.1.12 by enumerating those permutations that can be sorted with k *pop-stacks in series*. We note that by their interpretation, when the entire set of letters currently in the i-th pop-stack is popped, it is pushed onto the $(i+1)$-th pop-stack. To enumerate the objects, Avis and Newborn [63] proved that the set of permutations sortable by k pop-stacks in series can be characterized by a finite set of forbidden *classical* patterns. Even though the obtained formulas are rather complex, it is interesting that such enumeration can be done, taking into account the complexity of the case of (usual) stacks in series. See [58] for more results in this direction.

To complete the story, the situation for *pop-sorting in parallel* is as follows. Atkinson and Sack [56] proved that the set of permutations sortable with k *pop-stacks in parallel* is also (like in the series case) characterized by a finite set of forbidden *classical* patterns. For example, the following theorem holds.

Theorem 2.1.13. *([56])* $\mathcal{S}_n(3214, 2143, 24135, 41352, 14352, 13542, 13524)$ *is the set of permutations sortable with 2 pop-stacks in parallel. The number s_n of such permutations is defined by the conditions $s_1 = 1$, $s_2 = 2$, $s_3 = 6$, and the recurrence $s_n = 6s_{n-1} - 10s_{n-2} + 6s_{n-3}$.*

Moreover, Atkinson and Sack [56] conjectured that for all k, these permutations have a *rational* generating function. This conjecture was proved in [729] by Smith and Vatter.

Finally, the n-permutations sortable with k pop-stacks in parallel can be recognized in linear time [56].

2.1.5   A generalization of stack sorting permutations

In an (r, s)-stack defined by Atkinson [44], one is allowed to push into any of the first r positions and pop from any of the s positions at the top end of the stack. Notice that a usual stack corresponds to the case $r = s = 1$. Figure 2.9 gives an example of a (2,1)-stack sortable permutation (4231), which, by the way, is not stack sortable, and Figure 2.10 gives an example of a permutation (2341) that is not sortable with the (2,1)-stack. Notice that at Stage 3 in Figure 2.9 we intend to push 3 into the second position from the top instead of popping 2 out taking advantage of the new rules. Similarly, we used this trick (twice) at Stages 2 and 4 in Figure 2.10.

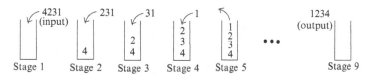

Figure 2.9: (2,1)-stack sorting the permutation 4231.

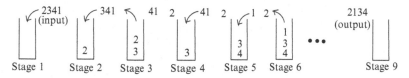

Figure 2.10: (2,1)-stack sorting the permutation 2341.

We conclude the section by stating several results (without proofs) on (r, s)-stack sortable permutations.

Proposition 2.1.14. *([44]) There is a one-to-one correspondence between (r,s)-stack sortable permutations and (s,r)-stack sortable permutations.*

Theorem 2.1.15. *([44]) A permutation is $(r,1)$-stack sortable if and only if it avoids all $r!$ patterns of the form $p_1p_2\cdots p_r(r+2)1$. Also, a permutation is $(1,s)$-stack sortable if and only if it avoids all $s!$ patterns of the form $2p_1p_2\cdots p_s1$.*

Notice that Theorem 2.1.15 is the reason for the permutation 4231 being $(2,1)$-sortable in Figure 2.9, and for the permutation 2341 being not $(2,1)$-sortable in Figure 2.10.

Proposition 2.1.16. *([44]) The set of $(r,1)$-stack sortable permutations, like the set of sortable permutations, has a* closure *property: any subsequence of an $(r,1)$-stack sortable permutation is $(r,1)$-stack sortable.*

Theorem 2.1.17. *([44]) If $n \le r$ then there are $n!$ $(r,1)$-stack sortable n-permutations, while otherwise, this number is the coefficient of x^{n-r+2} in*

$$-\frac{(r-1)!}{2}\sqrt{(r-1)^2x^2 - 2(r+1)x + 1}.$$

Theorem 2.1.18. *([44]) Asymptotically, the number of $(r,1)$-stack sortable permutations is*

$$\frac{1}{2}(r-1)!\sqrt{r^{1/2}/(\pi n^3)}(1+\sqrt{r})^{2n-2r+3}.$$

Theorem 2.1.19. *([44]) A permutation is $(2,2)$-stack sortable if and only if it avoids all of the following 8 patterns:* 23451, 23541, 32451, 32541, 245163, 246153, 425163, *and* 426153.

2.2 Planar maps, trees, bipolar orientations

In Section 2.1 we have already mentioned the fact that proving formula (2.1) for the number of 2-stack sortable permutations combinatorially involved several objects. In this section, we will learn more about these objects and see that they build a layer in a hierarchy related to permutation patterns and studied in [260] in connection with embeddings of certain structures into $\beta(0,1)$-*trees* (to be defined in Subsection 2.2.2). The variety of different classical combinatorial objects related to a single pattern class hierarchy is rather striking. Both the permutation patterns theory and the other structures involved benefit from the connection: for example, the number of 2-stack sortable permutations is obtained this way in a combinatorial fashion; an equidistribution result on non-separable permutations is obtained, to be discussed in Subsection 2.2.3; an alternative proof for the number of *planar bipolar orientations* defined in Subsection 2.2.1 is given via *Baxter permutations* discussed in Subsection 2.2.4.

In Subsections 2.2.1–2.2.6 we provide definitions of our objects of interest and their properties. Then we summarize all the connections in Subsection 2.2.7 (see Figure 2.32).

2.2.1 Planar maps and plane bipolar orientations

Definition 2.2.1. A *planar map* is a connected graph embedded in the plane with no edge-crossings. Such embeddings are considered up to continuous deformation. A map has *vertices* (points), *edges*, and faces (disjoint simply connected domains). The *outer face* is unbounded, the *inner faces* are bounded.

The two graphs below are the same as graphs, but they are different as planar maps since no continuous deformation transforms the first graph to the second one:

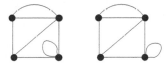

The maps we are dealing with are classical planar maps considered, for example, by Tutte [770] who founded the enumeration theory of planar maps in a series of papers in the 1960s (see [273] for references).

Definition 2.2.2. A *cut vertex* in a map is a vertex whose deletion disconnects the map. A *loop* is an edge whose endpoints coincide. A map is *non-separable* if it has no loops and no cut vertices.

The maps considered by us are *rooted*, meaning that a directed edge is distinguished as the root. Without loss of generality, we can assume that the root is incident to the outer face, and the outer face lies on its right side while following the root orientation. For such an orientation, the outer face will be the *root-face*. In general, the root face of a planar map is the face adjacent to the root that lies to the right of it while following the root direction. Also, the vertex from which the root comes out is called the *root-vertex*.

All rooted non-separable planar maps on 4 edges are given in Figure 2.11.

The number of rooted non-separable planar maps on $n + 1$ edges was first determined by Tutte [769] and it is given by

(2.5)
$$\frac{4(3n)!}{n!(2n+2)!},$$

Figure 2.11: All rooted non-separable planar maps on 4 edges.

which was also proved differently by Brown [190].

Definition 2.2.3. A planar map is *cubic* if all its vertices are of degree 3. A cubic planar map is *bicubic* if it is bipartite, that is, if its vertices can be colored using two colors, say, black and white, so that each edge is incident to different colors.

The simplest cubic non-separable map is the map with two vertices and three edges joining them. It is a well-known fact that the faces of a bicubic map can be colored using three colors so that adjacent faces have distinct colors, say, colors 1, 2, and 3, in a counterclockwise order around white vertices. We can assume that the root vertex is black and the root face has color 3. All bicubic planar maps on 6 edges are given in Figure 2.12.

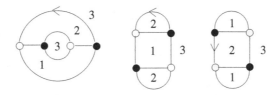

Figure 2.12: All bicubic planar maps on 6 edges.

The number of bicubic planar maps with $3n$ edges was given by Tutte [769]:

$$(2.6) \qquad \frac{3 \cdot 2^{n-1}(2n)!}{n!(n+2)!}.$$

Definition 2.2.4. In a directed graph, a *source* is a vertex with no incoming edges and a *sink* is a vertex with no outgoing edges. A *plane bipolar orientation O* is an acyclic orientation of a planar map with a unique source s and a unique sink t, both located on the outer face. One of the oriented paths going from s to t has the outer face on its right: this path is the *right border* of O, and its length (the number of edges) is the *right outer degree* of O. The *left outer degree* can be defined similarly.

See the rightmost picture in Figure 2.23 for an example of a plane bipolar orientation with right degree 2 and left degree 3. The vertices s and t are called the *poles* of O.

The coefficient of $x^1 y^0$ in the *Tutte polynomial* $T_M(x, y)$ of a non-separable planar map M having a fixed size, is the number of bipolar orientations of M [415, 431]. This number, up to a sign, is also the derivative of the *chromatic polynomial* of M, evaluated at 1 [555].

2.2.2 Description trees and skew ternary trees

Definition 2.2.5. A $\beta(1, 0)$-*tree* is a rooted plane tree labeled with positive integers such that

1. Leaves have label 1.

2. The root has label equal to the sum of its children's labels.

3. Any other node has a label no greater than the sum of its children's labels.

All $\beta(1, 0)$-trees on 3 edges are presented in Figure 2.13.

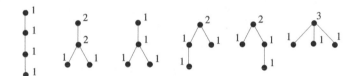

Figure 2.13: All $\beta(1, 0)$-trees on 4 nodes.

To state some of the upcoming results, we first need to define several statistics on $\beta(1, 0)$-trees. These are given in Table 2.1.

For the $\beta(1, 0)$-tree T in Figure 2.14, the values of the statistics are as follows: leaves$(T) = 6$, int$(T) = 5$, root$(T) = 4$, sub$(T) = 3$, lpath$(T) =$ rpath$(T) = 2$, stem$(T) = 1$, lsub$(T) = 2$, and rsub$(T) =$ beta$(T) = 1$. For another example, the tree second from the left in Figure 2.13 has leaves$(T) =$ int$(T) =$ root$(T) = 2$, sub$(T) = 1$, lpath$(T) =$ rpath$(T) =$ stem$(T) = 2$, lsub$(T) =$ rsub$(T) = 1$, and beta$(T) = 2$.

A $\beta(1, 0)$-tree T on at least two nodes is *indecomposable* if sub$(T) = 1$, that is, if the root of T has exactly one child; otherwise, T is *decomposable*. A $\beta(1, 0)$-tree

Statistic	Description in a $\beta(1,0)$-tree T
leaves(T)	# leaves;
int(T)	# internal nodes (or nonleaves);
root(T)	root's label;
sub(T)	# children of (subtrees coming out from) the root;
lpath(T)	# edges from the root to the leftmost leaf = length of the leftmost path (left-path);
rpath(T)	# edges from the root to the rightmost leaf = length of the rightmost path (right-path);
stem(T)	# internal nodes common to the left- and the right-paths;
lsub(T)	# 1's below the root on the left-path;
rsub(T)	# 1's below the root on the right-path;
beta(T)	Let $\ell_1,\ldots,\ell_m$ be the leaves from left to right. If no node on the path from ℓ_1 to the root, except for ℓ_1, has label 1, reduce the labels on all nodes on that path by 1. Now look at ℓ_2 and repeat the process, until we come to a leaf ℓ_i whose path to the root has a node (other than ℓ_i) that now has label 1. Then beta(T) = i.

Table 2.1: Statistics on $\beta(1,0)$-trees as described in [262].

Figure 2.14: A $\beta(1,0)$-tree.

T on at least two nodes is *right-indecomposable* if rsub(T) = 1, that is, if the right-path has exactly one 1 below the root; otherwise, T is *right-decomposable*. The idea of the involution called h on $\beta(1,0)$-trees, defined in [262], is to turn $\beta(1,0)$-tree decompositions into right-decompositions, and vice versa. A recursive description of h is shown schematically in Figure 2.15: as the base case, we map the 1 node tree to itself. In the case of an indecomposable tree, we remove the top edge to get A (the root may need to be adjusted), apply h recursively to get $h(A)$, and adjoin the removed edge to the proper place on the right path of $h(A)$ (to make the *rpath* statistic in the obtained tree equal to x, the value of the *root* statistic in the original tree) increasing all the labels above the rightmost leaf by 1. One the other hand, if our tree is decomposable, we locate its rightmost subtree B (which includes the root of the original tree), apply h recursively on it to get $h(B)$, and glue its rightmost leaf with the root of $h(A)$ obtained recursively (the glue node will receive label 1). See Figure 2.16 for an example of applying the involution h together with some of the steps involved in the recursive procedure.

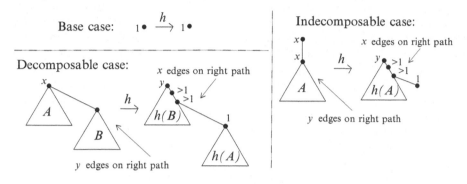

Figure 2.15: A schematic description of the involution h.

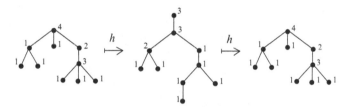

Some of the steps involved in the recursive calculations above:

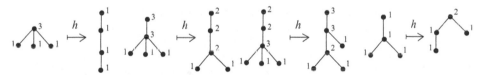

Figure 2.16: An example of applying the involution h together with some of the steps involved in the recursive procedure.

Theorem 2.2.6. *([262]) The map h is an involution (h^2 is the identity map) that sends the first tuple of statistics below to the second one (we refer to [262] for the definition of the gamma statistic; also, we can recall the notion of equidistribution of statistics in Definition 1.0.34):*

(leaves, int,   root,   rpath, sub, rsub, stem,   gamma)
(int,   leaves, rpath, root, rsub, sub, gamma, stem   ).

Another interesting property of h is that when restricted to $\beta(1,0)$-trees with all nodes labeled 1 (except for the root), which can be checked to be closed under h, the involution induces an involution on *unlabeled rooted plane trees*, very classical objects counted by the Catalan numbers (we can erase all labels from such $\beta(1,0)$-trees and reconstruct them, if needed). This involution is new and it is the subject of current studies in [261] by Claesson et al. One of the results that is a direct corollary to Theorem 2.2.6 is the following equidistribution fact; see Section 8.8 for more information on the subject.

Theorem 2.2.7. *([261]) On (unlabeled) rooted plane trees there is an automorphism (a bijection from the set of such trees to itself) that sends the first tuple of statistics below to the second one (in this case, the statistic* rpath *is identical to* rsub, *and* root *is identical to* sub*):*

$$(\text{ leaves, } \quad \text{int, } \quad \text{rpath, } \quad \text{sub })$$
$$(\text{ int, } \quad \text{leaves, } \quad \text{sub, } \quad \text{rpath }).$$

Definition 2.2.8. A $\beta(0,1)$-*tree* is defined on non-negative integers in a similar way to $\beta(1,0)$-trees:

1. Leaves have label 0.

2. The root has label equal to 1 + the sum of its children's labels.

3. Any other node has label no greater than 1 + the sum of its children's labels.

All $\beta(0,1)$-trees on 3 edges are presented in Figure 2.17.

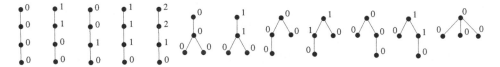

Figure 2.17: All $\beta(0,1)$-trees on 4 nodes.

Definition 2.2.9. A *ternary tree* is a rooted tree whose nodes have at most one son of each of the following three types: *left, middle,* and *right.* Vertices of ternary trees are labelled as follows: the root is labelled 0 and the non-root vertices take the label of their father, to which is added $+1$, $+0$, -1 when they are left, middle, or right sons, respectively. A ternary tree is *skew* if its labels are nonnegative.

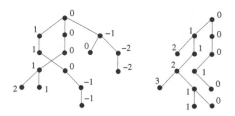

Figure 2.18: An example of a ternary tree and a skew ternary tree.

See Figure 2.18 for an example of a ternary tree and a skew ternary tree.

All but one statistics on skew ternary trees appearing below are straightforward to define: *"number of even labels"*, *"number of odd labels"*, and *"number of zeros"*. The non-straightforward statistic is called *"number of first zeros"* in [476] and it is defined as the maximum number of vertices in the sequence of middle sons with label 0 starting from the root. For example, for the tree to the right in Figure 2.18, the number of even labels is 7, the number of odd labels is 6, the number of zeros is 5, and the number of first zeros is 3.

2.2.3 Relevant pattern-avoidance

A combinatorial proof of West's former conjecture (that the number of 2-stack sortable permutations is given by (2.1)) presented in [324] connects rooted non-separable planar maps with 2-stack sortable permutations through eight different sets of permutations (see Figure 2.19). These sets are in bijection, either because they have isomorphic *generating trees* (out of four generating trees involved only two are identical) or because they can be obtained from each other by applying one of the trivial bijections (r, c, or i). Each of these eight permutation classes could enter Table 2.3 below, and the more general picture in Figure 2.32. However, we are including there only the set of permutations $\mathrm{Av}(2413, 41\overline{3}52) = \mathrm{Av}(2413, 3\underline{1}\underline{4}2)$ that is connected directly to rooted non-separable planar maps and is called the set of *non-separable permutations*. A special interest of this particular set is that the reverse of these permutations was studied in [262] by Claesson et al. in connection with rooted non-separable planar maps, where another bijection, preserving more statistics, was found.

Theorem 2.2.10. *([262]) There is a bijection showing that the first tuple below has the same distribution on $\beta(1, 0)$-trees as the second tuple does on $\mathrm{Av}(3142, 2\underline{41}3)$*

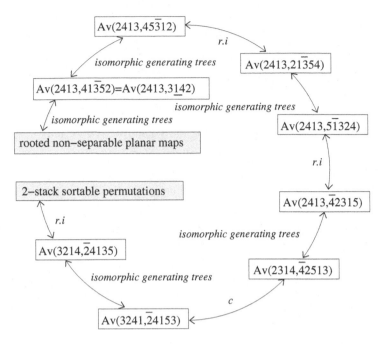

Figure 2.19: A connection between rooted non-separable planar maps and 2-stack sortable permutations through eight classes of pattern-restricted permutations.

(see Tables 2.1 and A.1 for definitions):

$$(\quad \text{sub}, \quad \text{leaves}, \quad \text{root}, \quad \text{lpath}, \quad \text{rpath}, \quad \text{lsub}, \quad \text{beta} \quad)$$
$$(\quad \text{comp}, \quad 1 + \text{asc}, \quad \text{lmax}, \quad \text{lmin}, \quad \text{rmax}, \quad \text{ldr}, \quad \text{lir} \quad).$$

The idea of the bijection proving Theorem 2.2.10 is close to the idea of generating trees: one wants to show that the objects in question can be generated in similar ways. More precisely, irreducible $\beta(1,0)$-trees (having the statistic sub equal to 1) are mapped into irreducible permutations avoiding the patterns (that have the statistic comp equal to 1). To achieve this a non-trivial procedure was found on permutations based on inserting the new maximum letter in proper places of smaller permutations and rearranging the parts to the left and to the right of this letter keeping the same relative orders (see [262] for the actual construction). As a matter of fact, two procedures to do the task may be found in [262], but one of them preserves more statistics of interest than the second one.

Using Theorem 2.2.6 together with the bijection proving Theorem 2.2.10, the following two equidistribution results were obtained (in proving the second equidistribution result the mirror image on $\beta(1,0)$-trees is involved as well).

Theorem 2.2.11. *([262]) The following pairs of tuples of statistics are equidistributed on the set* $\mathrm{Av}(3142, 24\underline{1}3)$*, that is, there is a bijection (automorphism) from* $\mathrm{Av}(3142, 24\underline{1}3)$ *to itself sending the first tuple of statistics to the second one in each pair:*

$$
\begin{array}{llll}
(&\text{asc,}&\text{lmax,}&\text{rmax}&)\\
(&\text{des,}&\text{rmax,}&\text{lmax}&)
\end{array}
$$

and

$$
\begin{array}{llllll}
(&\text{asc,}&\text{lmax,}&\text{lmin,}&\text{comp,}&\text{ldr}&)\\
(&\text{des,}&\text{lmin,}&\text{lmax,}&\text{ldr,}&\text{comp}&).
\end{array}
$$

Note that the first equidistribution result in Theorem 2.2.11, unlike the second one, is trivial on the set of *all* permutations: all one needs to do is to apply the reverse operation to the set of permutations. However, proving the same result on $\mathrm{Av}(3142, 24\underline{1}3)$ was unsuccessful for a long time before the involution h was invented. A direct (combinatorial) proof of results in Theorem 2.2.11 would be desirable. The diagrams in Figure 2.20 created by Anders Claesson may be of some help in solving the problem, though it is definitely not sufficient just to use them. The diagrams show translations of relevant statistics under compositions of two trivial bijections. We can use the fact that $i.r = c.i$, which is rather easy to prove, to obtain translations under compositions that are not present in the diagrams. The idea to approach finding a combinatorial proof of Theorem 2.2.11 is to use the fact that the set $\mathrm{Av}(3142, 24\underline{1}3)$ is closed under compositions of two trivial bijections, although it is not closed under any single such bijection.

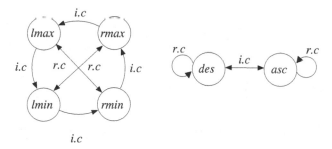

Figure 2.20: Translation of the statistics lmax, rmax, lmin, rmin, asc, and des under compositions involving two trivial bijections. To complete the picture, we can use $i.r = c.i$.

If one wishes to code rooted non-separable planar maps (equivalently, $\beta(1,0)$-trees) by permutations, it seems that the set Av$(3142, 2\underline{41}3)$ (equivalently for non-separable permutations, Av$(2413, 3\underline{14}2)$) is a better choice than 2-stack sortable permutations. Indeed, Av$(3142, 2\underline{41}3)$ is more symmetric (it is closed under compositions of two trivial bijections while the set of 2-stack sortable permutations is not closed under any combination of trivial bijections), and Av$(3142, 2\underline{41}3)$, under known bijections, captures better the structure of rooted non-separable planar maps by keeping track of more statistics than in the case of 2-stack sortable permutations. In either case, the following conjecture appears in [262] by Claesson et al. on relations between Av$(3142, 2\underline{41}3)$ and 2-stack sortable permutations.

Conjecture 2.2.12. ([262]) The quadruple $(comp, asc, ldr, rmax)$ has the same distribution on $\mathcal{S}_n(3142, 2\underline{41}3)$ as it has on 2-stack sortable permutations of length n.

Remark 2.2.13. It is known [323] that the pair of statistics $(asc, lmax)$ on the class Av$(3142, 2\underline{41}3)$ is equidistributed with the pair $(des, rmax)$ on 2-stack sortable permutations; this fact also follows from Table 2.3 below. If Conjecture 2.2.12 is true, then it would strengthen the result in [323].

In discussing the coding of planar maps/description trees by permutations, we would like to mention [127] by Bóna who enumerated Av(1342) (see Table 6.2). This was the first case of enumeration of non-monotonic patterns of length more than 3 and this result is relevant to the hierarchy we will discuss in Subsection 2.2.7 (see Figure 2.32).

Theorem 2.2.14. *([127]) The following three sets of objects are in one-to-one correspondence:*

- Av(1342);

- *Plane forests of $\beta(0,1)$-trees;*

- *Ordered collections of (rooted) bicubic planar maps.*

Using Theorem 2.2.14 and a known enumerative result from [769], the following theorem was proved.

Theorem 2.2.15. *([127]) One has the following enumeration results for Av(1342):*

- *The generating function for $s_n(1342)$ is*

$$\sum_{n\geq 0} s_n(1342)x^n = \frac{32x}{-8x^2 + 12x + 1 - (1-8x)^{3/2}};$$

- *The exact formula for $s_n(1342)$ is*

$$(-1)^{n-1}\frac{7n^2 - 3n - 2}{2} + 3\sum_{i=2}^{n}(-1)^{n-i}2^{i+1}\frac{(2i-4)!}{i!(i-2)!}\binom{n-i+2}{2};$$

- $\sqrt[n]{s_n(1342)}$ *converges to 8 when $n \to \infty$.*

We conclude this subsection with another result relevant to the hierarchy to be discussed in Subsection 2.2.7.

Theorem 2.2.16. *([260]) For $n \geq 0$, $s_n(24\underline{1}3) = s_n(34\underline{1}2)$. Moreover, there is a bijection between the sets that sends a permutation in $S_n(24\underline{1}3)$ with k occurrences of $34\underline{1}2$ to a permutation in $S_n(34\underline{1}2)$ with k occurrences of $24\underline{1}3$.*

Proof. For this proof, we let $P_1 = 24\underline{1}3$ and $P_2 = 34\underline{1}2$.

If a permutation avoids P_1 and P_2, we map it to itself.

Now suppose that an n-permutation π avoids P_1 and it contains at least one occurrence of P_2. We consider the leftmost pair, say xy, of consecutive letters, depicted in Figure 2.21 by the solid circles, that play the role of 4 and 1 in an occurrence of the pattern P_2; notice that this pair is well-defined.

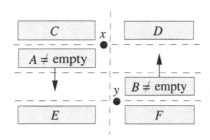

Figure 2.21: Sending a permutation from $S_n(P_1)$ to a permutation in $S_n(P_2)$.

One immediately realizes that if we restrict ourselves to the values of π between x and y, then in order to avoid P_1, everything to the left of xy must be larger than everything to the right of xy, which is shown schematically by the non-empty rectangles A and B in Figure 2.21. Notice that xy contributes $|A| \cdot |B|$ occurrences of P_2, where $|X|$ denotes the number of letters in X. The letters in A, C and E can be shuffled somehow, but they (together) do not contain an occurrence of P_1 or a pair of consecutive letters having the properties that xy above has (xy is the leftmost one with such properties). Also, D, B and F can be shuffled somehow in such a way that they (together) do not contain P_1.

We now decrease each letter of A by $|B|$ and increase each letter of B by $|A|$ thus turning each occurrence of P_2 involving xy into an occurrence of P_1 involving xy. We denote the resulting permutation π'.

Our claim is that no new occurrences of P_2 are introduced and no (new) occurrences of P_1, beyond those involving xy, are introduced after the described procedure is done. This follows easily from the fact that the elements of A (resp., B) do not change their relative position with respect to the elements of C and E (resp., D and F).

We can now proceed with π' and find $x'y'$, if there is one (otherwise we do not need to do anything), having the properties of xy. One then changes all the occurrences of P_2 involving $x'y'$ into occurrences of P_1 involving $x'y'$, in the way we did it above. There is only one difference between considering π and π': π contains no P_1, whereas π' does. However, the occurrences of P_1 in π' will not be affected by the procedure, again, because of the properties of A, C, and E. Indeed, either such an occurrence of P_1 is entirely in A, or in C, or in E, in which case it cannot disappear, or the occurrence has the letters corresponding to 2, 4, and 1 in P_1 either in C or in E, which, again, cannot cause the occurrence to disappear.

Thus we can go through all pairs xy from left to right and change all occurrences of P_2 to occurrences of P_1. The process terminates because of the fact that $x'y'$ is strictly to the right of xy. The map is easily seen to be injective and reversible, and it is easy to see that it sends a permutation in $S_n(P_1)$ with k occurrences of P_2 to a permutation in $S_n(P_2)$ with k occurrences of P_1. $\qquad\square$

Remark 2.2.17. It is straightforward to see from the proof of Theorem 2.2.16 that the bijection there preserves (sends to themselves) an enormous number of permutation statistics, which includes (but is not limited to!) the following statistics (see Table A.1 for definitions): maj, lmax, lmin, rmax, rmin, des, peak, last $.i$, head $.i$, ldr, lir, rdr, rir, comp, ddes, and dasc.

Remark 2.2.18. It follows from Proposition 1.3.7 and Theorem 2.2.16 that

$$s_n(2\underline{41}3) = s_n(3\underline{41}2) = s_n(21\overline{3}54).$$

2.2.4 Baxter permutations

In 1964, Glen Baxter [98] introduced the following class of permutations that now bears his name.

Definition 2.2.19. A permutation $\pi = \pi_1\pi_2\ldots\pi_n$ is a *Baxter permutation* if there are no four indices $i < j < k < \ell$ such that

- $k = j + 1$;

- $\pi_i \pi_j \pi_k \pi_\ell$ is an occurrence of the pattern 2413 or 3142.

In the language of vincular patterns, $\mathrm{Av}(2\underline{41}3, 3\underline{14}2)$ is the set of Baxter permutations.

Example 2.2.20. 25314 is a Baxter permutation, whereas 5327146 is not a Baxter permutation as it contains an occurrence of the pattern $3\underline{14}2$ (the subsequence 5274).

Gire [325, 419] showed that $\mathrm{Av}(25\overline{3}14, 41\overline{3}52)$ is exactly the set of Baxter permutations. The motivation for introducing the permutations defined in Definition 2.2.19 was the following problem in analysis on commuting continuous functions.

Suppose f and g are continuous functions from $[0,1]$ to $[0,1]$ that commute, that it, $g(f(x)) = f(g(x))$. We let $h(x)$ denote $g(f(x)) = f(g(x))$. Suppose h has finitely many fixed points $x_1 < x_2 < \cdots < x_n$. A fixed point x_i of $h(x)$ may have one of the following three types:

1. x_i is *up-crossing* if the sign of $h(x) - x$ changes from negative to positive in the neighborhood of x_i;

2. x_i is *down-crossing* if the sign of $h(x) - x$ changes from positive to negative in the neighborhood of x_i;

3. x_i is *touching* if $h(x) - x$ does not change sign in the neighborhood of x_i.

Example 2.2.21. In Figure 2.22, d is an up-crossing fixed point, b and f are down-crossing fixed points, and a, c, and e are touching fixed points.

Theorem 2.2.22. *([98, 167]) For the objects defined above, we have the following facts ($f(x)$ can be substituted by $g(x)$ throughout):*

- *The function f maps the fixed points $x_1, x_2, \ldots, x_n$ of h bijectively onto themselves;*

- *The fixed point $f(x_i)$ has the same type as x_i has;*

- *The permutation of the up-crossing fixed points is determined by the permutation of the down-crossing fixed points;*

- *The permutation of the down-crossing fixed points is a Baxter permutation.*

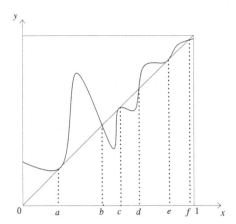

Figure 2.22: Fixed points of different types.

Baxter permutations are a widely studied class of permutations (see, for example, [98, 151, 167, 158, 272, 221, 257, 250, 379, 326, 419, 436, 620, 581, 787]). It is known that Baxter permutations of length n are equinumerous with several combinatorial objects, for example, with certain *rectangulations* with n points on the diagonal ([4, 379]) and with *plane bipolar orientations with n edges* (discussed in Subsection 2.2.1). Since the last set of objects is connected to maps, which is of special interest to us, we would like to sketch the idea of the bijection in [151] between Baxter permutations and plane bipolar orientations. We will explain the idea on the example in Figure 2.23 (we refer to [151] for a detailed description of the bijection).

Given a Baxter permutation, we start by drawing its permutation matrix using black circles. Next we add two white rectangles representing the poles, and we add white circles in certain places right after each ascent position (whenever we are coming from a smaller letter to a larger one while going from left to right) as shown in Figure 2.23. Then, starting from the source rectangle, we connect rectangles/circles (referred to as *nodes* in what follows) in the following way. Given a node x, draw an arrow from it to *each* node that is *visible directly* from x in the North-East direction. For example, from the leftmost white circle in Figure 2.23 we can see directly three black circles corresponding to the letters 7, 5, and 4 in the permutation 37568412. Thus, each directed path from the source to the sink is an alternation of black and white circles starting and ending with black circles. As the final step, we remove the black circles making the arrows going through them continuous as shown in Figure 2.23. The resulting object is a plane bipolar orientation.

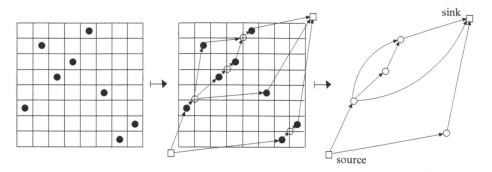

Figure 2.23: The Baxter permutation 37568412 and its transformation into a plane bipolar orientation under the bijection given in [151].

It is a funny coincidence, pointed out by Bonichon et al. in [151], that another Baxter, the physicist *Rodney* Baxter, in his studies [99] came across objects equinumerous with Baxter permutations without realizing the fact that the numbers he was dealing with are known and moreover, they bear his name! Baxter, the physicist, studied the sum of the Tutte polynomials $T_M(x, y)$ of non-separable planar maps M having a fixed size (see [99]). He found the coefficient of $x^1 y^0$ in $T_M(x, y)$, summed over all rooted non-separable planar maps M having $n + 1$ edges, $m + 2$ vertices, root-face of degree $i + 1$ and a root-vertex of degree $j + 1$. As was already mentioned in Subsection 2.2.1, this coefficient counts plane bipolar orientations of M (equivalently, Baxter permutations).

We will close our discussion of Baxter permutations for now by stating a few enumerative results on them. However, we will see Baxter permutations later in the book, in connection with pattern-avoidance in so-called *partial permutations*.

Theorem 2.2.23. *([250]) The number of Baxter permutations of length n is given by*

(2.7)
$$\sum_{i=0}^{n-1} \frac{\binom{n+1}{i}\binom{n+1}{i+1}\binom{n+1}{i+2}}{\binom{n+1}{1}\binom{n+1}{2}}.$$

While the proof of Theorem 2.2.23 given by Chung et al. [250] is analytical, Viennot [787] provided a bijective proof of formula (2.8). The following theorem is a refinement of Theorem 2.2.23.

Theorem 2.2.24. *([581]) The number of Baxter permutations of length n having m ascents, i left-to-right maxima and j right-to-left maxima (see Definition A.1.1)*

is given by
(2.8)
$$\frac{ij}{n(n+1)}\binom{n+1}{m+1}\left[\binom{n-i-1}{n-m-2}\binom{n-j-1}{m-1} - \binom{n-i-1}{n-m-1}\binom{n-j-1}{m}\right].$$

Review Definition 1.0.18 for the notion of alternating permutations.

Definition 2.2.25. A permutation is *doubly alternating* if it is alternating *and* its inverse is alternating.

Example 2.2.26. 13254 and 354612 are examples of doubly alternating (in fact, Baxter) permutations, whereas the permutation 24153 is not doubly alternating, as its inverse, $i(24153) = 31524$, is not alternating (it starts with a descent, not an ascent).

Theorem 2.2.27. *([272]) The number of alternating Baxter permutations of length $2n$ and $2n+1$ is given by C_n^2 and $C_n C_{n+1}$, respectively, where $C_n = \frac{1}{n+1}\binom{2n}{n}$ is the n-th Catalan number.*

Theorem 2.2.28. *([436]) The number of doubly alternating Baxter permutations of length $2n$ or $2n+1$ is given by $C_n = \frac{1}{n+1}\binom{2n}{n}$, the n-th Catalan number.*

The research done in [620] by Mansour and Vajnovszki is of the type discussed in Problem 1.7.13, namely, one restricts the set of objects (Baxter permutations in our case) by some conditions (in this case, the permutations must avoid the pattern 123) and then additional avoidance or containment constraints are considered. We provide here just two theorems proved in [620]. For more results of this type see Subsections 6.1.5 and 7.1.4.

Theorem 2.2.29. *([620]) The generating function for the number of 123-avoiding Baxter permutations is given by*

$$\frac{(1-x)^2}{1 - 3x + 2x^2 - x^3}.$$

In other words, the number of 123-avoiding n-permutations is given by the $(3n+3)$-th Padovan number.

Theorem 2.2.30. *([620]) The number of 123-avoiding Baxter permutations containing exactly r occurrences of the vincular pattern $\underline{132}$ (or $\underline{213}$) is given by*

$$\sum_{i=0}^{n-3r} 2^{n-3r-i}\binom{i+r-2}{r-2}\binom{n-3r-i+r}{r}.$$

Finally, another result related to Problem 1.7.13 is the following theorem dealing with Baxter involutions, that is, Baxter permutations whose (usual group-theoretical) square is the identity permutation.

Theorem 2.2.31. *([151]) The number of fixed-point-free Baxter involutions of length* $2n$ *is*

$$\frac{3 \cdot 2^{n-1}}{(n+1)(n+2)}\binom{2n}{n}.$$

2.2.5 Separable permutations

Definition 2.2.32. Suppose $\pi = \pi_1 \pi_2 \ldots \pi_m \in \mathcal{S}_m$ and $\sigma = \sigma_1 \sigma_2 \ldots \sigma_n \in \mathcal{S}_n$. We define the *direct sum* (or simply, *sum*) $\oplus$, and the *skew sum* $\ominus$ by building the permutations $\pi \oplus \sigma$ and $\pi \ominus \sigma$ as follows:

$$(\pi \oplus \sigma)_i = \begin{cases} \pi_i & \text{if } 1 \leq i \leq m, \\ \sigma_{i-m} + m & \text{if } m+1 \leq i \leq m+n, \end{cases}$$

$$(\pi \ominus \sigma)_i = \begin{cases} \pi_i + n & \text{if } 1 \leq i \leq m, \\ \sigma_{i-m} & \text{if } m+1 \leq i \leq m+n. \end{cases}$$

Example 2.2.33. For example, $14325 \oplus 4231 = 143259786$ and $14325 \ominus 4231 = 587694231$. This example is best understood by looking at the permutation matrices in Figure 2.24 of the permutations involved.

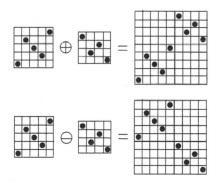

Figure 2.24: Permutation matrices illustration of the fact that $14325 \oplus 4231 = 143259786$ and $14325 \oplus 4231 = 587694231$.

Definition 2.2.34. The *separable permutations* are those which can be built from the permutation 1 by repeatedly applying the $\oplus$ and $\ominus$ operations.

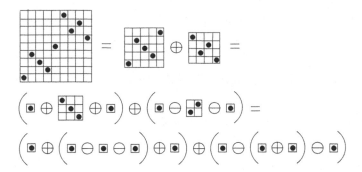

Figure 2.25: Decomposition of 143259786 using $\oplus$ and $\ominus$ operations.

Example 2.2.35. The permutations 143259786 and 587694231 appearing in Figure 2.24 are both separable. Figure 2.25 illustrates a step-by-step procedure to see that 143259786 is separable. All permutations of length 3 are separable, and only two permutations of length 4, 2413 and 3142, are not separable.

Bose et al. [155] introduced the notion of separable permutation in 1998, but the following well-known result is folkloric.

Theorem 2.2.36. *(folklore)* Av(2413, 3142) *is the set of all separable permutations.*

Remark 2.2.37. So far we introduced, in Definition 2.2.34, the class of separable permutations, Av(2413, 3142), and the class of non-separable permutations, Av(3142, 2413), in Subsection 2.2.3. We note that Av(2413, 3142) $\subseteq$ Av(3142, 2413) and thus each separable permutation is a non-separable one in our definitions (which sounds contradictory, but this is just a matter of names) but not vice versa. The word "separable" in the case of separable permutations came from the "process of separation," or decomposing permutations, whereas the word "non-separable" in the case of non-separable permutations came from a plain connection of the permutations to rooted non-separable planar maps, which has nothing to do with any separability on permutations themselves.

Throughout this book, by a "permutation class" we simply mean a set of permutations. In several places, however, to be mentioned explicitly, this expression has a stronger sense.

Definition 2.2.38. For permutations σ and π, we write $\sigma \leq \pi$ if σ occurs in π as a pattern (there exits a subsequence in π of the same length as σ that is order-isomorphic to σ). Thus we can define the *containment order* on the set of all permutations. Sets of permutations which are closed downward under this order are called *permutation classes* (or just *classes*). In other words, $\mathcal{C}$ is a class if for any $\pi \in \mathcal{C}$ and any $\sigma \leq \pi$, we have $\sigma \in \mathcal{C}$.

Example 2.2.39. It is easy to see that for any set P of classical patterns, $\text{Av}(P)$ is a (permutation) class, whereas if other patterns are involved, that does not have to be the case. Indeed, consider the permutation $23154 \in \text{Av}(12\underline{43})$. $\text{red}(2354) = 1243 \leq 23154$ and $1243 \notin \text{Av}(12\underline{43})$. Thus, $\text{Av}(12\underline{43})$ is not a permutation class.

Definition 2.2.40. For two sets (classes) of permutations $\mathcal{C}$ and $\mathcal{D}$ we let

$$\mathcal{C} \oplus \mathcal{D} = \{\pi \oplus \sigma \,|\, \pi \in \mathcal{C}, \sigma \in \mathcal{D}\},$$
$$\mathcal{C} \ominus \mathcal{D} = \{\pi \ominus \sigma \,|\, \pi \in \mathcal{C}, \sigma \in \mathcal{D}\}.$$

We would like to discuss just a couple of results on separable permutations. For more results on them, consult [27, 338, 546, 734].

Proposition 2.2.41. *([27]) The class of separable permutations* $\text{Av}(2413, 3142)$ *is the smallest nonempty class* $\mathcal{C}$ *which satisfies both* $\mathcal{C} \oplus \mathcal{C} \subseteq \mathcal{C}$ *and* $\mathcal{C} \ominus \mathcal{C} \subseteq \mathcal{C}$.

Since each of the patterns 2413 and 3142 contains every length 3 non-monotone pattern, all four of the classes $\text{Av}(132)$, $\text{Av}(213)$, $\text{Av}(132)$ and $\text{Av}(312)$ are contained in $\text{Av}(2413, 3142)$, and each of these has a characterization similar to one given by the following proposition.

Proposition 2.2.42. *([27]) The class* $\text{Av}(231)$ *is the smallest nonempty class* $\mathcal{C}$ *which satisfies both* $\mathcal{C} \oplus \mathcal{C} \subseteq \mathcal{C}$ *and* $1 \ominus \mathcal{C} \subseteq \mathcal{C}$.

One more similar result deals with so-called *skew-merged permutations* defined below in Definition 6.1.7.

Proposition 2.2.43. *([27]) The class* $\text{Av}(2143, 2413, 3142, 3412)$ *of separable skew-merged permutations is the smallest nonempty class* $\mathcal{C}$ *which contains* $\mathcal{C} \oplus 1$, $1 \oplus \mathcal{C}$, $\mathcal{C} \ominus 1$ *and* $1 \ominus \mathcal{C}$.

The following theorem is the main result in [27] by Albert et al.

Theorem 2.2.44. *([27]) If* $\mathcal{C}$ *is a subclass of the separable permutations that does not contain any of* $\text{Av}(132)$, $\text{Av}(213)$, $\text{Av}(231)$ *or* $\text{Av}(312)$ *then* $\mathcal{C}$ *has a rational generating function.*

It was conjectured by Shapiro and Getu and, for the first time, proved by West [801] that the set of separable permutations of length n, $\mathcal{S}_n(3142, 2413)$, is counted by the $(n-1)$-th Schröder number. The proof involves studying the generating tree for the restricted permutations and it uses a well-known relation between the Schröder numbers S_n and the Catalan numbers C_n:

$$(2.9) \qquad\qquad S_n = \sum_{i=0}^{n} \binom{2n-i}{i} C_{n-i}.$$

Theorem 2.2.45. *([801, 734, 27, 260]) The set of separable permutations of length* n, $\mathcal{S}_n(3142, 2413)$, *is counted by the* $(n-1)$-*th Schröder number.*

West asked for a more natural proof of the enumerative result, which was provided by Stankova [734]. In a recent paper, Albert et al. [27] provided a rather simple proof of the same result using decompositions involving the $\oplus$ and $\ominus$ operations. Using the same approach, the authors also gave an alternative proof of the fact that $\mathrm{Av}(231)$ is counted by the Catalan numbers. Two other proofs of the same result are presented in [260] by Claesson et al. In those bijective proofs, the Schröder paths (counted by the Schröder numbers) are mapped bijectively to plane rooted trees where some of the leaves may be marked, and then two bijections are found between the marked trees and $\mathrm{Av}(3142, 2413)$, simply by showing how to generate the permutations using relation (A.4) in two different ways.

The presentation in the rest of the subsection is based on [260].

Formula (2.9) is a standard one for calculating the Schröder numbers, but we can use another formula, which appears in [679]:

$$(2.10) \qquad\qquad S_n = \sum_{k=0}^{n} 2^k C_{n,k},$$

where $C_{n,k}$ is the number of Dyck paths of length $2n$ with k peaks (see Subsection A.2.2 for definitions). Indeed, if one takes a Dyck path of length $2n$ with k peaks then there are 2^k ways to decide which of the peaks will be turned into a double horizontal step, hh, thus ending up with a Schröder path of length $2n$. This procedure is obviously reversible.

There is an easy and standard correspondence between plane rooted trees with k leaves and Dyck paths with k peaks: one traverses a tree from the root (located, say, on top) using the *leftmost depth first algorithm*, and each step down in the tree corresponds to an up step in the Dyck path, whereas each step up in the tree corresponds to a down step in the Dyck path. See Figure 2.26 for an example of this correspondence.

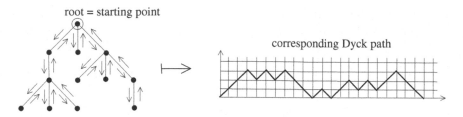

Figure 2.26: An example of a correspondence between plane rooted trees and Dyck paths.

To adapt the correspondence above for Schröder paths, we mark some of the leaves (maybe none, or all) in a tree with a star, which, once a marked leaf is reached, will instruct us to make a double horizontal step instead of creating a peak in the corresponding Schröder path. We call such trees *marked trees*. See Figure 2.27 for an example of the correspondence between Schröder paths and marked trees.

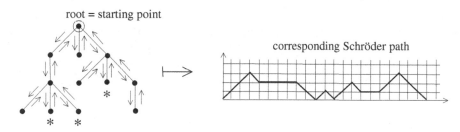

Figure 2.27: An example of a correspondence between plane rooted trees with marked leaves and Schröder paths.

We now interpret formula (A.4) ($S = 1 + hhS + uSdS$) generating the Schröder paths, as a generating relation for marked trees. Indeed, either a tree has one node (which cannot be marked by definition), or its root r has as its leftmost child a marked leaf (giving term hhS), or the leftmost child of r is the root of a tree (possibly a single node tree) and removing this tree leaves a tree with root r (this corresponds to the term $uSdS$ in (A.4)).

Using the interpretation above, we can easily see that all marked trees on n nodes can be generated from smaller marked trees using two operations: $\gamma_t^*(T)$ which adjoins to the tree T a marked leaf as the leftmost child of the root, and the $\oplus_t$ operation taking two trees as arguments and making the root of the left tree be the leftmost child of the root of the right tree (this adds an extra edge). In Figure 2.28, we show how to generate all marked trees on 3 nodes using the operations $\oplus_t$ and γ_t^*.

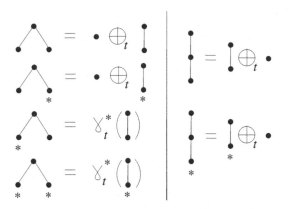

Figure 2.28: Generating all marked trees on 3 nodes.

The induced operation $\gamma_s^*(P)$ on Schröder paths corresponding to $\gamma_t^*(T)$ adjoins hh to the left of P. The operation $\oplus_s$ on paths corresponding to $\oplus_t$ on trees is defined as follows: for paths P_1 and P_2, $P_1 \oplus_s P_2$ is the Schröder path obtained by beginning with an up-step, then following P_1, then making a down-step and, finally, following P_2.

We distinguish two types of marked trees: trees of *type* 1 have the leftmost leaf marked, and all other trees are of *type* 2. Clearly, γ_t^* produces type 1 trees, while the type of $T_1 \oplus_t T_2$ is determined by T_1. Note that the induced definition for the Schröder paths is that if the leftmost increasing run of up-steps ends with a horizontal step (in particular, if a path begins with a horizontal step) then we have a *type* 1 *Schröder path*; otherwise we deal with a *type* 2 *Schröder path*. The number of type 1 trees/paths is easily seen to be the same as the number of type 2 trees/paths through a trivial bijection (involution) removing/adding a mark on the leftmost leaf for trees, and changing the leftmost peak to a double horizontal step and vice versa for paths. Thus the number of objects of each type is given by the *small Schröder numbers* (see Subsection A.2.1 for definition).

The statistic lpath(T), the length of the leftmost path, for a plane rooted tree T is defined as for the $\beta(1, 0)$-trees (see Table 2.1). We slightly change this definition for marked trees to define the statistic *lpath*$^*(T)$, the number of *non-marked* nodes on the leftmost path of a marked tree T below the root. Note that this statistic corresponds to the *length of the leftmost increasing run, lirun*, on the Schröder paths, that is, to the maximal number of the consecutive up-steps beginning a path. For the tree T in Figure 2.27, $lpath^*(T) = 3$ which matches the value $lirun(P) = 3$ for the corresponding Schröder path P, whereas in Figure 2.29, $lpath^*(T) = lirun(P) = 2$.

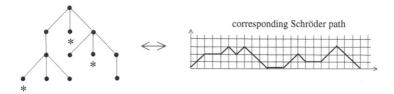

Figure 2.29: A marked tree and the Schröder path corresponding to it.

Another statistic of interest is 1 plus the number of marked leaves directly connected to the root in a marked tree, which we call $comp_t$, the *number of components*. This statistic obviously corresponds to 1 plus the number of hh steps on the ground level (x-axis) on the Schröder paths, which we call $comp_s$. For the tree T and the corresponding Schröder path P in Figure 2.27, we have $comp_t(T) = comp_s(P) = 1$, whereas in Figure 2.29, $comp_t(T) = comp_s(P) = 2$.

We can see that any $\pi \in \mathcal{S}_n(2413, 3142)$ has the following structure (see also Figure 2.30 for a schematic view of corresponding permutation matrices):

$$(2.11) \qquad \pi = L_1 L_2 \cdots L_m n R_m R_{m-1} \cdots R_1$$

where

- for $1 \leq i \leq m$, L_i and R_i are non-empty, with a possible exception for L_1 and R_m, separable permutations which are intervals in π;

- $L_1 < R_1 < L_2 < R_2 < \cdots < L_m < R_m$, where $A < B$, for two permutations A and B, means that each letter of A is less than every letter of B. In particular, L_1, if it is not empty, contains 1.

For example, if $\pi = 215643$ then $L_1 = 21$, $L_2 = 5$, $R_1 = 43$ and $R_2 = \emptyset$.

If $\pi \in \mathcal{S}_n(2413, 3142)$ and $\pi = L_1 L_2 \cdots L_m n R_m R_{m-1} \cdots R_1$, then the number of *left components* $lcomp(\pi) = comp(red(L_1 L_2 \cdots L_m n))$, that is, lcomp gives the number of components to the left of the largest letter in a separable permutation. For example, $lcomp(21376854) = 3$ which is the number of components in $red(21376) = 21354 = 21 - 3 - 54$. Note that if the largest letter is the leftmost letter in a permutation, then lcomp is 0. In the bijection we will describe below, lcomp will correspond to $lpath^*$ and to lirun.

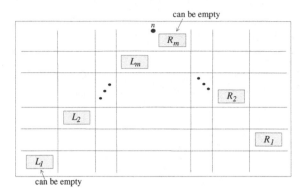

Figure 2.30: Schematic view of permutation matrices corresponding to separable permutations. Each L_i and R_j is a separable permutation.

We say that a separable permutation is of *type 1* if $L_1 = \emptyset$ in its decomposition (2.11), and it is of *type 2* otherwise. Clearly, by applying the reverse operation r, we see that the number of separable permutations of type 1 is the same as that of type 2 and is thus given by the small Schröder numbers.

Theorem 2.2.46. *([260]) There is a bijection between the separable permutations in $\mathcal{S}_{n+1}(2413, 3142)$ and the Schröder paths of length $2n$ (resp. plane rooted marked trees with n edges) such that the statistic* lcomp *on permutations corresponds to* lirun *on paths (resp.,* lpath* *on trees).*

Proof. We follow formula (A.4) to show a way to generate all separable permutations so that type 1 and type 2 permutations will correspond to type 1 and type 2 trees and paths, respectively. Also, from the generation it will be clear that lcomp on separable permutations will correspond to lpath* on marked trees and lirun on paths.

For separable permutations an analogue of the operations γ_t^*/γ_s^* on trees/paths, denoted γ_p^*, inserts the new largest letter in front of a given separable permutation. For example, $\gamma_p^*(21543) = 621543$. Clearly, this operation does not introduce any of the prohibited patterns. Also, it turns any permutation into a type 1 permutation. Finally, lcomp$(\gamma_p^*(\pi)) = 0$ for any permutation π which agrees well with the behavior of lpath* and lirun under γ_t^* and γ_s^*, respectively.

We now introduce $\oplus_p$, an analogue of the $\oplus_t/\oplus_s$ operations, on separable permutations. Suppose $\pi = KnP$ and $\sigma = LR$ are two separable permutations where n is the largest letter in π and $L \neq \emptyset$ is the leftmost irreducible component of σ (it is possible that $R = \emptyset$). Then, by definition,

$$\pi \oplus_p \sigma = K L^+ n^+ R^+ P,$$

where L^+ and R^+ are obtained from L and R, respectively, by increasing all of their letters by $|\pi| - 1$ and $n^+ = |\pi| + |\sigma|$. For example, $23154 \oplus_p 21543 = 23165(10)9874$.

We now make a few remarks. First of all, the operation $\oplus_p$ on separable permutations does not introduce occurrences of the prohibited patterns since the resulting permutation has structure (2.11). Second, the operation is proper with respect to lengths. Indeed, we would like permutations of length $(n+1)$ to correspond to trees with n edges; if T_1 and T_2 are trees corresponding to permutations π and σ, respectively, then $T_1 \oplus_t T_2$ has one more edge than the number of edges in T_1 and T_2, which is consistent with the fact that $|\pi \oplus_p \sigma| = |\pi| + |\sigma|$ (π (resp., σ) has one more element than the number of edges in T_1 (resp., T_2)). Moreover, obviously $\mathrm{lcomp}(\pi \oplus_p \sigma) = \mathrm{lcomp}(\pi) + 1$ which agrees well with the $\oplus_t$ operation on trees and $\oplus_s$ on paths: for instance, for trees, this operation applied to trees T_1 and T_2 produces a tree with lpath* statistic one more than lpath*(T_1). Finally, it is not hard to see that $\oplus_p$ is reversible like $\oplus_t$ and $\oplus_s$ are on trees and paths, respectively. $\square$

Remark 2.2.47. Based on computer experiments, we get that the result in Theorem 2.2.46 is maximal with respect to statistics, in the sense that no additional statistics from Table A.1, and their variations under trivial bijections, can be preserved if we require the statistics lcomp, lpath*, and lirun to correspond to each other. However, we can modify the bijection in Theorem 2.2.46 by generating the separable permutations differently, to prove Theorem 2.2.48 dealing with other, even more natural statistics, and again, providing a maximal result with respect to statistics in Table A.1.

Theorem 2.2.48. *([260]) There is a bijection between the separable permutations in $\mathcal{S}_{n+1}(2413, 3142)$ and the Schröder paths of length $2n$ (resp. plane rooted marked trees with n edges) such that the statistic* comp *on permutations corresponds to* comp$_s$ *on paths (resp.,* comp$_t$ *on trees).*

Proof. Notice how the statistics comp$_s$/comp$_t$ on the Schröder paths/marked trees (counting hh steps on the ground level/marked leaves directly connected to the root) change while generating the objects: comp$_s(\gamma_s^*(P)) = 1 + \mathrm{comp}_s(P)$, comp$_s(P_1 \oplus_s P_2) = \mathrm{comp}_s(P_2)$, comp$_t(\gamma_t^*(T)) = 1 + \mathrm{comp}_t(T)$, and comp$_t(T_1 \oplus_t T_2) = \mathrm{comp}_t(T_2)$.

We now introduce the following modifications to γ_p^* and $\oplus_p$ defined in the proof of Theorem 2.2.46 which we will call γ_p^{**} and $\oplus_p'$, respectively. The operation γ_p^{**} inserts the new largest letter at the end of a given separable permutation. For example, $\gamma_p^{**}(21543) = 215436$. Notice that γ_p^{**} increases the number of components by 1 (the largest letter is a component by itself) as it is supposed to mimic the behavior of comp_s and comp_t.

Next, for two separable permutations $\pi \neq \emptyset$ and $\sigma = LnR \neq \emptyset$ (n is the maximum letter in σ) we define

$$\pi \oplus'_p \sigma = \begin{cases} L\pi^+n^+R & \text{if } (n-1) \in R, \\ Ln^+\pi^+R & \text{otherwise,} \end{cases}$$

where π^+ is obtained from π by increasing each of its letters by $|LR|$, while $n^+ = 1 + |L\pi R|$. In particular, if $\sigma = 1$, we use the second line in the definition. For example,

$$312 \oplus'_p 1423 = 1645723 \quad \text{and} \quad 312 \oplus'_p 3412 = 3764512.$$

The outcome of the $\oplus'_p$ operation for two separable permutations is a separable permutation (which is easy to see since the structure is proper) with at least one letter to the right of the largest letter. Moreover, if $n-1$ was to the right of n in σ then the next largest letter is to the left of the largest letter in $\pi \oplus'_p \sigma$, and it is to the right of the largest letter in the sum otherwise. Thus, given $\tau \in \text{Av}(2413, 3142)$, we can either conclude that it was obtained using γ_p^{**} if the largest element is the rightmost letter, or, depending on the position of the next largest letter we can easily find π and σ such that $\pi \oplus'_p \sigma = \tau$.

Finally, we can see that $\text{comp}(\pi \oplus'_p \sigma) = \text{comp}(\sigma)$ as desired. $\square$

2.2.6   Schröder permutations

Kremer [546] showed that $s_n(1243, 2143)$ is given by the $(n-1)$-th Schröder number. For this reason, Egge and Mansour [338] called the set $\text{Av}(1243, 2143)$ the *Schröder permutations*. However, Kremer [546] actually proved that ten inequivalent (modulo trivial bijections) classes of permutations are counted by the Schröder numbers. Representatives from these classes are given in Table 2.2.

Since we have

$$\text{Av}(1324, 2314) = c(\text{Av}(4231, 3241)) \quad \text{and} \quad \text{Av}(1324, 2314) = i.r(\text{Av}(4231, 2431)),$$

we can see that class II contains the input- and output-restricted deque permutations introduced in Definition 2.1.10. Thus, one could call class II *deque-restricted permutations*. Class X represents the separable permutations. However, it is not so clear why class I should be called *the* Schröder permutations (class I is not any better than any other of the not yet mentioned classes counted by the Schröder numbers). In either case, since in the literature only classes I, II, and X seem to be studied, the following definition is justified.

I. Av(1234, 2134)	II. Av(1324, 2314)	III. Av(1342, 2341)
IV. Av(3124, 3214)	V. Av(3142, 3214)	VI. Av(3412, 3421)
VII. Av(1324, 2134)	VIII. Av(3124, 2314)	IX. Av(2134, 3124)
	X. Av(2413, 3142)	

Table 2.2: Classes of permutations counted by the Schröder numbers.

Definition 2.2.49. A permutation is a *Schröder permutation* if it avoids the patterns 1243 and 2143. Thus, Av(1243, 2143) is the class of all Schröder permutations.

Example 2.2.50. The permutation 264315 is a Schröder permutation, whereas 263514 is not as it contains an occurrence of the pattern 1243 (the subsequence 2354).

Remark 2.2.51. Once it comes to considering other, not yet studied classes of permutations counted by the Schröder numbers, one could invent something like the *Schröder permutation of the i-th kind*, where i is the class number in Table 2.2; following this scenario, for example, the *Schröder permutations of the first kind* would be simply the *Schröder permutations*.

Let us state a couple of results related to Schröder permutations coming from [338] by Egge and Mansour. Some of the open questions related to work in [338] are answered by Reifegerste [679] using so-called *essential sets* in permutation diagrams. We refer to [338, 679] for more results/details on that.

Recall from Chapter 1 that, for a pattern p, $p(\pi)$ denotes the number of occurrences of p in π. In particular, $12\cdots k(\pi)$ denotes the number of increasing subsequences of length k in π.

Theorem 2.2.52. ([338])

$$\sum_{\pi \in \mathrm{Av}(1243,2143)} \prod_{k \geq 1} x_k^{12\cdots k(\pi)} = 1 + \cfrac{x_1}{1 - x_1 - \cfrac{x_1 x_2}{1 - x_1 x_2 - \cfrac{x_1 x_2^2 x_3}{1 - x_1 x_2^2 x_3 - \cdots}}}.$$

One of the specializations of the variables x_i leads to the following result.

Theorem 2.2.53. ([338]) For $k \geq 1$,

$$\sum_{n \geq 0} s_n(1243, 2143, 12\cdots k) x^n = 1 + \cfrac{x}{1 - x - \cfrac{x}{1 - x - \cfrac{x}{1 - x - \cdots}}},$$

where the continued fraction has $k - 1$ denominators.

Example 2.2.54. If $k = 2$ in Theorem 2.2.53 then we get

$$\sum_{n \geq 0} s_n(1243, 2143, 12)x^n = 1 + \frac{x}{1 - x} = 1 + x + x^2 + x^3 + \ldots$$

which makes sense as for each n there is only one permutation, the decreasing permutation $n(n - 1) \ldots 1$, that avoids the pattern 12 (the other two prohibited patterns will be avoided automatically).

The formal power series in Theorem 2.2.53 admits another expression, in terms of the Chebyshev polynomials of the second kind (see Definition B.2.1), as shown in the following theorem.

Theorem 2.2.55. *([338]) For $k \geq 1$,*

$$\sum_{n \geq 0} s_n(1243, 2143, 12 \cdots k)x^n = 1 + \frac{\sqrt{x}U_{k-2}\left(\frac{1-x}{2\sqrt{x}}\right)}{\left(U_{k-1}\frac{1-x}{2\sqrt{x}}\right)}.$$

A similar result is recorded in the following theorem.

Theorem 2.2.56. *([338]) For $k \geq 1$,*

$$\sum_{n \geq 0} s_n(1243, 2143, 2134 \cdots k)x^n = 1 + \frac{\sqrt{x}U_{k-2}\left(\frac{1-x}{2\sqrt{x}}\right)}{U_{k-1}\left(\frac{1-x}{2\sqrt{x}}\right)}.$$

From Theorems 2.2.55 and 2.2.56,

$$\{1243, 2143, 12 \cdots k\} \sim \{1243, 2143, 2134 \cdots k\}$$

(these sets are Wilf-equivalent).

As particular cases of much more general theorems, the following theorem on restricted Schröder permutations is obtained.

Theorem 2.2.57. *([338, 679]) For $n \geq 2$,*

$$s_n(1243, 2143, 231) = (n + 2)2^{n-3},$$

and for $n \geq 1$,

$$s_n(1243, 2143, 321) = n + 2\binom{n}{3}.$$

For a final example of results in [338], we state the following theorem.

Theorem 2.2.58. *([338])*

$$\sum_{\pi} x^{|\pi|} = \frac{x(1+x)(1-x)^2}{\left(U_{k-1}\left(\frac{1-x}{2\sqrt{x}}\right)\right)^2}$$

where the sum on the left is over all permutations in $\mathrm{Av}(1243, 2143)$ *which contain exactly one occurrence of the pattern* $2134\cdots k$.

We close the subsection by mentioning a result related to Class II in Table 2.2. A reason to do this is that Bandlow et al. [73] define "A Schröder permutation is a permutation that is both 4132- and 4231-avoiding". As a matter of fact, $\mathrm{Av}(4132, 4231) = r.c(\mathrm{Av}(4231, 3241))$, and $\mathrm{Av}(4231, 3241)$ is defined by us as the set of the input-restricted deque permutations (a particular case of deque-restricted permutations); thus, in our terminology, $\mathrm{Av}(4132, 4231)$ belongs to Class II, not to Class I as the permutations' name suggests in [73]. Keeping this little inconsistency in mind, we now describe the result.

As it is defined in Table A.1, the *inversion* statistic for permutations, denoted inv, is the number of pairs $i < j$ such that $\pi_i > \pi_j$ in a permutation $\pi = \pi_1\pi_2\ldots\pi_n$. For example, if $\pi = 42513$ then $\mathrm{inv}(\pi) = 6$ since each of the 2-letter subsequences 42, 41, 43, 21, 51, and 53 contributes 1 to the total value of the statistic. Any permutation statisic that is equidistributed with inv is said to be *Mahonian*. The generating function for the inversion statistic on $\mathcal{S}_n(4231, 4132)$ is defined as

$$I_n(q) = \sum_{\pi \in \mathcal{S}_n(4231,4132)} q^{\mathrm{inv}(\pi)}.$$

Given a Schröder path P, the *area* statistic, $a(P)$, is the number of full squares and "upper" triangles (equivalently, triangles whose sides do not coincide with a double horizontal step in P) that lie below the path and above the x-axis. The definition is best understood by looking at the path P in Figure 2.31 and convincing yourself that $a(P) = 27$.

Figure 2.31: Illustration of the area statistic on a Schröder path.

Assuming S_n denotes the set of all Schröder paths on $2n$ steps, the generating function for the area statistic on Schröder paths is given by

$$\sum_{P \in S_n} q^{a(P)} = S_n(q)$$

and is known as the *Schröder polynomial* [152]. Specializing $q = 1$ in the Schröder polynomials gives usual Schröder numbers. Thus, the Schröder polynomials is a *q-analogue* to the Schröder numbers.

Using rather technical machinery, Barcucci et al. [81] show that

$$I_{n+1}(q) = S_n(q).$$

Bandlow et al. [73] give a constructive bijection from Schröder paths to Av(4231, 4132) that takes the area statistic on Schröder paths to the inversion number on permutations in $A(4231, 4132)$.

2.2.7 A hierarchy of permutation classes

Let us first summarize our knowledge of the following equinumerous objects: rooted non-separable planar maps, $\beta(1, 0)$-trees, skew ternary trees, 2-stack sortable permutations, and non-separable permutations. Everything but the last row in Table 2.3 essentially came from the corresponding table in [476]: we refer to this paper and references therein for further details (all but one of the statistics for objects involved are defined above and in Table A.1). The last row came from [262] as a particular case of Theorem 2.2.10 stated in Subsection 2.2.3 when the reverse operation is applied to get non-separable permutations from Av(3142, 2413). Note that "nodes" has the same meaning as "vertices" in Table 2.3 as opposed to the corresponding table in [476] where "nodes" actually means "non-leaf vertices". Finally, in Table 2.3, there is dependence between n, i, and j: $n = i + j + 1$.

We now let the equinumerous objects in Table 2.3 form a layer in a hierarchy (by set inclusion) of sets of permutations avoiding vincular patterns based on the permutations 2413 and 3142 (see Figure 2.32). This hierarchy is considered in [260] by Claesson et al. and its basic idea is as follows. Consider the set of permutations Av(3$\underline{14}$2). We can make the restriction 3$\underline{14}$2 stronger either by removing the underline thus arriving at the set Av(3142) $\subseteq$ Av(3$\underline{14}$2), or by adding an extra pattern to avoid, say, 2$\underline{41}$3 thus arriving at the set Av(2$\underline{41}$3, 3$\underline{14}$2) $\subseteq$ Av(3$\underline{14}$2). Then we can build other sets of permutations shown in Figure 2.32 in the same way. We would have a slightly different picture if instead of Av(3$\underline{14}$2) we started with Av(2$\underline{41}$3): instead of the chain

$$\text{Av}(2413, 3142) \subseteq \text{Av}(3142, 2\underline{41}3) \subseteq \text{Av}(3142) \subseteq \text{Av}(3\underline{14}2)$$

presented in Figure 2.32, we would get the chain

$$\text{Av}(2413, 3142) \subseteq \text{Av}(2413) \subseteq \text{Av}(2\underline{41}3).$$

rooted non-sep. planar maps	# edges $= n + 1$	# nodes $= i + 2$	# faces $= j + 2$	# cut-vert. after remov. root $= m$	# edges on outer face $= k + 1$
$\beta(1,0)$-trees	# nodes $= n + 1$	leaves $= i + 1$	int $= j + 1$	sub $= m + 1$	root $= k$
skew ternary trees	# nodes $= n$	even labels $= i + 1$	odd labels $= j$	first zeros $= m + 1$	zeros $= k$
2-stack sort. permutations	length $= n$	des $= i$	asc $= j$	see [423] for definition	rmax $= k$
non-separable permutations	length $= n$	des $= i$	asc $= j$	comp. r $= m + 1$	rmax $= k$

Table 2.3: Statistics translated under bijections between rooted non-separable planar maps, $\beta(1,0)$-trees, skew ternary trees, 2-stack sortable permutations, and non-separable permutations.

Once the hierarchy on sets of permutations is built, we can add to each layer other combinatorial objects related to the pattern-restricted classes and discussed in this section to see a "big" picture of relations between objects and to enjoy the variety of structures involved. In Figure 2.32 we use "∼" to show that one class of objects is equinumerous with another one, while "=" is used to show that the objects are actually the same. Finally, note that Figure 2.32 could accommodate more relevant objects, e.g. certain rectangulations equinumerous with the Baxter permutations and studied in [4, 379] (not discussed in this book), the deque-restricted permutations mentioned in Subsection 2.2.6, or any class of permutations, other than I, II, and X, in Table 2.2.

2.3 Schubert varieties and Kazhdan-Lusztig polynomials

We start by sketching several algebraic notions appearing in this section. However, if needed, one should consult other sources for precise definitions, for example, the book "Singular loci of Schubert varieties" by Billey and Lakshmibai [117].

A *Kazhdan-Lusztig polynomial* $P_{x,w}$ is a member of a family of integral polynomials introduced by Kazhdan and Lusztig [497] in 1979 (see [180] by Brenti for an introduction to the polynomials). These polynomials play an important role in Lie theory. Kazhdan and Lusztig originally defined the polynomials in terms of a complicated recurrence relation. While there are many uses for, and interpretations of,

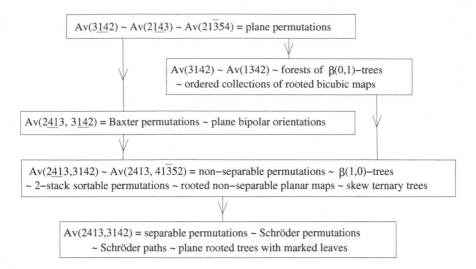

Figure 2.32: A hierarchy of permutation classes and related combinatorial objects.

Kazhdan-Lusztig polynomials, their combinatorial structure is not yet understood; in particular, there has been limited success in finding non-recursive formulas for them. However, for particular x and w, explicit formulas for Kazhdan-Lusztig polynomials can be obtained (see, e.g. [120] by Billey and Warrington for an overview of relevant results). One such particular case related to our pattern-avoidance is considered in Subsection 2.3.2.

An *algebraic variety* is the set of solutions of a system of polynomial equations. More precisely an algebraic variety is a space that is locally a set of solutions of a system of polynomial equations. Algebraic varieties are one of the central objects of study in algebraic geometry.

A *singularity* is a point at which a given object, e.g. a function, is not defined. *Local rings* are certain rings that are comparatively simple, and are used to describe what is called "local behaviour", in the sense of functions defined on varieties. The notion of local rings was introduced by Wolfgang Krull in 1938 under the name *Stellenringe*.

Below we will consider special subsets of the *flag variety* called *Schubert varieties*. A *flag* in $\mathbb{C}^n$ is an increasing sequence of subspaces in $\mathbb{C}^n$,

$$F_\bullet = \{0\} \subset F_1 \subset F_2 \subset \cdots \subset F_{n-1} \subset F_n = \mathbb{C}^n,$$

such that $\dim F_i = i$. The flag variety $\mathcal{F}l_n(\mathbb{C})$ is the set of all such flags. There is also an alternative description of this set which goes as follows: Consider the general linear group Gl_n consisting of all invertible $n \times n$ matrices and let B be the subset of invertible upper triangular matrices. Then given a matrix M from Gl_n we can construct a flag by letting F_i be the span of the first i columns. Two matrices M_1 and M_2 will correspond to the same flag if and only if there is a matrix $N \in B$ such that $M_2 = N \cdot M_1$. Thus $\mathcal{F}l_n(\mathbb{C}) = \mathrm{Gl}_n/B$. For the next definition we assume we have fixed a basis $e_1, e_2, \ldots, e_n$ of $\mathbb{C}^n$ and fixed a *reference flag* $E_\bullet$ such that E_i is the span of the first i basis vectors.

Definition 2.3.1. For each permutation $\pi \in S_n$ we define a subset of flags

$$X_\pi = \{F_\bullet \,|\, \dim(F_p \cap E_q) \geq \#\{i \leq p \,|\, \pi(i) \leq q \text{ for } 1 \leq p \leq q \leq n\}, \forall p, q\},$$

called a *Schubert variety.*

Example 2.3.2. Consider now the flag variety $\mathcal{F}l_3(\mathbb{C})$ and the Schubert variety X_{231}. The only non-trivial dimension condition becomes

$$\dim(F_1 \cap E_2) \geq 1.$$

Geometrically, this implies that the line F_1 should lie in the plane E_2. This implies that $X_{231} \cong \mathbb{C}^2$ and is therefore two-dimensional. Note that the permutation 231 has two inversions and it is a general fact that the number of inversions equals the dimension of the corresponding variety.

Schubert varieties form one of the most important and best-studied classes of algebraic varieties, and they are often used to test conjectures about more general varieties. A certain measure of the singularity of Schubert varieties is provided by Kazhdan-Lusztig polynomials, which encode their local *Goresky-MacPherson intersection cohomology*. These varieties are indexed by permutations and many properties of the varieties are encoded in the patterns that the permutations either contain or avoid. We will discuss relevant results in the next subsection.

2.3.1 Schubert varieties

Theorem 2.3.3. *([706, 551, 807, 745]) For $\pi \in S_n$, the variety X_π is smooth (i.e., has no singularities) if and only if $\pi \in \mathrm{Av}(4231, 3412)$.*

Following [160] by Bousquet-Mélou and Butler, which is influenced by Theorem 2.3.3, we give the following definition.

Definition 2.3.4. Any permutation in $\mathrm{Av}(4231, 3412)$ is called a *smooth permutation.*

A recurrence relation for counting smooth permutations is obtained by Stankova in [734] and the corresponding generating function is given by the following theorem.

Theorem 2.3.5. *([130, 160, 444]) The g.f. for $s_n(4231, 3412)$, the number of smooth n-permutations, is given by*

$$\frac{1 - 5x + 3x^2 + x^2\sqrt{1 - 4x}}{1 - 6x + 8x^2 - 4x^3}.$$

The following result was obtained by Bóna [130].

Theorem 2.3.6. *([130]) One has the following equinumeration result for five inequivalent (modulo trivial bijections) classes for $n \geq 0$:*

$$s_n(4231, 3412) = s_n(2431, 1342) = s_n(2431, 1423) = s_n(2431, 4132) = s_n(4231, 3142).$$

The generating function for these classes is given by Theorem 2.3.5 and there are no other inequivalent pairs of patterns that are Wilf-equivalent to the five above.

A weakening of smoothness is the notion of a *factorial* variety, which means that the local rings are unique factorization domains. Bousquet-Mélou and Butler [160] proved a conjecture by Yong and Woo on a characterization of factorial varieties. We state this result in the following theorem, where we use Proposition 1.3.7 to turn the barred pattern into the vincular one.

Theorem 2.3.7. *([160]) For $\pi \in \mathcal{S}_n$, the Schubert variety X_π is factorial if and only if*

$$\pi \in \mathrm{Av}(4231, 45\overline{3}12) = \mathrm{Av}(4231, 3\underline{41}2).$$

Remark 2.3.8. As it was remarked in [160] by Bousquet-Mélou and Butler, results of Cortez [284], and independently of Manivel [583], show that avoidance of 4231 and $45\overline{3}12$ characterizes *generically locally factorial Schubert varieties*, where *generic* has the following meaning: The variety is smooth at almost all points but has a closed subset Y_π where it is not smooth, and in that closed subset it is factorial at *almost* all points.

Remark 2.3.9. $\mathrm{Av}(4231, 45\overline{3}12) = \mathrm{Av}(4231, 3\underline{41}2)$ coincides with the class of *forest-like permutations* studied in [160]. We will not provide the original definition of forest-like permutations here, just saying that the definition is based on permutation matrices and certain drawings on them. Looking at the prohibited patterns, one sees that the class of smooth permutations is a subclass of the class of forest-like permutations (which is reminiscent of relations between some of the objects considered in Figure 2.32). However, there are three other subclasses of forest-like permutations, namely, *path-like permutations*, *tree-like permutations*, and *rooted tree-like permutations* (the last one is also a subclass of smooth permutations) all of which were enumerated in [160].

Theorem 2.3.10. *([160]) The g.f. for* $\mathrm{Av}(4231, 3\underline{41}2)$, *the forest-like permutations, is given by*

$$F(x) = \frac{(1-x)(1-4x-2x^2) - (1-5x)\sqrt{1-4x}}{2(1-5x+2x^2-x^3)}.$$

Gasharov and Reiner [414] defined a subclass of the factorial varieties that they name *defined by inclusions*. They described these varieties with a geometric condition and also with pattern-avoidance of four classical patterns (4231, 35142, 42513 and 351624). Úlfarsson and Woo [773] have shown that a relaxation of these conditions gives the Schubert varieties that are *local complete intersections*.

A further weakening is to only require that the local rings of X_π be *Gorenstein local rings*, in which case we say that X_π is a *Gorenstein variety*. Woo and Yong [809] gave a characterization of such varieties in terms of certain *Bruhat restrictions* with additional constraints. They also gave a characterization in terms of the avoidance of interval patterns [810]. However, Úlfarsson [772] provided a characterization of Gorenstein varieties in terms of bivincular patterns. To state the respective result (in Theorem 2.3.11), we need to define two infinite families, $\mathcal{G}_1$ and $\mathcal{G}_2$, of bivincular patterns.

- The family $\mathcal{G}_1$ is defined as

$$\mathcal{G}_1 = \left(\frac{1\overline{2345}}{53241}, \frac{1\overline{234567}}{7432651}, \frac{1\overline{23456789}}{954328761}, \cdots \right)$$

The general member of this family is of the form

$$\frac{1\overline{2\cdots\cdots\cdots k}}{k\ell\cdots2\cdots\ell+11},$$

where $\ell = (k-3)/2$.

- The family $\mathcal{G}_2$ is defined as

$$\mathcal{G}_2 = \left(\frac{\overline{12345}}{52431}, \frac{\overline{1234567}}{7326541}, \frac{\overline{123456789}}{943287651}, \cdots \right).$$

The general member of this family is of the form

$$\frac{\overline{12\cdots\cdots\cdots k}}{k\ell\cdots2\cdots\ell+11},$$

where $\ell = (k-1)/2$.

Note that the two families are the reverse complement of each other.

Theorem 2.3.11. *([772]) For $\pi \in \mathcal{S}_n$, the Schubert variety X_π is Gorenstein if and only if*

- *each Grassmannian permutation associated with π (see [772] for definitions) avoids every bivincular pattern in the families $\mathcal{G}_1$ and $\mathcal{G}_2$ defined above;*

- *π avoids the bivincular patterns $\frac{1 2 \overline{3} \overline{4} 5}{3 \underline{5} \underline{1} 4 2}$ and $\frac{1 \overline{2} \overline{3} 4 5}{4 2 \underline{5} \underline{1} 3}$.*

2.3.2   Kazhdan-Lusztig polynomials

Definition 2.3.12. Permutations in

$$\mathrm{Av}(321, 46718235, 46781235, 56718234, 56781234)$$

are called *321-hexagon-avoiding permutations*. The reason for this name is that if the *heap* of such a permutation is calculated, it does not contain a hexagon [120].

As we have already mentioned in the introduction to the section, the Kazhdan-Lusztig polynomials $P_{x,w}$ are defined in a complicated way, and finding explicit formulas for these polynomials for various x and w is a challenging task. Deodhar [295] proposes a combinatorial framework for determining the Kazhdan-Lusztig polynomials for an arbitrary *Coxeter group*. However, the algorithm is impractical for routine computations. On the other hand, the algorithm can be utilized efficiently to calculate $P_{x,w}$ in some cases, in particular, in the case of 321-hexagon-avoiding n-permutations w, as is shown in [120] by Billey and Warrington – an explicit description of the polynomials is obtained in these cases (we skip here most of the definitions and related results instead referring to the original source, [120]):

$$P_{x,w} = \sum q^{d(\sigma)},$$

where w is 321-hexagon-avoiding, $x \leq w$, $d(\sigma)$ is the *defect statistic*, the sum is over all *masks* σ on $\mathbf{a}$ whose product is x, and $\mathbf{a} = s_{i_1} s_{i_2} \cdots s_{i_r}$ is a reduced expression for $w \in \mathcal{S}_n$.

Definition 2.3.13. For the pattern 3412, the *height* of its occurrence in a permutation is the difference between the first and the last letters.

Example 2.3.14. There are four occurrences of the pattern 3412 in the permutation 461523: 4612, 4613, 4623 and 4523 of heights 2, 1, 1 and 1, respectively.

As is shown by Deodhar [294], $P_{id,w} = 1$ (for $\mathcal{S}_n$; id denotes the identity permutation) if and only if the Schubert variety X_w is smooth, and, more generally, $P_{u,w}(q) = 1$ if and only if X_w is smooth over the *Schubert cell* X_u°. The following theorem involving patterns was proved in [808] by Woo.

Theorem 2.3.15. *([808]) The Kazhdan-Lusztig polynomial for w satisfies $P_{id,w}(1) = 2$ if and only if the following two conditions are both satisfied:*

- *The* singular locus *of X_w has exactly one irreducible component;*

- *The permutation w avoids the patterns* 653421, 632541, 463152, 526413, 546213 *and* 465132.

More precisely, when these conditions are satisfied, $P_{id,w}(q) = 1 + q^h$ where h is the minimum height of a 3412 occurrence, with $h = 1$ if no such occurrence exists.

Finally we note that Billey and Postnikov [118] showed that pattern-avoidance can be extended to all Coxeter groups. This is done in terms of root subsystems and flattening maps. See also [115] by Billey and Braden.

For other materials relevant to this subsection, see [115, 121, 284, 496, 584].

2.4 A link to computational biology

In the last few decades, much has been done in the study of *genome evolution*, a research direction in *computational biology*. We refer to [164, 239] for some references on the biological aspects related to this section; we provide here almost no details on these. One of the many models for genome evolution, which take into account various biological phenomena, is the *tandem duplication-random loss model*, or simply the *duplication-loss model*. In this model, genomes are represented by permutations, that can evolve through *duplication-loss steps* representing the biological phenomenon that duplicates fragments of genomes, and then loses one copy of every duplicated gene. This model is well-studied in the biology literature, where it has been shown to be perhaps the most important rearrangement process in the case of *animal mitochondrial genomes*. For more on the biological motivation to study the duplication-loss model see [239] by Chaudhuri et al.

In this model, permutations can be modified by duplication-loss steps. Each of these steps is composed of two elementary operations, which are, for a given permutation π, as follows:

1. A factor (a fragment of consecutive letters) of π is duplicated, and the newly created factor is inserted immediately after the original copy; this is *tandem duplication*.

2. *Random loss* then takes place, which removes (exactly) one copy of *every* duplicated letter, resulting in a permutation.

For any duplication-loss step, the number of letters that are duplicated (the length of the duplicated factor) is called the *width* of the step.

Example 2.4.1. One step of a tandem duplication-random loss of width 3 applied to the permutation 123456 is as follows:

$$12 \underbrace{345}\ \underbrace{345}\ 6 \quad \mapsto \quad 1234\underline{53}4\underline{5}6 \quad \mapsto \quad 123546$$
$$\text{(tandem duplication)} \qquad \text{(random loss)}$$

As is mentioned in [166, 164], the duplication-loss model can be viewed as a particular case of *permuting machines* that sort and generate permutations, and they are defined in [19] by Albert et al. We will consider permutations that are obtained from the permutation $12 \cdots n$ after a given number r of duplication-loss steps.

Various duplication-loss models can be defined depending on a given so-called *cost function c*. Although it is intuitively clear that the cost of a duplication should be some non-decreasing function of the length of the duplication, it is not clear what exactly this function should be. We assume that the cost $c(k)$ of a duplication-loss step is dependent only on the width k of the step. In the original model of Chaudhuri et al. [239], $c(k) = \alpha^k$, for a parameter $\alpha \geq 1$. In [166] by Bouvel and Rossin, the cost function is defined by $c(k) = 1$ if $k \leq K$, $c(k) = \infty$ otherwise, for a parameter $K \in \mathbb{N}\backslash\{0,1\}$. In the model of [164] by Bouvel and Pergola, for all k, one has $c(k) = 1$, which is a special case of both the model of [239] (the case $\alpha = 1$) and the model of [166] (the case $K = \infty$). The model with $c(k) = 1$ is called the *whole genome duplication-random loss model*, which is motivated by the following argument: Since any step has cost 1, regardless of its width, we can, without loss of generality, assume that the whole permutation is duplicated at any step.

Pattern-avoidance is used in [166] to describe the set of permutations obtainable from an identity permutation after a number of duplication-loss steps of bounded width and we discuss it briefly in the following subsection. On the other hand, in the description of obtainable permutations in the whole genome duplication-random loss model, descents in permutations are involved, which are occurrences of the vincular (consecutive) pattern $\underline{21}$. Besides, consecutive 4-patterns are involved in a characterization of *minimal permutations with width d* (see Definition 2.4.7), objects involved in describing obtainable permutations in the whole genome duplication-random loss model which we discuss in Subsection 2.4.2. We refer to [239, 166] for algorithmic aspects related to the duplication-loss model. Finally, we refer to Definition 2.2.38 for the notion of a permutation class and to Section 8.1 for the notion of the *basis* of a permutation class.

2.4.1 Duplication-loss steps of bounded width

The main object of interest in [166] is given by the following definition.

Definition 2.4.2. Let $C(K, r)$ denote the class of all permutations obtainable from $12 \cdots n$ (for any n) after r duplication-loss steps of width at most K, for some constant parameters r and K.

Remark 2.4.3. Following remarks in [239, 166], the duplication-loss steps are not reversible and thus $C(K, r)$ is *not* the class of permutations that can be sorted to $12 \cdots n$ in r duplication-loss steps of width at most K.

Theorem 2.4.4. *([166]) $C(K, 1)$ is a class of pattern-avoiding permutations $\mathrm{Av}(B)$ whose basis B is finite of size $2^{K-1} + 3$. More precisely, $B = \{321, 3142, 2143\} \cup D$, where D is the set of all permutations in $\mathcal{S}_{K+1}$ that do not start with 1 nor end with $K + 1$, and contain exactly one descent.*

In the general case, Bouvel and Rossin [166] obtained the following result for $C(K, r)$.

Theorem 2.4.5. *([166]) $C(K, r)$ is a class of pattern-avoiding permutations whose basis is finite and contains patterns of size at most $(Kr + 2)^2 - 2$.*

2.4.2 The whole genome duplication-random loss model

Bouvel and Pergola [164] proved the following characterization theorem.

Theorem 2.4.6. *([164]) The permutations that can be obtained in at most r steps in the whole genome duplication-random loss model are exactly those whose number of descents is at most $2^r - 1$.*

Definition 2.4.7. A permutation is *minimal with d descents* if removing any of its letters and taking the reduced form gives a permutation with fewer descents.

Example 2.4.8. The permutation $\pi = 31254$ has 2 descents but it is not minimal with 2 descents as we can remove the letter 2 obtaining the permutation 2143 still having two descents. On the other hand, the permutation 642197385 is minimal with 6 descents.

Theorem 2.4.9. *([166, 164]) The class of permutations obtainable in at most r steps in the whole genome duplication-random loss model is a class of pattern-avoiding permutations whose basis is finite and is composed of the minimal permutations with 2^r descents.*

Taking into account the importance of minimal permutations with a specified number of descents in Theorem 2.4.9, we provide selected known facts on these permutations (consult [164] for more facts). A characterization of such permutations involving consecutive patterns is given in [164] by Bouvel and Pergola.

Theorem 2.4.10. *([164]) A permutation π is minimal with d descents if and only if it has exactly d descents and its ascents $\pi_i \pi_{i+1}$ are such that $2 \leq i \leq n-2$ and $\pi_{i-1} \pi_i \pi_{i+1} \pi_{i+2}$ forms an occurrence of either the pattern $\underline{2143}$ or the pattern $\underline{3142}$.*

Theorem 2.4.11. *([164]) The minimal permutations with d descents and of size*

- *$2d$ are enumerated by the d-th Catalan number $C_d = \frac{1}{d+1}\binom{2d}{d}$;*

- *$d+2$ are enumerated by $2^{d+2} - (d+1)(d+2) - 2$.*

Further studies of minimal permutations with d descents are carried out in [163] by Bouvel and Ferrari.

Chapter 3

More motivation points

In Chapter 2 we discussed how the (permutation) patterns theory is connected to other areas, such as sorting problems, maps and orientations, trees, Baxter permutations, Schröder structures and certain algebra problems. In this section, we discuss other important connections, namely, to statistical mechanics, to certain partially ordered sets and equinumerous objects, to the study of permutation statistics, to encoding combinatorial structures, and to proving combinatorial identities.

3.1 Applications to statistical mechanics

3.1.1 PASEP

The *Partially Asymmetric Simple Exclusion Process* (*PASEP*) is a paradigmatic model in *non-equilibrium statistical mechanics*. The model was introduced in the late 1960s by a mathematician, Spitzer [733] and by biologists MacDonald, Gibbs, and Pipkin [577]. The PASEP is a generalization of the *TASEP* (*Totally Asymmetric Simple Exclusion Process*), a very rich and well-studied *1D gas model* introduced by physicists [296] (see [168] for more references). The PASEP consists of black particles entering a row of n cells, each of which is either vacant or occupied by a black particle. A particle may enter the system from the left-hand side, hop to the right or to the left and leave the system from the right-hand side. Each cell contains at most one particle. It is convenient to think of the empty cells being filled with white particles ∘, and a *basic configuration* (a row with a number of cells, each containing either a black • or a white ∘) can be viewed as a word over the alphabet $\{\circ, \bullet\}$. We let $\mathcal{B}_n$ be the set of all basic configurations on n particles.

More formally, the PASEP can be defined as a discrete-time Markov chain P

on $\mathcal{B}_n$ with the transition probabilities α, β, and q (see [314]). Given that the system is in state X at time t, the probability $P_{X,Y}$ of finding the system in state Y at time $t + 1$ is defined by

- If $X = A \bullet \circ B$ and $Y = A \circ \bullet B$ then $P_{X,Y} = \dfrac{1}{n+1}$ (particle hops to the right)
  and $P_{Y,X} = \dfrac{q}{n+1}$ (particle hops to the left);

- If $X = \circ B$ and $Y = \bullet B$ then $P_{X,Y} = \dfrac{\alpha}{n+1}$ (particle enters from the left-hand side);

- If $X = B\bullet$ and $Y = B\circ$ then $P_{X,Y} = \frac{\beta}{n+1}$ (particle exits to the right-hand side);

- Otherwise, $P_{X,Y} = 0$ for $Y \neq X$ and $P_{X,X} = 1 - \sum\limits_{X \neq Y} P_{X,Y}$.

Example 3.1.1. See Figure 3.1 for the state diagram of the PASEP in the case $n = 2$.

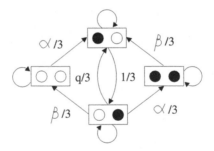

Figure 3.1: The Markov chain P for $n = 2$.

Remark 3.1.2. Note that the probabilities of the loops in Figure 3.1 are determined from the fact that the sum of the probabilities of all outgoing arrows of a given state must be 1. Thus, the probabilities of the loops starting from the top-most one and going clockwise are $\frac{2}{3}$, $\frac{3-\beta}{3}$, $\frac{3-q-\alpha-\beta}{3}$, and $\frac{3-\alpha}{3}$, respectively.

In the limit, the system reaches a steady state where all the probabilities $P_n(x_1, x_2, \ldots, x_n)$ of finding the system in configurations $(x_1, x_2, \ldots, x_n)$ (where $x_i =$

1 if cell i is occupied by a particle, and $x_i = 0$ otherwise) are stationary, that is, they satisfy

$$\frac{d}{dt}P_n(x_1, x_2, \ldots, x_n) = 0.$$

The stationary distribution is unique [297], and the question is then to solve for the probabilities $P_n(x_1, x_2, \ldots, x_n)$. The *"matrix ansatz"* was used by Derrida et al. [297] to obtain an exact expression for all the $P_n(x_1, x_2, \ldots, x_n)$, which is recorded in the following theorem.

Theorem 3.1.3. *([297]) Suppose that D and E are matrices, V and W are a column vector and a row vector, respectively, such that the following conditions hold:*

$$DE - qED = D + E,$$

$$DV = \frac{1}{\beta}V,$$

$$WE = \frac{1}{\alpha}W.$$

Then

$$P_n(x_1, x_2, \ldots, x_n) = \frac{W\left(\prod_{i=1}^{n}(x_i D + (1 - x_i)E)\right)V}{Z_n}$$

where

$$Z_n = W(D + E)^n V.$$

Reasons for PASEP's fame, apart of its solvability, include a number of features such as *boundary driven phase transitions, spontaneous symmetry breaking, and phase separation*. The PASEP is a primitive model for *biopolymerization* [577], *formation of shocks* [298], and *traffic flow* [715]. Moreover, it appears in a sort of *sequence alignment problem* in *computational biology* [194]. Also, the PASEP has relations to *orthogonal polynomials* [711], and to some interesting combinatorics of which the most important for our purposes is a connection to *permutation tableaux* (and via them to permutation patterns) to be defined in Subsection 3.1.2. In [283], the most general version of the PASEP was connected to *doubly signed permutations* counted by the formula $4^n n!$, that is, permutations where each letter is decorated by two signs: $++$, $+-$, $-+$, or $--$; the corresponding bijection appeared in [280].

Remark 3.1.4. The expression *ASEP* is used in the literature to abbreviate PASEP. However, one should pay attention to the context as some authors use ASEP to abbreviate TASEP. A short summary of the relevant terminology is as follows:

- *TASEP* means totally asymmetric exclusion process, that is, the parameter q representing the rate of hopping left is equal to 0;

- *SSEP* means symmetric exclusion process, that is, the rate of hopping left is equal to the rate of hopping right, in which case, typically $q = 1$;

- *PASEP* means that the rates of hopping left and hopping right do not have to be the same: typically hopping right is set to 1, which is not a loss of generality, and hopping left is a variable rate q.

3.1.2   Permutation tableaux

Corteel and Williams [281, 282] gave connections between PASEP and *permutation tableaux* defined and discussed in this subsection.

Definition 3.1.5. A *partition* $\lambda = (\lambda_1, \lambda_2, \ldots, \lambda_k)$ is a weakly decreasing sequence of nonnegative integers.

Example 3.1.6. For example, $\lambda = (7, 4, 4, 2, 1, 1)$ is a partition of 19 (the sum of all the entries of the partition).

Definition 3.1.7. The *Young diagram* Y_λ *of shape* λ for a partition λ having $\sum_{1 \leq i \leq k} i = m$, is a left-justified diagram of m boxes, with λ_i boxes in the i-th row counted from the top.

Example 3.1.8. For $\lambda = (5, 4, 2, 2, 1)$, the Young diagram Y_λ is

Definition 3.1.9. A *permutation tableau* T is a partition λ together with a filling of the boxes of Y_λ with 0s and 1s such that the following two properties hold:

1. Each column of the shape contains at least one 1;

2. There is no 0 which has a 1 above it in the same column *and* a 1 to its left in the same row. Equivalently, reading the columns of T from right to left, whenever we see a 0 in a column which lies beneath a 1, all the entries to the left of the 0 (in the same row) must be 0s.

For the purpose of keeping track of statistics, it is convenient to make a rectangular shape from a permutation tableau by filling missing cells with 2s as shown in Example 3.1.10.

Example 3.1.10. See Figure 3.2 for an example of a permutation tableau represented in two ways.

```
0 0 1 1 0 0 0 1 0        0 0 1 1 0 0 0 1 0
1 0 1 1 1 0              1 0 1 1 1 0 2 2 2
0 0 0 0 0 0              0 0 0 0 0 0 2 2 2
1 0 1 1                  1 0 1 1 2 2 2 2
0 1 1 1                  0 1 1 1 2 2 2 2
```

Figure 3.2: A permutation tableau represented in two ways.

Remark 3.1.11. Permutation tableaux come from the enumeration of the *totally positive Grassmannian cells* [656, 806].

The following connection between permutation tableaux and permutations was established in [274] by Corteel, where the length of a tableau is defined in Definition 3.1.14.

Theorem 3.1.12. *([274]) There exists a bijection between n-permutations with k descents and permutation tableaux of length n with k columns.*

A more sophisticated bijection than that in Theorem 3.1.12 appears in [753] by Steingrímsson and Williams. This bijection involves patterns and we state it in the following theorem, where $p(\pi)$ denotes the number of occurrences of the pattern p in the permutation π, and, more generally, $[p_1 + p_2 + \cdots + p_k](\pi)$ is the total number of occurrences of all patterns from the set $\{p_1, p_2, \ldots, p_k\}$ in π.

Theorem 3.1.13. *([753]) Let $T(k, a, b, c)$ be the set of permutation tableaux with k rows and $(n - k)$ columns, which are filled with precisely a 0s, b 1s and c 2s. Let $P(k, a, b, c)$ be the set of all permutations $\pi \in \mathcal{S}_n$, such that*

- $k - 1 = \operatorname{des}(\pi)$,

- $a = [\underline{3}1\underline{2} + \underline{2}1\underline{3} + \underline{3}\underline{2}\underline{1}](\pi) - \binom{\operatorname{des}(\pi)}{2}$,

- $b = 2\underline{3}\underline{1}(\pi) + n - 1 - \operatorname{des}(\pi)$,

- $c = [1\underline{3}\underline{2} + \underline{3}\underline{2}\underline{1}](\pi) - \binom{\operatorname{des}(\pi)}{2}$.

Then $|T(d, a, b, c)| = |P(d, a, b, c)|$.

We refer to [279] by Corteel and Nadeau for two more bijections between permutation tableaux and permutations that involve, in particular, the permutation statistic rmin and occurrences of the pattern 3̲1̲2. Chen and Liu [244] introduced the *inversion number* of a permutation tableau T which corresponds to occurrences of the vincular pattern 3̲2̲1 in the reverse complement of the image of T under a bijection of Corteel and Nadeau [279]. Also, Chen and Liu [244] show that permutation tableaux without inversions coincide with *L-Bell tableaux* introduced by Corteel and Nadeau [279].

For some follow up work related to [753], see [199, 197, 789]. In particular, in [789], Viennot connected certain permutation tableaux, called *Catalan tableaux*, to TASEP.

Definition 3.1.14. The *length* of a tableau is the sum of the number of parts and the largest part of the partition λ. A zero in a permutation tableau is *restricted* if there is a one above it in the same column. A row is *unrestricted* if it does not contain a restricted zero. A restricted zero is a *rightmost* restricted zero if it is restricted and it has no restricted zero to the right of it in the same row. A one is *superfluous* if it has a one above it in the same column.

Example 3.1.15. The permutation tableau in Figure 3.2 has length 14, 7 superfluous ones, and 3 unrestricted rows.

We close the subsection by mentioning some of the main results in [277] by Corteel and Hitczenko on the expected values of several statistics on permutation tableaux.

Theorem 3.1.16. *([277]) Let n be a fixed integer.*

1. *The expected number of rows in a tableau of length n is $(n+1)/2$.*

2. *The expected number of unrestricted rows in a tableau of length n is $H_n = 1 + \frac{1}{2} + \frac{1}{3} + \cdots + \frac{1}{n}$, the n-th harmonic number.*

3. *The expected number of ones in the first row in a tableau of length n is $H_n - 1$.*

4. *The expected number of superfluous ones in a tableau of length n is*

$$\frac{(n-1)(n-2)}{12}.$$

3.1.3   PASEP and pattern-avoidance

Brak et al. [168] showed that many properties of the PASEP, in particular, the *normalization Z_n for the stationary distribution* (introduced in Theorem 3.1.3 above)

k	The number of n-permutations with k occurrences of $2\underline{13}$
0	$\frac{1}{n+1}\binom{2n}{n}$
1	$\binom{2n}{n-3}$
2	$\frac{n}{2}\binom{2n}{n-4}$
3	$\frac{(n+1)(n+2)}{6}\binom{2n}{n-5}$
4	$\frac{-36-100n-13n^2+4n^3+n^4}{24(n+6)}\binom{2n}{n-5}$
5	$\frac{(n+4)(-108-192n+3n^2+8n^3+n^4)}{120(n+7)}\binom{2n}{n-6}$
6	$\frac{20160+44448n+548n^2-4196n^3-565n^4+67n^5+17n^6+n^7}{720(n+7)(n+8)}\binom{2n}{n-6}$
7	$\frac{(n+5)(40320+67824n-20180n^2-7556n^3-5n^4+211n^5+25n^6+n^7)}{5040(n+8)(n+9)}\binom{2n}{n-7}$
8	$\frac{f(n)+3647724n^3-416320n^4-249417n^5-19971n^6+2646n^7+576n^8+39n^9+n^{10}}{40320(n+8)(n+9)(n+10)}\binom{2n}{n-7}$

Table 3.1: Counting occurrences of a vincular pattern in permutations, from [251, 264, 650]. Here, $f(n) = -7983360 - 12956832n + 10475400n^2$.

for certain probabilities, can be determined by counting weighted lattice paths. The distribution was connected by Parviainen [650] to counting occurrences of the vincular pattern $2\underline{13}$ in permutations. As a result of this study, closed forms for the number of permutations containing k occurrences of the pattern $2\underline{13}$, for $k \leq 8$, were obtained in [650], which extended results of Claesson and Mansour [264] dealing with the cases $1 \leq k \leq 3$ (a bijective proof of the case $k = 1$ appears in [279]); the case $k = 0$ was considered by Claesson [251]. We summarize these results in Table 3.1, where $f(n) = -7983360 - 12956832n + 10475400n^2$.

Theorems 3.1.17 and 3.1.20 below can be viewed as theorems on the pattern $2\underline{13}$ because of the complement operation (note that descents become ascents under taking the complement) – we have chosen to use the pattern $2\underline{31}$ in Theorems 3.1.17 and 3.1.20 that appears in the original publications.

Theorem 3.1.17. *([275, 753]) The number of permutations in $\mathcal{S}_n$ with $k-1$ descents and m occurrences of the pattern $2\underline{31}$ is equal to the coefficient of q^m in*

$$\hat{E}_{k,n}(q) = q^{-k^2}\sum_{i=0}^{k-1}(-1)^i(1+q+q^2+\cdots+q^{k-i-1})^n q^{ki}\left(\binom{n}{i}q^{k-i}+\binom{n}{i-1}\right).$$

Remark 3.1.18. The polynomials $\hat{E}_{k,n}(q)$ were first introduced in [806] by Williams.

Remark 3.1.19. The formula for $\hat{E}_{k,n}(q)$ is the only known polynomial expression which gives the *complete* distribution of a permutation pattern of length greater than 2. Regarding length 2 patterns the following is known: For the pattern $\underline{12}$ (equivalently, for $\underline{21}$) the distribution is given by the *Eulerian numbers* (see Subsection A.2.1 for definition):

$$\sum_{n\geq 0}\frac{x^n}{n!}\sum_{\sigma\in\mathcal{S}_n}q^{\underline{12}(\sigma)} = \sum_{n\geq 0}\frac{x^n}{n!}\sum_{\sigma\in\mathcal{S}_n}q^{\underline{21}(\sigma)} = \frac{(1-q)e^{x(1-q)}}{1-qe^{x(1-q)}},$$

and for the pattern 12 (equivalently, for 21) the distribution on $\mathcal{S}_n$ is given by the coefficients of

$$(1+q)(1+q+q^2)\cdots(1+q+q^2+\cdots+q^{n-1}).$$

Theorem 3.1.20. *Let $\hat{E}(q,x,y) := \displaystyle\sum_{n,k\geq 0}\hat{E}_{k,n}(q)y^kx^n$. Then*

1. *according to [806, 753], $\hat{E}(q,x,y)$ is given by*

$$\sum_{i=0}^{\infty}\frac{y^i(q^{2i+1}-y)}{q^{i^2+i+1}(q^i-q^{i+1}(1+q+q^2+\cdots+q^{i-1})x+(1+q+q^2+\cdots+q^{i-1})xy)},$$

2. *and, according to [275],*

$$\hat{E}(q,x,y) = \cfrac{1}{1-b_0x-\cfrac{\lambda_1x^2}{1-b_1x-\cfrac{\lambda_2x^2}{1-b_2x-\cfrac{\lambda_3x^2}{\ddots}}}},$$

where $b_n = y(1+q+q^2+\cdots+q^n)+(1+q+q^2+\cdots+q^{n-1})$ and $\lambda_n = y(1+q+q^2+\cdots+q^{n-1})^2$.

The following theorem gives the expected number of occurrences of the pattern $\underline{312}$ which is equivalent to that of the pattern $2\underline{13}$ because of the reverse operation.

Theorem 3.1.21. *([277]) The expected value of the number of occurrences of the vincular pattern $\underline{312}$ in a random n-permutation is equal to*

$$\frac{(n-1)(n-2)}{12}.$$

The connection by Parviainen [650] mentioned above was almost simultaneously expanded by Corteel and Williams to permutation tableaux [281, 282], which were shown to carry information about many important aspects of the PASEP, including distributional results significant from the physics point of view. For example, Corteel and Williams [282] show how the solution (D_1, E_1) to the matrix ansatz (see Theorem 3.1.3) leads to a natural connection between the PASEP model and permutation tableaux, and hence to permutations.

Further aspects of the PASEP solution can be understood by studying the joint distribution of occurrences of the pattern $21\underline{3}$ and cycles on permutations [651]. Not that many publications consider cycles in connection with permutation patterns [330, 487, 652]. An example of an enumerative result appearing in [652] is the following theorem that is a refinement of the standard continued fraction representation of the Eulerian numbers (note that $\mathrm{des}(\pi) = \underline{21}(\pi)$ and $\mathrm{asc}(\pi) = \underline{12}(\pi)$).

Theorem 3.1.22. *([652]) Denoting by* $\mathrm{cyc}(\pi)$ *the number of cycles in a permutation* π *and by* $|\pi|$ *the number of letters in* π, *we have*

$$\sum_{\pi \in \mathcal{S}} p^{\mathrm{des}(\pi)} q^{\mathrm{asc}(\pi)} x^{\mathrm{cyc}(\pi)} z^{|\pi|} =$$

$$\cfrac{1}{1 - xz - \cfrac{qxz^2}{1 - (q+p+x)z - \cfrac{2q(p+x)z^2}{1 - (2q+2p+x)z - \cfrac{3q(2p+x)z^2}{1 - (3q+3p+x)z - \cdots}}}}.$$

3.2 Posets, diagrams, and matrices

The main object of interest in this section is the *unlabeled (2+2)-free posets* (to be defined in Subsection 3.2.1). These posets attract significant attention in the literature, in particular, because, as was shown by Fishburn [385], they are in one-to-one correspondence with the well-known and much-studied *interval orders*.

Definition 3.2.1. A poset $P = (X, \le)$ is an *interval order* provided that we can assign to each $x \in X$ a real line interval $[x_L, x_R]$ such that $x_R < y_L$ in the real numbers if and only if $x < y$ in P.

In this section we link the (2+2)-free posets with so-called *ascent sequences* (see Subsection 3.2.2), *Stoimenow's diagrams* (see Subsection 3.2.3), certain upper triangular matrices (see Subsection 3.2.4), and, of course, with pattern-avoidance in permutations (avoiding the bivincular pattern $\overline{1}\overline{2}3 \atop 231$; see Subsection 3.2.5). Among the enumerative results, in Subsection 3.2.6 we will discuss how a conjecture of Pudwell on the barred pattern $3\overline{1}52\overline{4}$, as well as a conjecture of Jovovic on certain binary upper triangular matrices were solved using studies on $(2+2)$-free posets.

3.2.1   Unlabeled $(2+2)$-free posets

Definition 3.2.2. A *partial order* is a binary relation $\leq$ over a set P which is *reflexive*, *antisymmetric*, and *transitive*, that is, for all a, b, and c in P, we have that:

- $a \leq a$ (reflexivity);

- if $a \leq b$ and $b \leq a$ then $a = b$ (antisymmetry);

- if $a \leq b$ and $b \leq c$ then $a \leq c$ (transitivity).

Definition 3.2.3. A set P with a partial order $\leq$ on it is called a *partially ordered set*, or *poset*, and is denoted $(P, \leq)$.

We use Hasse diagrams to represent posets as in Figure 3.3 where we draw two posets.

Definition 3.2.4. An element x in a poset P is called *minimal* if no element in P is covered by x. An element $x \in P$ is called *maximal* if x is not covered by any element in P.

Definition 3.2.5. A poset is $(2+2)$-free if it does not contain an induced subposet that is isomorphic to $2+2$, the union of two disjoint 2-element chains. That is, in a $(2+2)$-free poset we would never find four elements, $a < b$ and $c < d$ such that a is incomparable to both c and d, and b is incomparable to both c and d.

Example 3.2.6. The poset P_1 in Figure 3.3 contains a $(2+2)$ configuration shown in bold. On the other hand, the poset P_2 in Figure 3.3 is $(2+2)$-free.

Definition 3.2.7. For an element $x \in P$, its *strict down-set* is defined as $D(x) = \{y \in P | y < x\}$, and its *strict up-set* is defined as $U(x) = \{y \in P | y > x\}$.

We will be interested in *unlabeled* $(2+2)$-free posets in which two elements, x and y, are *indistinguishable* if $D(x) = D(y)$ and $U(x) = U(y)$. For example, in Figure 3.3, the minimal elements x and y are indistinguishable, as well as the minimal elements a and b.

The following result is easy to prove (e.g., see [161] for a proof).

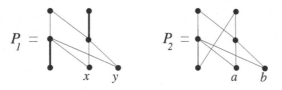

Figure 3.3: The poset P_1 contains a 2+2 configuration, whereas P_2 is (2+2)-free.

Theorem 3.2.8. *(folklore) A poset is (2+2)-free if and only if its collection of strict down-sets can be linearly ordered by inclusion. Therefore, each element x in a (2+2)-free poset can be assigned the level $\ell(x)$ depending on the size of the down-set $D(x)$ (level 0 corresponds to the minimal elements having no elements in their down-sets; level 1 is assigned to elements having next smallest down-set; etc.)*

Example 3.2.9. Even though we are dealing with unlabeled posets, it is convenient to label the elements of the posets in question to be able to refer to a particular element. In Figure 3.4, we give two drawings of the same poset. The drawing on the left is a standard Hasse diagram; on the right, we redraw the poset with respect to its levels. For this picture, we have, $D_0 = D(a) = D(e) = D(h) = D(j) = \emptyset$, $D_1 = D(b) = D(c) = D(d) = \{a\}$, $D_2 = D(g) = \{a, c, d\}$, $D_3 = D(f) = \{a, b, c, d\}$, $D_4 = D(i) = \{a, b, c, d, e, f, g, h\}$, and

$$\emptyset = D_0 \subset D_1 \subset D_2 \subset D_3 \subset D_4.$$

For example, $\ell(c) = 1$ and $\ell(f) = 3$. Note that j is both minimal and maximal element.

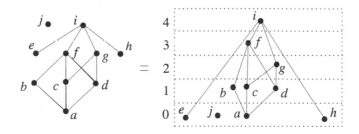

Figure 3.4: A (2+2)-free poset redrawn according to its levels.

Definition 3.2.10. ([743, p.100]) For labeled or unlabeled posets, the *ordinal sum* of two posets $(P, \leq_P)$ and $(Q, \leq_Q)$ is the poset $(P \oplus Q, \leq_{P \oplus Q})$ on the union $P \cup Q$ such that $x \leq_{P \oplus Q} y$ if $x \leq_P y$, or $x \leq_Q y$, or $x \in P$ and $y \in Q$.

Example 3.2.11. See Figure 3.5 for an example of the ordinal sum of two posets. In particular, one way to draw the sum of two posets is to draw one poset above the other making sure that each minimal element of the top poset covers every maximal element of the bottom poset.

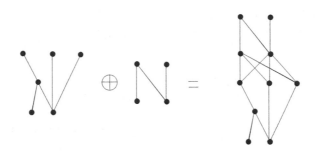

Figure 3.5: An example of the ordinal sum of two posets.

Remark 3.2.12. It is easy to see that for posets P and Q, $P \oplus Q$ is (2+2)-free if and only if both P and Q are (2+2)-free.

Definition 3.2.13. We say that a poset P has k *components* if it is the ordinal sum of k, but not $k + 1$, nonempty posets.

Definition 3.2.14. Given a poset P, we let size(P) be the number of elements of P and min(P) (resp., max(P)) denote the number of minimal (resp., maximal) elements of P. $\ell(P)$ denotes the number of levels in P, and $\ell^*(P)$ denote the level of a *minimal maximal element* (that is, of a maximal element from the lowest possible level). Also, we let comp$_p(P)$ denote the number of components of P.

Example 3.2.15. For the poset in Figure 3.4, size$(P) = 10$, min$(P) = 4$, max$(P) = 2$, $\ell(P) = 5$, $\ell^*(P) = 0$, and comp$_p(P) = 1$.

Definition 3.2.16. A (2+2)-free poset is *primitive* if it contains no pair of indistinguishable elements.

Definition 3.2.17. Let maxindist(P) be the maximum number of indistinguishable elements of P. Also, we define the statistic rep(P) to be the minimum number of elements of P that need to be removed to create a primitive poset.

Example 3.2.18. For the poset in Figure 3.4, maxindist$(P) = 2$ (the elements c and d are indistinguishable, as well as the elements e and h). For the same poset, rep$(P) = 2$ (to get a primitive poset, we need to remove one element in $\{c, d\}$, and one element in $\{e, h\}$).

Remark 3.2.19. Note that maxindist$(P) = 1$ if and only if rep$(P) = 0$.

Definition 3.2.20. We let $\mathcal{P}_n$ denote the set of all (unlabeled) (2+2)-free posets on n elements, and $\mathcal{P} = \cup_{n \geq 0} \mathcal{P}_n$. Moreover, $\mathcal{P}_n^{(k)}$ denotes the set of all (2+2)-free posets for which maxindist(P) is at most k. In particular, $\mathcal{P}_n^{(1)}$ is the set of primitive (2+2)-free posets on n elements.

3.2.2 Ascent sequences

The sequences defined below play an important role in the study of (2+2)-free posets. In particular, they allow us to enumerate these posets (with respect to several statistics) as discussed in Subsection 3.2.6. Recall that asc(x) denotes the number of ascents in a sequence x (see Table A.1 for definition).

Definition 3.2.21. A sequence $(x_1, x_2, \ldots, x_n) \in \mathbb{N}^n$ is an *ascent sequence of length* n if and only if it satisfies $x_1 = 0$ and $x_i \in [0, 1 + \text{asc}(x_1, \ldots, x_{i-1})]$ for all $2 \leq i \leq n$.

Definition 3.2.22. Let $\mathcal{A}_n$ be the set of ascent sequences of length n and $\mathcal{A}$ be the set of all ascent sequences. If $a \in \mathcal{A}_n$ then we will write $|a| = n$.

Example 3.2.23. For example, $(0, 1, 0, 1, 3, 1, 0, 1, 4)$ is an ascent sequence in $\mathcal{A}_9$. Also, there are 15 ascent sequences of length 4: $(0, 0, 0, 0)$, $(0, 0, 0, 1)$, $(0, 0, 1, 0)$, $(0, 0, 1, 1)$, $(0, 0, 1, 2)$, $(0, 1, 0, 0)$, $(0, 1, 0, 1)$, $(0, 1, 0, 2)$, $(0, 1, 1, 0)$, $(0, 1, 1, 1)$, $(0, 1, 1, 2)$, $(0, 1, 2, 0)$, $(0, 1, 2, 1)$, $(0, 1, 2, 2)$, and $(0, 1, 2, 3)$.

Definition 3.2.24. A *right-to-left maximum* of an ascent sequence a is a letter with no larger letter to its right. We let rmax(a) denote the number of right-to-left maxima in a. This definition is consistent with the definition of rmax on permutations (see Table A.1).

Example 3.2.25. For example rmax$(0, 1, 0, 2, 3, 1, 2, 1, 2, 0, 1, 1) = 5$, where the right-to-left maxima are in bold.

Definition 3.2.26. For an ascent sequence a, we let zeros(a) denote the number of 0s in a, and last(a) denote the rightmost letter of a. Also, asc(a), the number of ascents in a, and comp(a), the number of components in a, are defined as in the case of permutations (see Table A.1).

Example 3.2.27. For the ascent sequence $a = (0, 1, 0, 2, 3, 2, 3)$, we have zeros$(a) = $ comp$(a) = 2$, last$(a) = 3$, and asc$(a) = 4$.

Definition 3.2.28. A *run* in an ascent sequence is a maximal consecutive factor of equal letters. Let $\mathcal{A}^{(k)}$ be the set of ascent sequences whose runs have length at most k, and let $\mathcal{A}_n^{(k)}$ be those $a \in \mathcal{A}^{(k)}$ that have $|a| = n$. A *primitive* ascent sequence is a sequence with no runs of length greater than 1. Thus, $\mathcal{A}_n^{(1)}$ is the set of all primitive ascent sequences of length n.

Definition 3.2.29. Given $a = (a_1, a_2, \ldots, a_n) \in \mathcal{A}_n$, we call a pair (a_i, a_{i+1}) with $a_i = a_{i+1}$ an *equal pair* of the sequence. We denote the number of equal pairs in a sequence a by $epairs(a)$.

Example 3.2.30. For example, $epairs(0, 0, 0, 0, 1, 1, 1, 2, 2, 1, 1) = 7$ since $(a_1, a_2) = (a_2, a_3) = (a_3, a_4) = (0, 0)$, $(a_5, a_6) = (a_6, a_7) = (1, 1)$, $(a_8, a_9) = (2, 2)$, and $(a_{10}, a_{11}) = (1, 1)$.

Remark 3.2.31. The equal consecutive pairs introduced in Definition 3.2.29 are exactly the same as the statistic *"level,"* the number of levels on words, that appears often in the literature, and also in this book. However, in this section we want to distinguish levels on sequences from levels on $(2+2)$-free posets; also, we decided to follow the original notation in [318].

Theorem 3.2.32. *([161]) There is a one-to-one correspondence between the sets $\mathcal{P}_n$ and $\mathcal{A}_n$.*

The basic idea to prove Theorem 3.2.32 is to decompose a given $(2+2)$-free poset by removing one element at a time and recording the level from which that element was removed. At each step, one removes a minimal maximal element which is marked with an asterisk in Figure 3.6 illustrating the process. Three cases are possible: two easy cases and one complicated one. In the first easy case, a minimal maximal element x to be removed is not alone on its level: simply remove x together with edges coming out from it (if any) to go to the next step after recording x's level. The other easy case is when the current poset has just one maximal element (the maximum element). In this case, proceed like in the other easy case. For a proper description of the complicated case, we refer to [161], however, the basic idea is that on top of what is to be done in an easy case, we have to watch for each element y (marked with the $\#$ symbol in Figure 3.6 illustrating the process) incomparable to x (to be removed) but covered by the elements from the level, say i, one higher than x's level. Each such y will become a new maximal element in the next step, and additionally, every element that was less than x (before removing x), must become less than the elements from the (former) i-th level (level i after removing x will become level $i - 1$). Once the last element is removed (it has to be from level 0), we obtain the corresponding ascent sequence by reading the levels from which elements were removed in the reverse order (starting from the last removed 0).

Example 3.2.33. In Figure 3.6 we provide the steps in the decomposition of the poset in Figure 3.4. All steps except Step 5 are easy cases (note that we record Steps 3 and 4 and Steps 8 and 9 together: in these cases, we deal with indistinguishable elements and the order in which they are removed is irrelevant). In Step 5, one element is marked with $\#$. This element is not comparable to the element to be removed from level 2, but it is below the element on level 3. The element marked with # becomes a maximal element on the next step, and no extra rearrangement must be done as all the elements less than the element from level 2 are already less than the element from level 3 (which does not have to be the case in general). The corresponding ascent sequence obtained by reading the numbers on top of arrows in reverse order is $(0, 1, 1, 2, 1, 2, 0, 0, 4, 0)$.

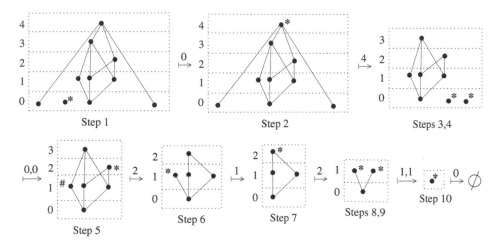

Figure 3.6: Steps in decomposition of the poset in Figure 3.4 to obtain the corresponding ascent sequence (0,1,1,2,1,2,0,0,4,0).

Remark 3.2.34. Once one is faced with solving a problem on (2+2)-free posets, one should check if the problem can be rephrased, using the bijection in Theorem 3.2.32, in terms of ascent sequences. This may be an easier way to the solution of the problem as dealing with ascent sequences is normally much easier than dealing with posets. In this way several enumerative problems were solved [161, 527, 318] which will be discussed in Subsection 3.2.6.

To conclude the subsection, we introduce the notion of a *modified ascent sequence* which is important in settling a conjecture of Pudwell (discussed in Subsection 3.2.6) and in translating statistics between equinumerous sets of relevant objects (discussed in Subsection 3.2.5).

Suppose $a = (a_1, a_2, \ldots, a_n)$ is an ascent sequence and $AP(a)$ is the (ordered) list of positions where an ascent occurs:

$$AP(a) = (i \mid i \in [n-1] \text{ and } a_i < a_{i+1}).$$

In particular, $asc(a) = |AP(a)|$. We describe an algorithm that takes a as an input and produces its *modified ascent sequence* $\widehat{a}$ (of the same length).

```
for i ∈ AP(a):
    for j ∈ [i − 1]:
        if aⱼ ≥ aᵢ₊₁ then aⱼ := aⱼ + 1
```

The resulting sequence is $\widehat{a}$.

Example 3.2.35. For $a = (0, 1, 0, 1, 2, 0, 2)$, $AP(x) = (1, 3, 4, 6)$ and the algorithm computes the modified ascent sequence $\widehat{a}$ in the following steps:

$$
\begin{aligned}
a = \; & 0\ \mathbf{1}\ 0\ 1\ 2\ 0\ 2 \\
& 0\ 1\ 0\ \mathbf{1}\ 2\ 0\ 2 \\
& 0\ 2\ 0\ 1\ \mathbf{2}\ 0\ 2 \\
& 0\ 3\ 0\ 1\ 2\ 0\ \mathbf{2} \\
& 0\ 4\ 0\ 1\ 3\ 0\ 1 = \widehat{a}
\end{aligned}
$$

Definition 3.2.36. An ascent sequence a is *self-modified* if $\widehat{a} = a$.

Example 3.2.37. $a = (0, 0, 1, 0, 2, 3, 0)$ is an example of a self-modified ascent sequence.

3.2.3   Stoimenow's diagrams

Definition 3.2.38. A *matching* of $[2n] = \{1, 2, \ldots, 2n\}$ is a partition of that set into blocks of size 2.

Example 3.2.39. For $n = 5$, an example of a matching is

$$M = \{(1, 5), (2, 8), (3, 4), (6, 10), (7, 9)\},$$

which can be represented by the diagram in Figure 3.7 (called a *chord diagram*), where an *arc* connects i and j precisely when $(i, j) \in M$.

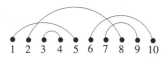

Figure 3.7: A diagram representing the matching M in Example 3.2.39.

Remark 3.2.40. A matching of $[2n]$ can be seen as a fixed-point-free involution on $[2n]$. For example, the matching in Example 3.2.39 corresponds to the involution $58431(10)9276$.

Definition 3.2.41. A *nesting* of a matching M is a pair of arcs (i, ℓ) and (j, k) with $i < j < k < \ell$. Such a nesting is called a *left-nesting* (resp., *right-nesting*) if $j = i+1$ (resp., $\ell = k + 1$).

Example 3.2.42. The matching in Example 3.2.39 represented by the chord diagram in Figure 3.7 has a right-nesting, the pair of arcs $(1, 5)$ and $(3, 4)$, and a pair of arcs $(6, 10)$ and $(7, 9)$ which is both a right- and a left-nesting.

Definition 3.2.43. A *Stoimenow's diagram*, also called a *regular linearized chord diagram* (*RLCD*), is a matching without *neighbour nestings*, that is, a matching with neither left-nestings nor right-nestings. We let $\mathcal{D}_n$ denote the set of Stoimenow's diagrams on $[2n]$, and $\mathcal{D} = \cup_{n \geq 0} \mathcal{D}_n$.

Stoimenow [754] used the linearized chord diagrams to give an upper bound on the dimension of the *space of Vassiliev's knot invariants* of a given degree. A couple of years later, Zagier [812] derived the generating function for the number of such diagrams with respect to size, which coincides with the formula derived by Bousquet-Mélou et al. in [161] for the number of (2+2)-free posets, and we state it in Subsection 3.2.6.

Definition 3.2.44. For an arc in a Stoimenow's diagram M, call the left endpoint the *opener* and the right endpoint the *closer*. We can then talk about runs (maximal factors) of openers and closers. Label each arc with the number of runs of closers that precede it. Let iopeners(M) denote the number of elements in the *initial run of openers* (equivalently, this is the number of arcs labeled 0). Let lclosers(M) denote the number of elements in the last run of closers. We let $\ell_d(M)$ denote the maximum arc label.

Definition 3.2.45. For a Stoimenow's diagram on $[2n]$, the arc having the closer $2n$ is called the *maximum arc*. $\ell_d^*(M)$ is the label of the arc whose closer is one plus the maximum arc opener. Finally, the number of components in M, $\text{comp}_d(M)$, is the number of cuts in the following procedure: read M from left to right starting at 1, and cut at each place when the number of openers is equal to that of closers (there are no arcs beginning to the left of a cut point and ending to the right of it).

Example 3.2.46. For the Stoimenow's diagram M in Figure 3.8, runs of openers are marked by rectangles. We have iopeners(M) = 3 (given by 1, 2, and 3), lclosers(M) = 3 (given by 12, 13, and 14), $\ell_d(M) = 2$, $\ell_d^*(M) = 1$ (given by the label of the arc $[8, 12]$), and $\text{comp}_d(M) = 2$ (we can make a cut between 6 and 7, and a cut to the right of 14).

Definition 3.2.47. If (i, j) and $(i+1, j+1)$ are arcs in a matching M, we say that they are *similar*. Let echords(M) be the minimum number of arcs in M one has to

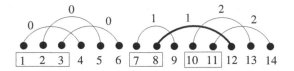

Figure 3.8: A Stoimenow's diagram illustrating defined statistics.

remove to obtain a matching without similar arcs. Let $\mathcal{D}_n^{(k)}$ be the set of matchings $M \in \mathcal{D}_n$ such that for no pair i and j do all of $(i,j), (i+1,j+1), \ldots, (i+k,j+k)$ belong to M.

Theorem 3.2.48. *([161]) There is a one-to-one correspondence between the sets $\mathcal{P}_n$ and $\mathcal{D}_n$.*

3.2.4   Upper triangular matrices

Definition 3.2.49. Let $\mathcal{M}_n$ be the set of upper triangular matrices of non-negative integers such that

- no row or column contains all zero entries, and

- the sum of the entries is n.

Let $\mathcal{M} = \cup_{n \geq 0} \mathcal{M}_n$, and $\mathcal{M}_n^{(k)}$ be the set of matrices in $\mathcal{M}_n$ that have no entries exceeding k. In particular, $\mathcal{M}_n^{(1)}$ is the set of binary matrices satisfying the conditions.

Example 3.2.50. For example, $\mathcal{M}_3$ consists of the following 5 matrices:

$$(3), \begin{pmatrix} 2 & 0 \\ 0 & 1 \end{pmatrix}, \begin{pmatrix} 1 & 1 \\ 0 & 1 \end{pmatrix}, \begin{pmatrix} 1 & 0 \\ 0 & 2 \end{pmatrix}, \begin{pmatrix} 1 & 0 & 0 \\ 0 & 1 & 0 \\ 0 & 0 & 1 \end{pmatrix}.$$

Definition 3.2.51. For any $A \in \mathcal{M}_m$, we let $\dim(A)$ be the dimension of A (the number of rows = the number of columns), $|A|$ be the sum of the entries in A, extra(A) be $|A|$ minus the number of non-zero entries in A, rsum$_i(A)$ be the sum of entries in row i, csum$_i(A)$ be the sum of entries in column i, index(A) be the smallest value of i such that $A_{i,\dim(A)}$ is non-zero, and blocks(A) be the number of *diagonal blocks* in A. A diagonal block in A is a square submatrix A' of A that is built by consecutive rows and columns, whose diagonal elements are A's diagonal elements, and there are no non-zero elements above, below, to the left, or to the right of A'.

Example 3.2.52. Let

$$A = \begin{pmatrix} 1 & 3 & 0 & 0 \\ 0 & 0 & 2 & 0 \\ 0 & 0 & 0 & 4 \\ 0 & 0 & 0 & 2 \end{pmatrix}.$$

Then $\dim(A) = 4$, $|A| = 1+3+2+4+2 = 12$, $\text{extra}(A) = 12-5 = 7$, $\text{rsum}_1(A) = 4$, $\text{csum}_4(A) = 6$, $\text{blocks}(A) = 1$, and $\text{index}(A) = 3$ since the first non-zero entry in the final column is in the third row.

Theorem 3.2.53. *([321]) There is a one-to-one correspondence between the sets $\mathcal{P}_n$ and $\mathcal{M}_n$.*

3.2.5 Connections between the objects and pattern-avoidance

The set $\mathcal{S}_n(\overline{\substack{123\\231}})$ can be described as follows:

$$\{\pi_1 \pi_2 \cdots \pi_n \in \mathcal{S}_n \mid \text{if } \text{red}(\pi_i \pi_j \pi_k) = 231 \text{ then } j \neq i+1 \text{ or } \pi_i \neq \pi_k + 1\}.$$

In the terminology of [161], $\mathcal{S}_n(\overline{\substack{123\\231}}) = \mathcal{S}_n(\boxed{\,\substack{\bullet\\ \bullet}\,})$.

Note that in our terminology, $\mathcal{S}_n(\overline{\substack{123\\231}}, r(\overline{\substack{12\cdots k+1\\12\cdots k+1}}))$ is the subset of $\mathcal{S}_n(\overline{\substack{123\\231}})$ such that there do not exist integers i and m with $\pi_i = m$, $\pi_{i+1} = m-1, \ldots,$ $\pi_{i+k} = m-k$ (together with the pattern $\overline{\substack{123\\231}}$, we avoid the reverse of the bivincular pattern $\overline{\substack{12\cdots k+1\\12\cdots k+1}}$). In particular, for $k = 1$, $\mathcal{S}_n(\overline{\substack{123\\231}}, r(\overline{\substack{12\\12}}))$ are those permutations in $\mathcal{S}_n(\overline{\substack{123\\231}})$ that do not contain a factor $m(m-1)$ for some m.

See Table A.1 for definitions of the permutation statistics lmin, asc, rmax, and comp we will need below. One more statistic we need, $b(\pi)$, is defined as follows.

Definition 3.2.54. For a permutation $\pi \in \mathcal{S}_n(\overline{\substack{123\\231}})$ its *active sites* are positions in which we can insert the element $(n+1)$ to get a permutation in $\mathcal{S}_{n+1}(\overline{\substack{123\\231}})$. Label active sites from left to right by $0, 1, 2$, etc. Then $b(\pi)$ is the label immediately to the left of the maximum element n in π.

Example 3.2.55. For $\pi = 31764825 \in \mathcal{S}_n(\overline{\substack{123\\231}})$, its active sites are $_0 3_1 1_2 7 6 4_3 8 4 2_5 5_6$, and $b(\pi) = 3$.

Theorems 3.2.32, 3.2.48, and 3.2.53 can be refined considering statistics and are connected to pattern-avoidance as stated in the following theorem. Most of the results on Stoimenow's diagrams are implicit in [254].

Theorem 3.2.56. *([161, 254, 321]) There are bijections between the set $\mathcal{P}_n$ of unlabeled (2+2)-free posets on n elements and the set $\mathcal{A}_n$ of ascent sequences of length n, the set $\mathcal{M}_n$ of upper triangular matrices with the sum of entries n (with no row or column containing all zero entries), the set $\mathcal{D}_n$ of Stoimenow's diagrams on $[2n]$, and the set $\mathcal{S}_n(\overline{123}\!\!\;_{231})$ of $\overline{123}\!\!\;_{231}$-avoiding permutations. Moreover, assuming that $P \in \mathcal{P}_n$, $a \in \mathcal{A}_n$, $A \in \mathcal{M}_n$, $M \in \mathcal{D}_n$, and $\pi \in \mathcal{S}_n(\overline{123}\!\!\;_{231})$ correspond to one another under the bijections, the following statistics are preserved ($i(\pi)$ below is the inverse of π, usually denoted π^{-1}, $\mathrm{asc}\,.i(\pi) = \mathrm{asc}(i(\pi))$, and "perms" stands for "permutations"):*

posets		sequences		matrices		diagrams		perms
$\min(P)$	=	$\mathrm{zeros}(a)$	=	$\mathrm{rsum}_1(A)$	=	$\mathrm{iopeners}(M)$	=	$\mathrm{lmin}(\pi)$
$\max(P)$	=	$\mathrm{rmax}(\hat{a})$	=	$\mathrm{csum}_{dim(A)}(A)$	=	$\mathrm{lclosers}(M)$	=	$\mathrm{rmax}(\pi)$
$\mathrm{comp}_p(P)$	=	$\mathrm{comp}(\hat{a})$	=	$\mathrm{blocks}(A)$	=	$\mathrm{comp}_d(M)$	=	$\mathrm{comp}(\pi)$
$\ell(P)$	=	$\mathrm{asc}(a)$	=	$\dim(A) - 1$	=	$\ell_d(M)$	=	$\mathrm{asc}\,.i(\pi)$
$\ell^*(P)$	=	$\mathrm{last}(a)$	=	$\mathrm{index}(A) - 1$	=	$\ell_d^*(M)$	=	$b(\pi)$

Actually references [161] and [321] contain results generalizing Theorem 3.2.56. For example, under the bijections, the number of elements on level i in a (2+2)-free poset P, is the same as the number of copies of i in the modified ascent sequence $\hat{a}$, which is the same as the number of letters in the permutation π between the i-th and $(i + 1)$-th active sites. Even more can be said. Given a (2+2)-free poset $P \in \mathcal{P}_n$, let us label by i the element removed at step i in the decomposition process, for $i = 1, 2, \ldots, n$. We can obtain the permutation corresponding to P by arranging elements on each level in decreasing order, and placing them in a line starting from the bottom level. See Figure 3.9, based on Figure 3.6, for an example.

We would also like to explain how to obtain the permutation $\pi \in \mathcal{S}_n(\overline{123}\!\!\;_{231})$ corresponding to an ascent sequence $a \in \mathcal{A}_n$ following the bijection in Theorem 3.2.56. Read $a = (a_1, a_2, \ldots, a_n)$ from left to right and build the corresponding permutation inserting letters $1, 2, 3$, etc., depending on the symbol read: if $a_i = j$, then insert i in the j-th active site of the $(i - 1)$-permutation constructed so far. We illustrate this process by example. Suppose $a = (0, 0, 1, 0, 2, 1)$. The corresponding permutation is obtained as follows:

$$_0 \rightarrow {_0}1_1 \rightarrow {_0}21_1 \rightarrow {_0}21_1 3_2 \rightarrow {_0}421_1 3_2 \rightarrow {_0}421_1 3_2 5_3 \rightarrow 421635.$$

One more illustration of the bijections in Theorem 3.2.56 is given in Figure 3.10, where it is shown how to get the (2+2)-free poset corresponding to a

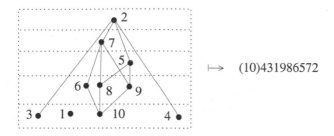

$\longmapsto$ (10)431986572

Figure 3.9: Encoding a (2+2)-free poset, based on Figure 3.6, by a permutation.

given Stoimenow's diagram. The basic idea is to view a Stoimenow's diagram as an interval order (see Definition 3.2.1).

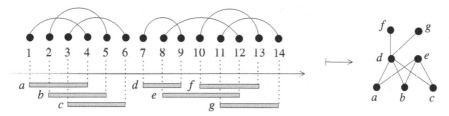

Figure 3.10: Obtaining the (2+2)-free poset corresponding to the Stoimenow's diagram $M = \{(1,4),(2,5),(3,6),(7,9),(8,12),(10,13),(11,14)\}$ through interval orders.

Another use of the bijections in Theorem 3.2.56 is shown by Dukes et al. in [318] and is recorded in the following theorem.

Theorem 3.2.57. *([318]) We have*

$$\mathcal{P}_n^{(k)} = \mathcal{A}_n^{(k)} = \mathcal{M}_n^{(k)} = \mathcal{D}_n^{(k)} - \mathcal{S}_n(\overline{\tfrac{123}{231}}, r(\overline{\tfrac{12\cdots k+1}{12\cdots k+1}})).$$

The set of all permutations $\mathcal{S}_n$ is a natural superset of $\mathcal{S}_n(\overline{\tfrac{123}{231}})$, and $|\mathcal{S}_n| = n!$. Claesson and Linusson [263] considered a superset of Stoimenow's diagrams by removing the restriction to avoid right-nestings.

Theorem 3.2.58. *([263]) There are $n!$ matchings of $[2n]$ with no left-nestings.*

Since Stoimenow's diagrams are in one-to-one correspondence with (2+2)-free posets, a natural question to ask is "Which superset of the set of (2+2)-free posets on n elements is counted by $n!$?" Let us describe the *factorial posets* introduced in [263] which give an answer to the question.

Definition 3.2.59. A labeling of a poset $(P, <_P)$ is *natural*, if for two elements in P labeled by i and j, $i <_P j$ implies $i < j$ (in the usual order).

Example 3.2.60. The poset to the left in Figure 3.11 is naturally labeled, whereas the poset to the right is not — the element labeled 3 in the poset is larger than the element labeled 4.

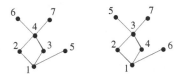

Figure 3.11: The poset to the left is naturally labeled; the poset to the right is not.

One is interested in the following specialization of naturally labeled posets.

Definition 3.2.61. A naturally labeled poset P on $[n]$ is *factorial*, if $i <_P k$ whenever $i < j <_P k$ for some $j \in [n]$.

Example 3.2.62. All factorial posets on $\{1, 2, 3\}$ are in Figure 3.12.

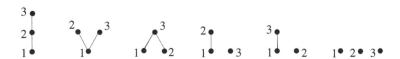

Figure 3.12: All factorial posets on $\{1, 2, 3\}$.

Theorem 3.2.63. *([263]) Factorial posets are (2+2)-free and there are $n!$ factorial posets on n elements.*

Another superset of unlabeled (2+2)-free posets is considered by Claesson et al. in [255] in connection with *composition matrices* generalizing our upper triangular matrices and defined in Definition 3.2.64 below. However, we do not provide any further details on [255], instead referring to the source.

Definition 3.2.64. Let X be a finite subset of $\{1, 2, \dots\}$. A *composition matrix* on X is an upper triangular matrix over the powerset of X satisfying the following properties:

(i) each column and row contain at least one non-empty set;

(ii) the non-empty sets partition X.

Example 3.2.65. Here is an example of a composition matrix:

$$\begin{pmatrix} \{2,11\} & \emptyset & \{7\} & \{3,9\} \\ \emptyset & \{4\} & \{5,8,12\} & \{1\} \\ \emptyset & \emptyset & \emptyset & \{13\} \\ \emptyset & \emptyset & \emptyset & \{6,10\} \end{pmatrix}$$

3.2.6 Enumeration and solving two conjectures

As was mentioned above, Zagier [812] derived the generating function for the number of Stoimenow's diagrams, which enumerates at once all objects of interest in this section using the bijections in [161, 321]. However, [161] by Bousquet-Mélou et al. contains an independent derivation of this formula using ascent sequences.

Theorem 3.2.66. *([161, 321]) We have*

$$\sum_{n \geq 0} |\mathcal{P}_n| x^n = \sum_{n \geq 0} |\mathcal{A}_n| x^n = \sum_{n \geq 0} |\mathcal{M}_n| x^n = \sum_{n \geq 0} |\mathcal{D}_n| x^n = \sum_{n \geq 0} s_n(\overline{231}) x^n$$

$$(3.1) \quad = \sum_{n \geq 0} \prod_{i=1}^{n} \left(1 - (1-x)^i\right) = 1 + x + 2x^2 + 5x^3 + 15x^4 + 53x^5 + 217x^6 + \cdots.$$

Remark 3.2.67. In fact, there are other enumerative results in the literature attempting to find the number of (2+2)-free posets. Khamis [499] used a recursive description of (2+2)-free posets to derive a pair of functional equations that define the corresponding generating function, but he did not solve these equations. Haxell, McDonald and Thomasson [452] provided an algorithm, based on a complicated recurrence relation, to produce the first few values of $|\mathcal{P}_n|$.

Theorem 3.2.66 was refined in [527] when the number of minimal elements in (2+2)-free posets, and thus the corresponding statistics on other objects in Theorem 3.2.56 are taken into account.

Theorem 3.2.68. *([527]) Let $p_{n,k}$ be equal to the number of (2+2)-free posets of size n with k minimal elements. Thus, by Theorem 3.2.56, $p_{n,k}$ = the number of ascent sequences of length n with k zeros = the number of upper triangular $n \times n$ matrices with the first row sum k = the number of Stoimenow's diagrams on $[2n]$*

with initial run of openers of length $k =$ *the number of* $\overline{\underline{\underline{2}}}\underline{3}\underline{1}$ *-avoiding n-permutations with* k *left-to-right minima. Then*

$$(3.2) \qquad P(x, z) = \sum_{n,k \geq 0} p_{n,k} x^n z^k = 1 + \sum_{n \geq 0} \frac{zx}{(1 - zx)^{n+1}} \prod_{i=1}^{n} (1 - (1 - t)^i).$$

However, Kitaev and Remmel conjectured a different form of formula (3.2). This conjecture was proved independently by three people and we record this result in the next theorem, though we state references only to two papers that are available (the paper by Jelínek is still in preparation).

Theorem 3.2.69. *([557, 811]) For $P(x, z)$ defined in Theorem 3.2.68, we have*

$$(3.3) \qquad P(x, z) = \sum_{n \geq 0} \prod_{i=1}^{n} (1 - (1 - x)^{i-1}(1 - zx)).$$

Remark 3.2.70. Note the striking similarity between formulas (3.1) and (3.3).

To state more enumerative results, we need Definitions 3.2.71 and 3.2.73.

Definition 3.2.71. In the process of decomposing a (2+2)-free poset P, one will reach a point where the remaining poset is an *antichain*, which is one or more pairwise incomparable elements. We define $lds(P)$ to be the maximum size of such an antichain, which is also equal to the size of the down-set of the last removed element that has a non-trivial down-set. By definition, the value of lds on an antichain is 0 as there are no non-trivial down-sets for such a poset.

Example 3.2.72. For the poset P in Figure 3.6, $lds(P) = 1$, whereas for the third poset in Figure 3.12 (if the labels are disregarded), the value of lds is 2.

To provide a counterpart for the poset statistic lds (defined in Definition 3.2.71) on ascent sequences and permutations avoiding the bivincular pattern $\overline{\underline{\underline{2}}}\underline{3}\underline{1}$, we state the following definition. The corresponding statistics can be defined on upper triangular matrices and Stoimenow's diagrams as well (based on the bijections in Theorem 3.2.56), but we do not do this here.

Definition 3.2.73. For an ascent sequence a, we let $\mathrm{zrun}(a)$ be the number of 0s in the maximum initial run of 0s in a. By definition, $\mathrm{zrun}(0, 0, \ldots, 0) = 0$. For a permutation π, we let $\mathrm{decr}(\pi) \in S_n$ be the length of the maximum decreasing factor of the form $i(i - 1) \cdots 1$ in π. By definition, $\mathrm{decr}(n(n - 1) \cdots 1) = 0$.

Example 3.2.74. $\mathrm{zrun}((0, 0, 0, 1, 2, 1, 0, 3)) = 3$ and $\mathrm{decr}(6432175) = 4$.

For $k \geq 1$, we let

$$
\begin{aligned}
\delta_k &= u - (1-t)^k(u-1) \\
\gamma_k &= u - (1-zt)(1-t)^{k-1}(u-1) \\
\bar{\delta}_k &= \delta_k|_{u=uv} = uv - (1-t)^k(uv-1) \\
\bar{\gamma}_k &= \gamma_k|_{u=uv} = uv - (1-zt)(1-t)^{k-1}(uv-1)
\end{aligned}
$$

and we set $\delta_0 = \gamma_0 = \bar{\delta}_0 = \bar{\gamma}_0 = 1$.

The following theorem is a refinement of Theorems 3.2.68 and 3.2.69.

Theorem 3.2.75. *([527]) For* $\mathcal{X} = \mathrm{Av}(\frac{\overline{123}}{231})$, *we have*

$$
\begin{aligned}
G(x,u,v,z,t) &= \sum_{P \in \mathcal{P}} x^{\mathrm{size}(P)} u^{\ell(P)} v^{\ell^*(P)} z^{\min(P)} t^{\mathrm{lds}(P)} \\
&= \sum_{a \in \mathcal{A}} x^{|a|} u^{\mathrm{asc}(a)} v^{\mathrm{last}(a)} z^{\mathrm{zeros}(a)} t^{\mathrm{zrun}(a)} \\
&= \sum_{\pi \in \mathcal{X}} x^{|\pi|} u^{\mathrm{asc}.i(\pi)} v^{b(\pi)} z^{\mathrm{lmin}(\pi)} t^{\mathrm{decr}(\pi)}
\end{aligned}
$$

$$
= \frac{1}{(1-xz)} + \frac{x^2 z t u}{(1-xzt)(v\delta_1 - 1)}\left(v(v-1) + \right.
$$

$$
x(1-u)(z(v-1)-v)\sum_{s \geq 0} \frac{u^s(1-x)^s}{\delta_s \delta_{s+1} \prod_{i=1}^{s+1}\gamma_i} + uv^3 t(1-uv)\sum_{s \geq 0} \frac{(uv)^s(1-x)^s}{\bar{\delta}_s \bar{\delta}_{s+1} \prod_{i=1}^{s+1}\bar{\gamma}_i}\left.\right).
$$

We can use Mathematica, for example, to compute

$$
\begin{aligned}
&G(x,u,v,z,t) \\
&= 1 + zx + \left(uvtz + z^2\right)x^2 + \left(uvtz + u^2v^2tz + utz^2 + uvt^2z^2 + z^3\right)x^3 \\
&\quad + \left(uvtz + u^2vtz + 2u^2v^2tz + u^3v^3tz + utz^2 + u^2tz^2 + u^2vtz^2\right. \\
&\quad + u^2v^2tz^2 + uvt^2z^2 + u^2v^2t^2z^2 + utz^3 + ut^2z^3 + uvt^3z^3 + z^4\right)x^4 \\
&\quad + \left(uvtz + 3u^2vtz + u^3vtz + 3u^2v^2tz + 2u^3v^2tz + 4u^3v^3tz + u^4v^4tz\right. \\
&\quad + utz^2 + 3u^2tz^2 + u^3tz^2 + 3u^2vtz^2 + u^3vtz^2 + 2u^2v^2tz^2 + 2u^3v^2tz^2 + 3u^3v^3tz^2 \\
&\quad + uvt^2z^2 + u^2vt^2z^2 + 2u^2v^2t^2z^2 + u^3v^3t^2z^2 + utz^3 + 3u^2tz^3 + u^2vtz^3 + u^2v^2tz^3 \\
&\quad + ut^2z^3 + u^2t^2z^3 + u^2vt^2z^3 + u^2v^2t^2z^3 + uvt^3z^3 + u^2vt^3z^3 + utz^4 \\
&\quad + ut^2z^4 + ut^3z^4 + uvt^4z^4 + z^5\right)x^5 + \ldots.
\end{aligned}
$$

For instance, the 3 ascent sequences corresponding to the term $3u^2v^2tzx^5$ are $(0,1,1,1,2)$, $(0,1,1,2,2)$, and $(0,1,2,2,2)$.

More enumerative results appear in [318] by Dukes et al.

Theorem 3.2.76. *([318]) We have*

$$\sum_{n\geq 0}|\mathcal{P}_n^{(k)}|x^n=\sum_{n\geq 0}|\mathcal{A}_n^{(k)}|x^k=\sum_{n\geq 0}|\mathcal{M}_n^{(k)}|x^k=\sum_{n\geq 0}|\mathcal{D}_n^{(k)}|x^k=\sum_{n\geq 0}s_n(\overline{\underset{231}{123}},r(\overline{\underset{12\cdots k\text{:}1}{12\cdots k\text{+}1}}))x^k$$

$$=\sum_{n\geq 0}\prod_{i=1}^{n}\left(1-\left(\frac{1-x}{1-x^{k+1}}\right)^{i}\right).$$

As is pointed out in [318], a particular case of Theorem 3.2.76 (when $k=1$), solves a conjecture of Jovovic stating that the generating function for the number of upper triangular binary matrices with no zero rows and zero columns, by the sum of entries, is equal to

$$\sum_{n\geq 0}\prod_{i=1}^{n}\left(1-\frac{1}{(1+x)^{i}}\right).$$

Another conjecture that was solved while dealing with (2+2)-free posets and the equinumerous objects is that of Pudwell on $\mathrm{Av}(3\overline{1}52\overline{4})$. We state this result in Theorem 3.2.77. The enumeration of $3\overline{1}52\overline{4}$-avoiding permutations was done by observing that $\mathrm{Av}(3\overline{1}52\overline{4})$ is exactly the set of permutations corresponding to self-modified ascent sequences (see Definition 3.2.36) under the respective bijection in Theorem 3.2.56, and it is rather easy to enumerate the self-modified ascent sequences.

Theorem 3.2.77. *([161]) The number of n-permutations avoiding the pattern $3\overline{1}52\overline{4}$ is*

$$\sum_{k=1}^{n}\binom{\binom{k+1}{2}+n-k-1}{n-k}.$$

Moreover, the k-th term of this sum counts those permutations that have k right-to-left minima, or, equivalently, $k-1$ ascents. The corresponding generating function is

$$\sum_{n\geq 0}s_n(3\overline{1}52\overline{4})x^n=\sum_{k\geq 1}\frac{x^k}{(1-x)^{\binom{k+1}{2}}}.$$

As a matter of fact, we can see[1] that the self-modified ascent sequences, under the bijection in Theorem 3.2.56, correspond to the posets that are both (2+2)- and *N-avoiding* (N-avoidance can be defined similarly to (2+2)-avoidance). Thus, an alternative way to prove the conjecture of Pudwell is to use the connection between $3\overline{1}52\overline{4}$-avoiding permutations and (2+2)- and N-avoiding posets via self-modified ascent sequences and to use the formula derived in [100].

---

[1]This fact is an unpublished observation by Claesson, Dukes, Kitaev and Parviainen, 2009.

3.3 A classification of the Mahonian statistics

This section discusses the original motivation to introduce vincular patterns in [64], namely, a classification of the *Mahonian statistics* on permutations (these statistics also appear in different contexts, not to be considered here, such as the study of *Motzkin paths, orthogonal polynomials, rook theory,* and more). Most of our presentation is based on [64] by Babson and Steingrímsson. In Section 3.4, we consider other applications of vincular patterns to encode various combinatorial objects.

The notion of a Mahonian statistic was introduced in Subsection 2.2.6, but we redefine it here.

Definition 3.3.1. A permutation statistic is *Mahonian* if it has the same distribution as the inversion statistic inv, the number of inversions in permutations, which, in turn, is equivalent to the number of occurrences of the pattern 21, that is, $inv(\pi) = 21(\pi)$ for all permutations π.

Definition 3.3.2. The q-analogue of n is

$$[n]_q = [n] := 1 + q + q^2 + \cdots + q^{n-1} = \frac{1 - q^n}{1 - q},$$

and the q-analogue of $n!$ is

$$[n]_q! = [n]! := [n][n-1] \cdots [1].$$

It is rather easy to see, and was first proved by Rodriguez [699] in 1839, that the distribution of inv is given by the generating function

$$(3.4) \qquad \sum_{\pi \in \mathcal{S}_n} q^{inv(\pi)} = [n]!.$$

One way to prove (3.4) is by induction on n with the trivial base case $n = 1$: assuming the statement is true for $(n-1)$-permutations, we note that adding a letter i to the right of an $(n-1)$-permutation (increasing all other letters greater than or equal to i by 1, and thus creating an n-permutation) introduces $i-1$ inverses. Moreover, a way to generate all n-permutations is to consider all possible extensions of $(n-1)$-permutations to the right. Thus

$$\sum_{\pi \in \mathcal{S}_n} q^{inv(\pi)} = (1+q+q^2+\cdots+q^{n-1}) \sum_{\pi \in \mathcal{S}_{n-1}} q^{inv(\pi)} = (1+q+q^2+\cdots+q^{n-1})[n-1]! = [n]!.$$

Remark 3.3.3. An even easier way to prove (3.4) is to insert n in all possible places in an $(n-1)$-permutation and to register changes of the statistic inv.

3.3.1 Mahonian statistics in the literature

It turns out that almost all Mahonian statistics in the literature are either defined as a linear combination of vincular patterns or can be defined in this way. See Subsection 3.3.2 for an example of a Mahonian statistic on permutations, defined in [653], whose pattern definition is unknown, though this may not be impossible to find. In this subsection we provide pattern definitions of Mahonian statistics including those appearing in the classification of the *Mahonian 3-functions* (*d-functions* are defined in Definition 3.3.11 in Subsection 3.3.3 below).

Definition 3.3.4. The statistic *major index* of a permutation, maj, is defined as the sum of the descent positions in the permutation.

Example 3.3.5. $\mathrm{maj}(3251476) = 1 + 3 + 6 = 10$, since the permutation has descents in positions 1, 3 and 6.

MacMahon [578] showed that the statistic maj is Mahonian. Further, it was shown in [64] by Babson and Steingrímsson that

$$\mathrm{maj}(\pi) = (1\underline{32} + 2\underline{31} + 3\underline{21} + \underline{21})(\pi),$$

where recall from Remark 1.0.36 that the sum of patterns denotes the total number of occurrences of the patterns from the set $\{1\underline{32}, 2\underline{31}, 3\underline{21}, \underline{21}\}$ in π. Indeed, a way to compute $\mathrm{maj}(\pi)$ is to count, for each descent in π, the letters in π preceding the descent bottom (the second letter in the descent). If a letter thus preceding a descent is smaller than both letters in the descent it will be counted by the pattern $1\underline{32}$. If the letter is between the descent letters in size, it will be counted by the pattern $2\underline{31}$, and if it is larger than both the descent letters, then it will be counted by $3\underline{21}$. Finally, one needs to count the first letter in the descent which gives an occurrence of the pattern $\underline{21}$ in π.

Another Mahonian statistic, introduced by Foata and Zeilberger in [400], is mak, which is shown in [64] to equal the following combination of patterns:

$$\mathrm{mak}(\pi) = (2\underline{31} + \underline{32}1 + 1\underline{32} + \underline{21})(\pi).$$

Yet another Mahonian statistic mad introduced in [269] by Clarke et al., can be defined as

$$\mathrm{mad}(\pi) = (2\underline{31} + 2\underline{31} + 3\underline{12} + \underline{21})(\pi),$$

as was mentioned in [64]. One more Mahonian statistic in [269] can be written in terms of patterns as follows:

$$\mathrm{makl}(\pi) = (1\underline{32} + 3\underline{12} + \underline{32}1 + \underline{21})(\pi).$$

Simion and Stanton [722] defined 16 different Mahonian statistics, each of which is a combination of the patterns $2\underline{31}$, $\underline{31}2$, $\underline{21}$, and $\underline{12}$. One of these statistics is mad, and the other statistics, up to trivial symmetries, are as follows:

$$(2 \cdot \underline{13}2 + 2\underline{13} + \underline{21})(\pi);$$

$$(2 \cdot \underline{13}2 + 2\underline{31} + \underline{21})(\pi);$$

$$(\underline{13}2 + 2 \cdot 2\underline{31} + \underline{21})(\pi),$$

where, for a pattern p, $2 \cdot p = p + p$.

By induction on the length of a permutation, the following statistics can be shown to be Mahonian (see [64])

$$\text{stat}(\pi) = (\underline{13}2 + \underline{21}3 + \underline{32}1 + \underline{21})(\pi);$$

$$\text{stat}'(\pi) = (\underline{13}2 + \underline{31}2 + \underline{32}1 + \underline{21})(\pi);$$

$$\text{stat}''(\pi) = (1\underline{32} + 2\underline{13} + 3\underline{21} + \underline{21})(\pi).$$

The following statistics were conjectured in [64] to be Mahonian, which was confirmed by Foata and Zeilberger [401]:

$$(\underline{13}2 + \underline{21}3 + 3\underline{21} + \underline{21})(\pi);$$

$$(1\underline{32} + 2 \cdot 2\underline{31} + \underline{21})(\pi);$$

$$(\underline{23}1 + 2 \cdot \underline{31}2 + \underline{21})(\pi).$$

Remark 3.3.6. It was shown in [64] using Corollary 3.3.22 stated in Subsection 3.3.3 below, that the *only* possible Mahonian statistics (modulo trivial bijections) based on vincular patterns of length at most 3 are the fourteen statistics mentioned above (if one takes into account the inv statistic).

All of the Mahonian statistics mentioned above, except for *inv*, are *descent-based* meaning that they are defined in terms of descents or ascents in a permutation. However, there are *excedance-based* Mahonian statistics in the literature (the statistic *exc*, the number of excedances, is defined in Table A.1 as the number of positions i in a permutation $\pi = \pi_1\pi_2 \cdots \pi_n$ such that $\pi_i > i$). *Denert's statistic*, den, introduced by Denert in [291] was the first known excedance-based Mahonian statistic. It seems that all known excedance-based Mahonian statistics in the literature can be translated into descent-based Mahonian statistics via the bijection in [269] by Clarke et al. In particular, an excedance-based statistic of Haglund [442], called hag in [64], can be translated into a descent-based statistic dag that can be written, as it is shown in [64], as the following combination of patterns:

$$
\begin{aligned}
\mathrm{dag}(\pi) = \quad & (\underline{21} + \underline{312} + \underline{321} + \underline{132} + \\
& 2 \cdot \underline{3142} + 2 \cdot \underline{3241} + \underline{4312} + \underline{4321} + \underline{4123} + \\
& \underline{4213} + \underline{1432} + \underline{1342} + \underline{1243} + \underline{2143})(\pi).
\end{aligned}
$$

The following statistics were conjectured in [64] to be Mahonian, which was confirmed by Foata and Randrianarivony [398]:

$$
\begin{aligned}
S_1(\pi) &= (1\underline{32} + 2\underline{13} + \underline{321} + \underline{\lfloor 21})(\pi); \\
S_2(\pi) &= (1\underline{32} + 2\underline{13} + 321 + \underline{\lfloor 21})(\pi); \\
S_3(\pi) &= (1\underline{32} + 2\underline{31} + \underline{321} + \underline{\lfloor 21})(\pi); \\
S_4(\pi) &= (1\underline{32} + 2\underline{31} + 321 + \underline{\lfloor 21})(\pi).
\end{aligned}
$$

3.3.2   A Mahonian statistic

In this subsection we give an example of a Mahonian statistic on permutations (see Definition 3.3.9), defined by Petersen in [653], for which no pattern definition is known.

Definition 3.3.7. A *transposition* $t_{i,j}$, for $i \neq j$, is an n-permutation, $n \geq 2$, having j in the i-th place, i in the j-th place, and every other letter k in the k-th place.

Example 3.3.8. $t_{2,5} = 1534267$ and $t_{3,4} = 1243$ are examples of transpositions.

Definition 3.3.9. For a permutation π, the *sorting index*, $\mathrm{sor}(\pi)$, is defined as

$$
\mathrm{sor}(\pi) = \sum_{s=1}^{k} (j_s - i_s),
$$

where k, and the i_ss and j_ss come from the unique expression of π,

$$
(3.5) \qquad\qquad \pi = t_{i_1,j_1} t_{i_2,j_2} \cdots t_{i_k,j_k},
$$

as a product of transpositions with $j_1 < j_2 < \cdots < j_k$.

Example 3.3.10. The permutation $2431756 = t_{1,2} t_{2,4} t_{5,6} t_{5,7}$ and thus

$$
\mathrm{sor}(2431756) = (2-1) + (4-2) + (6-5) + (7-5) = 6.
$$

The transpositions in factorization (3.5) are exactly the transpositions used in the *"straight selection sort"* algorithm described in [653].

3.3.3 Mahonian statistics and combinations of patterns of arbitrary length

In this subsection we provide an explicit numerical description of the combinations of patterns a Mahonian statistic must have, depending on the maximal length of its patterns.

Definition 3.3.11. A *k-pattern* is a function from the set of all permutations $\mathcal{S}_\infty$ to $\mathbb{N}$ that counts the number of occurrences of a vincular pattern of length k (that is, having k letters). A *pattern function* is a linear combination of patterns and a *d-function* is a linear combination of patterns that have length at most d.

Example 3.3.12. Examples of 5-patterns are $\underline{25431}$ and $3\underline{124}5$. The following combination of patterns is an example of a 4-function:

$$(1\underline{423}\rfloor + \underline{2134} + 2 \cdot \underline{312}\rfloor + 12)(\pi).$$

Definition 3.3.13. Let p be a vincular pattern with the underlying permutation $p_1 p_2 \cdots p_k$. There are $k + 1$ positions associated with p: before each p_i, $1 \leq i \leq k$, and after p_k. We number the positions $0, 1, \ldots, k$ from left to right. Position 0 is *restricted* if p begins with a hook. Position k is *restricted* if p ends with a hook. For any other position i, it is *restricted* if p_i and p_{i+1} are underlined by the same line segment. We call the pattern p a (k, i)-*pattern*, if p has i unrestricted positions.

Example 3.3.14. For the pattern $\underline{2134}$, positions 0, 3, and 4 are unrestricted, thus the pattern is a $(4, 3)$-pattern. On the other hand, $\lfloor\underline{231}\rfloor$ is a $(3, 0)$-pattern.

Remark 3.3.15. *Unrestricted positions* for vincular patterns in our terminology are *dashes* for generalized patterns in the terminology of Babson and Steingrímsson [64]. See Remark 1.3.4 discussing translations between the GP and VP notations.

Proposition 3.3.16. *([64]) Any d-function, when restricted to $\mathcal{S}_n$ for $n \geq d$, can be written as a linear combination of d-patterns.*

To illustrate Proposition 3.3.17, assume that a 4-function contains the vincular pattern $\underline{213}$, then, as it is shown in [64], it can be rewritten as the following combination:

$$\underline{213}(\pi) = (\underline{2143} + \underline{2134} + \underline{3124} + \underline{3214} + \lfloor\underline{213}\rfloor)(\pi).$$

However, since $\lfloor\underline{213}\rfloor(\pi) = 0$ for all π except for the permutation $\pi = 213 \in \mathcal{S}_3$, we have, that for $|\pi| \geq 4$,

$$\underline{213}(\pi) = (\underline{2143} + \underline{2134} + \underline{3124} + \underline{3214})(\pi).$$

Such rewriting as above of the pattern $\underline{213}$ is called *upgrading* in [64]; upgrading never increases the number of unrestricted positions in a vincular pattern.

Proposition 3.3.17. *([64]) The statistic inv, when restricted to $\mathcal{S}_n$ for $n \geq d$, can be written as a combination of d-patterns, of which $d!/2$ have three unrestricted positions, $(d-2)d!/2$ have two unrestricted positions, and $\binom{d-1}{2}d!/2$ have one unrestricted position.*

Definition 3.3.18. The *weight* on $\mathcal{S}_n$ of a function f is the sum $\displaystyle\sum_{\pi \in \mathcal{S}_n} f(\pi)$.

It is not hard to show [64] that the weight of the *inv* statistic is $\frac{n!}{2}\binom{n}{2}$. More generally, the following proposition is true, where we define $\binom{m}{-1}$ to be 1 if $m = -1$ and 0 otherwise.

Proposition 3.3.19. *([64]) The weight on $\mathcal{S}_n$ of a d-pattern with $k+1$ unrestricted positions is given by*

$$W_n(d,k) = \frac{n!}{d!}\binom{n-d+k}{k}.$$

In particular, the weight of a d-pattern with no unrestricted positions (the case $k = -1$) is 1 if $n = d$ and 0 otherwise.

Since two functions with the same distribution must have the same weight, we have the following proposition.

Proposition 3.3.20. *([64]) The weight of a Mahonian function on $\mathcal{S}_n$ is $\frac{n!}{2}\binom{n}{2}$, the weight of inv.*

As a particular corollary to the following theorem we have that the number of Mahonian d-functions is finite for each d.

Theorem 3.3.21. *([64]) Let f be an arbitrary Mahonian d-function, written so that all of its k-patterns, for $k < d$, have no unrestricted positions. Then f has $d!/2$ $(d,3)$-patterns, $(d-2)d!/2$ $(d,2)$-patterns, and $\binom{d-1}{2}d!/2$ $(d,1)$-patterns.*

Finally, we state the following corollary to Theorem 3.3.21 that was used in [64] to classify the Mahonian 3-functions (see Remark 3.3.6 above).

Corollary 3.3.22. *([64]) Let f be a Mahonian d-function whose patterns all have at least two unrestricted positions. Then f can be written as a sum of $k!/2$ patterns of length k with two unrestricted positions, for $2 \leq k < d$, and $d!/2$ patterns of length d with three unrestricted positions.*

3.4   Encoding combinatorial structures

In previous chapters/sections, we have already seen several links between pattern-restricted permutations and various combinatorial objects, that can be seen as encoding combinatorial objects by restricted permutations. In this section, we focus on combinatorial structures appearing in the consideration of vincular patterns. The first systematic analysis of vincular patterns of length 3 is done by Claesson in [251]. Our strategy is to define objects of interest in this section that are not defined in Section A.2, and to state several connections between vincular patterns and other combinatorial structures in Table 3.2. Note that Table 3.2 is not necessary comprehensive though we attempted to include in it as many facts as possible.

In Subsection 3.4.1, we provide all necessary definitions (beyond those defined in Section A.2) and results to be able to list connections between vincular patterns and other combinatorial objects. In Subsection 3.4.2, we provide examples of bijections involved in Table 3.2.

3.4.1   Objects of interest and the table with connections

Definition 3.4.1. A *partition* of a set S is a collection $P = \{A_1, A_2, \ldots, A_k\}$ of pairwise disjoint non-empty subsets of S such that $S = \cup_{1 \leq i \leq k} A_i$. A_i is called a *block* of P.

Set partitions are counted by the *Bell numbers* (see Subsection A.2.1).

Definition 3.4.2. Two blocks A and B of a partition P *overlap* if

$$\min A < \min B < \max A < \max B.$$

A partition is *non-overlapping* if no pair of blocks overlaps.

Example 3.4.3.

$$P = \{\{1, 3, 10\}, \{2\}, \{4, 8, 9\}, \{5, 6, 7\}, \{11, 14\}, \{12, 13\}\}$$

is a non-overlapping partition whose pictorial representation is given in Figure 3.13.

The n-th *Bessel number* is introduced by Flajolet and Schott in [386] as the number of non-overlapping partitions over n elements. The reason for the numbers B_n^* to have this name is a close relation with *Bessel functions*. The sequence of Bessel numbers starts with

$$1, 1, 2, 5, 14, 43, 143, 509, 1922, 7651, 31965, 139685, 636712, 3020203, \ldots,$$

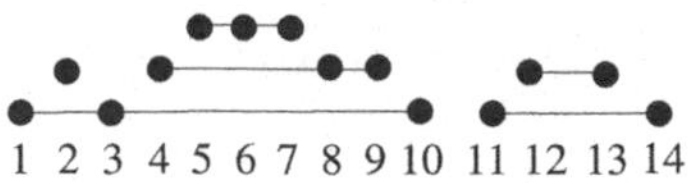

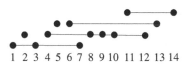

Figure 3.13: A pictorial representation of the non-overlapping set partition P in Example 3.4.3.

and the corresponding (ordinary) generating function can be written as a continued fraction expansion as follows:

$$\sum_{n\geq 0} B_n^* x^n = \cfrac{1}{1 - 1\cdot x - \cfrac{x^2}{1 - 2\cdot x - \cfrac{x^2}{1 - 3\cdot x - \cfrac{x^2}{\ddots}}}}.$$

Using the expression above, the asymptotic formula for B_n^* was derived in [386]:

$$B_n^* \sim \sum_{k\geq 0} \frac{k^{n+2}}{(k!)^2}.$$

Definition 3.4.4. Let P be a partition whose non-singleton blocks $\{A_1, A_2, \ldots, A_k\}$ are ordered so that for all $1 \leq i \leq k-1$, $\min A_i > \min A_{i+1}$. If $\max A_i > \max A_{i+1}$ for all $1 \leq i \leq k-1$, then P is a *monotone partition*.

Example 3.4.5.

$$P = \{\{1,3,7\}, \{2\}, \{4,8,9,10,12\}, \{5\}, \{6,13\}, \{11,14\}\}$$

is a monotone partition whose pictorial representation is given in Figure 3.14.

Figure 3.14: A pictorial representation of the monotone set partition P in Example 3.4.5.

Proposition 3.4.6. *([251]) Monotone partitions of $[n]$ are in one-to-one correspondence with non-overlapping partitions of $[n]$. Hence, the monotone partitions are counted by the Bessel numbers.*

Definition 3.4.7. An *involution* is a permutation which is its own inverse.

Example 3.4.8. 341256 and 12746538 are involutions.

Remark 3.4.9. We can see that a permutation π is an involution if and only if each cycle of π is of length one or two.

Definition 3.4.10. A walk in a rooted tree in which each parent node is traversed before its children is called a *pre-order walk*. In an *increasing rooted tree*, nodes are numbered and the numbers increase as we move away from the root. An increasing rooted tree has *increasing leaves* if its leaves, taken in pre-order, are increasing.

Increasing rooted trees are considered in [743, Chapter 1] and they are fundamental in, e.g. [360] by Elizalde and Noy in the study of consecutive patterns to be discussed in Chapter 5.

Definition 3.4.11. A *trimmed tree* is a tree where no node has a single leaf as a child (every leaf has a sibling).

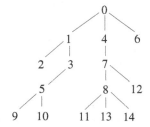

Figure 3.15: The increasing rooted trimmed tree corresponding to the permutation $29(10)531(11)(13)(14)8(12)746$ under a bijection in [503].

Example 3.4.12. See Figure 3.15 for an example of an increasing rooted trimmed tree. This tree corresponds to the permutation $29(10)531(11)(13)(14)8(12)746 \in S_{14}(\underline{132},\underline{21})$ under the corresponding bijection in [503]. Figure 3.16 shows two increasing rooted trees: the one to the left has increasing leaves while the one to the right does not.

Theorem 3.4.13. *([219]) Let $a_n = s_n(1\underline{234})$. Then $a_n = \sum_{k=1}^{n} a_{n,k}$, where*

$$a_{n,k} = a_{n-1,k-1} + k \sum_{j=k}^{n-1} a_{n-1,j}$$

for $1 \leq k \leq n$, with initial conditions $a_{0,0} = 1$ and $a_{n,0} = 0$ for $n \geq 1$.

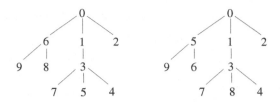

Figure 3.16: The increasing rooted tree to the left has increasing leaves, while the other tree does not.

Theorem 3.4.14. *([503]) The e.g.f. for the number of increasing rooted trimmed trees is given by*

$$\frac{e^{-x^2/2}}{1 - \int_0^x e^{-t^2/2}dt} - x - 1.$$

In Table 3.2 we have *set partitions with one marked block*. For example,

$$\{\overline{\{1,3\}}, \{2,5\}, \{4\}\}, \ \{\{1,3\}, \overline{\{2,5\}}, \{4\}\} \text{ and } \{\{1,3\}, \{2,5\}, \overline{\{4\}}\}$$

are three different partitions of [5] with one marked block.

For Table 3.2 we need the following theorem that turns out to collect under one roof five connections between restricted Dyck paths (and, actually *standard Young tableaux* though a simple bijection) and permutations restricted by vincular patterns.

Theorem 3.4.15. *([433]) The number of Dyck paths of length 2n that start with at least k up-steps, end with at least p down-steps, and touch the x-axis somewhere between the first up-step and the last down-step is given by*

$$\frac{k+p+1}{n+1}\binom{2n-k-p}{n} = \frac{d+1}{n+1}\binom{2n-d}{n},$$

where d = p + k.

Definition 3.4.16. A *compact directed rooted animal* of size n with k roots is a subset of n lattice points in the plane satisfying the following conditions. All of the k roots having integer coordinates are on the line $x + y = k - 1$. The remaining $n - k$ points having integer coordinates are added one by one so that at each step we add a point which is immediately to the East or North of a point we added before. The order in which we add the points is irrelevant.

Example 3.4.17. All 14 compact directed rooted animals with 3 consecutive roots of size 5 are in Figure 3.17. Starting from size 3, the number of the animals begins with

$$1, 4, 14, 45, 140, 427, 1288, 3858, 11505, \ldots$$

(see [726, A005775]).

Figure 3.17: Compact directed rooted animals with three consecutive roots of size 5.

Theorem 3.4.18. *([425]) The (ordinary) generating function for the compact directed rooted animals with three consecutive roots is given by*

$$\frac{x^2 + x - 1 + (x^2 - 3x + 1)\sqrt{\frac{1+x}{1-3x}}}{2x^2}.$$

3.4.2 Examples of bijections involved in Table 3.2

Two of our examples come from [251] by Claesson and one from [503] by Kitaev. Our choice of examples is based on easiness of presentation: some of the bijections in question are much more complicated like, for example, the two bijections in [262] connecting (3142, 24̲1̲3)-avoiding permutations with rooted non-separable planar maps and 2-stack sortable permutations through $\beta(1,0)$-trees (see Section 2.11).

Proposition 3.4.19. *([251]) There is a bijection between partitions of $[n]$ and $\mathcal{S}_n(1\underline{32})$.*

Proof. Let P be a partition of $[n]$. We define a *standard representation* of P by requiring that

1. The elements of a block are written in increasing order;

Patterns	Objects	Counting	Ref.
$1\underline{23}$	set partitions	e.g.f. e^{e^x-1}	[251]
$1\underline{32}$	set partitions	e.g.f. e^{e^x-1}	[251]
$2\underline{13}$	Dyck paths of length $2n$	$C_n = \frac{1}{n+1}\binom{2n}{n}$	[251]
$1\underline{234}$	increasing rooted trees with increasing leaves (in preorder)	see the recurrence in Theorem 3.4.13	[219]
$1\underline{23}$, $\underline{123}$	non-overlapping partitions of $[n]$	B_n^*	[251]
$1\underline{23}$, $1\underline{32}$	involutions	e.g.f. $e^{x+\frac{x^2}{2}}$	[251]
$1\underline{23}$, $1\underline{32}$	Motzkin paths	g.f. $\frac{1-x-\sqrt{1-2x-3x^2}}{2x^2}$	[251]
$1\underline{32}$, $\underline{12}1$	increasing rooted trimmed trees	e.g.f. Theorem 3.4.14	[503]
$1\underline{32}$, $2\underline{1}1$	partitions of $[n-1]$ with one block marked	$\displaystyle\sum_{i=0}^{n-2}\binom{n-1}{i}B_i$	[503]
3142, $2\underline{41}3$	rooted non-separable planar maps; see Section 2.11	$\frac{2(3n)!}{(2n+1)!(n+1)!}$	[262]
123, $1\underline{2}1$	certain Dyck paths (case $d=2$ in Theorem 3.4.15)	$\frac{3}{n+1}\binom{2n-2}{n}$	[433]
$123(\pi)=1$ & $1\underline{23}(\pi)=1$	certain Dyck paths (case $d=4$ in Theorem 3.4.15)	$\frac{5}{n+2}\binom{2n-2}{n+1}$	[433]
$123(\pi)=1$	certain Dyck paths (case $d=5$ in Theorem 3.4.15)	$\frac{6}{n+3}\binom{2n-1}{n+2}$	[433]
$123(\pi)=1$ & $1\underline{23}(\pi)=0$	certain Dyck paths (case $d=6$ in Theorem 3.4.15)	$\frac{7}{n+3}\binom{2n-2}{n+2}$	[433]
$123(\pi)=1$ & $1\underline{23}(\pi)=0$ & $\underline{123}(\pi)=0$	certain Dyck paths (case $d=7$ in Theorem 3.4.15)	$\frac{8}{n+3}\binom{2n-3}{n+2}$	[433]
$132(\pi)=1$ & $1\underline{23}(\pi)=0$	compact directed rooted animals with three consecutive roots	g.f. Theorem 3.4.18	[654]

Table 3.2: Permutations restricted by vincular patterns and other combinatorial objects. Here set partitions are counted by the Bell numbers B_n; C_n is the n-th Catalan number, and B_n^* is the n-th Bessel number. A single pattern p stands for avoidance: $p(\pi) = 0$.

2. The blocks are written in decreasing order of their least element, and with dashes separating the blocks.

The permutation π corresponding to P is obtained by writing P in standard form and erasing the dashes.

It is easy to see that π avoids $1\underline{32}$. Conversely, P can be recovered from π uniquely by inserting a dash in between each descent in π. □

Example 3.4.20. The bijection in Proposition 3.4.19 sends the set partition

$$P = \{\{1,4,6\}, \{2,3,9\}, \{5\}, \{7,8\}\}$$

to the permutation $\pi = 785239146 \in \mathcal{S}_9(1\underline{32})$ through P's standard form

$$78 - 5 - 239 - 146.$$

Proposition 3.4.21. *([251]) There is a bijection between $\mathcal{S}_n(\underline{12}3, 1\underline{32})$ and involutions in $\mathcal{S}_n$.*

Proof. Let π be an involution of length n (each cycle of π is of length one or two). We define a *standard form* for writing π in cycle notation by requiring that

1. Each cycle is written with the least element first;

2. The cycles are written in decreasing order of their least element.

The permutation σ corresponding to π is obtained from π by writing it in standard form and erasing the parentheses separating the cycles.

Note that $\sigma = \sigma_1\sigma_2\cdots\sigma_n$ avoids $\underline{12}3$. Indeed, assuming $\sigma_i < \sigma_{i+1}$ one gets that $(\sigma_i\sigma_{i+1})$ is a cycle in π and, by construction of σ, σ_i is a left-to-right minimum in σ. Thus, σ_i and σ_{i+1} cannot be involved in an occurrence of the pattern $\underline{12}3$, and σ avoids $\underline{12}3$.

Also, the permutation σ avoids $1\underline{32}$. Indeed, assuming $\sigma_i > \sigma_{i+1}$, σ_{i+1} must be the smallest element of a cycle leading σ_{i+1} to be a left-to-right minimum in σ. Thus, σ_i and σ_{i+1} cannot be involved in an occurrence of the pattern $1\underline{32}$, and σ avoids $1\underline{32}$.

Conversely, if σ is a ($\underline{12}3$, $1\underline{32}$)-avoiding permutation then the involution π is given by the condition: $(\sigma_i\sigma_{i+1})$ is a cycle in π if and only if $\sigma_i < \sigma_{i+1}$. □

Example 3.4.22. The bijection in Proposition 3.4.21 sends the involution $\pi = 15432876$ to the permutation $\sigma = 76834251 \in \mathcal{S}_8(\underline{12}3, 1\underline{32})$ through π's standard form

$$(7)(68)(34)(25)(1).$$

Proposition 3.4.23. *([503]) There is a bijection between increasing rooted trimmed trees (IRTTs) on $n+1$ nodes and $\mathcal{S}_n(\underline{132},\underline{|21})$.*

Proof. We describe a bijective function f from $\mathcal{S}_n(\underline{132},\underline{|21})$ to IRTTs on $n+1$ nodes.

Suppose $\pi \in \mathcal{S}_n(\underline{132},\underline{|21})$ and $\pi = B_0 a_0 B_1 a_1 \ldots B_k a_k$, where the a_is are the right-to-left minima of π and the B_js are (possibly empty) factors of π. We construct $T = f(\pi)$ as follows. The root of T is labelled by 0 and $a_0, a_1, \ldots, a_k$ are the labels of the root's children if we read them from left to right. Then we let the right-to-left minima of B_i be the labels of the children of a_i, and so on. It is easy to see that, since π avoids the pattern 132 and begins with the pattern 12, T avoids *limbs* of length 2 (no node in T has a single leaf as a child). Also, T is an increasing rooted tree and hence T is an IRTT. For example, $f(29(10)531(11)(13)(14)8(12)746)$ is shown in Figure 3.15.

It is easy to see that f is an injection. To see that f is a surjection, we show how to construct the permutation $\pi \in \mathcal{S}_n(\underline{132},\underline{|21})$ corresponding to a given IRTT T. If a_i and a_j are siblings and $a_i < a_j$, then the labels of the nodes of the subtree below a_j are all the letters in π between a_i and a_j, that is, $a_{i+1}, a_{i+2}, \ldots, a_{j-1}$. If a_i is a single child, then the labels of the nodes of the subtree below a_i appear immediately to the left of a_i in π. That is, if there are k nodes in the subtree below a_i then the k corresponding labels form the factor $a_{i-k}a_{i-k+1} \ldots a_{i-1}$. We now start from the first level of T consisting of the root's children and apply the instructions above. Then we consider the second level, and so on. The fact that T is an IRTT ensures that π avoids the pattern 132 and begins with the pattern 12. Thus, f is a bijection. $\qquad\square$

3.5   Several structures under one roof

This section is dedicated to POPs (partially ordered patterns) defined in Section 1.5.

The POPs were introduced in [509] as an auxiliary tool to study the maximum number of non-overlapping occurrences of consecutive VPs (vincular patterns, see Definition 1.3.2), that is, VPs whose occurrences in permutations form contiguous factors. However, the most useful property of POPs known so far is their ability to "encode" certain sets of VPs which provides a convenient notation for those sets and often gives an idea of how to treat them. For example, the original proof of the fact that $s_n(\underline{123},\underline{132},\underline{213}) = \binom{n}{\lfloor n/2 \rfloor}$ was 3 pages long ([505]); on the other hand, if one notices that $s_n(\underline{123},\underline{132},\underline{213}) = s_n(acb)$, where the relations between the letters

a, b and c are given by the following poset

that is, $a < b$ and c is incomparable to both a and b, then the result is easy to see. Indeed, we may use the property that the letters in odd and even positions of a "good" permutation do not affect each other because of the form of acb. Thus we choose the letters in the odd positions in $\binom{n}{\lfloor n/2 \rfloor}$ ways, and we must arrange them in decreasing order. We then must arrange the letters in the even positions in decreasing order too.

As a matter of fact, some POPs appeared in the literature before they were actually introduced. Thus the notion of a POP allows us to collect under one roof (to provide a uniform notation for) several combinatorial structures such as *peaks*, *valleys*, *modified maxima* and *modified minima* in permutations, *horse permutations*, non-consecutive occurrences of patterns, *V- and Λ-patterns* and *p-descents* in permutations discussed in Subsections 3.5.1–3.5.5.

On the other hand, similarly to VPs (see Table 3.2), POPs can be used to encode certain combinatorial objects by restricted permutations. Examples of that are given in Subsection 3.5.6.

3.5.1   Co-unimodal patterns

Recall that for a permutation $\pi = \pi_1 \pi_2 \cdots \pi_n \in \mathcal{S}_n$, the statistic $\mathrm{inv}(\pi)$ (called the *inversion index* of π) is the number of ordered pairs (i, j) such that $1 \leq i < j \leq n$ and $\pi_i > \pi_j$. Also, recall from Definition 3.3.4, that the major index, $\mathrm{maj}(\pi)$, is the sum of all i such that $\pi_i > \pi_{i+1}$. Suppose σ is a consecutive POP and

$$\mathrm{place}_\sigma(\pi) = \{i \mid \pi \text{ has an occurrence of } \sigma \text{ starting at } \pi_i\}.$$

Let $\mathrm{maj}_\sigma(\pi)$ be the sum of the elements of $\mathrm{place}_\sigma(\pi)$.

If a consecutive POP σ is *co-unimodal*, meaning that $k = \sigma_1 > \sigma_2 > \cdots > \sigma_j < \cdots < \sigma_k$ for some $2 \leq j \leq k$ (see Figure 3.18 for the corresponding poset in the case $j = 3$ and $k = 5$), then the following holds [123]:

$$(3.6) \qquad \sum_{\pi \in \mathcal{S}_n} t^{\mathrm{maj}_\sigma(\pi^{-1})} q^{\mathrm{maj}(\pi)} = \sum_{\pi \in \mathcal{S}_n} t^{\mathrm{maj}_\sigma(\pi^{-1})} q^{\mathrm{inv}(\pi)}.$$

If $k = 2$ we deal with the usual descents. Thus a co-unimodal pattern can be viewed as a generalization of the notion of a descent. This could be a reason why a co-unimodal pattern p is called a *p-descent* in [123]. Also, setting $t = 1$ in (3.6) we get a well-known result by MacMahon [578] on the equidistribution of maj and inv mentioned in Subsection 3.3.1.

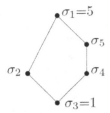

Figure 3.18: The poset for the co-unimodal pattern in the case $j = 3$ and $k = 5$.

3.5.2   Peaks and valleys in permutations

Definition 3.5.1. A permutation $\pi = \pi_1\pi_2\cdots\pi_n$ has k *peaks* (resp., *valleys*), also known as *local maxima* (resp., *local minima*), if $|\{j \mid \pi_j > \max\{\pi_{j-1}, \pi_{j+1}\}\}| = k$ (resp., $|\{j \mid \pi_j < \min\{\pi_{j-1}, \pi_{j+1}\}\}| = k$).

Example 3.5.2. In the permutation 248165397, peaks are in positions 3, 5 and 8, and valleys are in positions 4 and 7.

Clearly, an occurrence of a peak in a permutation is an occurrence of the consecutive POP $\underline{1'21''}$, where relations in the poset are $1' < 2$ and $1'' < 2$. Similarly, occurrences of valleys correspond to occurrences of the consecutive POP $\underline{2'12''}$, where $2' > 1$ and $2'' > 1$. See Figure 3.19 for the posets corresponding to peaks and valleys. So, research done on the peak or valley statistics can be regarded as part of the research on (consecutive) POPs (e.g. see [796]).

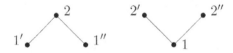

Figure 3.19: Posets corresponding to peaks and valleys.

Definition 3.5.3. For a permutation $\sigma_1\sigma_2\cdots\sigma_n$ we say that σ_i is a *modified maximum* if $\sigma_{i-1} < \sigma_i > \sigma_{i+1}$ and a *modified minimum* if $\sigma_{i-1} > \sigma_i < \sigma_{i+1}$, for $i = 1, 2, \ldots, n$, where we assume $\sigma_0 = \sigma_{n+1} = 0$ for modified maxima, and $\sigma_0 = \sigma_{n+1} = \infty$ for modified minima.

Example 3.5.4. The permutation 4265173 has modified maxima in positions 1, 3 and 6, and modified minima in positions 2, 5 and 7.

The results related to modified maxima and modified minima can be viewed as results on consecutive POPs. Indeed, the study of the distribution of modified maxima (resp., minima) is the same as the study of the pattern function $(\underline{ac} + 1'21'' + \underline{ca})(\pi)$ (resp., $(\underline{ca} + 2'12'' + \underline{ac})(\pi))$ where $a < c$ and the relations between the other letters are taken from Figure 3.19.

A specific result in this direction is recorded in Theorem 3.5.8. We need the following definition to state it.

Definition 3.5.5. We say that σ_i is a *double rise* (resp., *double fall*) if $\sigma_{i-1} < \sigma_i < \sigma_{i+1}$ (resp., $\sigma_{i-1} > \sigma_i > \sigma_{i+1}$). Clearly, occurrences of double rises and double falls are exactly the occurrences of the vincular patterns $\underline{123}$ and $\underline{321}$, respectively. The notion of double rises and double falls can be generalized to that of *increasing runs* and *decreasing runs*, respectively, when increasing/decreasing factors of length more than 2 are considered.

Remark 3.5.6. Double rises and double falls are also known as *double ascents* and *double descents*, respectively.

Example 3.5.7. The permutation 314795268 has double rises in positions 3, 4 and 8, and a double fall in position 6. The factors 147 and 1479 form increasing runs of lengths 3 and 4, respectively, while 952 is decreasing run of length 3.

Theorem 3.5.8. *([422]) The number of permutations in $\mathcal{S}_n$ with i_1 modified minima, i_2 modified maxima, i_3 double rises, and i_4 double falls is*

$$
(3.7) \qquad \left[u_1^{i_1} u_2^{i_2-1} u_3^{i_3} u_4^{i_4} \frac{x^n}{n!} \right] \frac{e^{\alpha_2 x} - e^{\alpha_1 x}}{\alpha_2 e^{\alpha_1 x} - \alpha_1 e^{\alpha_2 x}}
$$

where $\alpha_1 \alpha_2 = u_1 u_2$, $\alpha_1 + \alpha_2 = u_3 + u_4$. Formula (3.7) means the coefficient to $u_1^{i_1} u_2^{i_2-1} u_3^{i_3} u_4^{i_4} \frac{x^n}{n!}$ in the expansion of

$$
\frac{e^{\alpha_2 x} - e^{\alpha_1 x}}{\alpha_2 e^{\alpha_1 x} - \alpha_1 e^{\alpha_2 x}}.
$$

As a corollary to a much more general result (see Theorem 5.1.2), an explicit bivariate generating function for the *distribution of peaks* (equivalently, *distribution of valleys*) in permutations is obtained in [511] by Kitaev:

Theorem 3.5.9. *([511])*

$$
\sum_{n \geq 0} \sum_{\pi \in \mathcal{S}_n} y^{\text{peak}(\pi)} \frac{x^n}{n!} = 1 - \frac{1}{y} + \frac{1}{y} \sqrt{y-1} \cdot \tan\left(x\sqrt{y-1} + \arctan\left(\frac{1}{\sqrt{y-1}} \right) \right).
$$

Remark 3.5.10. We remark that thanks to the study of POPs, the entire distribution of peaks, a classical object in the literature, was found (see Theorem 3.5.9). An alternative formula for the distribution was obtained by Elizalde and Noy [360], and by Mendes and Remmel in [629]:

$$\sum_{n\geq 0}\sum_{\pi\in\mathcal{S}_n} y^{\mathrm{peak}(\pi)}\frac{x^n}{n!} = \frac{\sqrt{y-1}}{\sqrt{y-1}-\tan(x\sqrt{y-1})},$$

leading to a combinatorial proof of the identity

$$\frac{\sqrt{y-1}}{\sqrt{y-1}-\tan(x\sqrt{y-1})} = 1 - \frac{1}{y} + \frac{1}{y}\sqrt{y-1}\cdot\tan\left(x\sqrt{y-1}+\arctan\left(\frac{1}{\sqrt{y-1}}\right)\right).$$

Theorem 3.5.9 is an analogue of a result in [362] where the circular case of permutations is considered, that is, when the first letter of a permutation is thought to be to the right of the last letter in the permutation. In [362] it is shown that if $M(n,k)$ denotes the number of circular permutations in $\mathcal{S}_n$ having k maxima, then

$$\sum_{n\geq 1}\sum_{k\geq 0} M(n,k)y^k\frac{x^n}{n!} = \frac{zx(1-z\tanh xz)}{z-\tanh xz}$$

where $z = \sqrt{1-y}$.

3.5.3   Patterns containing the □ symbol

Hou and Mansour [465] study simultaneous avoidance of the patterns 132 and 1□23.

Definition 3.5.11. A permutation $\pi = \pi_1\pi_2\cdots\pi_n$ avoids 1□23 if there is no $\pi_i < \pi_j < \pi_{j+1}$ such that $1 \leq i \leq j-2$. Thus the □ symbol has the same meaning as an *unrestricted position* in a pattern (see Definition 3.3.13) except that □ does not allow the letters it separates to be adjacent in an occurrence of the corresponding pattern.

Example 3.5.12. The permutation 24783651 contains two occurrences of the pattern 1□23, namely, the subsequences 278 and 236. Note that 247 and 478 are not occurrences of 1□23 since the letters 2 and 4, and 4 and 7, respectively, are adjacent.

In the POP language, 1□23 is the pattern 1*a*23, or 1*a*23, or 1*a*23, where *a* is incomparable to the letters 1, 2 and 3 which, in turn, are ordered naturally: $1 < 2 < 3$.

Definition 3.5.13. The permutations avoiding 132 and 1□23 are called *horse permutations*.

The reason for the name in Definition 3.5.13 came from the fact that these permutations are in one-to-one correspondence with *horse paths*, which are the lattice paths from $(0,0)$ to (n,n) containing the steps $(0,1)$, $(1,1)$, $(2,1)$, and $(1,2)$ and not passing the line $y = x$. According to [465], the generating function for the horse permutations is

$$\frac{1 - x - \sqrt{1 - 2x - 3x^2 - 4x^3}}{2x^2(1 + x)}.$$

Moreover, in [465] the generating functions for horse permutations avoiding, or containing (exactly) once, certain patterns are given.

In [381], patterns of the form $xy\square z$ are studied, where $xyz \in \mathcal{S}_3$ (the position between x and y in the pattern is unrestricted; see Definition 3.3.13). Such a pattern can be written in the POP-notation as, for example, $xyaz$ where a is not comparable to x, y, and z. A bijection between permutations avoiding the pattern $12\square 3$, or $21\square 3$, and the set of *odd-dissection convex polygons* is given in [381] (see Subsection 3.5.6). Moreover, generating functions for permutations avoiding $13\square 2$ and certain additional patterns are obtained in [381].

3.5.4 Avoiding non-consecutive occurrences of patterns

Claesson [253] and Callan [216] study permutations avoiding *non-consecutive occurrences* of a classical pattern p: if p occurs in a permutation then this occurrence must be a factor in the permutation. This research is similar in nature to that discussed in Subsection 3.5.3. We state here results in [253, 216] first making an explicit connection to POPs in Proposition 3.5.14 that is easy to prove (the proof is omitted).

Proposition 3.5.14. *Suppose $p = p_1p_2 \cdots p_k$ is a classical pattern. The number of permutations avoiding non-consecutive occurrences of p is the same as that avoiding simultaneously all of the partially ordered patterns in the following set, where the letter a is incomparable to each p_i, $1 \leq i \leq k$:*

$$\{p_1ap_2p_3\ldots p_k, p_1p_2ap_3\ldots p_k, \ldots, p_1p_2p_3\ldots p_{k-1}ap_k\}.$$

Using the reverse and complement operations, from Theorems 3.5.15, 3.5.17 and 3.5.18 one gets enumerative results on non-consecutive avoidance of all permutation patterns on 2 and 3 letters.

Theorem 3.5.15. *([216]) The number of permutations in $\mathcal{S}_n$ avoiding non-consecutive occurrences of the pattern 21 is given by the $(n+1)$-th Fibonacci number.*

Theorem 3.5.15 can be refined as follows.

Theorem 3.5.16. *([253]) The number of permutations in S_n avoiding non-consecutive occurrences of the pattern 21 and having exactly k occurrences of the pattern $\underline{21}$ is given by $\binom{n-k}{k}$.*

Theorem 3.5.17. *([216]) The g.f. for the number of permutations in S_n avoiding non-consecutive occurrences of the pattern 321 is given by*

$$\frac{(1+x^2)(C(x)-1)}{1+x+x^2-xC(x)},$$

where $C(x) = \frac{1-\sqrt{1-4x}}{2x}$ is the g.f. for the Catalan numbers. The counting sequence, for $n \geq 0$, begins

$$1, 1, 2, 6, 18, 56, 182, 607, 2064, \ldots.$$

Theorem 3.5.18. *([216]) The g.f. for the number of permutations in S_n avoiding non-consecutive occurrences of the pattern 132 is given by*

$$C(x+x^3) = \frac{1-\sqrt{1-4x-4x^3}}{2(x+x^3)}.$$

The counting sequence, for $n \geq 0$, begins

$$1, 1, 2, 6, 18, 57, 190, 654, 2306, \ldots.$$

Definition 3.5.19. A *bicolored Dyck path* is a Dyck path in which each up-step is assigned one of two colors, say, red or green. The *height* of a step in a (bicolored) Dyck path is the height above the x-axis of its left point.

Example 3.5.20. The marked up-step of the Dyck path in Figure 3.20 is of height 2.

Figure 3.20: The marked up-step is of height 2.

Theorems 3.5.21 and 3.5.22 below provide combinatorial interpretations for the permutations in question, and they are also refinements of Theorems 3.5.17 and 3.5.18, respectively.

Theorem 3.5.21. *([253]) The permutations in $\mathcal{S}_n$ having k occurrences of the pattern* $\underline{321}$ *and avoiding non-consecutive occurrences of the pattern 321 are in one-to-one correspondence with bicolored Dyck paths of length $2n - 4k$ with k red up-steps, each of height less than 2. The number of such permutations is given by*

$$\sum_{i \geq 0} \frac{2k + i + 1}{n - k + i + 1} \binom{k - 1}{k - i} \binom{2n - 4k + i}{n - 3k}.$$

Theorem 3.5.22. *([253]) The permutations in $\mathcal{S}_n$ having k occurrences of the pattern* $\underline{132}$ *and avoiding non-consecutive occurrences of the pattern 132 are in one-to-one correspondence with bicolored Dyck paths of length $2n - 4k$ with k red up-steps. The number of such permutations is given by*

$$\binom{n - 2k}{k} C_{n-2k}.$$

3.5.5   V- and Λ-patterns

In [524] the notion of V- and Λ-patterns, generalizing the concepts of valleys, peaks, and increasing and decreasing runs (see Subsection 3.5.2 for definitions), is defined.

Definition 3.5.23. A factor $a_{i-k}a_{i-k+1} \ldots a_i a_{i+1} \ldots a_{i+\ell}$ of a permutation $a_1 a_2 \cdots a_n$ is an occurrence of the pattern $V(k, \ell)$ (resp., $\Lambda(k, \ell)$) if $a_{i-k} > a_{i-k+1} > \cdots > a_i < a_{i+1} < \cdots < a_{i+\ell}$ (resp., $a_{i-k} < a_{i-k+1} < \cdots < a_i > a_{i+1} > \cdots > a_{i+\ell}$).

Example 3.5.24. The permutation $\pi = 643198235$ has two occurrences of the pattern $V(2, 1)$, namely, the factors 4319 and 9823. Also, π has one occurrence of the pattern $\Lambda(1, 2)$ (the factor 1982) and one occurrence of the pattern $V(3, 1)$ (the factor 64319).

It is clear that $V(k, \ell)$ and $\Lambda(k, \ell))$ patterns are particular cases of consecutive POPs similar to the co-unimodal patterns considered in Subsection 3.5.1 (the difference is that $V(k, \ell)$ does not require the first letter to be the largest one in the pattern and $\Lambda(k, \ell)$ is obtained from $V(k, \ell)$ by taking the complement).

Below we list enumerative results on avoiding $V(k, \ell)$ and $\Lambda(k, \ell))$ patterns mostly coming from [524]. We note that a general approach leading to solving systems of differential equations, to find the e.g.f. for permutations avoiding a $V(k, \ell)$ pattern, is offered in [524].

Theorem 3.5.25. *([505]) The number of n-permutations avoiding the pattern $V(1, 1)$ is given by 2^{n-1}. The corresponding e.g.f. is $\frac{e^{2x}+1}{2}$.*

Theorem 3.5.26. *([524]) The e.g.f. for the number of permutations avoiding $V(2,1)$ is given by*

$$1 + \exp\left(\frac{3x}{2}\right) \sec\left(\frac{\sqrt{3}x}{2} + \frac{\pi}{6}\right) \int_0^x \exp\left(-\frac{3u}{2}\right) \cos\left(\frac{\sqrt{3}u}{2} + \frac{\pi}{6}\right) du.$$

Expanding the e.g.f. one gets the initial values for the number of the permutations in question for $n \geq 1$:

$$1, 2, 6, 21, 90, 450, 2619, 17334, \ldots.$$

The following theorem is a direct corollary to the well known fact that the e.g.f. for the number of (reverse) alternating permutations is given by $\tan x + \sec x$.

Theorem 3.5.27. *([524]) The e.g.f. for the number of permutations avoiding all four V- and Λ-patterns of length 4 ($V(1,2)$, $V(2,1)$, $\Lambda(1,2)$, and $\Lambda(2,1)$) is given by*

$$2(\tan x + \sec x) + 2e^x - x^2 - 3x - 3.$$

Theorem 3.5.28. *([524]) The e.g.f. for the number of permutations simultaneously avoiding the patterns $V(1,2)$, $V(2,1)$, and $\Lambda(1,2)$ is given by*

$$\frac{1}{2}(e^x + (\tan x + \sec x)(e^x + 1) - (1 + 2x + x^2)).$$

The initial values in question, for $n \geq 1$, are

$$1, 2, 6, 15, 47, 178, 791, 4025, 23057, \ldots.$$

Theorem 3.5.29. *([524]) The e.g.f. for the number of permutations simultaneously avoiding the patterns $V(1,2)$ and $\Lambda(1,2)$ is given by*

$$1 + x + (\tan x + \sec x - 1)(e^x - 1).$$

The initial values in question, for $n \geq 1$, are

$$1, 2, 6, 18, 60, 232, 1036, 5278, 30240, \ldots.$$

Theorem 3.5.30 below was proved using the fact shown in [360] that the e.g.f. for the number of permutations with no double rises is given by

$$\frac{\sqrt{3}}{2} \exp\left(\frac{x}{2}\right) \sec\left(\frac{\sqrt{3}}{2}x + \frac{\pi}{6}\right).$$

Theorem 3.5.30. *([524]) The e.g.f. for the number of permutations simultaneously avoiding the patterns $V(1, 2)$ and $\Lambda(2, 1)$ is given by*

$$\frac{\sqrt{3}}{2} \exp\left(\frac{x}{2}\right) \sec\left(\frac{\sqrt{3}}{2}x + \frac{\pi}{6}\right) + e^x - \left(1 + x + \frac{x^2}{2}\right).$$

The initial values in question, for $n \geq 1$, are

$$1, 2, 6, 18, 71, 350, 2018, 13359, 99378, \ldots.$$

Theorem 3.5.31. *([524]) The number of n-permutations simultaneously avoiding $V(1, 2)$ and $V(2, 1)$ is given by*

$$A_n = \sum_{i=1}^{n} \sum_{\substack{j = 1 \\ n - i - j \text{ is odd}}}^{n-i+1} A_{i,j}^n$$

with

$$A_{i,j}^n = \begin{cases} \binom{n-1}{i-1} & \text{if } n \geq i \geq 1 \text{ and} \\ & n - i - j = -1, \\ \binom{n}{i}\binom{n-i}{j} - \binom{n-1}{i-1} - \binom{n-1}{i} - \binom{n-1}{i+1} & \text{if } n - i - j = 1, \\ \binom{n}{i}\binom{n-i}{j} E_{n-i-j} - A_{i+2,j}^n - A_{i,j+2}^n - A_{i+2,j+2}^n & \text{if } n - i - j \geq 3 \text{ is odd} \end{cases}$$

where E_n is the n-th Euler number (the number of alternating n-permutations). The initial values for A_n, for $n \geq 1$, are given by

$$1, 2, 6, 18, 66, 252, 1176, 5768, 34216, \ldots.$$

Theorem 3.5.31 was proved by observing that a permutation avoiding at the same time the patterns $V(1, 2)$ and $V(2, 1)$ consists of three blocks (the second of which may be empty): it begins with an increasing run followed by a reverse alternating factor of odd length, and ends with a decreasing run, as shown schematically in the picture below.

Theorem 3.5.32. *([524]) The e.g.f. for the number of permutations avoiding the patterns* $\underline{213}$ *and* $V(1,2)$ *is given by*

$$1 + \int_0^x e^{2t+\frac{t^2}{2}}\, dt.$$

The initial values in question, for $n \geq 1$, *are*

$$1, 2, 5, 14, 43, 142, 499, 1850, 7193, \ldots.$$

3.5.6 Encoding combinatorial objects by POPs

Similarly to Section 3.4 dealing with encoding combinatorial objects by vincular pattern-restricted permutations, we first define objects of interest providing several results, and then we state connections in Table 3.3. We also provide examples of bijections involved in the table (see Theorems 3.5.38 and 3.5.39 below). We refer to [204] for additional information on encoding by POP restricted permutations. In particular, a general approach (different from a standard one involving the Online Encyclopedia of Integer Sequences [726]; see Section 1.7) for looking for connections between restricted permutations and other combinatorial objects, is suggested in [204]. Its idea is to consider those structures that can be "controlled," that is, for which we can find a set of patterns that force our permutations to have the prescribed structure. See Subsection 3.1 in [204] for details.

Definition 3.5.33. Let G_n be a *convex n-gon* in the plane $\mathbb{R}^2$ with vertices labeled $1, 2, \ldots n$ clockwise. A *dissection* of G_n is a partition of vertices of G_n into k polygons $G^1, G^2, \ldots, G^k$ by non-crossing diagonals of G_n. An *odd-dissection* of G_n is a dissection $G^1, G^2, \ldots, G^k$ such that G^i is not a $(2m)$-gon $(m > 1)$ for $1 \leq i \leq k$. Let O_n be the set of all odd-dissections of a convex $(n+2)$-gon.

Example 3.5.34. See Figure 3.21 for the set O_4 of all odd-dissections of the convex 6-gon.

Theorem 3.5.35. *([381]) Let* c *be incomparable to any of* $1 < 2 < 3$. *Then*

1. *There is a bijection between the set* $\mathcal{S}_n(\underline{12}c3)$ *and* O_n;

2. *There is a bijection between the set* $\mathcal{S}_n(\underline{21}c3)$ *and* O_n;

3. *The corresponding generating function is*

$$\sum_{n \geq 0} |O_n| x^n = 1 + \sum_{n \geq 1} \left(\sum_{k \geq 0} \frac{1}{n-k} \binom{2n-2k}{n-1-2k} \binom{n-k}{k} \right) x^n.$$

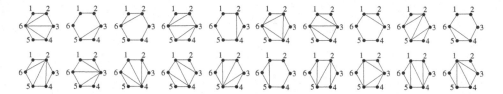

Figure 3.21: The set O_4 of all odd-dissections of the convex 6-gon.

Definition 3.5.36. The *corona* of a graph G is the graph G' constructed from G by adding for each vertex v a new vertex v' and the edge (v, v'). Let K_n denote the complete graph on n nodes and K'_n denote the corona of K_n. A *matching* or *independent edge set* in a graph is a set of edges without common nodes. Let M_n denote the number of matchings in K'_n including the empty one.

Example 3.5.37. There are two examples of matchings in K'_n in Figure 3.22:

$$\{(1, 5), (2, 2'), (3, 3')\} \text{ and } \{(1, 3), (2, 2')\}$$

are matchings in K'_5 and in K'_4, respectively.

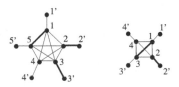

Figure 3.22: Examples of matchings in K'_5 and K'_4.

Theorem 3.5.38. *([524]) The set* $\mathcal{S}_{n+1}(\underline{213}, V(1, 2))$ *is in one-to-one correspondence with the set of all matchings of* K'_n.

We now turn our attention to *flat posets* built on $k + 1$ elements $a, a_1, \ldots, a_k$ with the only relations $a < a_i$ for all i (a_i and a_j are incomparable for $i \neq j$). A Hasse diagram for the flat poset is in Figure 3.23.

The following theorem generalizes Proposition 6 in [251] (see Proposition 3.4.21). Indeed, letting $k = 2$ in Theorem 3.5.39 we deal with involutions and permutations avoiding $1\underline{23}$ and $1\underline{32}$.

Theorem 3.5.39. *([511]) The permutations in* $\mathcal{S}_n$ *having cycles of length at most* k *are in one-to-one correspondence with permutations in* $\mathcal{S}_n(\underline{a a_1 a_2 \cdots a_k})$.

Figure 3.23: A flat poset.

Proof. Let $\pi \in \mathcal{S}_n$ be a permutation with cycles of length at most k. A standard form for writing π in cycle notation is requiring that

1. Each cycle is written with its least element first;

2. The cycles are written in decreasing order of their least element.

Let $\hat{\pi}$ be the permutation obtained from π by writing it in standard form and erasing the parentheses separating the cycles. The permutation $\hat{\pi}$ avoids $a\underline{a_1 a_2 \cdots a_k}$. Indeed, the number of letters between two left-to-right minima in $\hat{\pi}$ does not exceed $k-1$ because of the restriction on the cycle lengths. Thus if $\hat{\pi}$ contains $a\underline{a_1 a_2 \cdots a_k}$ then among the letters of $\hat{\pi}$ corresponding to $a_1 a_2 \cdots a_k$ there is at least one left-to-right minimum, say m, and the letter in $\hat{\pi}$ corresponding to a must be less than m. This contradicts the definition of a left-to-right minimum as a can be chosen to be a left-to-right minimum.

Conversely, if $\hat{\pi}$ is an $a\underline{a_1 a_2 \cdots a_k}$-avoiding permutation then there are at most $k-1$ letters between any of $\hat{\pi}$'s consecutive left-to-right minima, since otherwise we would have an occurrence of the pattern $a\underline{a_1 a_2 \cdots a_k}$ starting at a left-to-right minimum preceding a factor of length at least k that does not contain other left-to-right minima. The left-to-right minima of $\hat{\pi}$ define cycles of π. $\square$

The following proposition is easy to prove.

Proposition 3.5.40. *([511]) One has* $\mathcal{S}_n(a\underline{a_1 a_2 \cdots a_k}) = \mathcal{S}_n(\underline{a a_1 a_2} \cdots a_k).$

In what follows in the rest of the subsection, the relations between POP letters are $1 < 2 < 3$, $1' < 2'$ and the letters with a prime are not comparable with the other letters.

Proposition 3.5.41. *([510]) The $(n+1)$-permutations avoiding $\underline{12'}21'$ are in one-to-one correspondence with different walks of n steps between lattice points, each in a direction North, South, East or West, starting from the origin and remaining in the positive quadrant.*

$$s_n(\underline{12'21'}) = \binom{n-1}{\lfloor (n-1)/2 \rfloor}\binom{n}{\lfloor n/2 \rfloor}.$$

Proposition 3.5.42. *([204]) For $n \geq 3$, there is a bijection between n-permutations avoiding the patterns $\underline{11'22'}$ and $\underline{22'11'}$ and the set of all $(n+1)$-step walks on the x-axis with the steps $a = (1,0)$ and $\bar{a} = (-1,0)$ starting from the origin but not returning to it.*

$$s_n(\underline{11'22'}, \underline{22'11'}) = 2\binom{n}{\lfloor n/2 \rfloor}.$$

Definition 3.5.43. A *general graph* (*pseudograph*) is a graph in which both graph loops and multiple edges are permitted. A *homeomorphically irreducible* general graph is a graph with multiple edges and loops and without nodes of degree 2. We let 2-*HIGGs* stands for homeomorphically irreducible general graphs on two labeled nodes. Assuming L and R are the nodes of a 2-HIGG with n edges, we let x_L and x_R be the number of loops of L and R respectively, and x_{LR} be the number of edges between L and R.

Example 3.5.44. All six 2-HIGGs on three edges are in Figure 3.24.

It is easy to see [204] that the number of 2-HIGGs with $n \geq 3$ edges is

$$\binom{n+2}{2} - 4 = \frac{n^2 + 3n - 6}{2}.$$

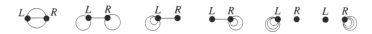

Figure 3.24: All six 2-HIGGs on three edges.

Proposition 3.5.45. *For $n \geq 3$, there is a bijection between $\mathcal{S}_n(\underline{121'2'}, \underline{211'2'}, \underline{121'3})$ and the set of 2-HIGGs with n edges such that either $x_{LR} > 0$ or $x_{LR} = 0$ and $x_L = n$.*

Definition 3.5.46. A permutation $\pi = \pi_1 \pi_2 \cdots \pi_n$ is *semi-alternating* if and only if for each $i \geq 1$, $\pi_i > \pi_{i+1}$ exactly when $\pi_{i+2} < \pi_{i+3}$ and $\pi_i < \pi_{i+1}$ exactly when $\pi_{i+2} > \pi_{i+3}$. Hence, descents and ascents of π alternate at exactly every other position.

Example 3.5.47. 479312865, 3562147 and 27513(10)6489 are examples of semi-alternating permutations.

Definition 3.5.48. A set partition is *bicolored* if each of its blocks is assigned one of two colors, say, red or green.

For Table 3.3 we need the poset in Figure 3.25.

Figure 3.25: A poset.

Theorem 3.5.49. *([448]) Permutations in $\mathcal{S}_{n+1}(y\underline{12}z, y(k-1)(k-2)\cdots 1z)$ are in one-to-one correspondence with bicolored set partitions with block size less than k.*

Remark 3.5.50. In [448] a different notation, say for the POP $y\underline{12}z$ is used, namely $3\underline{12}3$. That is, incomparable letters are denoted by the same letter, while the order relations between the letters is natural. The notation in [448] is very convenient and intuitively clear for the *permutation* patterns under consideration. However, this notation does not work for POPs on words bringing a confusion whether the same letters in a pattern stand for incomparable letters in an occurrence of the pattern or for the same letters (in an occurrence of a word POP, repetitions of the same letter can be required). Thus, to have a more general set up, we introduced the poset in Figure 3.25 and built some patterns in Table 3.3 based on it.

Definition 3.5.51. For a group G, a G-labeled set (S, α) is a set S together with a mapping $\alpha : S \to G$. Two G-labeled sets (S, α) and (S, β) are *equivalent* if there is $g \in G$ such that $\alpha = g\beta$. We let $[S, \alpha]$ denote the *equivalence class* containing (S, α). A *partial G-partition* of a set S is a set $\{[A_1, \alpha_1], [A_2, \alpha_2], \ldots, [A_k, \alpha_k]\}$ where the A_is are disjoint subsets of S, for $1 \leq i \leq k$.

Dowling [311] showed that partial G-partitions of a set S form a lattice, for every group G and set S. We are interested in $G = (\mathbb{Z}_2, +)$.

Example 3.5.52. ([448]) A partial $(\mathbb{Z}_2, +)$-partition of the set [9]:

$$A_1 = \{1, 6\}, \alpha_1(1) = 0, \alpha_1(6) = 1;$$

$$A_2 = \{7\}, \alpha_2(7) = 0;$$

$$A_3 = \{2, 4, 5, 9\}, \alpha_3(2) = 0, \alpha_3(4) = 0, \alpha_3(5) = 1, \alpha_3(9) = 0.$$

Theorem 3.5.53. *([448]) Permutations in $\mathcal{S}_{n+1}(xy\underline{1}zw)$ are in one-to-one correspondence with partial G-partitions on the set $[n]$, where G is the two-element group $(\mathbb{Z}_2, +)$.*

Patterns	Objects	Counting	Ref.
$\underline{12}c3$	odd-dissection convex polygons	g.f. Theorem 3.5.35	[381]
$\underline{12'21'}$	certain lattice walks (Prop. 3.5.40)	$\binom{n-1}{\lfloor(n-1)/2\rfloor}\binom{n}{\lfloor n/2\rfloor}$	[510]
$y\underline{12}z$	bicolored set partitions	e.g.f. $1 + \int_0^x e^{2(e^t-1)}dt$	[448]
$xy\underline{1}zw$	partial $(\mathbb{Z}_2, +)$-partitions of $[n]$	$1 + \int_0^x \exp\left(\frac{e^{2t}+2t-1}{2}\right) dt$	[448]
$a a_1 a_2 \cdots a_k$	perms with cycles of length $\leq k$	e.g.f. $\exp(\sum_{i=1}^{k} \frac{x^i}{i})$	[511]
132, $1c\underline{23}$	horse paths (see Subsection 3.5.3)	g.f. $\frac{1-x-\sqrt{1-2x-3x^2-4x^3}}{2x^2(1+x)}$	[465]
$\underline{21}3$, $V(1,2)$	matchings of the corona of K_n	e.g.f. $1 + \int_0^x e^{2t+\frac{t^2}{2}} dt$	[524]
$\underline{11'22'}, \underline{22'11'}$	certain lattice walks (Prop. 3.5.42)	$2\binom{n}{\lfloor n/2\rfloor}$	[204]
$\underline{121'2'}, \underline{212'1'}$	semi-alternating permutations	implicit e.g.f. [204]	[204]
$y\underline{12}z$ and $y(k-1)\cdots\underline{1}z$	bicolored set partitions with block size $\leq k$	?	[448]
$\underline{121'2'}, \underline{211'2'}$ and $\underline{121'3}$	homeomorphically irreducible graphs on 2 labeled nodes (2-HIGGs)	$\frac{n^2+3n-6}{2}$	[204]

Table 3.3: Permutations restricted by POPs and other combinatorial objects. Here "perms" stands for "permutations"; c is incomparable to any other letter; K_n is the complete graph on n nodes; relations between a and a_is are as in Figure 3.23; relations between the other letters are given by the poset in Figure 3.25.

There are other relations between POP restricted permutations and different combinatorial objects that do not appear in Table 3.3. For example, [204] links certain POP restricted permutations to the following combinatorial objects (note connections to classical pattern-avoidance):

1. $\pi \in \mathrm{Av}(1324)$ such that $\underline{21}(\pi) = 1$ ([726, A000292]).

2. Permutations π such that $132(\pi) = 1$ and $123(\pi) = 1$ ([726, A001815]).

3. Acute triangles made from the vertices of a regular n-polygon ([726, A007290]).

4. Lines drawn through the points of intersections of straight lines in a plane, no two of which are parallel, and no three of which are concurrent ([726, A050534]).

5. Number of 1s in all palindromic compositions ([726, A057711]). E.g. there are five palindromic compositions of 6 which contain 1s, namely 111111, 11211, 2112, 1221, and 141, containing a total of sixteen 1s.

6. Binary bitonic sequences. A bitonic sequence of length n is $a_1 \leq a_2 \leq \cdots \leq a_h \geq a_{h+1} \geq \cdots \geq a_{n-1} \geq a_n$ or $a_1 \geq a_2 \geq \cdots \geq a_h \leq a_{h+1} \leq \cdots \leq a_{n-1} \leq a_n$ ([726, A014206]).

7. Binary strings with no singletons, that is, where any 0 has a 0 standing next to it, and any 1 has a 1 standing next to it ([726, A006355]). E.g. there are six such strings of length 5: 00000, 00011, 00111, 11000, 11100, and 11111.

Remark 3.5.54. Finding bijections between the objects above and the corresponding permutations is left in [204] as open problems.

3.6   Proving identities combinatorially

In Remark 3.5.10 we show how patterns on permutations are used to obtain a combinatorial proof of a non-trivial identity.

Using formulas (3.2) and (3.3), one gets a combinatorial proof of the following identity:

$$1 + \sum_{n \geq 0} \frac{zx}{(1 - zx)^{n+1}} \prod_{i=1}^{n} (1 - (1 - t)^i) = \sum_{n \geq 0} \prod_{i=1}^{n} (1 - (1 - x)^{i-1}(1 - zx)).$$

The proofs of the identity above, even though not appearing in terms of permutations, can be translated into a proof on $\overline{231}$-avoiding permutations using the appropriate bijections in [161].

In [503], considerations of pattern-avoiding permutations led to a combinatorial proof (using *marked set partitions*, though) of the following identity involving the Bell numbers B_n and the Stirling numbers of the second kind $S(n, k)$:

$$\sum_{i=0}^{n-1} \binom{n}{i} B_i = \sum_{i=0}^{n} i \cdot S(n, i).$$

Yet another example that uses a structure of (POP) restricted permutations is given in [204] where the following simple identity is proved combinatorially:

$$3 \binom{n}{4} = \left(\binom{n-1}{2} \atop 2 \right).$$

Barcucci et al. [81] considered the multi-variable generating function

$$\sum_{\sigma \in \text{Av}(4231,4132)} s^{a(\sigma)} x^{\text{length}(\sigma)} y^{\text{rmin}(\sigma)} q^{\text{inv}(\sigma)}$$

for what they call "Schröder permutations", which in our terminology are the reverse of "deque-restricted permutations" (see class II in Subsection 2.2.6). Here $a(\sigma)$ is the number of active sites in σ (see Definition 6.1.5). Combined with the results on Schröder paths in [81] (see Subsection A.2.2 for the definition), the following q-identity is obtained:

$$\frac{\sum_{n \geq 0} (-1)^n \frac{x^{n+1} y q^{n(n+1)}}{(xy, q)_{n+1} (q, q)_n}}{\sum_{n \geq 0} (-1)^n \frac{x^n q^{n^2}}{(xy, q)_n (q, q)_n}} = \frac{\sum_{n \geq 0} (-1)^n \frac{x^{n+1} y q^{\frac{n(n+1)}{2}}}{(q, q)_n} \prod_{k=0}^{n-1} (y + q^{k+1})}{\sum_{n \geq 0} (-1)^n \frac{x^n q^{\frac{n(n-1)}{2}}}{(q, q)_n} \prod_{k=0}^{n-1} (y + q^{k+1})},$$

where we denote $(a, q)_n = \prod_{i=0}^{n-1} (1 - aq^i)$.

Kitaev and Remmel [525] studied the problem of counting descents (occurrences of the pattern $\underline{21}$) according to whether the first or the second element in a descent pair is equivalent to 0 mod k for $k \geq 2$. This extension of the notion of a pattern is considered briefly in Section 9.1. The following identity is obtained as a result of studies in [525]:

$$\sum_{r=0}^{s}(-1)^{s-r}\binom{(k-1)n+j+r}{r}\binom{kn+j+1}{s-r}\prod_{i=0}^{n-1}(r+1+j+(k-1)i) =$$
$$\sum_{r=0}^{n-s}(-1)^{n-s-r}\binom{(k-1)n+j+r}{r}\binom{kn+j+1}{n-s-r}\prod_{i=1}^{n}(r+(k-1)i)$$

Moreover, as a corollary to a theorem on the extended notion of patterns, Kitaev and Remmel [525] obtained combinatorial proofs for two special cases of *Saalschütz's identity:*

$$\binom{n}{s}^{2}=\sum_{r=0}^{s}(-1)^{s-r}\binom{n+r}{r}^{2}\binom{2n+1}{s-r};$$

$$\binom{n}{s}\binom{n+1}{s+1}=\sum_{r=0}^{s}(-1)^{s-r}\binom{n+r+1}{r}\binom{n+r+1}{r+1}\binom{2n+2}{s-r}.$$

Other identities proved combinatorially using the extended notion of patterns can be found in [525].

Finally, we provide one more example, Theorem 3.6.2, coming from study of patterns in words [203].

We need the following definition to state Theorem 3.6.2.

Definition 3.6.1. The *Möbius function* is defined by

$$\mu(n)=\begin{cases} 0, & \text{if } n \text{ has one or more repeated prime factors,} \\ 1, & \text{if } n=1, \\ (-1)^{k}, & \text{if } n \text{ is a product of } k \text{ distinct primes,} \end{cases}$$

so $\mu(n) \neq 0$ indicates that n is squarefree. The first few values are

$$1, -1, -1, 0, -1, 1, -1, 0, 0, 1, -1, 0, \ldots.$$

The following theorem is proved combinatorially. The idea of the proof in [203] is to find, in two ways, the cardinality (number of elements) of a minimal *unavoidable set of prohibitions* for so-called *pattern words* (in the context, a set is unavoidable if any infinite pattern word contains a factor from the set).

Theorem 3.6.2. *([203])*

$$\sum_{i=1}^{m}\sum_{a_1+a_2+\cdots+a_i=n}\binom{n}{a_1,a_2,\ldots,a_i}=$$

$$\sum_{i|n} \sum_{j=0}^{\min(i,m)-1} (-1)^j \binom{\min(i,m)-1}{j} \sum_{d|i} \mu(d)(\min(i,m)-j)^{\frac{i}{d}},$$

where the second sum in the left-hand side is over all weak compositions of n, that is, $a_j > 0$, for $1 \leq j \leq i$.

Chapter 4

Bijections between 321- and 132-avoiding permutations

In Proposition 2.1.3 we proved that the number of 231-avoiding n-permutations is given by the n-th Catalan number, $C_n = \frac{1}{n+1}\binom{2n}{n}$. Using the reverse and complement operations, and their composition, one sees that all of the patterns in the set $\{132, 231, 213, 312\}$ are Wilf-equivalent (see Definition 1.0.30). Also, the patterns in $\{123, 321\}$ are Wilf-equivalent by simply applying the reverse operation. It turns out that the second set of permutations is also counted by the Catalan numbers, which was first established by MacMahon [578] in 1915. However, there is no easy way to see directly why any two patterns from different sets above have to be Wilf-equivalent.

This chapter presents a 30+ year history of bijections between permutations avoiding a pattern from $\{132, 231, 213, 312\}$ and permutations avoiding a pattern from $\{123, 321\}$. The presentation is based on a recent work of Claesson and Kitaev [259]. The first such bijection appeared in Knuth [540, pp.242–243] in 1968, and since then researchers have been discovering new, sometimes rather sophisticated bijections between the sets. A question raised in [259] was if there is any simple way to get from one bijection between 321- and 132-avoiding permutations[1] and another one. In particular, are there structural relationships between any of these bijections which appear very different on the surface? Another question considered was of a philosophical nature that goes far beyond the scope of this book:

---

[1]Because of the trivial bijections, any bijection between permutations avoiding a pattern from $\{132, 231, 213, 312\}$ and permutations avoiding a pattern from $\{123, 321\}$ can easily be extended to yield a bijection between 321- and 132-avoiding permutations. Therefore, some of our bijections are not always described directly between these sets. For example, the Knuth-Richards bijection is presented as a bijection between 132- and 123-avoiding permutations. Our choice of representatives from these sets when describing a particular bijection is normally based on the original source.

"Given two bijections between equinumerous sets of objects, which one is better?" Perhaps, a reasonable and widely acceptable answer to this question would be "The one that is easier to describe". While this is a good quality and often we are looking for easily describable bijections between sets, another important aspect is how well the bijection reflects the structural similarity of the respective sets.

A natural measure of structural similarity between two equinumerous sets of permutations is how many statistics (in our case, permutation statistics) on one set correspond to statistics on the other. This gives a way to compare two bijections saying that one bijection is "better" than the other one if it respects more statistics. Of course, one bijection could be better on one set of statistics, whereas another could be better on a different set of statistics. What seems to be fair in comparison of two bijections is to include all (or as close to all as possible) permutation statistics typically appearing in the literature on permutations. More details on this can be found in Section 4.2 where a classification of bijections between 321- and 132-avoiding permutations is presented. That is, the bijections will be partitioned into equivalent groups, modulo trivial bijections, and the number of statistics each group respects will be indicated.

It should be noted that the research in [259] would not be possible without some powerful software created by Anders Claesson. The idea of this software can be described as follows. Given two equinumerous sets, S_1 and S_2, of combinatorial objects (which could be permutations, trees, maps, etc) with lists $L_1 = \{s_1, s_2 \ldots\}$ and $L_2 = \{s_1', s_2', \ldots\}$ of statistics on sets S_1 and S_2, respectively, we can produce a conjecture that a tuple of statistics $(s_{i_1}, s_{i_2}, \ldots, s_{i_k})$ on S_1 is equinumerous with $(s_{j_1}', s_{j_2}', \ldots, s_{j_k}')$ on S_2 and these tuples of statistics are not extendable by adding extra statistics from L_1 and L_2. This is done by exhausting all possibilities for a reasonably large (finite) collection of objects. Once such a conjecture is obtained, it normally becomes much easier to find a bijection from S_1 to S_2 since one gets a hint as to which statistic on S_1 corresponds to a given statistic on S_2. In practice, assuming large enough objects are considered, all such conjectures produced by the software turn out to be true.

More relevant to this chapter is the following feature of the software mentioned above. Given two bijections, f_1 and f_2, between the sets S_1 and S_2, we can use the software not only to see if f_1 is equivalent to f_2 modulo trivial bijections, but also to see how many statistics from the lists L_1 and L_2 are preserved. This method of analysis was employed in [259]: first conjectures were obtained by the software and then proofs were given confirming the conjectures. Note that without using the software the results in [259] would be extremely painful to obtain, since for a human being it would be too difficult to see/guess all the connections. In either case, there are other publications [124, 217, 710], not being considered in this chapter, related to analysis of several bijections of interest. The set up in [259] is on a much more general scale than that in the other publications.

In Section 4.1 we follow [259] to introduce known bijections between permutations avoiding a pattern from $\{132, 231, 213, 312\}$ and permutations avoiding a pattern from $\{123, 321\}$, and in Section 4.2 we provide their classification. However, there are bijections appearing in the literature that were not considered under the analysis done in [259] either because the authors in [259] were not aware of them, or because such bijections appeared after [259] was published. Some of these "non-classified" bijections, which will not be considered in this chapter, appear in [52, 66, 124, 696], and hopefully, one day, these bijections will be compared with those considered in [259].

It is worth mentioning that there are three generalizations of bijections between permutations avoiding a pattern from $\{132, 231, 213, 312\}$ and permutations avoiding a pattern from $\{123, 321\}$. None of these generalizations will be discussed in detail in this chapter, but we record these results in Theorems 4.0.3, 4.0.5, and 4.0.7 below. One of the generalizations is due to Backelin et al. [66], and it is stated in the following theorem.

Theorem 4.0.3. *([66]) $s_n(12 \cdots kP) = s_n(k(k-1) \cdots 1P)$ for any permutation P on the set $\{k+1, k+2, \ldots, \ell\}$.*

Remark 4.0.4. The case $k = 2$ in Theorem 4.0.3 was first proved in [799] by West, and the case $k = 3$ in [65] by Babson and West.

A transformation ϕ^* used to prove Theorem 4.0.3, recursively replaces all occurrences of the pattern $k(k-1) \cdots 1$ in a permutation π by occurrences of the pattern $(k-1)(k-2) \cdots 21k$ until no occurrences of $k(k-1) \cdots 1$ remain, which induces a bijection from $\mathcal{S}_n((k-1)(k-2) \cdots 1k)$ onto $\mathcal{S}_n(k(k-1) \cdots 1)$. This transformation is defined not only on permutations, but also on more general objects called *full rook placements on a Ferrers shape* considered in [162]. Another property of ϕ^* shown in [162] is that $\phi^*(\pi^{-1}) = (\phi^*(\pi))^{-1}$, that is, the transformation ϕ^* commutes with applying the inverse operation.

Stankova and West [737] obtained the following result.

Theorem 4.0.5. *([737]) $s_n(231P) = s_n(312P)$ for any permutation P on the set $\{4, 5, \ldots, k\}$.*

One may find it convenient to think of the permutations involved in Theorems 4.0.3 and 4.0.5 as permutation matrices. A schematic way to state these theorems in matrix form is given in Figures 4.1 and 4.2, respectively.

Remark 4.0.6. The bijections in Theorems 4.0.3 and 4.0.5 were generalized in [257] by Claesson et al. where instead of permutations one considers *partial permutations*, but still deals with usual classical patterns. We discuss this in Section 9.5.

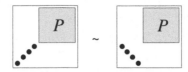

Figure 4.1: A schematic "matrix way" to state Theorem 4.0.3.

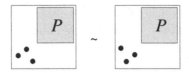

Figure 4.2: A schematic "matrix way" to state Theorem 4.0.5.

As a special case of a more general bijection showing equidistribution of patterns $\tau_1\tau_2\cdots\tau_{k-2}k(k-1)$ and $\tau_1\tau_2\cdots\tau_{k-2}(k-1)k$ over the set of all permutations, Reifegerste [678] proved that when $k = 3$, the bijection coincides with another bijection provided by Simion and Schmidt [721] which we describe in Subsection 4.1.3.

Theorem 4.0.7. *([678]) There is a one-to-one correspondence between permutations avoiding each of the patterns of the form $\tau_1\tau_2\cdots\tau_{k-2}k(k-1) \in \mathcal{S}_k$ and those avoiding each of the patterns of the form $\tau_1\tau_2\cdots\tau_{k-2}(k-1)k \in \mathcal{S}_k$.*

There are two main points in this chapter:

- Considering bijections between permutations avoiding a pattern from the set $\{132, 231, 213, 312\}$ and permutations avoiding a pattern from $\{123, 321\}$ is a great way to illustrate a variety of different approaches to tackle a single problem in the pattern-avoidance theory. The approaches in question could be of help in solving other similar problems.

- The idea of ranking bijections between two equinumerous sets based on the number of statistics they preserve is of general importance and can be applied in many situations when dealing with bijections.

4.1   Bijections between $\mathrm{Av}(321)$ and $\mathrm{Av}(132)$

Some of the bijections described in this section use Dyck paths (see Subsection A.2.2 for definition) as auxiliary objects, and there is a natural bijection, occurring often in the literature, from 132-avoiding permutations to Dyck paths. We call this bijection the *standard bijection* and we describe it below. The standard bijection can be obtained by reformulating the bijection from 312-avoiding permutations to "stack words," given by Knuth [540, pp.242–243]. Also, this bijection is the same as the non-recursive bijection given by Krattenthaler [543] (see Subsection 4.1.6).

Let $\pi = \pi_L n \pi_R$ be a 132-avoiding n-permutation. Each letter of π_L is larger than any letter of π_R, or else a 132 pattern involving n would be formed. Let $\pi'_L = \text{red}(\pi_L)$. We define the standard bijection f recursively by $f(\pi) = uf(\pi'_L)df(\pi_R)$ and $f(\varepsilon) = \varepsilon$, where ε denotes the empty word/permutation. Thus, under the standard bijection, the position of n in π determines the first return to the x-axis and vice versa. For instance,

$$f(6753412) = uf(1)df(53412) = uuddudf(3412) = uuddududf(1)df(12) =$$

$$uudduduudduf(1)d = uudduduudduudd =$$

4.1.1   Knuth's bijection, 1973

Knuth [539, pp.60–61] gives a bijection from 321-avoiding permutations to Dyck paths that we describe below. This bijection can be combined with the standard one to obtain a bijection between 321- and 132-avoiding permutation which is called *Knuth's bijection* in [259].

Given a 321-avoiding permutation, start by applying the *Robinson-Schensted-Knuth correspondence* (*RSK-correspondence*). This classic correspondence gives a bijection between permutations π of length n and pairs (P, Q) of *standard Young tableaux* of the same shape $\lambda \vdash n$. It is well-known that the length of the longest decreasing subsequence in π corresponds to the number of rows in P (or Q), and thus, for 321-avoiding permutations, the tableaux P and Q have at most two rows.

The *insertion tableau* P is obtained by reading $\pi = a_1 a_2 \cdots a_n$ from left to right and, at each step, inserting a_i into the partial tableau obtained thus far. Assume that $a_1, a_2, \ldots, a_{i-1}$ have been inserted. If a_i is larger than all the elements in the first row of the current tableau, place a_i at the end of the first row. Otherwise, let m be the leftmost element in the first row that is larger than a_i. Place a_i in the square that is occupied by m, and place m at the end of the second row. The *recording tableau* Q has the same shape as P and is obtained by placing i, for i from 1 to n, in the position of the square that in the construction of P was created at step i (when a_i was inserted).

Example 4.1.1. The pair of tableaux corresponding to the 321-avoiding permutation 23514687 and the steps involved in applying RSK is shown in Figure 4.3.

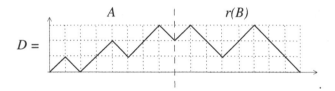

$$23514687 \;\rightarrow$$

$$\left(\boxed{2}, \boxed{1}\right) \;\rightarrow\; \left(\boxed{2\,3}, \boxed{1\,2}\right) \;\rightarrow\; \left(\boxed{2\,3\,5}, \boxed{1\,2\,3}\right) \;\rightarrow\; \left(\begin{smallmatrix}\boxed{1\,3\,5}\\\boxed{2}\end{smallmatrix}, \begin{smallmatrix}\boxed{1\,2\,3}\\\boxed{4}\end{smallmatrix}\right) \;\rightarrow\; \left(\begin{smallmatrix}\boxed{1\,3\,4}\\\boxed{2\,5}\end{smallmatrix}, \begin{smallmatrix}\boxed{1\,2\,3}\\\boxed{4\,5}\end{smallmatrix}\right) \;\rightarrow$$

$$\left(\begin{smallmatrix}\boxed{1\,3\,4\,6}\\\boxed{2\,5}\end{smallmatrix}, \begin{smallmatrix}\boxed{1\,2\,3\,6}\\\boxed{4\,5}\end{smallmatrix}\right) \;\rightarrow\; \left(\begin{smallmatrix}\boxed{1\,3\,4\,6\,8}\\\boxed{2\,5}\end{smallmatrix}, \begin{smallmatrix}\boxed{1\,2\,3\,6\,7}\\\boxed{4\,5}\end{smallmatrix}\right) \;\rightarrow\; \left(\begin{smallmatrix}\boxed{1\,3\,4\,6\,7}\\\boxed{2\,5\,8}\end{smallmatrix}, \begin{smallmatrix}\boxed{1\,2\,3\,6\,7}\\\boxed{4\,5\,8}\end{smallmatrix}\right) = \left(P, Q\right)$$

Figure 4.3: The RSK-correspondence applied to the 321-avoiding permutation 23514687.

The pair of tableaux (P, Q) is then translated into a Dyck path D. The first half, A, of the Dyck path we get by recording, for i from 1 to n, an up-step if i is in the first row of P, and a down-step if it is in the second row. Let B be the word obtained from Q in the same way but interchanging the roles of u and d. Then $D = Ar(B)$ where $r(B)$ is the reverse of B. For the permutation 23514687 whose pair of tableaux (P, Q) is given in Figure 4.3, the corresponding Dyck path D is

$$D =$$

Elizalde and Pak [361] use this bijection together with a slight variant of the standard bijection to give a combinatorial proof of a generalization of the result by Robertson et al. [697] that fixed points have the same distribution on $\mathrm{Av}(123)$ and $\mathrm{Av}(132)$. The modification they use is to reflect the Dyck path obtained from the standard bijection with respect to the vertical line crossing the path in the middle (in other words, they use the reverse of the path). We apply the same modification. We can check that the image of the 132-avoiding permutation 67435281 under the standard bijection is the reverse of the path above. Thus the image of $23514687 \in \mathcal{S}_8(321)$ under Knuth's bijection is $67435281 \in \mathcal{S}_8(132)$.

4.1.2   Knuth-Rotem's bijection, 1975

Rotem [704] gives a bijection between 321-avoiding permutations and Dyck paths, described below. Combining it with the standard bijection gives a bijection from 321- to 132-avoiding permutations, which is called *Knuth-Rotem's bijection* in [259].

Let $\mathcal{B}_n$ be the set of sequences $b_1 b_2 \cdots b_n$ satisfying the two conditions

1. $b_1 \leq b_2 \leq \cdots \leq b_n$;

2. $0 \leq b_i \leq i - 1$, for $i = 1, 2, \ldots, n$.

Let $\pi = \pi_1 \pi_2 \cdots \pi_n \in \mathcal{S}_n(321)$. From it we construct a sequence $b_1 b_2 \cdots b_n \in \mathcal{B}_n$: Let $b_1 = 0$, and, for $i = 2, \ldots, n$, let $b_i = b_{i-1}$ if π_i is a left-to-right maximum in π, and let $b_i = \pi_i$ otherwise.

Example 4.1.2. For the permutation $\pi = 2513476$ we obtain the sequence 0013446. We can represent this sequence by a bar-diagram, which in turn can be viewed as a lattice path from $(0,0)$ to $(7,7)$ with steps $(1,0)$ and $(0,1)$, as shown in Figure 4.4.

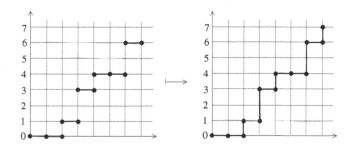

Figure 4.4: The bar-diagram representing 0013446 and the corresponding lattice path.

Rotating the path obtained in Example 4.1.2 (Figure 4.4) counter clockwise by $3\pi/4$ radians we get

which corresponds to the 132-avoiding permutation 7564213 under the standard bijection. Thus the image of $2513476 \in \mathcal{S}_7(321)$ under Knuth-Rotem's bijection is $7564213 \in \mathcal{S}_7(132)$.

4.1.3   Simion-Schmidt's bijection, 1985

Consider the following algorithm:

> Input: A permutation $\pi = \pi_1 \pi_2 \cdots, \pi_n \in \mathcal{S}_n(123)$.
> Output: A permutation $\tau = c_1 c_2 \ldots c_n \in \mathcal{S}_n(132)$.
>
> 1 $c_1 := \pi_1;\ x := \pi_1$
> 2   for $i = 2, \ldots, n$:
> 3   if $\pi_i < x$:
> 4 $c_i := \pi_i;\ x := \pi_i$
> 5   else:
> 6 $c_i := \min\{\, k \mid x < k \le n,\ k \ne c_j \text{ for all } j < i \,\}$

The map sending π to τ is the *Simion-Schmidt bijection* [721]. It can be seen from the algorithm description that the bijection preserves the values and positions of the left-to-right minima, normally rearranging other elements. As a matter of fact, if positions and values of a left-to-right minima sequence are specified, then there exists exactly one such 123-avoiding permutation, and exactly one such 132-avoiding permutation. As an example, $6743152 \in \mathcal{S}_7(123)$ maps to $6743125 \in \mathcal{S}_7(132)$ under the Simion-Schmidt bijection; note that, in both permutations, 6, 4, 3, and 1 is the sequence of left-to-right minima occupying the positions 1, 3, 4, and 5.

We remark that Fulmek [407] has given a simple diagrammatic description of what is essentially the Simion-Schmidt bijection. To be more precise, the bijection Fulmek describes is the composition of maps $c \circ$ Simion-Schmidt $\circ\, c$, where c is the complement operation.

4.1.4   Knuth-Richards's bijection, 1988

Richards's bijection [691] from Dyck paths to 123-avoiding permutations is given by the following algorithm:

> Input: A Dyck path $P = b_1 b_2 \cdots b_{2n}$.
> Output: A permutation $\pi = \pi_1 \pi_2 \cdots \pi_n \in \mathcal{S}_n(123)$.
>
> 1 $r := n + 1;\ s := n + 1;\ j := 1$
> 2   for $i = 1, \ldots, n$:
> 3   if b_j is an up-step:
> 4   repeat $s := s - 1;\ j := j + 1$ until b_j is a down-step
> 5 $\pi_s := i$
> 6   else:
> 7   repeat $r := r - 1$ until π_r is unset
> 8 $\pi_r := i$
> 9 $j := j + 1$

As a matter of fact, the algorithm above can be described differently than was presented in [259]. We provide the alternative description before defining the Knuth-Richards bijection.

Consider the Dyck path $P = uudduududuuddd$ of semi length $n = 7$. Let us index its up- and down-steps 1 through n:

$$P = u_1 u_2 d_1 d_2 u_3 u_4 d_3 u_5 d_4 u_6 u_7 d_5 d_6 d_7.$$

From this path we shall construct a permutation $\pi = \pi_1 \pi_2 \cdots \pi_n$. Scan down-steps of P from left to right: if d_i is preceded by an up-step u_j, then let $\pi_{n+1-j} = i$; otherwise, let j be the largest value for which π_j is unset, and let $\pi_j = i$. In our example, this leads to:

1. d_1 is preceded by the up-step u_2; let $\pi_{8-2} = \pi_6 = 1$.

2. d_2 is preceded by a down-step; let $j = 7$ and $\pi_7 = 2$.

3. d_3 is preceded by the up-step u_4; let $\pi_{8-4} = \pi_4 = 3$.

4. d_4 is preceded by the up-step u_5; let $\pi_{8-5} = \pi_3 = 4$.

5. d_5 is preceded by the up-step u_7; let $\pi_{8-7} = \pi_1 = 5$.

6. d_6 is preceded by a down-step; let $j = 5$ and $\pi_5 = 6$.

7. d_7 is preceded by a down-step; let $j = 2$ and $\pi_2 = 7$.

The resulting permutation is $\pi = 5743612$. Lines 3, 4 and 5 of the algorithm cover the case when d_i is preceded by an up-step; lines 6, 7 and 8 cover the case when d_i is preceded by a down-step. Plainly speaking, if d_i is preceded by an up-step u_i

then $u_i d_j$ is a peak in P. Moreover, $\pi_{n+1-j} = i$ is a left-to-right minimum in the corresponding permutation.

The *Knuth-Richards bijection*, from $\mathcal{S}_n(132)$ to $\mathcal{S}_n(123)$, is defined by the composition

$$\text{Knuth-Richards} = \text{Richards} \circ \text{standard},$$

where "standard" is the standard bijection f from $\mathrm{Av}(132)$ to Dyck paths, and "Richards" is the algorithm just described. As an example, applying Knuth-Richards's bijection to $6743125 \in \mathcal{S}_7(132)$ yields $5743612 \in \mathcal{S}_7(123)$:

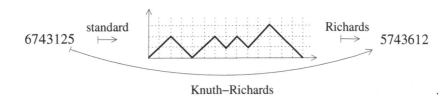

standard
$6743125 \quad \longmapsto$

Richards
$\longmapsto \quad 5743612$

Knuth–Richards

Actually, as was shown in [259], the Knuth-Richards bijection accepts a recursive description which we call the Claesson-Kitaev bijection and provide here.

Recall that a permutation π is *irreducible* if $\mathrm{comp}(\pi) = 1$. If we define the sum $\oplus_p$ on permutations by $\sigma \oplus_p \tau = \sigma\tau'$, where τ' is obtained from τ by adding $|\sigma|$ to each of its letters, then a permutation is irreducible if it cannot be written as the sum of two nonempty permutations. We shall describe, separately for 231- and 321-avoiding permutations, how to generate the irreducible permutations, thus inducing the Claesson-Kitaev bijection.

For a permutation of length n to be 231-avoiding everything to the left of n has to be smaller than anything to the right of n. Clearly, if there is at least one letter to the left of n, then the permutation is reducible (everything to the right of n, including n, would form the last component). Thus a 231-avoiding permutation of length n is irreducible if and only if it starts with n. To build an irreducible 231-avoiding permutation of length n from a 231-avoiding permutation of length $n - 1$ we simply prepend n. Let us call this map α. For instance, $\alpha(2134) = 52134$.

Given k irreducible 231-avoiding permutations $\pi_1, \pi_2, \ldots, \pi_k$, we build the corresponding permutation by summing: $\pi_1 \oplus_p \pi_2 \oplus_p \cdots \oplus_p \pi_k$. Given k irreducible 321-avoiding permutations $\pi_1, \pi_2, \ldots, \pi_k$ we build the corresponding permutation by summing in reverse order: $\pi_k \oplus_p \pi_{k-1} \oplus_p \cdots \oplus_p \pi_1$.

Here is how we build an irreducible 321-avoiding permutation π' of length n from a 321-avoiding permutation π of length $n - 1$:

$$\pi \; = \quad 2 \quad 4 \quad 1 \quad 3 \quad 5 \quad \boxed{7} \quad 6 \quad \boxed{9} \quad 8$$

$$2 \quad 4 \quad \boxed{10} \quad 1 \quad 3 \quad 5 \quad \boxed{7} \quad 6 \quad \boxed{9} \quad 8$$

$$\pi' \; = \quad 2 \quad 4 \quad \boxed{7} \quad 1 \quad 3 \quad 5 \quad \boxed{9} \quad 6 \quad \boxed{10} \quad 8 \quad .$$

In the first row we box the left-to-right maxima to the right of 1 that are not right-to-left minima. Here, those are 7 and 9. In the second row we insert a new largest letter, 10, immediately to the left of 1 and box it. Finally, in the third row, we cyclically shift the sequence of boxed letters one step to the left, thus obtaining π'. Let us call this map β.

The induced map Claesson-Kitaev, between 231- and 321-avoiding permutations is then formally defined by

$$\text{Claesson-Kitaev}(\varepsilon) = \varepsilon;$$

$$\text{Claesson-Kitaev}(\alpha(\sigma)) = \beta(\text{Claesson-Kitaev}(\sigma));$$

$$\text{Claesson-Kitaev}(\sigma \oplus_p \tau) = \text{Claesson-Kitaev}(\tau) \oplus_p \text{Claesson-Kitaev}(\sigma).$$

As an example, consider the permutation 5213476 in $\mathcal{S}_6(231)$. Decompose it using $\oplus_p$ and α:

$$5213476 = 52134 \oplus_p 21 = \alpha(2134) \oplus_p \alpha(1) = \alpha(\alpha(1) \oplus_p 1 \oplus_p 1) \oplus_p \alpha(1).$$

Reverse the order of summands and change each α to β:

$$\beta(1) \oplus_p \beta(1 \oplus_p 1 \oplus_p \beta(1)) = 21 \oplus_p \beta(1243) = 21 \oplus_p 41253 = 2163475.$$

In conclusion, Claesson-Kitaev(5213476) = 2163475.

4.1.5   West's bijection, 1995

West's bijection [801] is induced by an isomorphism between *generating trees*. The two isomorphic trees generate 123- and 132-avoiding permutations, respectively. We give a brief description of that bijection: Given a permutation $\pi = \pi_1 \pi_2 \cdots \pi_{n-1}$ and a positive integer $i \le n$, let

$$\pi^i \; = \; \pi_1 \cdots \pi_{i-1} \; n \; \pi_i \cdots \pi_{n-1};$$

we call this *inserting n into site i*. With respect to a fixed pattern τ, we call site i of $\pi \in \mathcal{S}_{n-1}(\tau)$ *active* if the insertion of n into site i creates a permutation in $\mathcal{S}_n(\tau)$. We have already met this notion in Chapter 3.

Avoiding 123	Avoiding 132	Signature
$_11_2$	$_11_2$	2
$_11_22$	$_112_2$	22
$_13_21_32$	$_13_212_3$	322
$_13_21_342$	$_134_212_3$	3322
$_15_23_31_442$	$_15_234_3124$	43322
$_15_23_36142$	$_15346_212_3$	343322

Table 4.1: Justification that $536142 \in \mathcal{S}_6(123)$ and $534612 \in \mathcal{S}_6(132)$ have the same signature 343322. Active sites are numbered from left to right.

For $i = 0, 1, \ldots, n-1$, let a_i be the number of active sites in the permutation obtained from π by removing the i largest letters. The *signature* of π is the word $a_0 a_1 \cdots a_{n-1}$.

West [801] showed that for 123-avoiding permutations, as well as for 132-avoiding permutations, the signature determines the permutation uniquely. This induces a natural bijection between the two sets. For example, $536142 \in \mathcal{S}_6(123)$ corresponds to $534612 \in \mathcal{S}_6(132)$—both have the same signature, 343322, as justified in Table 4.1.

4.1.6 Krattenthaler's bijection, 2001

Krattenthaler's bijection [543] uses Dyck paths as intermediate objects, and permutations mapped to the same Dyck path correspond to each other under this bijection.

The first part of Krattenthaler's bijection is a bijection from Av(123) to Dyck paths. Reading from right to left, let the right-to-left maxima in π be $m_1, m_2, \ldots,$ m_s, so that

$$\pi = w_s m_s \cdots w_2 m_2 w_1 m_1,$$

where w_i is the factor of π in between m_{i+1} and m_i. Since π is 123-avoiding, the letters in w_i are in decreasing order. Moreover, all letters of w_i are smaller than those of w_{i+1}.

To define the bijection, read π from right to left. With the convention $m_0 = 0$, any right-to-left maximum m_i is translated into $m_i - m_{i-1}$ up-steps. Any factor w_i is translated into $|w_i| + 1$ down-steps, where, as usual, $|w_i|$ denotes the number of letters of w_i. Finally, the resulting path is reversed.

The second part of Krattenthaler's bijection is a bijection from 132-avoiding permutations to Dyck paths. Read $\pi = \pi_1 \pi_2 \cdots \pi_n$ in $\mathcal{S}_n(132)$ from left to right and generate a Dyck path. When π_j is read, adjoin to the path obtained thus far as many up-steps as necessary (perhaps none) to reach height $h_j + 1$, followed by a down-step to height h_j (measured from the x-axis). Here, h_j is the number of letters in $\pi_{j+1}\pi_{j+2}\cdots\pi_n$ which are larger than π_j. This procedure can be shown to be equivalent to the standard bijection from 132-avoiding permutations to Dyck paths.

For instance, Krattenthaler's bijection sends the permutation $536142 \in \mathcal{S}_6(123)$ to the permutation $452316 \in \mathcal{S}_6(132)$—both map to the same Dyck path

4.1.7   Billey-Jockusch-Stanley-Reifegerste's bijection, 2002

Figure 4.5 illustrates the *Billey-Jockusch-Stanley-Reifegerste bijection* as it was presented by Reifegerste in [678]. It pictures, using permutation matrices, the 321-avoiding permutation $\pi = 13256847$ and the corresponding 132-avoiding permutation $\pi' = 78564213$.

	1	2	3	4	5	6	7	8
1							∘	
2					□			∘
3					∘			
4				□		∘		
5			□	∘				
6	□	∘						
7	∘							
8			∘					

Figure 4.5: Correspondence between $13256847 \in \mathcal{S}_8(321)$ and $78564213 \in \mathcal{S}_8(132)$ as it is described by Reifegerste [678].

Let $\pi = \pi_1\pi_2\cdots\pi_n \in \mathcal{S}_n(321)$, and let E be the set of pairs

$$E = \{\, (i, \pi_i) \mid i \text{ is an excedance} \,\}.$$

(Recall from Table A.1 that a position i in π is an *excedance* if $\pi_i > i$.) For each pair (i, π_i) in E, we place a square, called an *E-square*, in position $(i, n + 1 - \pi_i)$ (that is, in row i from above and in column $n + 1 - \pi_i$ from left to right) in an $n \times n$ permutation matrix. (We can see that E uniquely determines π.) Next we shade each square (a, b) of the matrix where there are no E-squares in the region $\{(i, j) \mid i \geq a, \ j \geq b\}$, thus obtaining a *Ferrers diagram*. Finally, we get the 132-avoiding permutation π' corresponding to π by placing circles row by row starting from the first row, in the leftmost available shaded square such that no two circles lie in the same column or row. If (i, j) contains a circle, then $\pi'(i) = j$.

Reifegerste [678] showed that the map $\pi \mapsto \pi'$ is the composition of a bijection (between 321-avoiding permutations and Dyck paths) given by Billey, Jockusch and Stanley [116] and the standard bijection (from 132-avoiding permutations to Dyck paths). Therefore we call $\pi \mapsto \pi'$ the Billey-Jockusch-Stanley-Reifegerste bijection.

4.1.8   Elizalde-Deutsch's bijection, 2003

Let $\mathcal{D}_n$ be the set of all Dyck paths on $2n$ steps. A bijection given by Elizalde and Deutsch [358] can be outlined as follows: Map 321- and 132-avoiding permutations bijectively to Dyck paths, use an automorphism Ψ on Dyck paths, and match permutations with equal paths.

We start by describing the automorphism Ψ. Let $P \in \mathcal{D}_n$. Each up-step of P has a corresponding down-step in the sense that the path between the up-step and the down-step forms a proper Dyck path. For example, in the following path we show the corresponding steps by the same label:

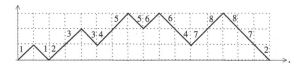

Match such pairs of steps. Let $\sigma = \sigma_1 \sigma_2 \cdots \sigma_{2n} \in \mathcal{S}_{2n}$ be the permutation defined by $\sigma_i = (i + 1)/2$ if i is odd, and $\sigma_i = 2n + 1 - i/2$ otherwise. That is,

$$\sigma = 1(2n)2(2n - 1)3(2n - 2)4 \cdots (n - 1)(n + 2)n(n + 1).$$

For i from 1 to $2n$, consider the σ_i-th step of P. If the corresponding matching step has not yet been read, define the i-th step of $\Psi(P)$ to be an up-step, otherwise let it be a down-step. For example, Ψ applied to the path above gives the path $uududuuddduduudd$, while

$$\Psi(uuduududududdud) = uuuddduduuuddud.$$

The bijection ψ from 321-avoiding permutations to $\mathcal{D}_n$ is defined as follows: Any permutation π in $\mathcal{S}_n$ can be represented as an $n \times n$ array with crosses in the squares (i, π_i). Here by the square (a, b) we mean the square in the a-th row from the top and b-th column from the left. Note that this array is obtained from the permutation matrix of π (see Definition 1.0.14) by rotating it $\pi/2$ radians clockwise. Given the array of π in $\mathcal{S}_n(321)$, consider the path with *down-* and *right*-steps along the edges of the squares that goes from the upper-left corner to the lower-right corner of the array leaving all the crosses to the right and remaining always as close to the main diagonal as possible. Then the corresponding Dyck path is obtained from this path by reading an up-step every time the path moves down, and a down-step every time the path moves to the right. For example,

$$\psi(23147586) = uuudddduduuduuddd$$

as illustrated in Figure 4.6.

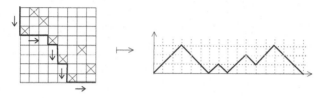

Figure 4.6: Justification for $\psi(23147586) = uuudddduduuduuddd$.

The bijection ϕ from 132-avoiding permutations to $\mathcal{D}_n$ is the standard bijection followed by a reflection of the path with respect to a vertical line through the middle of the path. For example,

$$\phi(7432516) = uuduududuuddud.$$

The *Elizalde-Deutsch bijection*, from $\mathcal{S}_n(321)$ to $\mathcal{S}_n(132)$, is defined by the following composition:

$$\text{Elizalde-Deutsch} = \phi^{-1} \circ \Psi^{-1} \circ \psi.$$

As an example, the Elizalde-Deutsch bijection sends $2314657 \in \mathcal{S}_7(321)$ to $7432516 \in \mathcal{S}_7(132)$ as shown in detail in Figure 4.7.

4.1.9   Mansour-Deng-Du's bijection, 2006

For $i = 1, 2, \ldots, n-1$, let s_i denote the adjacent transposition $(i \ \ i+1)$ in $\mathcal{S}_n$. For any permutation π of length n, the *canonical reduced decomposition* of π is

$$\pi = \sigma_1 \sigma_2 \cdots \sigma_k,$$

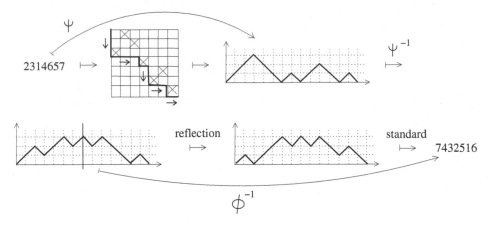

Figure 4.7: $2314657 \in \mathcal{S}_7(321)$ maps to $7432516 \in \mathcal{S}_7(132)$ under Elizalde-Deutsch's bijection.

where $\sigma_i = s_{h_i}s_{h_i-1}\cdots s_{t_i}$, $h_i \geq t_i$, $1 \leq i \leq k$ and $1 \leq h_1 < h_2 < \cdots < h_k \leq n-1$. For example, $415263 = (s_3s_2s_1)(s_4s_3)(s_5)$.

Mansour, Deng and Du [606] use canonical reduced decompositions to construct a bijection between $\mathcal{S}_n(321)$ and $\mathcal{S}_n(231)$. They show that a permutation is 321-avoiding precisely when $t_i \geq t_{i-1} + 1$ for $2 \leq i \leq k$. They also show that a permutation is 231-avoiding precisely when $t_i \geq t_{i-1}$ or $t_i \geq h_{i-j} + 2$ for $2 \leq i \leq k$ and $1 \leq j \leq i - 1$. Using these two theorems they build their bijection, which is composed of two bijections: one from $\mathcal{S}_n(321)$ to $\mathcal{D}_n$, and one from $\mathcal{S}_n(231)$ to $\mathcal{D}_n$, where $\mathcal{D}_n$ denotes the set of all Dyck paths of length $2n$.

For a Dyck path P, we define the $(x+y)$-labeling of P as follows: each cell in the region enclosed by P and the x-axis, whose corner points are (i,j), $(i+1,j-1)$, $(i+1,j+1)$ and $(i+2,j)$ is labeled by $(i+j)/2$. If $(i-1,j-1)$ and (i,j) are starting points of two consecutive up-steps, then we call the cell with leftmost corner (i,j) an *essential cell* and the up-step $((i-1,j-1),(i,j))$ its *left arm*. We say that two cells are *connected* if they share at least one point.

We define the *zigzag strip* of P as follows: If there is no essential cell in P, then the zigzag strip is empty. Otherwise, the zigzag strip is the longest run of connected cells (from left to right) that ends at the rightmost essential cell and is such that each cell in the run shares an edge with P (it lies on the border of P). For example, the zigzag strip of the Dyck path $uuduuuudddduddduduuuddud$ in Figure 4.8 is the shaded cell labeled by 9, while for the Dyck path $uuduuuududdd$ (obtained from that in Figure 4.8 by ignoring the steps 15 to 22) the zigzag strip is the run of shaded connected cells labeled by 4, 5, 5, 5 and 6.

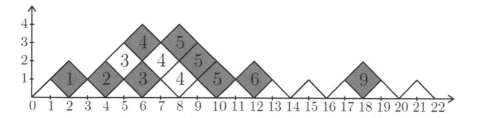

Figure 4.8: The zigzag decomposition of the path *uuduuududdddudduduuddud*.

Let $P_{n,k}$ be a Dyck path of semi-length n containing k essential cells. We define its *zigzag decomposition* as follows: The zigzag decomposition of $P_{n,0}$ is the empty set. The zigzag decomposition of $P_{n,1}$ is the zigzag strip. If $k \geq 2$, then we decompose $P_{n,k} = P_{n,k-1}Q$, where Q is the zigzag strip of $P_{n,k}$ and $P_{n,k-1}$ is the Dyck path obtained from P by deleting Q. Reading the labels of Q from left to right, ignoring repetitions, we get a sequence of numbers $\{i, i+1, \ldots, j\}$, and we associate Q with the sequence of adjacent transpositions $\sigma_k = s_j s_{j-1} \cdots s_i$. For $P_{n,i}$ with $i \leq k - 1$ repeat the above procedure to get $\sigma_{k-1}, \sigma_{k-2}, \ldots, \sigma_1$. The zigzag decomposition of $P_{n,k}$ is then given by $\sigma = \sigma_1 \sigma_2 \cdots \sigma_k$.

From the zigzag decomposition we get a 321-avoiding permutation π whose canonical reduced decomposition is σ. For the Dyck path $P_{11,4}$ in Figure 4.8 we have

$$\pi = (s_3 s_2 s_1)(s_4 s_3)(s_6 s_5 s_4)(s_9) = 41572368(10)9(11) \in \mathcal{S}_{11}(321).$$

We will now describe a map from Dyck paths to 231-avoiding permutations. For a Dyck path P, we define the $(x - y)$-labeling of P as follows (this labeling seems to be considered for the first time in [74]): each cell in the region enclosed by P and the x-axis, whose corner points are (i, j), $(i+1, j-1)$, $(i+1, j+1)$ and $(i+2, j)$ is labeled by $(i - j + 2)/2$.

Let us say that a cell is on the x-axis if it shares a point with the x-axis. We define the *trapezoidal strip* of P as follows: If there is no essential cell in P, then the trapezoidal strip is empty. Otherwise, the trapezoidal strip is the longest run of connected cells on the x-axis (from left to right) that ends at the rightmost essential cell on the x-axis. For example, the trapezoidal strip of the Dyck path *uuduuududdddudduduuddud* in Figure 4.9 is the shaded cell labeled by 9, while for the Dyck path *uuduuududddd* (obtained from that in Figure 4.9 by ignoring the steps 15 to 22) the zigzag strip is the downmost shaded strip with labels 1, 2, 3, 4, 5 and 6.

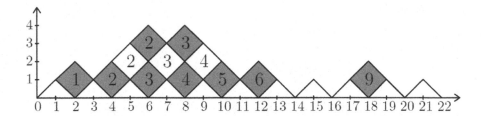

Figure 4.9: The trapezoidal decomposition of the path *uuduuududdddudduduuddud*.

Let $P_{n,k}$ be a Dyck path of semi-length n containing k essential cells. We define its *trapezoidal decomposition* as follows: The trapezoidal decomposition of $P_{n,0}$ is the empty set. The trapezoidal decomposition of $P_{n,1}$ is the trapezoidal strip. If $k \geq 2$, then we decompose $P_{n,k}$ into $P_{n,k} = Q_1 u Q_2 d$, where u is the left arm of the rightmost essential cell that touches the x-axis, d is the last down step of $P_{n,k}$, and Q_1 and Q_2 carry the labels in $P_{n,k}$. Reading the labels of the trapezoidal strip of $P_{n,k}$ from left to right we get a sequence $\{i, i+1, \ldots, j\}$, and we set $\sigma_k = s_j s_{j-1} \cdots s_i$. Repeat the above procedure for Q_1 and Q_2. Suppose the trapezoidal decomposition of Q_1 and Q_2 are σ' and σ'' respectively, then the trapezoidal decomposition for $P_{n,k}$ is $\sigma = \sigma' \sigma'' \sigma_k$.

From the trapezoidal decomposition we get a 231-avoiding permutation π whose canonical reduced decomposition is σ. For the Dyck path $P_{11,4}$ in Figure 4.9 we have

$$\pi = (s_3 s_2)(s_4 s_3 s_2)(s_6 s_5 s_4 s_3 s_2 s_1)(s_9) = 71542368(10)9(11) \in \mathcal{S}_{11}(231).$$

Together, the two maps involving Dyck paths described in this subsection induce a bijection from 321-avoiding to 231-avoiding permutations, which we call *Mansour-Deng-Du's bijection*.

4.2   Classification of bijections between 321- and 132-avoiding permutations

As mentioned in the beginning of the chapter, our intention is to see how many permutation statistics, out of those appearing most frequently in the literature, each of the bijections preserves. These statistics substitute our "base" set of statistics and they are recorded, together with definitions, in Table A.1. To do a more comprehensive analysis, we include in our list of statistics those that are obtained from our "basic" statistics by applying to them the trivial bijections on permutations: the reverse r, the complement c, the inverse i, and their compositions[2]. Moreover, for each statistic stat, we consider two other statistics: $n - \text{stat}(\pi)$ and $m - \text{stat}(\pi) = n + 1 - \text{stat}(\pi)$, where n is the length of the permutation π. The meaning of $n - \text{stat}$ or $m - \text{stat}$ is often "non-stat"; for example, $n - fp$ counts non-fixed-points in a permutation.

Thus, each basic statistic gives rise to 24 statistics in our extended list. We could also extend the basic set and include all the modifications of each basic statistic. However, many statistics in the obtained set are actually equivalent statistics. For example, des $= \text{asc}.r = \text{asc}.c$, and peak $= \text{peak}.r = \text{valley}.c = \text{valley}.c.r$, where a dot is used to denote composition of functions. Choosing one representative from each of the classes of equal statistics results in a final set of 148 statistics which we call *STAT*.

In the theorems below, the statistics presented are linearly independent. An example of a non-trivial linear dependence among the statistics over 132-avoiding permutations is lmin $-$ lmax $\text{I} (n - \text{des}) - \text{head} = 0$. The results below are also maximal in the sense that they cannot be non-trivially extended using statistics from STAT. That is, adding one more pair of equidistributed statistics from STAT to any of the results would create a linear dependency among the statistics.

The main results of [259] are presented in the following three theorems.

Theorem 4.2.1. *([259]) The following results are maximal in the sense that adding one more pair of equidistributed statistics from STAT to any of the results would create a linear dependency among the statistics. Also, we indicate the sets between which a bijection acts. The numbers in parentheses show the number of statistics respected.*

(11) **Knuth-Richards**, $\text{Av}(132) \to \text{Av}(123)$

| valley.i | valley | lmin | ldr.i | head.i | comp.r | rank | ldr | lir.i |
| valley | valley.i | lmin | ldr | head | comp.r | rank | ldr.i | slmax.c |

and, also, we have
| lir | rmax |
| slmax.$i.r$ | head.$i.r$ |

(11) **Simion-Schmidt**, $\text{Av}(123) \to \text{Av}(132)$

---

[2]As one can notice, the statistics list in Table A.1 already contains results of applying trivial bijections to other statistics, and thus it is not a minimal possible base set. However, our main interest is in an extension of this set, called STAT, so the issue of providing the minimal base set is not important here.

valley   valley.*i*   lmin   ldr   head   comp.*r*   rank   ldr.*i*   slmax.*c*
valley   valley.*i*   lmin   ldr   head   comp.*r*   rank   ldr.*i*   lir

and, also, we have   slmax.*i.r*   head.*i.r*
                  lir.*i*   rmin

(11) **Krattenthaler**, Av(123) → Av(132)

peak.*i*   peak   rmax   zeil   last.*i.r*   comp.*r*   rank.*r.c*   rdr   slmax.*r.i*
valley   valley.*i*   lmin   ldr   head   comp.*r*   rank   ldr.*i*   lir

and, also, we have   slmax.*r*   last
                  lir.*i*   rmin

(11) **Mansour-Deng-Du**, Av(321) ↠ Av(231)

valley   peak.*i*   rmin   rir   last   comp   rank.*r*   lir.*i*   slmax.*c.r*
valley   peak.*i*   rmin   rir   last   comp   rank.*r*   lir.*i*   rdr

and, also, we have   slmax.*i*   head.*i*
                  ldr.*i*   lmin

(9) **Knuth-Rotem**, Av(321) → Av(132)

valley.*i*   peak   exc   slmax   head   slmax.*r.c.i*   rir.*i*   lir   last.*i*
valley.*i*   valley   des   rdr   ldr.*i*   zeil   lmax   rmin $m -$ ldr

(9) **Billey-Jockusch-Stanley-Reifegerste**, Av(321) → Av(132)

valley   peak.*i*   exc   slmax.*i*   head.*i*   slmax.*r.c*   rir   lir.*i*   last
valley   valley.*i*   des   zeil   ldr   rdr   rmin   lmax $m -$ ldr.*i*

(7) **West**, Av(123) → Av(132)

valley.*i*   exc.*r*   slmax.*i.r*   slmax.*c*   ldr   ldr.*i*   head
valley.*i*   asc   lir.*i*   comp   rmax   ldr.*i*   head

(5) **Knuth**, Av(321) → Av(132)

exc   fp   lir.*i*   lir   lis
exc   fp   rmin   lmax $n -$ rank

(1) **Elizalde-Deutsch**, Av(321) → Av(132)

fp
fp

Example 4.2.2. In Subsection 4.1.1, we see that $\pi_1 = 23514687 \in \mathcal{S}_n(321)$ is mapped to $\pi_2 = 67435281$ under Knuth's bijection. Theorem 4.2.1 suggests that five statistics should be preserved under this map. Indeed, $\mathrm{exc}(\pi_1) = \mathrm{exc}(\pi_2) = 4$ (excedances in π_1 are 1, 2, 3 and 7, whereas excedances in π_2 are also 1, 2, 3 and 7); $\mathrm{fp}(\pi_1) = \mathrm{fp}(\pi_2) = 1$ (6 and 5 are the fixed points in π_1 and π_2, respectively); $\mathrm{lir}.i(\pi_1) = \mathrm{rmin}(\pi_2) = 1$ (1 is the largest i such that $12 \cdots i$ is a subsequence in π_1, whereas the sequence of right-to-left minima in π_2 is 1); $\mathrm{lir}(\pi_1) = \mathrm{lmax}(\pi_2) = 3$ (235 is the longest increasing run in π_1, whereas the sequence of left-to-right maxima in π_2 is 6, 7, and 8); finally, $\mathrm{lis}(\pi_1) = n - \mathrm{rank}(\pi_2) = 5$ (a longest increasing subsequence in π_1 is 2, 3, 5, 6, and 8, whereas $\mathrm{rank}(\pi_2) = 3$ and $n = 8$).

A few samples of arguments involved in proving Theorem 4.2.1 are discussed at the end of this section.

It turns out that bijections in Theorem 4.2.1 with the same number of statistics preserved are related via "trivial" bijections. The next theorem makes this precise.

Theorem 4.2.3. *([259]) The following relations among bijections between 321- and 132-avoiding permutations hold (r and i stand for "reverse" and "inverse," respectively):*

$$r \circ \text{Claesson-Kitaev}^{-1}$$

$$= i \circ \text{Simion-Schmidt} \circ r$$

$$= i \circ \text{Krattenthaler} \circ r \circ i$$

$$= i \circ r \circ \text{Mansour-Deng-Du}$$

$$= \text{Knuth-Richards}^{-1} \circ r$$

and

$$\text{Billey-Jockusch-Stanley-Reifegerste} = i \circ \text{Knuth-Rotem} \circ i.$$

Also, there are no other relations among the bijections and their inverses via the trivial bijections that do not follow from the ones above.

The last statement of Theorem 4.2.3 is obtained from the following argument: The set STAT is by definition closed with respect to the trivial bijections, meaning that if f is a trivial bijection and *stat* is in STAT, then *stat.f* is in STAT too. Thus, two bijections related by trivial bijections will preserve the same number of statistics from STAT. Consequently, bijections preserving different numbers of statistics from STAT cannot be related by trivial bijections.

The main idea to prove the equalities in Theorem 4.2.3 is to control where the "basic elements" of a permutation π go under the two bijections to be compared. Such a basic element for $\pi \in \mathrm{Av}(132)$ is the sequence of positions and values of left-to-right minima π, which, as was mentioned above, determines π uniquely (and thus, there is no point in controlling other elements in π).

If we regard all bijections as bijections from 321- to 132-avoiding permutations— applying the transformations in Theorem 4.2.3—then we get the following condensed version of Theorem 4.2.1.

Theorem 4.2.4. *([259]) For bijections from 321- to 132-avoiding permutations we have the following equidistribution results. These results are maximal in the sense that adding one more pair of equidistributed statistics from STAT to any of the results would create a linear dependency among the statistics.*

(11) **Claesson-Kitaev, Knuth-Richards, Krattenthaler, Mansour-Deng-Du, Simion-Schmidt**

| valley | peak$.i$ | rmin | rir | last | comp | rank$.r$ | lir$.i$ | slmax$.c.r$ |
| valley | valley$.i$ | lmin | ldr | head | comp$.r$ | rank | ldr$.i$ | lir |

and, also, we have
$$\begin{array}{cc} \text{slmax}.i & \text{head}.i \\ \text{lir}.i & \text{rmin} \end{array}$$

(9) **Knuth-Rotem, Billey-Jockusch-Stanley-Reifegerste**

| valley | peak$.i$ | exc | slmax$.i$ | head$.i$ | slmax$.r.c$ | rir | lir$.i$ | last |
| valley | valley$.i$ | des | zeil | ldr | rdr | rmin | lmax | $m - ldr.i$ |

(7) **West**

| peak$.i$ | exc | slmax$.i$ | slmax$.r.c$ | rir | lir$.i$ | last |
| valley$.i$ | asc | lir$.i$ | comp | rmax | ldr$.i$ | head |

(5) **Knuth**

| exc | fp | lir$.i$ | lir | lis |
| exc | fp | rmin | lmax | $n - $rank |

(1) **Elizalde-Deutsch**

fp
fp

We would like to conclude this section by providing four samples of proofs (coming almost verbatim from [259]) of statements in Theorem 4.2.1 related to Simion-Schmidt's bijection. Our examples use the fact that under the bijection, positions and values of the sequence of left-to-right minima in a permutation $\pi \in$ Av(123) are preserved. We let $\pi' = $ Simion-Schmidt(π), and, for brevity, we write stat$_1 \simeq$ stat$_2$ below when, for all $\pi \in$ Av(123), we have stat$_1(\pi) =$ stat$_2(\pi')$.

- slmax$.i.r \simeq$ lir$.i$: The statistic slmax$.i.r$ is one less than the minimal i such that the letter i is to the left of the letter 1. Note that such an i in π must be a left-to-right minimum. Suppose slmax$.i.r(\pi) = i - 1$ and $i < n$ (the case slmax$.i.r(\pi) = n - 1$ is trivial). Then i and 1 are two consecutive left-to-right minima in π. Consequently, i and 1 are two consecutive left-to-right minima in π'. Thus the letters $2, 3, \ldots, i - 1$ must be to the right of 1 in π'. To avoid forming an occurrence of 132, those letters must also be in increasing order. Thus lir$.i(\pi') = i - 1$.

- ldr$.i \simeq$ ldr$.i$: By definition ldr$.i(\pi)$ is the largest i such that $i, i - 1, \ldots, 1$ is a subsequence in π. In particular, $i + 1$, if it exists, is to the right of i in π. Suppose that ldr$.i(\pi) = i$. Clearly, the letters $i, i - 1, \ldots, 1$ are consecutive left-to-right minima in π and $i + 1$ is a non-left-to-right minimum. Thus $i, i - 1, \ldots, 1$ is a subsequence in π' and $i + 1$ is to the right of i in π'; hence ldr$.i(\pi') = i$.

- head$.i.r \simeq$ rmin: We can see that head$.i.r(\pi) = n - i + 1$ where i is the position of 1 in π. Let $\pi' = \sigma 1 \tau$. The letter 1 is in the same position in π' as it is in π and so $|\tau| = n - i$. To avoid the pattern 132 the letters of τ must be in increasing order. The sequence of right-to-left minima in π' is thus simply 1τ and therefore head$.i.r(\pi) = \text{rmin}(\pi')$.

- rank $\simeq$ rank: Suppose that rank$(\pi) = k$ and π has a in position $k + 1$. We distinguish two cases based on whether a is a left-to-right minimum or not. If a is a non-left-to-right-minimum, then $k + 1$ is the left-to-right minimum closest to a in π from the left. Since positions and values of the left-to-right minima of π are preserved, we have rank$(\pi') = k$. On the other hand, if a is a left-to-right minimum, then $a \leq k + 1$ and π' will have the same left-to-right minimum, a, in position $k + 1$. Thus rank$(\pi') = k$ in this case as well.

Chapter 5

Consecutive patterns

Consecutive patterns are defined in Definition 1.3.2. Examples of consecutive patterns are $\underline{132}$ and $\underline{43251}$. There are many approaches in the literature to study consecutive patterns ranging from the symbolic method, the spectral approach, the symmetric functions approach, and the cluster method to inclusion-exclusion arguments and homological algebra, and this chapter provides an introduction to the methods. Also, we state most of the known results involving consecutive patterns and provide references to several other results. There are more than 60 references to different papers and books in this chapter.

5.1   The symbolic method

There is a direct correspondence between set-theoretic operations on combinatorial classes and algebraic operations on e.g.f.s. Let $\mathcal{A}$, $\mathcal{B}$, and $\mathcal{C}$ be classes of labeled combinatorial objects, and $A(x)$, $B(x)$, and $C(x)$ be their e.g.f.s, respectively. If $\mathcal{A} = \mathcal{B} \cup \mathcal{C}$ is the union of disjoint objects then $A(x) = B(x) + C(x)$. If $\mathcal{A} = \mathcal{B} \star \mathcal{C}$ is the labeled product, that is, the usual Cartesian product enriched with the relabeling operation, then $A(x) = B(x)C(x)$. If $\mathcal{A} = \Pi(\mathcal{B})$ is taking the set of all subsets of $\mathcal{B}$, then $A(x) = \exp(B(x))$; if $\mathcal{A} = \mathcal{B}^{\square} \star \mathcal{C}$ is the box product, that is, the subset of $\mathcal{B} \star \mathcal{C}$ formed by those pairs in which the smaller label lies in the $\mathcal{B}$ component, then $A(x) = \int_0^x (\frac{d}{dt}B(t)) \cdot C(t)dt$. The same holds if we have *bivariate generating functions* (*b.g.f.s*) discussed below instead of e.g.f.s. We summarize the facts above, used in [360] and [511] dealing with consecutive patterns, in Table 5.1.

We give an example of applying Table 5.1 to derive a well-known fact on permutations. This approach is referred to as the *symbolic method*; more on this method can be found in the book [387] by Flajolet and Sedgewick.

Construction		Operation on e.g.f.
Union	$\mathcal{A} = \mathcal{B} \cup \mathcal{C}$	$A(x) = B(x) + C(x)$
Labeled product	$\mathcal{A} = \mathcal{B} \star \mathcal{C}$	$A(x) = B(x)C(x)$
Set	$\mathcal{A} = \Pi(\mathcal{B})$	$A(x) = \exp(B(x))$
Box product	$\mathcal{A} = \mathcal{B}^{\square} \star \mathcal{C}$	$A(x) = \int_0^x (\frac{d}{dt} B(t)) \cdot C(t) dt$

Table 5.1: The basic combinatorial constructions and their translation into exponential generating functions.

Suppose we would like to enumerate all permutations in $\mathcal{S}_n$, that is, to find $s_n = |\mathcal{S}_n|$. We begin by considering a permutation π from the set $\mathcal{S}_\infty = \cup_{n \geq 0} \mathcal{S}_n$: either π is the empty permutation ε or $\pi = \pi_1 1 \pi_2$, where $\mathrm{red}(\pi_1)$ and $\mathrm{red}(\pi_2)$ are possibly empty permutations in $\mathcal{S}_\infty$. Thus, $\mathcal{S}_\infty$ satisfies the recursive definition

$$\mathcal{S}_\infty = \{\varepsilon\} + \{x\}^{\square} \star \mathcal{S}_\infty \star \mathcal{S}_\infty,$$

where $\{x\}$ represents the permutation 1, and the box indicates that the 1 is unchanged in the composed objects (after something to the left and to the right of it is added). Thus, the e.g.f. for the number of all permutations is

$$S(x) = 1 + \int_0^x S^2(t) dt,$$

which reduces to the differential equation $S'(x) = S^2(x)$ (the derivative is with respect to x) with the initial condition $S(0) = 1$, having the solution

$$S(x) = \frac{1}{1-x} = 1 + x + x^2 + x^3 + \cdots = 1 + \frac{1!}{1!}x + \frac{2!}{2!}x^2 + \frac{3!}{3!}x^3 + \cdots + \frac{n!}{n!}x^n + \cdots.$$

Consequently, $s_n = n!$ as expected.

In Subsection 5.1.1 we provide two general theorems on consecutive (partially ordered) patterns and some of their corollaries; the proofs are obtained by a direct application of the symbolic method. In Subsection 5.1.2, we state more general theorems on consecutive patterns and several of their consequences. The theorems in Subsection 5.1.2 were proved in [360] by Elizalde and Noy who first transferred the problem to *increasing binary trees* and then applied the symbolic method.

5.1.1   A direct approach

The theorems in this subsection are related to *flat posets* introduced in Subsection 3.5.6; see Figure 3.23 for the Hasse diagram related to the flat poset built on the elements $a, a_1, \ldots, a_k$ with the only relations $a < a_i$ for all i (a_i and a_j are incomparable for $i \neq j$). Theorem 5.1.1 deals with the distribution of the pattern $aa_1a_2 \cdots a_k$ appearing in Theorem 3.5.40, and Theorem 5.1.2 deals with the distribution of the pattern $a_1a_2 \cdots a_k aa_{k+1}a_{k+2} \cdots a_{k+\ell}$ which generalizes Theorem 5.1.1.

Theorem 5.1.1. *([511]) Let*

$$P := P(x, y) = \sum_{n \geq 0} \sum_{\pi \in S_n} y^{p(\pi)} \frac{x^n}{n!}$$

be the bivariate generating function (b.g.f.) for permutations where $p(\pi)$ is the number of occurrences of the consecutive POP $p = \underline{aa_1a_2 \cdots a_k}$ in π. Then P is the solution of the following partial differential equation (PDE):

$$(5.1) \qquad \frac{\partial P}{\partial x} = yP^2 + \frac{(1-y)(1-x^k)}{1-x}P$$

with initial condition $P(0, y) = 1$.

Proof. Let $\pi = \pi' 1 \pi''$ be a permutation. Then

$$p(\pi) = \begin{cases} p(\pi') + p(\pi'') + 1 & \text{if } |\pi''| \geq k, \\ p(\pi') & \text{if } |\pi''| < k \end{cases}$$

since an occurrence of p cannot start inside π' and end outside π'; also when π'' is of length at least k, it contributes one extra occurrence of p starting at 1.

Suppose

$$P_{<k} := P_{<k}(x, y) = \sum_{n=0}^{k-1} \sum_{\pi \in S_n} y^{p(\pi)} \frac{x^n}{n!} = \sum_{n=0}^{k-1} x^n = \frac{1-x^k}{1-x}.$$

In the rest of the proof we apply the symbolic method to see that

$$P' = P(y(P - P_{<k}) + P_{<k})$$

with the initial condition $P(0, y) = 1$; the desired result is then easy to obtain by plugging in $P_{<k}$ and rewriting the equation.

Let $\mathcal{P}_{<k}$ be the class of permutations of length less than k. With some abuse of notation, we introduce the parameter y in the equation for classes, meaning that it will be placed there when we write the corresponding differential equations for the b.g.f.s. With this notation and using the structure of $p(\pi)$, we can write

$$\mathcal{S}_\infty = \{\epsilon\} + \{x\}^\square \star \mathcal{S}_\infty \star [y(\mathcal{S} - \mathcal{P}_{<k}) + \mathcal{P}_{<k}].$$

We differentiate the corresponding equation for b.g.f.s to obtain the desired result. $\qquad \square$

Note that if $y = 0$ in Theorem 5.1.1 then the function $\exp(\sum_{i=1}^{k} x^i/i)$ from Table 3.3 would be the solution to (5.1), and it is. If $k = 1$ in Theorem 5.1.1, then as the solution to (5.1) we obtain the distribution of descents in permutations:

$$\frac{1-y}{e^{(y-1)x} - y}.$$

Thus Theorem 5.1.1 can be seen as a generalization of the descent distribution formula.

The following theorem generalizes Theorem 5.1.1. Indeed, Theorem 5.1.1 is obtained from Theorem 5.1.2 by plugging in $\ell = 0$ and observing that the patterns $aa_1 \cdots a_k$ and $a_1 \cdots a_k a$ are equidistributed.

Theorem 5.1.2. *([511]) Let*

$$P := P(x, y) = \sum_{n \geq 0} \sum_{\pi \in \mathcal{S}_n} y^{p(\pi)} \frac{x^n}{n!}$$

be the b.g.f. for permutations where $p(\pi)$ is the number of occurrences of the consecutive POP $p = a_1 a_2 \cdots a_k a a_{k+1} a_{k+2} \cdots a_{k+\ell}$ in π. Then P is the solution of the following PDE:

$$(5.2) \quad \frac{\partial P}{\partial x} = y \left(P - \frac{1 - x^k}{1 - x} \right) \left(P - \frac{1 - x^\ell}{1 - x} \right) + \frac{2 - x^k - x^\ell}{1 - x} P - \frac{1 - x^k - x^\ell + x^{k+\ell}}{(1 - x)^2}$$

with initial condition $P(0, y) = 1$.

If $y = 0$ in Theorem 5.1.2 then we get the following corollary:

Corollary 5.1.3. *([511]) The e.g.f. $E(x)$ for the number of permutations avoiding the consecutive POP $p = a_1 a_2 \cdots a_k a a_{k+1} a_{k+2} \cdots a_{k+\ell}$ satisfies the following differential equation with initial condition $E(0) = 1$:*

$$E'(x) = \frac{2 - x^k - x^\ell}{1 - x} E(x) - \frac{1 - x^k - x^\ell + x^{k+\ell}}{(1 - x)^2}.$$

The following corollaries to Corollary 5.1.3 are obtained by plugging in $k = \ell = 1$, and $k = 1$ and $\ell = 2$, respectively.

Corollary 5.1.4. *([505]) The e.g.f. for the number of permutations avoiding the pattern $a_1 a a_2$ is $(\exp(2x) + 1)/2$ and thus $s_n(a_1 a a_2) = 2^{n-1}$.*

Corollary 5.1.5. *([511]) The e.g.f. for the number of permutations avoiding the pattern $\underline{a_1aa_2a_3}$ is*

$$1 + \sqrt{\frac{\pi}{2}}(\operatorname{erf}(\frac{1}{\sqrt{2}}x + \sqrt{2}) - \operatorname{erf}(\sqrt{2}))e^{\frac{1}{2}x(x+4)+2}$$

where $\operatorname{erf}(x) = \frac{2}{\sqrt{\pi}}\int_0^x e^{-t^2}\,dt$ *is the error function.*

If $y = 0$ in Theorem 5.1.2 then we obtain Corollary 5.1.3. If $k = 1$ and $\ell = 1$ then our pattern $\underline{a_1aa_2}$ is simply the valley statistic. In [693] a recursive formula for the generating function of permutations with exactly k valleys is obtained; however, this does not seem to enable us to easily find the corresponding b.g.f.. As a corollary to Theorem 5.1.2 one obtains this b.g.f. by solving (5.2) for $k = 1$ and $\ell = 1$. The solution is recorded in Theorem 3.5.9 (the valley statistic is equivalent to the peak statistic by taking the complement).

Expanding the b.g.f. in Theorem 3.5.9 we can get, for example, the sequences A000431, A000487 and A000517 appearing in [726] for the number of permutations with exactly one, two and three valleys respectively. Note that we have already found the number of *valleyless* permutations in Corollary 5.1.4. The valleyless permutations were studied in [693].

5.1.2 Increasing binary trees

Let $\pi = \pi_1\pi_2\cdots\pi_n$ be a word on the alphabet of the natural numbers $\mathbb{N} = \{1, 2, \ldots\}$ with no repeated letters. Define the binary tree $T(\pi)$ corresponding to π as follows. If $\pi = \varepsilon$ then $T(\pi)$ is the empty tree. Otherwise, let i be the smallest letter of π and $\pi = \pi_1 i \pi_2$. Define $T(\pi)$ recursively as the tree with the root labeled i that has $T(\pi_1)$ and $T(\pi_2)$ as the left and right subtrees, respectively. This correspondence gives a bijection between $\mathcal{S}_n$ and the set of *increasing binary trees* on n vertices [743, Chapter 1]. In particular, the number of such trees is $n!$.

Example 5.1.6. Figure 5.1 shows the increasing binary tree corresponding to the permutation 427319586.

The theorems in this subsection, appearing in [360] by Elizalde and Noy, are obtained by representing permutations as binary increasing trees, and deriving (from the representation using the symbolic method) differential equations satisfied by the corresponding e.g.f.s. The following two theorems are very important results in the literature on consecutive patterns. They deal with the distribution on permutations of consecutive patterns of special types. In the rest of this subsection, all derivatives are taken with respect to x.

Figure 5.1: The increasing binary tree corresponding to the permutation 427319586.

Theorem 5.1.7. *([360]) Let m be a positive integer, let $p = 12\cdots(m+2)$, and let $P(u,x)$ be the b.g.f. of permutations where u marks the number of occurrences of p. Then, $P(u,x) = 1/w(u,x)$, where $w = w(u,x)$ is the solution of*

$$w^{(m+1)} + (1-u)(w^{(m)} + w^{(m-1)} + \cdots + w' + w) = 0$$

with $w(u,0) = 1$, $w'(u,0) = -1$, and $w^{(k)}(u,0) = 0$ for $2 \le k \le m$.

Theorem 5.1.8. *([360]) Let m and a be positive integers with $a \le m$ and let $p = 12\cdots a\tau(a+1)$, where τ is any permutation of $\{a+2, a+3, \ldots, m+2\}$. Let $P(u,x)$ be the b.g.f. for permutations where u marks the number of occurrences of p. Then $P(u,x) = 1/w(u,x)$, where $w = w(u,x)$ is the solution of*

$$w^{a+1} + (1-u)\frac{x^{m-a+1}}{(m-a+1)!}w' = 0$$

with $w(u,0) = 1$, $w'(u,0) = -1$ and $w^{(k)}(u,0) = 0$ for $2 \le k \le a$. In particular, the distribution does not depend on τ.

Note that as in the case of avoidance, the distribution of patterns of length 3 constitutes two equivalence classes: $\{\underline{123}, \underline{321}\}$ and $\{\underline{132}, \underline{231}, \underline{213}, \underline{312}\}$ (one can simply apply the trivial bijections). The following theorem is a corollary to Theorems 5.1.7 and 5.1.8 and it gives the distribution and, in particular, avoidance for all consecutive patterns of length 3.

Theorem 5.1.9. *([360]) Let $P(u,x)$ and $Q(u,x)$ be the b.g.f.s of permutations where u marks the number of occurrences of the patterns $\underline{123}$ and $\underline{132}$, respectively. Then*

$$P(u,x) = \frac{2e^{1/2(1-u+\sqrt{(u-1)(u+3)})x}\sqrt{(u-1)(u+3)}}{1+u+\sqrt{(u-1)(u+3)}+e^{\sqrt{(u-1)(u+3)}x}(-1-u+\sqrt{(u-1)(u+3)})} \quad \text{and}$$

$$Q(u,x) = \frac{1}{1 - \int_0^x e^{(u-1)t^2/2}dt}.$$

Thus, the e.g.f. for the number of permutations avoiding the pattern $\underline{123}$ *is given by*

$$E_{\underline{123}}(x) = P(0, x) = \frac{\sqrt{3}}{2} \frac{e^{x/2}}{\cos\left(\frac{\sqrt{3}}{2}x + \frac{\pi}{6}\right)},$$

while the e.g.f. for the number of permutations avoiding the pattern $\underline{132}$ *is given by*

$$E_{\underline{132}}(x) = Q(0, x) = \frac{1}{1 - \int_0^x e^{-t^2/2}dt}.$$

One can compare $s_n(\underline{123})$ and $s_n(\underline{132})$ which shows that it is "harder" to avoid the pattern $\underline{132}$ than $\underline{123}$:

Proposition 5.1.10. *([360]) For every $n \geq 4$, we have $s_n(\underline{123}) > s_n(\underline{132})$.*

Another result in the same vein deals with consecutive patterns of length 4:

Proposition 5.1.11. *([360]) For every $n \geq 7$, we have $s_n(\underline{1342}) > s_n(\underline{1243})$.*

When experimenting by computer, we often tend to believe that if a fact on permutations holds for large enough n, say, $n \leq 11$, then it is likely to hold for all n. One should be careful with such assumptions: for example, as is discussed in [360] by Elizalde and Noy, $s_n(\underline{1324}) \geq s_n(\underline{2143})$ for all $n \leq 11$, but $s_{12}(\underline{1324}) < s_{12}(\underline{2143})$. For $n \leq 32$, Aldred et al. [34] have found no further oddities comparing $s_n(p_1)$ and $s_n(p_2)$ for length 4 patterns p_1 and p_2.

Using the reverse and complement operations, we can partition the set of all consecutive patterns of length 4 into equivalence classes with respect to distribution on permutations:

$$\{\underline{1234}, \underline{4321}\}, \{\underline{2413}, \underline{3142}\}, \{\underline{2143}, \underline{3412}\}, \{\underline{1324}, \underline{4231}\}, \{\underline{1423}, \underline{3241}, \underline{4132}, \underline{2314}\},$$
$$\{\underline{1342}, \underline{2431}, \underline{4213}, \underline{3124}, \underline{1432}, \underline{2341}, \underline{4123}, \underline{3214}\}, \text{ and } \{\underline{1243}, \underline{3421}, \underline{4312}, \underline{2134}\}.$$

Theorems 5.1.7 and 5.1.8 can be used to find distributions for 3 of these equivalence classes which are recorded in the following theorem (we pick a representative from each equivalence class).

Theorem 5.1.12. *([360]) Let $P(u, x)$ be the b.g.f. of permutations where u marks the number of occurrences of the consecutive pattern in question. Then*

- *For the pattern* $\underline{1342}$,

$$P(u, x) = \frac{1}{1 - \int_0^x e^{(u-1)t^3/6}dt}$$

which, in the case of avoidance, gives the following e.g.f.:

$$P(0,x) = \frac{1}{1 - \int_0^x e^{-t^3/6} dt}.$$

- *For the pattern* 1234, $P(u,x) = 1/w$, *where* $w = w(u,x)$ *is the solution of*

$$w''' + (1-u)(w'' + w' + w) = 0$$

  with $w(u,0) = 1$, $w'(u,0) = -1$ *and* $w''(u,0) = 0$. *In the case of avoidance, the solution is the following e.g.f.:*

$$P(0,x) = \frac{2}{\cos x - \sin x + e^{-x}}.$$

- *For the pattern* 1243, $P(u,x) = 1/w$, *where* $w = w(u,x)$ *is the solution of*

$$w''' + (1-u)xw' = 0$$

  with $w(u,0) = 1$, $w'(u,0) = -1$ *and* $w''(u,0) = 0$.

Consideration of increasing binary trees together with the symbolic method can be used to deal with more than one pattern at the same time. The following theorem provides the joint distribution of two consecutive patterns, 123 and 231, namely the multivariate generating function

$$P = P(u,v,x) = \sum_{\pi \in \mathcal{S}_\infty} u^{123(\pi)} v^{231(\pi)} \frac{x^{|\pi|}}{|\pi|!}.$$

Theorem 5.1.13. *([360]) Let $K = K(u,v,x)$, $L = L(u,v,x)$ and $M = M(u,v,x)$ be defined similarly to P, except instead of the set $\mathcal{S}_\infty$ of all permutations, the sum is taken over all permutations avoiding* 12, *avoiding* 12, *and those that avoid both* 12 *and* 12, *respectively. Then the following system of differential equations holds:*

$$\begin{cases} P' = [L + v(P-L)][K + u(P-K)], \\ K' = 1 + [M + v(K-M) - 1][K + u(P-K)], \\ L' = [L + v(P-L)][M + u(L-M) - x], \\ M' = 1 + [M + v(K-M) - 1][M + u(L-M) - x], \end{cases}$$

with the initial conditions $P(u,v,0) = K(u,v,0) = L(u,v,0) = M(u,v,0) = 1$.

A corollary to Theorem 5.1.13 is the following result.

Corollary 5.1.14. *([360]) The e.g.f.* $A = A(x)$ *for* $s_n(\underline{123}, \underline{231})$ *is given by the solution to the following system of differential equations, where derivatives are with respect to* x *and* $B = B(x)$, $C = C(x)$ *and* $D = D(x)$:

$$\begin{cases} A' = CB, \\ B' = 1 + (D + x - 1)B, \\ C' = CD, \\ D' = (D + x - 1)D, \end{cases}$$

with the initial conditions $A(0) = B(0) = C(0) = D(0) = 1$.

As remarked in [360], an involved explicit form for $A(x)$ in Corollary 5.1.14 can be found in terms of integrals containing the error function $\mathrm{erf}(x) = \frac{2}{\sqrt{\pi}} \int_0^x e^{-t^2} dt$. As many initial terms for $s_n(\underline{123}, \underline{231})$ as desired can be obtained from the system in Corollary 5.1.14. The first numbers in that sequence are the following:

$$1, 2, 4, 11, 39, 161, 784, 4368, 27260, 189540, 1448860, 12076408, \dots .$$

5.2 An inclusion-exclusion approach

In this section we provide three theorems on consecutive patterns that were proved using an inclusion-exclusion approach. Two theorems are of a general type (Theorems 5.2.1 and 5.2.4), while Theorem 5.2.5 is on a consecutive partially ordered pattern of length 4. To demonstrate the approach, we provide a proof of Theorem 5.2.5; the other two theorems can be proved using similar considerations. The inclusion-exclusion approach allows us to get explicit formulas for the e.g.f. in terms of infinite series instead of having to solve differential equations as discussed in Section 5.1. However, in particular cases, we can use certain differential equations to simplify the series (see Remark 5.2.2).

Theorem 5.2.1. *([422]) Let* $E_k(x)$ *be the e.g.f. for the number of permutations avoiding the pattern* $p = \underline{12 \cdots k}$. *Then*

$$E_k(x) = \frac{1}{F_k(x)}$$

where

$$F_k(x) = \sum_{i \geq 0} \frac{x^{ki}}{(ki)!} - \sum_{i \geq 0} \frac{x^{ki+1}}{(ki+1)!}.$$

Remark 5.2.2. For some k it is possible to simplify the function $F_k(x)$ in Theorem 5.2.1. Indeed, $F_k(x)$ satisfies the differential equation $F_k^{(k)}(x) = F_k(x)$ with the

k initial conditions $F_k(0) = 1$, $F_k'(0) = -1$ and $F_k^{(i)}(0) = 0$ for $i = 2, 3, \ldots, k-1$. For instance, if $k = 4$, that is, if we deal with avoidance of the pattern $\underline{1234}$, then

$$E_4(x) = \frac{1}{2}(\cos x - \sin x + e^{-x}),$$

which is consistent with Theorem 5.1.12.

Remark 5.2.3. See Theorem 5.13.3 below for an alternative way to express $E_k(x)$ in Theorem 5.2.1.

Theorem 5.2.4. ([509]) Let k and a be positive integers with $a < k$, and $p = 12 \cdots a\tau(a+1)$, where τ is any permutation of the elements $\{a+2, a+3, \ldots, k+1\}$. That is, p is a consecutive pattern of length $k+1$. Let $E_{k,a}(x)$ be the e.g.f. for the number of permutations that avoid p. Further, let

$$F_{k,a}(x) = \sum_{i \geq 1} \frac{(-1)^{i+1}x^{ki+1}}{(ki+1)!} \prod_{j=2}^{i} \left(\frac{jk-a}{k-a} \right).$$

Then

$$E_{k,a}(x) = \frac{1}{1 - x + F_{k,a}(x)}.$$

If $k = 2$ and $a = 1$ in Theorem 5.2.4, corresponding to avoidance of the pattern $p = \underline{132}$, then by Theorem 5.2.4 the function $F_{2,1}(x)$ (which is the same for the patterns p, $\underline{231}$, $\underline{312}$ and $\underline{213}$ because of the trivial bijections) can be written as:

$$F_{2,1}(x) = \sum_{i \geq 1} \frac{(-1)^{i+1}x^{ki+1}}{i!(k!)^i(ki+1)} = x - \int_0^x e^{-t^2/2}\, dt.$$

That is

$$E_{2,1} = \frac{1}{1 - \displaystyle\int_0^x e^{-t^2/2}\, dt},$$

which is recorded in Theorem 5.1.9 and is a special case of Theorem 5.1.8.

One more result that was proved in [509] using the inclusion-exclusion approach is the following theorem on a consecutive POP.

Theorem 5.2.5. ([509]) Let $1 < 2$ and $1' < 2'$ be the only order relations on the set $\{1, 2, 1', 2'\}$. Let $E_p(x)$ be the e.g.f. for the number of permutations that avoid the POP $p = \underline{122'1'}$. Then

$$E_p(x) = \frac{1}{2} + \frac{1}{4}\tan x(1 + e^{2x} + 2e^x \sin x) + \frac{1}{2}e^x \cos x.$$

Proof. Let $b_n = s_n(122'1')$ and $a_n = s_n(122'1',\lfloor 21)$; that is, a_n is the number of n-permutations that avoid p and begin with an ascent. Let $A(x)$ be the e.g.f. for the numbers a_n. We set $b_0 = a_0 = a_1 = 1$. Suppose $\pi \in \mathcal{S}_{n+1}(122'1')$. For $n \geq 1$, there are three mutually exclusive possibilities:

1) $\pi = (n+1)\pi_2$;

2) $\pi = \pi_1(n+1)$;

3) $\pi = \pi_1(n+1)\pi_2$ and $\pi_1, \pi_2 \neq \varepsilon$.

Obviously, in 1) and 2) the letter $(n+1)$ does not affect the rest of the permutation π in the sense that $(n+1)$ cannot be involved in an occurrence of the pattern $122'1'$, and thus in each of these cases we have b_n permutations that avoid p. In 3), it is easy to see that if π_1 has more than one letter then π_1 must end with the pattern $21\rfloor$, while if π_2 has more than one letter then π_2 must begin with the pattern $\lfloor 12$. By applying the reverse operation, the number of n-permutations that avoid p and end with the pattern $21\rfloor$ is the same as the number of n-permutations that avoid p and begin with the pattern $\lfloor 12$. So if $|\pi_1| = i$ then we can choose the letters of π_1 in $\binom{n}{i}$ ways and then choose a permutation π_1 in a_i ways and a permutation π_2 in a_{n-i} ways, since the letters of π_1 and π_2 do not affect each other. From this we get

$$b_{n+1} = 2b_n + \sum_{i=1}^{n-1}\binom{n}{i}a_i a_{n-i} = 2b_n + \sum_{i=0}^{n}\binom{n}{i}a_i a_{n-i} - 2a_n.$$

We multiply both sides of the last equality by $x^n/n!$ to obtain

$$b_{n+1}\frac{x^n}{n!} = 2b_n\frac{x^n}{n!} + \sum_{i=0}^{n}\frac{a_i}{i!}x^i \frac{a_{n-i}}{(n-i)!}x^{n-i} - 2a_n\frac{x^n}{n!}.$$

Summing both sides over all natural numbers n we obtain:

(5.3) $$B'(x) = 2B(x) + A^2(x) - 2A(x).$$

To solve this differential equation with the initial condition $B(0) = 1$, we need to determine $A(x)$. We can observe that if a permutation π avoids p and begins with the pattern $\lfloor 12$ then π has the structure $\pi = x_1 y_1 x_2 y_2 x_3 y_3 \cdots$, where $x_i < y_i$ for all i. Moreover, if $y_1 < x_2$ then we must have $x_1 < y_1 < x_2 < y_2 < x_3 < \cdots$ since otherwise we obviously have an occurrence of the pattern p. Our first approximation is that $a_n = \binom{n}{2}a_{n-2}$, because we can choose $x_1 y_1$ in π in $\binom{n}{2}$ ways and then pick an arbitrary $(n-2)$-permutation that avoids p and begins with the pattern $\lfloor 12$, to be

$x_2 y_2 x_3 y_3 \cdots$, in a_{n-2} ways. But it is possible that $y_1 < x_2$ in which case $y_1 x_2 y_2 x_3$ can be an occurrence of p in π, and it is an occurrence of p unless $x_2 < y_2 < x_3 < \cdots$. So in order to avoid this we must subtract the number of permutations of the form $abcd\pi'$, where $a < b < c < d$ and π' is any $(n-4)$-permutation that avoids p, from the first approximation of a_n. Thus the second approximation is that $a_n = \binom{n}{2}a_{n-2} - \binom{n}{4}a_{n-4}$. We observe that in the second approximation we do not count the increasing permutation $12 \cdots n$. Moreover, among the permutations counted by $\binom{n}{4}a_{n-4}$, there are permutations that begin with 6 increasing letters. Except for the increasing permutation, such permutations are not counted by $\binom{n}{2}a_{n-2}$. We must therefore add the number of such permutations. So the third approximation is that $a_n = \binom{n}{2}a_{n-2} - \binom{n}{4}a_{n-4} + \binom{n}{6}a_{n-6}$, and so on. That is,

$$(5.4) \qquad a_n = \sum_{i \geq 1} (-1)^{i+1} \binom{n}{2i} a_{n-2i}.$$

We observe that if $n = 4k$ or $n = 4k+1$ then we do not count the increasing permutation in our sum. This, together with Equation (5.4), gives us

$$\sum_{i \geq 0} (-1)^i \binom{n}{2i} a_{n-2i} = \begin{cases} 1, & \text{if } n = 4k \text{ or } n = 4k+1, \\ 0, & \text{if } n = 4k+2 \text{ or } n = 4k+3. \end{cases}$$

Multiplying both sides of the equality by $x^n/n!$ and summing over all natural numbers n we get

$$\left(a_0 + a_1 x + \frac{a_2}{2!}x^2 + \cdots \right)\left(1 - \frac{x^2}{2!} + \frac{x^4}{4!} - \frac{x^6}{6!} + \cdots \right) = \sum_{k=0}^{\infty} \left(\frac{x^{4k}}{(4k)!} + \frac{x^{4k+1}}{(4k+1)!} \right).$$

The left hand side of this equality is equal to $A(x)\cos x$. Let $F(x)$ be the function in the right hand side of the equality. Then it is easy to see that $F(x)$ is the solution to the differential equation $F^{(4)}(x) = F(x)$ with the initial conditions $F(0) = F'(0) = 1$, $F^{(2)}(0) = F^{(3)}(0) = 0$. So $F(x) = \frac{1}{2}(\cos x + \sin x + e^x)$ and

$$A(x) = \frac{1}{2}\left(1 + \tan x + \frac{e^x}{\cos x}\right).$$

Now we solve the differential equation (5.3) and get

$$B(x) = \frac{1}{2} + \frac{1}{4}\tan x (1 + e^{2x} + 2e^x \sin x) + \frac{1}{2}e^x \cos x.$$

$\square$

Remark 5.2.6. The series expansion of $B(x)$ in Theorem 5.2.5 begins with

$$B(x) = 1 + x + x^2 + x^3 + \frac{3}{4}x^4 + \frac{11}{20}x^5 + \frac{7}{20}x^6 + \frac{7}{30}x^7 + \frac{103}{720}x^8 + \cdots.$$

That is, the initial values for B_n are $1, 2, 6, 18, 66, 252, 1176, 5768, \ldots$.

5.3 The symmetric functions approach

The approach discussed in this section originates from understanding a connection between symmetric functions and the permutation enumeration of $\mathcal{S}_n$. A proper description of the method can be found in the upcoming book "Symmetric Functions and Generating Functions for Permutations and Words" by Mendes and Remmel [627]. Several PhD theses deal with the approach [402, 562, 626, 642, 692]; for relevant papers see [403, 523, 630, 628, 629, 689] and references therein.

5.3.1 Symmetric functions

Let Λ denote the *ring of symmetric functions* over infinitely many variables $x_1, x_2, \ldots$ with coefficients in the field of complex numbers $\mathbb{C}$. The *n-th elementary symmetric function* e_n in the variables $x_1, x_2, \ldots$ is given by

$$E(t) = \sum_{n \geq 0} e_n t^n = \prod_{i \geq 1} (1 + x_i t)$$

and the *n-th homogeneous symmetric function* h_n in the variables $x_1, x_2, \ldots$ is given by

$$H(t) = \sum_{n \geq 0} h_n t^n = \prod_{i \geq 1} \frac{1}{1 - x_i t}.$$

In the case of a finite number of variables, say N, we have $1 \leq i \leq N$ in the products above.

Example 5.3.1. We have

$$e_3(x_1, x_2, x_3, x_4) = x_1 x_2 x_3 + x_1 x_2 x_4 + x_1 x_3 x_4 + x_2 x_3 x_4;$$
$$e_3(x_1, x_2, \ldots) = x_1 x_2 x_3 + x_1 x_2 x_4 + \cdots + x_1 x_3 x_4 + x_1 x_3 x_5 + \cdots + x_2 x_3 x_4 + x_2 x_3 x_5 + \cdots;$$
$$h_3(x_1, x_2, x_3) = x_1 x_2 x_3 + x_1^2 x_2 + x_1^2 x_3 + x_2^2 x_1 + x_2^2 x_3 + x_3^2 x_1 + x_3^2 x_2 + x_1^3 + x_2^3 + x_3^3.$$

Furthermore, let

$$P(t) = \sum_{n \geq 0} p_n t^n$$

where $p_n = \sum_{i \geq 1} x_i^n$ is the *n-th power symmetric function.*

It is well-known that

(5.5)
$$H(t) = \frac{1}{E(-t)}$$

and

(5.6)
$$P(t) = \frac{\sum_{n \geq 1} (-1)^{n-1} n e_n t^n}{E(-t)}.$$

Let $\lambda = (\lambda_1, \lambda_2, \ldots, \lambda_\ell)$ be an integer partition, where unlike Definition 3.1.5 we assume that $0 < \lambda_1 \leq \lambda_2 \leq \cdots \leq \lambda_\ell$. Let $\ell(\lambda)$ denote the number of integers in λ. If the sum of these integers is n, we say that λ is a partition of n and write $\lambda \vdash n$. For any partition $\lambda = (\lambda_1, \lambda_2, \ldots, \lambda_\ell)$, let $e_\lambda = e_{\lambda_1} e_{\lambda_2} \cdots e_{\lambda_\ell}$. The well-known fundamental theorem on symmetric functions [454] says that $\{e_\lambda : \lambda \text{ is a partition}\}$ is a basis for Λ, alternatively that $\{e_0, e_1, \ldots\}$ is an algebraically independent set of generators for Λ. Thus, we can specify a ring homomorphism φ on Λ by simply defining $\varphi(e_n)$ for all $n \geq 0$. For example, the following ring homomorphism (to be used below) which maps Λ to the ring $\mathbb{Q}[x]$ can be defined:

(5.7)
$$\varphi(e_n) = \frac{(-1)^n}{n!} \sum_{i=0}^{\infty} (-1)^{n-i} \mathcal{R}_{n-1,i,k-1} x^i,$$

where $\mathcal{R}_{n,i,k}$ is the number of rearrangements of i 0s and $n - i$ 1s such that k 0s never appear consecutively.

A large number of results on generating functions for various permutation statistics in the literature can be derived by applying appropriate homomorphisms on Λ to simple identities such as (5.5) and (5.6). See Subsection 5.3.3 for examples of such results.

5.3.2   Brick tabloids

Definition 5.3.2. A *brick tabloid* of shape (n) and type $\lambda = (\lambda_1, \lambda_2, \ldots, \lambda_k)$ is a filling of a row of n squares of cells with bricks of lengths $\lambda_1, \lambda_2, \ldots, \lambda_k$ such that bricks do not overlap.

Example 5.3.3. A brick tabloid of shape (12) and type $(1, 1, 2, 3, 5)$ is shown in Figure 5.2.

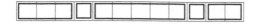

Figure 5.2: A brick tabloid of shape (12) and type $(1, 1, 2, 3, 5)$.

Let $\mathcal{B}_{\lambda,n}$ denote the set of all λ-brick tabloids of shape (n) and let $B_{\lambda,n} = |\mathcal{B}_{\lambda,n}|$. Through simple recursions using (5.5), Eğecioğlu and Remmel [370] proved that

$$(5.8) \qquad h_n = \sum_{\lambda \vdash n} (-1)^{n - \ell(\lambda)} B_{\lambda,n} e_\lambda.$$

Several theorems are proved in the literature using (5.8). To demonstrate the technique used to prove these, and similar theorems, we provide a theorem in [629], Theorem 5.3.4, with a proof (essentially taken from [629] by Mendes and Remmel).

Theorem 5.3.4. *([629])*

$$\sum_{n=0}^{\infty} \frac{t^n}{n!} \sum_{\sigma \in \mathcal{S}_n(k(k-1)\cdots 1)} x^{\operatorname{des}(\sigma)} = \left(\sum_{n=0}^{\infty} \frac{t^n}{n!} \sum_{n=0}^{\infty} (-1)^i \mathcal{R}_{n-1,i,k-1} x^i \right)^{-1}.$$

Proof. We will apply the homomorphism φ defined by (5.7) to $n! h_n$ and will describe the result by *decorating* brick tabloids. From (5.8), we have

$$n! \varphi(h_n) = n! \sum_{\lambda \vdash n} (-1)^{n - \ell(\lambda)} B_{\lambda,n} \varphi(e_\lambda)$$

$$= n! \sum_{\lambda \vdash n} (-1)^{n - \ell(\lambda)} B_{\lambda,n} \prod_{m=1}^{\ell(\lambda)} \frac{(-1)^{\lambda_m}}{\lambda_m!} \left(\sum_{i=0}^{\infty} (-1)^{\lambda_m - i} \mathcal{R}_{\lambda_m - 1, i, k-1} x^i \right)$$

$$(5.9) \qquad = \sum_{\lambda \vdash n} \binom{n}{\lambda_1, \lambda_2, \ldots, \lambda_{\ell(\lambda)}} (-1)^{\ell(\lambda)} B_{\lambda,n} \prod_{m=1}^{\ell(\lambda)} \left(\sum_{i=0}^{\infty} (-1)^{\lambda_m - i} \mathcal{R}_{\lambda_m - 1, i, k-1} x^i \right).$$

We now give a combinatorial interpretation for (5.9) in order to perform a sign-reversing, weight-preserving involution whose fixed points will correspond to permutations $\sigma \in \mathcal{S}_n(k(k-1)\cdots 1)$: Select a partition $\lambda \vdash n$ and use the $B_{\lambda,n}$ term in (5.9) to pick a brick tabloid of shape (n) and type λ. Using the multinomial coefficient, select λ_1 letters from $[n] = \{1, 2, \ldots, n\}$ to place in the first brick of length λ_1 in decreasing order, λ_2 of the remaining letters to place in a brick of length λ_2 in decreasing order, and so on, so that each letter appears once in the brick tabloid.

Now, for a part λ_m of λ, in (5.9) we have the factor $\sum_{i=0}^{\infty}(-1)^{\lambda_m-i}\mathcal{R}_{\lambda_m-1,i,k-1}x^i$. λ_m in the partition λ corresponds to a brick of length λ_m in the brick tabloid, and for this brick we do the following: Place a -1 in the terminal cell of the brick to use one power of (-1) from $(-1)^{\lambda_m-i}$; use the remaining (-1)s and select a rearrangement of i xs and λ_m-i-1 (-1)s that avoids $k-1$ consecutive xs. Place the rearrangement above the cells of the brick. Do the same with the other parts of λ. The only component in (5.9) we have not used so far is the factor of the form $(-1)^{\ell(\lambda)}$ which may be used to reverse the sign of the terminal -1 appearing in each block to obtain a decorated brick tabloid having the properties:

- each brick contains a subset of $[n]$ in decreasing order; no letter in a brick appears in another brick;

- the terminal cell in each brick is labeled with 1;

- the rest of the cells in each brick are labeled with a rearrangement of xs and (-1)s avoiding $k-1$ consecutive xs.

Since each brick ends with 1 and avoids $k-1$ consecutive xs, there are no $k-1$ consecutive xs in a decorated brick tabloid. Let $\mathcal{T}_\varphi$ be the set of decorated brick tabloids. See Figure 5.3 for an example of two elements of $\mathcal{T}_\varphi$ corresponding to $k=4$ and $n=12$.

Figure 5.3: Two decorated brick tabloids for $k=4$ and $n=12$.

Define the weight $w(T)$ of $T \in \mathcal{T}_\varphi$ to be the product of all of the powers of x and -1 in T. From our construction, $n!\varphi(h_n) = \sum_{T \in \mathcal{T}_\varphi} w(T)$. Note that the weight of a brick tabloid can be either positive or negative. We will define a sign-reversing involution preserving the powers of x on a brick tabloid to pair off every $T \in \mathcal{T}_\varphi$ with negative weight with another brick tabloid in $\mathcal{T}_\varphi$ with positive weight. Take $T \in \mathcal{T}_\varphi$ and scan it from left to right looking for the first case of either of the following two situations:

1. a cell labeled with -1, or

2. the integer in the last cell of a brick is larger than the integer in the cell to its right; thus, we have two consecutive bricks whose integer contents form a descent between the bricks.

If situation 1 appears first, break the brick labeled with -1 into two bricks immediately after the violation and change the -1 to 1. On the other hand, if situation 2 appears first, combine the two consecutive bricks and change the 1 label on the brick to a -1. This process flipping the sign on T is the desired involution. The images of the two decorated brick tabloids in Figure 5.3 under the involution are shown in Figure 5.4.

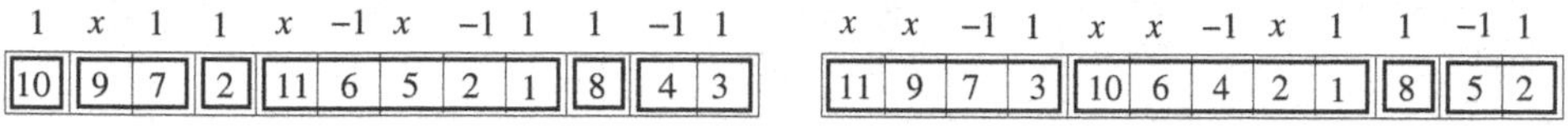

1		x	1	1		x	-1	x		-1	1		1		-1	1
10		9	7	2		11	6	5		2	1		8		4	3

x		x	-1	1		x	x	-1	x		1	1		-1	1
11		9	7	3		10	6	4	2		1		8		5	2

Figure 5.4: The images of the decorated brick tabloids in Figure 5.3 under the sign-reversing involution.

The fixed points of the involution are those $T \in \mathcal{T}_\varphi$ where no brick is longer than $k-1$ cells, there are no decreases between bricks, and there are no -1 labels. Clearly, each fixed point corresponds to a permutation in $\mathcal{S}_n(k(k-1)\cdots 1)$, and the number of x labels counts the number of descents in the corresponding permutation. An example of a fixed point is in Figure 5.5.

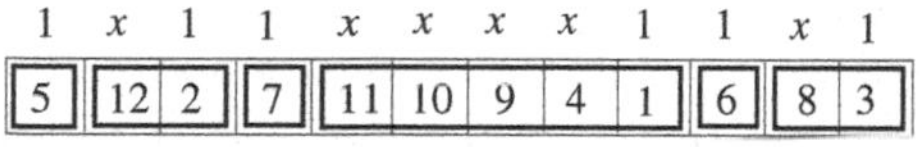

1		x	1	1		x	x	x	x	1	1		x	1
5		12	2	7		11	10	9	4	1		6		8	3

Figure 5.5: A fixed point under the sign-reversing involution.

What we have just shown is that

$$n!\varphi(h_n) = \sum_{T \in \mathcal{T}_\varphi} w(T) = \sum_{\substack{T \in \mathcal{T}_\varphi \text{ is a fixed point}}} w(T) = \sum_{\sigma \in \mathcal{S}_n(k(k-1)\cdots 1)} x^{\mathrm{des}(\sigma)}.$$

Using (5.5),

$$\sum_{n=0}^{\infty} \frac{t^n}{n!} \sum_{\sigma \in \mathcal{S}_n(k(k-1)\cdots 1)} x^{\mathrm{des}(\sigma)} = \varphi\left(\sum_{n=0}^{\infty} h_n t^n\right) = \varphi\left(\sum_{n=0}^{\infty} e_n(-t)^n\right)^{-1}$$

$$= \left(\sum_{n=0}^{\infty} \frac{t^n}{n!} \sum_{n=0}^{\infty} (-1)^i \mathcal{R}_{n-1,i,k-1} x^i\right)^{-1},$$

where in the last step we used the definition of φ and the fact that φ is a ring homomorphism. $\square$

Mendes and Remmel [629] show how to derive Theorem 5.2.1 from Theorem 5.3.4 by first substituting $x = 1$. Another corollary to Theorem 5.3.4 discussed in [629] is the following formula:

$$\sum_{n=0}^{\infty} \frac{t^n}{n!} \sum_{\sigma \in S_n(\underline{321})} x^{\mathrm{des}(\sigma)} = \frac{\exp(\frac{t}{2})}{\cos\left(\frac{t\sqrt{4x-1}}{2}\right) - \frac{1}{\sqrt{4x-1}} \sin\left(\frac{t\sqrt{4x-1}}{2}\right)}.$$

Compare the last formula with the following result that can be obtained by applying a suitable homomorphism to the identity (5.5):

$$\sum_{n=0}^{\infty} \frac{t^n}{n!} \sum_{\sigma \in S_n} x^{\mathrm{des}(\sigma)} = \frac{1 - x}{-x + \exp(t(x-1))}.$$

A q-analogue of the last formula is stated in Theorem 5.3.10.

5.3.3   Results obtained by using symmetric functions

In this subsection we state several results in the literature obtained by applying appropriate homomorphisms to simple identities like (5.5) and (5.6); more results on this will be mentioned in Chapter 9, where we will discuss other definitions of the notion of a pattern in permutations or words. Note that the results related to the statistic des are essentially results on the consecutive pattern $\underline{21}$ since $\mathrm{des}(\sigma) = \underline{21}(\sigma)$; also, $\mathrm{inv}(\sigma) = 21(\sigma)$ bringing the corresponding results to the domain of classical patterns. However, the formulas below are stated in terms of statistics inv, des, coinv, and others, to be consistent with the original sources.

See Definition 3.3.2 for the notions of $[n]_q$ and $[n]_q!$. We also need the standard q-analogue for the binomial coefficient $\binom{n}{k}$:

$$\begin{bmatrix} n \\ k \end{bmatrix}_q = \frac{[n]_q!}{[k]_q![n-k]_q!},$$

and p, q-analogues of n and $n!$, respectively:

$$[n]_{p,q} = \frac{p^n - q^n}{p - q} \quad \text{and} \quad [n]_{p,q}! = [n]_{p,q}[n-1]_{p,q} \cdots [1]_{p,q}.$$

By convention, $[0]_{p,q} = 0$ and $[0]_{p,q}! = 1$. A p, q-analogue for the exponential function is defined by

$$e_{p,q}^t = \sum_{n \geq 0} \frac{t^n}{[n]_{p,q}!} q^{\binom{n}{2}}.$$

In addition, let $(x, y; p, q)_0 = 1$ and

$$(x, y; p, q)_n = (x - y)(xp - yq) \cdots (xp^{n-1} - yq^{n-1}).$$

Definition 5.3.5. For permutations $\sigma, \tau \in \mathcal{S}_n$, the value of the statistic $\mathrm{comdes}(\sigma, \tau)$ is the number of positions in which *both* σ and τ have descents. Also, $\mathrm{coinv}(\sigma)$ is the number of pairs $1 \leq i < j \leq n$ such that $\sigma_i < \sigma_j$, that is, $\mathrm{coinv}(\sigma) = 12(\sigma)$.

Example 5.3.6. $\mathrm{coinv}(2413) = 3$, while $\mathrm{comdes}(2534176, 4317625) = 2$ since only positions 2 and 4 are common descent positions for the permutations.

Often, the statistics ris and asc (see Table A.1 for definitions) have the same definition; however, for this section, let $\mathrm{ris}(\sigma) = 1 + \mathrm{asc}(\sigma)$.

Definition 5.3.7. For a permutation σ, $\mathrm{comaj}(\sigma)$ is defined as the length of σ plus the sum of the positions in which ascents occur.

Example 5.3.8. $\mathrm{comaj}(4623715) = 7 + 1 + 3 + 4 + 6 = 21$ since ascents are in positions 1, 3, 4 and 6.

The following theorem was proved by Carlitz [223].

Theorem 5.3.9. *([223])*

$$\sum_{n=0}^{\infty} \frac{t^n}{(n!)^2} \sum_{(\sigma,\tau) \in \mathcal{S}_n \times \mathcal{S}_n} x^{\mathrm{comdes}(\sigma,\tau)} = \frac{1-x}{-x + \sum_{n=0}^{\infty} \frac{(t(x-1))^n}{(n!)^2}}.$$

The formulas in the following theorem are due to Stanley [739].

Theorem 5.3.10. *([739])*

$$\sum_{n=0}^{\infty} \frac{t^n}{[n]!} \sum_{\sigma \in \mathcal{S}_n} x^{\mathrm{des}(\sigma)} q^{\mathrm{inv}(\sigma)} = \frac{1-x}{-x + \sum_{n=0}^{\infty} \frac{(t(x-1))^n}{[n]_q!} q^{\binom{n}{2}}};$$

$$\sum_{n=0}^{\infty} \frac{t^n}{[n]!} \sum_{\sigma \in \mathcal{S}_n} x^{\mathrm{des}(\sigma)} q^{\mathrm{coinv}(\tau)} = \frac{1-x}{-x + \sum_{n=0}^{\infty} \frac{(t(x-1))^n}{[n]_q!}}.$$

Fédou and Rawlings [378] proved the following theorem.

Theorem 5.3.11. *([378])*

$$\sum_{n=0}^{\infty} \frac{t^n}{[n]_q! [n]_p!} \sum_{(\sigma,\tau) \in \mathcal{S}_n \times \mathcal{S}_n} x^{\mathrm{comdes}(\sigma,\tau)} q^{\mathrm{inv}(\sigma)} p^{\mathrm{inv}(\tau)} = \frac{1-x}{-x + \sum_{n=0}^{\infty} \frac{(t(x-1))^n}{[n]_q! [n]_p!} q^{\binom{n}{2}} p^{\binom{n}{2}}}.$$

The following theorem is due to Mendes and Remmel [630] (see Definition 3.3.4 for the notion of maj).

Theorem 5.3.12. *([630])*

$$\sum_{n\geq 0} \frac{t^n}{[n]_{p,q}!(y,x;v,u)_{n+1}} \sum_{\sigma\in S_n} x^{\mathrm{des}(\sigma)} y^{\mathrm{ris}(\sigma)} u^{\mathrm{maj}(\sigma)} v^{\mathrm{comaj}(\sigma)} q^{\mathrm{inv}(\sigma)} p^{\mathrm{coinv}(\sigma)}$$

$$= \sum_{k\geq 0} \frac{x^k}{y^{k+1} e_{p,q}^{-t(u/v)^0} e_{p,q}^{-t(u/v)^1} \cdots e_{p,q}^{-t(u/v)^k}}.$$

Among other results on consecutive patterns, Mendes and Remmel [629] used the symmetric functions approach to express the e.g.f. for the number of permutations avoiding an arbitrary consecutive pattern p:

$$\sum_{n\geq 0} s_n(p) \frac{t^n}{n!} = \frac{1}{1 - t + \sum_{n\geq 2} \frac{t^n}{n!} \sum_{w\in J_p,\ ||w||=n} (-1)^{\bar{\ell}(w)} |\mathcal{P}_w^p|},$$

where $||w|| = |p| + w_1 + w_2 + \cdots + w_n$, J_p is a certain collection of words associated with p, and $\mathcal{P}_w^p$ is a certain collection of permutations $\sigma \in S_{||w||}$; we refer to [629] for precise definitions. The main goal of [563] by Liese and Remmel is to show how to compute explicitly

$$\sum_{n\geq 2} \frac{t^n}{n!} \sum_{w\in J_p,\ ||w||=n} (-1)^{\bar{\ell}(w)} |\mathcal{P}_w^p|$$

in several cases. For example, Liese and Remmel [563] consider various classes of consecutive patterns which are shuffles of the increasing permutation $12\cdots n$ with an arbitrary permutation of the numbers $\{n+1, n+2, \ldots, n+m\}$. Moreover, the following theorem is proved in [563].

Theorem 5.3.13. *([563]) The e.g.f. for the number of permutations avoiding the pattern $\underline{1324}$ is given by $\frac{1}{1-t-f(t)}$, where*

$$f(t) = F\left(\frac{t\left(1 + 2t^2 + 2t^3 - \sqrt{1+4t^2}\right)}{2(1 + t + 2t^2 + t^3)}\right)$$

and F is the operator on formal power series $A(t) = \sum_{n\geq 0} a_n t^n$ such that $F(A(x)) = \sum_{n\geq 0} a_n \frac{t^n}{n!}$.

One more example of results in [563] is the following theorem, which is a special case of a more general result.

Theorem 5.3.14. *([563]) The e.g.f. for the number of permutations avoiding the pattern $\underline{1\tau 2}$, where τ is a permutation of $3, 4, \ldots, k+2$ is*

$$\frac{1}{1 - \int_0^t e^{-z^{k+1}/(k+1)!} dz}.$$

Remark 5.3.15. Theorem 5.3.14 includes the results of Elizalde and Noy in [360] on enumeration of $\mathrm{Av}(\underline{132})$ and $\mathrm{Av}(\underline{1342})$ (see Theorems 5.1.9 and 5.1.12, respectively).

5.4 The cluster and chain methods

For two consecutive patterns, p_1 and p_2, we say that $p_1 \leq p_2$ if p_1 occurs in p_2 viewed as a permutation (the underline is ignored). Clearly, this defines a partial order on the set of all consecutive patterns. In this subsection when avoiding a set P of consecutive patterns, without loss of generality, we assume that the patterns in P form an *antichain*. Indeed, if $p_1 \leq p_2$ and $p_1, p_2 \in P$, then we can remove p_2 from P without changing the set of P-avoiding permutations, since if a permutation avoids p_1 then it avoids p_2.

The *cluster method* of Goulden and Jackson [421, 645], based on the inclusion-exclusion approach discussed in Section 5.2, is a powerful tool to deal with consecutive patterns. The notion of a *cluster* is defined for both permutations and words, and roughly speaking it is a way to connect together several patterns to form a given set. The following definition deals with clusters in permutations.

Definition 5.4.1. Let P be a set of consecutive patterns and $\sigma = \sigma_1 \sigma_2 \cdots \sigma_n$ be a permutation. A *q-cluster* with respect to P and σ is a triple

$$(\sigma; \pi_1, \pi_2, \ldots, \pi_q; i_1, i_2, \ldots, i_q)$$

such that

- for every $j = 1, 2, \ldots, q$, $\mathrm{red}(\sigma_{i_j} \sigma_{i_j+1} \cdots \sigma_{i_j+\ell_j-1}) = \pi_j$, where ℓ_j is the length of π_j;

- $i_{j+1} > i_j$ and $i_{j+1} < i_j + \ell_j$ (the patterns π_j and π_{j+1} overlap in σ);

- $i_1 = 1$, and $i_q + \ell_q - 1 = n$ (σ is completely covered by the patterns π_1, $\pi_2, \ldots, \pi_q$).

Example 5.4.2. Let $P = \{\underline{123}, \underline{132}, \underline{321}, \underline{2143}, \underline{1423}\}$ and $\sigma = 42651873$. Then

- $(42651873; \underline{2143}, \underline{321}, \underline{132}, \underline{321}; 1, 3, 5, 6)$ is a 4-cluster and

- $(42651873; \underline{2143}, \underline{2143}, \underline{321}; 1, 4, 6)$ is a 3-cluster with respect to P.

Given a set of consecutive patterns P, let $a_{n,k} = a_{n,k}(P)$ denote the number of n-permutations with exactly k occurrences of patterns from P, and let $\mathrm{cl}_{n,q} = \mathrm{cl}_{n,q}(P)$ denote the number of q-clusters for n-permutations. Also, let

$$\Pi_P(x, t) = \sum_{n,k \geq 0} a_{n,k} \frac{x^n}{n!} t^k \quad \text{and} \quad \mathrm{Cl}_P(x, t) = \sum_{n,k \geq 0} \mathrm{cl}_{n,q} \frac{x^n}{n!} t^q.$$

The following theorem by Goulden and Jackson [421] shows a relation between $\Pi_P(x, t)$ and $\mathrm{Cl}_P(x, t)$. This theorem is a basis for the algorithmic considerations of consecutive pattern-avoidance in permutations in [641] by Nakamura.

Theorem 5.4.3. *([421]) For a set of consecutive patterns P, we have*

$$\Pi_P(x, t) = \frac{1}{1 - x - \mathrm{Cl}_P(x, t - 1)},$$

and thus, the e.g.f. for $\mathrm{Av}(P)$ is given by

(5.10)
$$\frac{1}{1 - x - \mathrm{Cl}_P(x, -1)}.$$

To improve formula (5.10), Dotsenko and Khoroshkin [310] introduced the notion of *q-chains* which, unlike *q*-clusters, take into account possible self-overlaps of patterns. The idea here is that in general, a permutation σ can occur in several *q*-clusters for different *q*s, which will result in cancelations in $\mathrm{Cl}(x, -1)$. This approach is influenced by the *homological algebra* considerations by Anick [40] for pattern-avoidance in words, and it is based on *free resolutions of Anick type* for *shuffle algebras* defined by generators and relations (see [702]).

Definition 5.4.4. We inductively define *q-chains* and their *tails* as follows:

- the empty permutation ε is the 0-chain and it coincides with its tail;

- the permutation 1 is a 1-chain and it also coincides with its tail;

- a *q*-chain is a permutation $\sigma = \sigma_1 \sigma_2$, where $\mathrm{red}(\sigma_1)$ is a $(q-1)$-chain and σ_2 is the tail of σ;

- if we denote the tail of σ_1 as σ_1' in the decomposition above, then there exists a "factorization" $\sigma_1' = \alpha\beta$ with $\mathrm{red}(\beta\sigma_2) \in P$, and the factor $\beta\sigma_2$ is the only occurrence of a pattern from P in $\sigma_1'\sigma_2$.

Basically, $(q+1)$-chains are *q*-clusters with additional restrictions:

- if π_i and π_j overlap in the permutation, then either $i = j+1$ or $j = i+1$ (only neighbors overlap);

- the permutation covered by $\pi_1, \pi_2, \ldots, \pi_q$ forms a *q*-chain;

- no proper left factor of the permutation (beginning of the permutation not coinciding with the whole permutation) forms a $(q+1)$-chain.

Example 5.4.5. If $P = \{\underline{12}\}$, the only *n*-chain, for $n \geq 1$, is $12\cdots(n+1)$. If $P = \{\underline{123}\}$, we can see that 123 is the only 1-chain, while 1234 is the only 2-chain. However, 12345 is *not* a 2-chain because it begins with the 2-chain 1234, and it is not a 3-chain because in the only tiling of 12345 by three copies of the patten in P, its first and third occurrences overlap:

$$1\,2\,\overbrace{3\,4}\,5.$$

Proposition 5.4.6. *([310]) If σ is an n-chain, then the choice of π_1, $\pi_2,\ldots$ is unique, that is, there is a unique way to cover σ by properly overlapping patterns from a given set of consecutive patterns.*

Let $\mathrm{Ch}_{n,k}(P)$ denote the number of k-chains in $\mathcal{S}_n$. The following theorem provides an alternative expression to (5.10) for consecutive pattern-avoidance.

Theorem 5.4.7. *([310]) For a set of consecutive patterns P, the e.g.f. for $\mathrm{Av}(P)$ is given by*

$$\frac{1}{\sum_{n,k\geq 0}(-1)^k\,\mathrm{Ch}_{n,k}(P)\frac{t^n}{n!}}.$$

Remark 5.4.8. In some cases, the number of n-chains is substantially smaller than the number of n-clusters. For example, this happens if the set of forbidden consecutive patterns P contains the pattern $\underline{12\cdots k}$. See [310] for examples.

Using the approach suggested by Dotsenko and Khoroshkin in [310], Khoroshkin and Shapiro [500] presented a sufficient condition for two sets of consecutive patterns to have the same distribution on permutations, which was discovered independently by Nakamura [641]. The idea here is to guarantee that the corresponding e.g.f.s match by exhibiting *length-preserving* bijections with certain properties between the sets of patterns. To state the main result in [500] precisely (see Theorem 5.4.12 below), we need the following definition.

Definition 5.4.9. Let τ_1 and τ_2 be consecutive patterns of length ℓ_1 and ℓ_2, respectively. A permutation $\sigma = \sigma_1\sigma_2\cdots\sigma_n$ is called a *linkage* of the (ordered) pair (τ_1,τ_2) if

1. $n < \ell_1 + \ell_2$;

2. $\mathrm{red}(\sigma_1\sigma_2\cdots\sigma_{\ell_1}) = \tau_1$ and $\mathrm{red}(\sigma_{n-\ell_2+1}\sigma_{n-\ell_2+2}\cdots\sigma_n) = \tau_2$.

Remark 5.4.10. Note that the notion of a linkage is related to the notion of a 2-cluster.

Example 5.4.11. The permutation 25341 is a linkage of $(\underline{132},\underline{4231})$.

Theorem 5.4.12. *([500, 641]) Two sets of patterns, P_1 and P_2, have the same distribution on permutations if there exists a bijection $\varphi : P_1 \to P_2$ preserving the following three properties:*

- (lengths) *For any $\tau \in P_1$, length$(\tau) =$ length$(\varphi(\tau))$;*

- (linkages) *For $\tau_1, \tau_2 \in P_1$, the pair (τ_1, τ_2) has a linkage of length n if and only if the pair $(\varphi(\tau_1), \varphi(\tau_2))$ has a linkage of length n;*

- (overlapping sets) *Let $p_1 = a_1 a_2 \cdots a_{\ell_1} \in P_1$ and $p_2 = b_1 b_2 \cdots b_{\ell_2} \in P_1$. For any $k \le \min(\ell_1, \ell_2)$, if $\mathrm{red}(a_{\ell_1-k+1} a_{\ell_1-k+2} \cdots a_{\ell_1}) = \mathrm{red}(\overline{b_1 b_2 \cdots b_k})$ then*

  - $\{a_{\ell_1-k+1}, a_{\ell_1-k+2}, \dots, a_{\ell_1}\} = \{\varphi(a_{\ell_1-k+1}), \varphi(a_{\ell_1-k+2}), \dots, \varphi(a_{\ell_1})\}$ *and*
  - $\{b_1, b_2, \dots, b_k\} = \{\varphi(b_1), \varphi(b_2), \dots, \varphi(b_k)\}$.

As illustrations of applications of Theorem 5.4.12, patterns from the following sets are shown in [500] to be strongly Wilf-equivalent (to have the same distribution):

$$\{\underline{1734526}, \underline{1735426}, \underline{1743526}, \underline{1745326}, \underline{1753426}, \underline{1754326}\}, \ \{\underline{143265987}, \underline{134265897}\},$$

$$\{\{\underline{145623}, \underline{13452}\}, \{\underline{145623}, \underline{13542}\}, \{\underline{146523}, \underline{13452}\}, \{\underline{146523}, \underline{13542}\}\}.$$

5.4.1 Applications of the methods

Cluster and chain methods are used in [310] by Dotsenko and Khoroshkin to rederive several known results, as well as to prove the following four theorems extending Theorem 5.1.12, in the case of avoidance, to the remaining cases of consecutive patterns of length 4. A common feature for all patterns of length 4, except for the patterns $\underline{1234}$ and $\underline{4321}$, is that they only have self-overlaps of lengths 1 and 2, so patterns in a cluster overlap only if they are neighbors. In this case, as is explained in [310], chains actually coincide with clusters. We note that applying the methods often involves dealing with linear extensions of posets.

Theorem 5.4.13. *([310]) The e.g.f. for* $\mathrm{Av}(\underline{1324})$ *is given by*

$$\frac{1}{1 - x - \sum_{n \ge 2, \ell \ge 1} \frac{\mathrm{cl}_{n,\ell}\, x^n (t-1)^\ell}{n!}},$$

where the cluster numbers $\mathrm{cl}_{n,\ell}$ *satisfy the following recurrence relation*

$$\mathrm{cl}_{n,\ell} = \sum_{4 \le 2k+2 \le n} \frac{1}{k+1} \binom{2k}{k} \mathrm{cl}_{n-2k-1,\ell-k}$$

with initial conditions $\mathrm{cl}_{1,0} = 1$, $\mathrm{cl}_{1,\ell} = 0$ *for* $\ell > 0$, *and* $\mathrm{cl}_{2,\ell} = \mathrm{cl}_{3,\ell} = 0$.

Remark 5.4.14. The e.g.f. for $\mathrm{Av}(\underline{1324})$ in Theorem 5.4.13 is given in Theorem 5.3.13 in a different form.

Theorem 5.4.15. *([310]) The e.g.f. for* $\mathrm{Av}(\underline{1423})$ *is given by*

$$\frac{1}{1 - x - \sum_{n \geq 2, \ell \geq 1} \frac{\mathrm{cl}_{n,\ell}\, x^n (t-1)^\ell}{n!}},$$

where the cluster numbers $\mathrm{cl}_{n,\ell}$ *satisfy the following recurrence relation*

$$\mathrm{cl}_{n,\ell} = \sum_{4 \leq 2k+2 \leq n} \binom{n-k-2}{k} \mathrm{cl}_{n-2k-1,\ell-k}$$

with initial conditions $\mathrm{cl}_{1,0} = 1$, $\mathrm{cl}_{1,\ell} = 0$ *for* $\ell > 0$, *and* $\mathrm{cl}_{2,\ell} = \mathrm{cl}_{3,\ell} = 0$.

Theorem 5.4.16. *([310]) The e.g.f. for* $\mathrm{Av}(\underline{2143})$ *is given by*

$$\frac{1}{1 - x - \sum_{n \geq 2, \ell \geq 1} \frac{\mathrm{cl}_{n,\ell}\, x^n (t-1)^\ell}{n!}},$$

where the cluster numbers $\mathrm{cl}_{n,\ell}$ *are given by*

$$\mathrm{cl}_{n,\ell} = \sum_{2 \leq p < n-2} \mathrm{cl}_{n,\ell}(p)$$

and the numbers $\mathrm{cl}_{n,\ell}(p)$ *satisfy the following recurrence relation*

$$\mathrm{cl}_{n,\ell}(p) = \sum_{4 \leq 2k+2 \leq q \leq n} \binom{q-p-1}{2k-2}(p-1)(n-q)\,\mathrm{cl}_{n-2k-1,\ell-k}(q-2k)$$

with initial conditions $\mathrm{cl}_{1,0}(1) = 1$, $\mathrm{cl}_{1,\ell}(p) = 0$ *for* $\ell \neq 0$ *or* $p \neq 1$, *and* $\mathrm{cl}_{2,\ell}(p) = \mathrm{cl}_{3,\ell}(p) = 0$.

Theorem 5.4.17. *([310]) The e.g.f. for* $\mathrm{Av}(\underline{2413})$ *is given by*

$$\frac{1}{1 - x - \sum_{n \geq 2, \ell \geq 1} \frac{\mathrm{cl}_{n,\ell}\, x^n (t-1)^\ell}{n!}},$$

where the cluster numbers $\mathrm{cl}_{n,\ell}$ *are given by the formula*

$$\mathrm{cl}_{n,\ell} = \sum_{1 < p < q-1 < n} \mathrm{cl}_{n,\ell}(p,q)$$

and the numbers $\mathrm{cl}_{n,\ell}(p,q)$ *satisfy the following recurrence relation*

$$\mathrm{cl}_{n,\ell}(p,q) = \sum_{r < p < s < q} \mathrm{cl}_{n-2,\ell-1}(r, s-1) + \sum_{p < r < s < q} (p-1)\,\mathrm{cl}_{n-3,\ell-1}(r-1, s-1)$$

$$+ \sum_{p < r < q < s} (p-1)\,\mathrm{cl}_{n-3,\ell-1}(r-1, s-2)$$

with initial conditions $\mathrm{cl}_{2,\ell}(p,q) = \mathrm{cl}_{3,\ell}(p,q) = 0$, $\mathrm{cl}_{4,1}(2,4) = 1$ *and* $\mathrm{cl}_{4,\ell}(p,q) = 0$ *if* $\ell \neq 1$ *or* $p \neq 2$ *or* $q \neq 4$.

5.4.2   Patterns without self-overlaps

Yet another application of the cluster method is in proving Theorem 5.4.23 below, which was conjectured by Elizalde in [353]. An alternative proof of Theorem 5.4.23 is due to Remmel and Duane [312] and it uses the symmetric functions approach discussed in Section 5.3.

Definition 5.4.18. A consecutive pattern p of length k is said to be without *self-overlaps* if there does not exist a permutation of length at most $2k - 2$ containing at least two occurrences of p. Such patterns are also called *minimal overlapping*.

Remark 5.4.19. Clearly, for any pattern p of length k there exists a permutation of length $2k - 1$ with at least two occurrences of p.

Example 5.4.20. $\underline{132}$, $\underline{1243}$, $\underline{12354}$, and, in general, $12 \cdots (n-2)n(n-1)$ are patterns without self-overlaps, while $\underline{1324}$ is not such a pattern, since the permutation 132546 of length $6 < 2 \cdot 4 - 1$ contains two occurrences of the pattern.

Definition 5.4.21. Given a consecutive pattern p of length k without self-overlaps, we say that a permutation $\sigma \in \mathcal{S}_{k+s(k-1)}$ is a *maximum packing* for p if and only if σ has occurrences of p starting at positions $1, k, 2k - 1, 3k - 2, \ldots, sk - (s-1)$. Let $\mathcal{MP}_{p,k+s(k-1)}$ be the set of $(k + s(k-1))$-permutations σ such that σ is a maximum packing for p and

$$MP_{p,k+s(k-1)}(q) = \sum_{\sigma \in \mathcal{MP}_{p,k+s(k-1)}} q^{\mathrm{inv}(\sigma)}.$$

Example 5.4.22. The permutation $\sigma = 1627354$ is a maximum packing for $\underline{132}$ ($k = 3$ and $s = 2$).

Theorem 5.4.23. *([310, 500, 312]) For a consecutive pattern $p = p_1 p_2 \cdots p_k$ without self-overlaps, the number of permutations of length n with m occurrences of p depends only on k, p_1 and p_k. That is, two minimal overlapping patterns of length k have the same distribution (in particular, they are Wilf-equivalent) if their first and last letters are the same.*

Another result on non-self-overlapping patterns, stated in the following theorem, is due to Remmel and Duane [312].

Theorem 5.4.24. *([312]) For a consecutive pattern p of length k without self-overlaps, we have*

$$\sum_{n \geq 0} \frac{t^n}{n!} \sum_{\sigma \in \mathcal{S}_n} x^{p(\sigma)} q^{\mathrm{inv}(\sigma)} = \frac{1}{1 - \left(t + \sum_{s \geq 0} \frac{t^{k+s(k-1)}}{[k+s(k-1)]_q!} (x-1)^{s+1} MP_{p,k+s(k-1)}(q) \right)}.$$

Setting $q = 1$ in Theorem 5.4.24 we obtain the e.g.f. for the number of permutations avoiding a consecutive pattern without self-overlaps:

$$\frac{1}{1 - \left(t + \sum_{s \geq 0} \frac{t^{k+s(k-1)}}{(k+s(k-1))!}(x-1)^{s+1}|\mathcal{MP}_{p,k+s(k-1)}|\right)}.$$

According to [312], $|\mathcal{MP}_{p,k+s(k-1)}|$ can be computed in many cases of consecutive patterns $p = p_1 p_2 \cdots p_k$ without self-overlaps. For example, if $p_1 = 1$ and $p_k = m$ then for $s \geq 1$,

$$|\mathcal{MP}_{p,k+s(k-1)}| = \prod_{i=1}^{s} \binom{k+i(k-1)-m}{k-m}.$$

5.5   Applications of Fédou's bijection

Fédou's insertion-shift bijection is described in Subsection 5.12.1 below in connection with a hierarchy on the consecutive patterns enumeration problems.

Definition 5.5.1. For a set P of consecutive patterns, let

$$A_{n,P}(q, \mathbf{t}) = \sum_{\sigma \in \mathcal{S}_n} q^{\text{inv}(\sigma)} \prod_{p \in P} t_p^{p(\sigma)},$$

where $\mathbf{t}$ is the vector of t_is indexed by permutations in P.

Rawlings [671] presented a new method, based on Fédou's bijection and a result of Goulden and Jackson, for finding $A_{n,P}(q, \mathbf{t})$. In particular, a number of q-analogues of several general theorems on consecutive patterns stated above were obtained. In this subsection we provide selected results from [671]. Review Definition 3.3.2 for the notions of $[n]_q$ and $[n]_q!$.

Theorem 5.5.2. *([671]) For $m \geq 3$ and $t = t_p$, the q-exponential generating function for permutations by the consecutive pattern $p = \underline{12 \cdots (m-2)m(m-1)}$ and by the inversion number is*

$$\sum_{n \geq 0} \frac{A_{n,p}(q,t)x^n}{[n]_q!} = \left(1 - \int \sum_{k \geq 0} \frac{(q(t-1)x^{m-1})^k \prod_{j=1}^{k-1}[j(m-1)+1]_q}{[k(m-1)]_q!} dx_q\right)^{-1}.$$

As a corollary to Theorem 5.5.2, we have the following formula, which, for $q = 1$, gives $Q(u, x)$ in Theorem 5.1.9.

$$\sum_{n \geq 0} \frac{A_{n,\underline{132}}(q,t)x^n}{[n]_q!} = \left(1 - \int \sum_{k \geq 0} \frac{(q(t-1)x^2)^k}{[2]_q[4]_q \cdots [2k]_q} dx_q\right)^{-1}.$$

Theorem 5.5.3. *([671]) For $m \geq 3$ and $t = t_p$, the q-exponential generating function for permutations by the consecutive pattern $p = \underline{1m(m-1)\cdots 2}$ and by the inversion number is*

$$\sum_{n \geq 0} \frac{A_{n,p}(q,t)x^n}{[n]_q!} = \left(1 - \int \sum_{k \geq 0} \frac{(q^{\binom{m-1}{2}}(t-1)x^{m-1})^k}{([m-2]_q!)^k \prod_{i=1}^{k}[(m-1)i]_q} dx_q\right)^{-1}.$$

The case $m = 4$ and $q = 1$ of Theorem 5.5.3 is due to Elizalde and Noy [360] (see the first case in Theorem 5.1.12 for an equivalent pattern).

Rawlings [671] also derived $A_{n,P}(q,\mathbf{t})$ for $P = \{\underline{12\cdots m}\}$ and $P = \{\underline{123},\underline{132}\}$.

5.6   Walks in graphs of overlapping patterns

In this section we provide yet another approach, suggested by Avgustinovich and Kitaev [61], to derive the distribution of a consecutive pattern. It is hard to say how useful this approach is, since no research seems to have been done in this direction. However, the approach has motivated the study of reconstructing permutations from their *walks* in graphs of overlapping patterns and we discuss it in this section. Just as with the cluster and chain methods, this method's reconstruction problem has an intimate connection to linear extensions of posets.

The approach is based on considering the *graph of overlapping patterns*, which is similar to the *de Bruijn graph* studied widely in the literature (see Definition 8.6.4), primarily in connection with *combinatorics on words* and *graph theory*. The graph of overlapping patterns is the graph $G_\emptyset$ of overlapping permutations mentioned in Subsection 5.7.2 (see Definition 5.7.15 and Example 5.7.16). We let $\mathcal{P}_k$ denote the graph of overlapping permutations of *order k*, that is, the nodes of $\mathcal{P}_k$ are permutations in $\mathcal{S}_k$.

Suppose we are interested in the number of occurrences of a consecutive pattern τ of length k in a permutation π of length n. To find this number, we scan π from left to right with a "window" of length k, that is, we consider factors $P_i = \pi_i\pi_{i+1}\cdots\pi_{i+k-1}$ for $i = 1, 2, \ldots, n-k+1$: if we meet an occurrence of τ, we register it. Each P_i forms a pattern of length k, and the procedure of scanning π gives us a *walk* (in the graph-theoretical terminology) in $\mathcal{P}_k$. Thus, for any n-permutation there is a walk in $\mathcal{P}_k$ of length $n-k+1$ corresponding to it. For example, if $k = 3$ then to the permutation 13542 there corresponds the walk $123 \to 132 \to 321$ in $\mathcal{P}_3$. It is also clear that every walk in the graph corresponds to at least one permutation.

The approach in [61] to study the distribution of a consecutive pattern τ of length k among n-permutations is to take $\mathcal{P}_k$ and to consider all walks of length

$n - k + 1$ passing through the node τ exactly ℓ times, where $\ell = 0, 1, \ldots, n - k + 1$. Then we can try to count the permutations corresponding to the walks. Similarly, for the "avoidance problem" we proceed as follows: given a set of patterns of length k to avoid, we remove the corresponding nodes and corresponding arcs from $\mathcal{P}_k$, then we consider all the walks of a certain length in the obtained graph, and then count the permutations of interest. The prohibited patterns could have length less than k as well, in which case we remove from $\mathcal{P}_k$ those permutations containing prohibited patterns.

However, a complication with this approach is that a permutation is not necessarily uniquely reconstructible from the walk corresponding to it. For example, the permutation 13542 above has the same walk in $\mathcal{P}_3$ corresponding to it as the permutations 23541 and 12543. Thus, different walks in $\mathcal{P}_k$ may have different contributions to the number of permutations with the required properties; in particular, some of the walks in $\mathcal{P}_k$ have exactly one permutation corresponding to them. Such permutations are called *uniquely k-determined* in [61] or just *k-determined* for brevity. The study of these permutations is the main concern of [61] by Avgustinovich and Kitaev, and it may be considered as the first step in understanding how to apply the approach to the distribution problem. Below, we discuss known results on uniquely k-determined permutations.

Two criteria for permutations to be uniquely k-determined are given in [61]. We state just one of them in Theorem 5.6.3 below.

Definition 5.6.1. Suppose $\pi = \pi_1 \pi_2 \cdots \pi_n$ is a permutation and $i < j$. The *distance* $d_\pi(\pi_i, \pi_j) = d_\pi(\pi_j, \pi_i)$ between the elements π_i and π_j is $j - i$.

Example 5.6.2. $d_{253164}(5, 6) = d_{253164}(6, 5) = 3$.

Theorem 5.6.3. *([61]) An n-permutation π is k-determined if and only if for each $1 \le i < n$, the distance $d_\pi(i, i + 1) \le k - 1$.*

Corollary 5.6.4. *([61]) An n-permutation π is not k-determined if and only if there exists a letter i, $1 \le i < n$, such that $d_\pi(i, i + 1) \ge k$.*

So, to determine if a given n-permutation is k-determined, all we need to do is to check the distances for $n - 1$ pairs of numbers: $(1, 2)$, $(2, 3)$,..., $(n - 1, n)$. Note that the language of determined k-permutations is *factorial* in the sense that if $\pi_1 \pi_2 \cdots \pi_n$ is k-determined, then so is $\mathrm{red}(\pi_i \pi_{i+1} \cdots \pi_j)$ for any $i \le j$. Coming back to the permutation 13542 above and using Corollary 5.6.4, we see why this permutation is not 3-determined ($k = 3$): the distance $d_{13542}(2, 3) = 3 = k$.

One has the following lower and upper bounds for the number $A_{k,n}$ of k-determined n-permutations.

Theorem 5.6.5. *([61]) We have $2((k-1)!)^{\lfloor n/k \rfloor} < A_{k,n} < 2(2(k-1))^n$.*

The set of k-determined n-permutations can be described by the language of prohibited patterns $\mathcal{L}'_{k,n}$ as follows. Using Theorem 5.6.3, we can describe the set of k-determined n-permutations by prohibiting patterns of the forms $xX(x+1)$ and $(x+1)Xx$, where X is a permutation on $\{1, 2, \ldots, |X|+2\} - \{x, x+1\}$ ($|X|$ is the number of letters in X), the length of X is at least $k-1$, and $1 \le x \le 1 + |X| < n$. We collect all such patterns in the set $\mathcal{L}'_{k,n}$; also, let $\mathcal{L}'_k = \cup_{n \ge 0} \mathcal{L}'_{k,n}$.

Definition 5.6.6. A prohibited pattern $X = aYb$ from $\mathcal{L}'_k$, where a and b are some consecutive elements and Y is a (possibly empty) word, is called *irreducible* if $\mathrm{red}(Yb)$ and $\mathrm{red}(aY)$ are not prohibited, in other words, if $\mathrm{red}(Yb)$ and $\mathrm{red}(aY)$ are k-determined permutations. Let $\mathcal{L}_k$ be the set consisting *only* of irreducible prohibited patterns in $\mathcal{L}'_k$.

Theorem 5.6.7. *([61]) Let k be fixed. The number of (irreducible) prohibitions in $\mathcal{L}_k$ is finite. Moreover, the longest prohibited patterns in $\mathcal{L}_k$ are of length $2k-1$.*

Theorem 5.6.7 allows us to use the *transfer matrix method* (see Subsection 8.4.1) to obtain the following result.

Theorem 5.6.8. *([61]) The generating function $A_k(x) = \sum_{n \ge 0} A_{k,n} x^n$ for the number of k-determined permutations is rational.*

5.7 Asymptotic enumeration

Elizalde [355] gives the following theorem that we provide with a proof.

Theorem 5.7.1. *([355]) Let p be a consecutive pattern of length $k \ge 3$.*

(i) There exist constants $0 < c, d < 1$ such that

$$c^n n! < s_n(p) < d^n n!$$

for all $n \ge k$.

(ii) There exists a constant $0 < w \le 1$ such that

$$\lim_{n \to \infty} \left(\frac{s_n(p)}{n!} \right)^{1/n} = w.$$

Note that c, d and w depend only on p. One can compare this result with the former conjecture of Warlimont settled by Ehrenborg et al. [346] (see Theorem 5.7.21 below).

Proof. Note that

$$(5.11) \qquad\qquad s_{m+n}(p) \le s_m(p)s_n(p)\binom{m+n}{n}.$$

Indeed, suppose $\pi = \pi_1\pi_2$ is a p-avoiding permutation of length $m + n$ and π_1 is of length m. Then $\text{red}(\pi_1)$ and $\text{red}(\pi_2)$ must be p-avoiding permutations. The inequality (5.11) is now easy to see taking into account that concatenation of two p-avoiding permutations is not necessarily a p-avoiding permutation. For example, if $p = \underline{12}$, $m = 3$, $n = 2$, $\text{red}(\pi_1) = 321$, and $\text{red}(\pi_2) = 21$, then the right-hand side of (5.11) counts, e.g. the permutation 43152 which is not p-avoiding.

By induction on $n \ge k$, one gets

$$s_{m+n}(p) < d^m m! d^n n! \binom{m+n}{n} = d^{m+n}(m+n)!$$

for some positive $d < 1$.

For the lower bound, let $p_1 p_2 p_3$ be the first (leftmost) three letters of p and $\tau = \text{red}(\underline{p_1 p_2 p_3})$. Clearly, $\mathcal{S}_n(\tau) \subseteq \mathcal{S}_n(p)$ for all n, since an occurrence of p in a permutation is necessarily an occurrence of τ. Thus, $s_n(\tau) \le s_n(p)$. Since τ is a consecutive pattern of length 3, Theorem 5.7.3 classifying asymptotic behavior of such patterns can be applied to obtain that

$$s_n(p) \ge s_n(\underline{132}) > c^n n!$$

for some $c > 0$.

To prove (ii), we can express (5.11) as

$$\frac{s_{m+n}(p)}{(m+n)!} \le \frac{s_m(p)}{m!}\frac{s_n(p)}{n!}$$

and apply *Fekete's lemma* [776, Lemma 11.6] to the function $\frac{n!}{s_n(p)}$ to conclude that

$$\lim_{n\to\infty}\left(\frac{s_n(p)}{n!}\right)^{1/n}$$ exists. Calling it w and applying part (i), we get

$$0.7839769\ldots = \lim_{n\to\infty}\left(\frac{s_n(\underline{132})}{n!}\right)^{1/n} \le w \le 1.$$

$\square$

Regarding another general result, Bóna [143] proved that the distribution of occurrences of the pattern $\underline{12\cdots k}$ in permutations is asymptotically *normal*. In the special case of $k = 2$, it is nothing else but the classic result stating that descents of permutations asymptotically have normal distribution (see [406] and references therein for various proofs of this fact, or [144] for a generalization).

5.7.1   A complex analysis result

Theorem 5.7.2. *([387, Chapter 4]) Let $A(x)$ be a* meromorphic *function on a domain of the complex plane including the origin, and let ρ be the unique pole of $A(x)$ such that $|\rho|$ is minimum. Asymptotically*

$$[x^n]A(x) \sim \gamma \cdot \rho^{-n}$$

where γ is the residue of $A(x)$ in ρ, and $[x^n]A(x)$ is the coefficient to x^n in the Taylor series expansion of $A(x)$ around 0.

From Theorems 5.1.9 and 5.7.2, the following result was obtained in [360] by Elizalde and Noy, which by the trivial bijections gives the asymptotics for avoidance of any consecutive pattern of length 3.

Theorem 5.7.3. *([360]) Asymptotically, the numbers $s_n(\underline{123})$ and $s_n(\underline{132})$ are given by*

$$s_n(\underline{123}) \sim \gamma_1 \cdot (\rho_1)^n \cdot n!, \qquad s_n(\underline{132}) \sim \gamma_2 \cdot (\rho_2)^n \cdot n!,$$

where $\rho_1 = \frac{3\sqrt{3}}{2\pi}$, $\gamma_1 = e^{3\sqrt{3}\pi}$, $(\rho_2)^{-1}$ is the unique positive root of $\int_0^x e^{-t^2/2}dt = 1$, and $\gamma_2 = \exp((\rho_2)^{-2}/2)$. The approximate values are as follows:

$$\rho_1 = 0.8269933, \qquad \gamma_1 = 1.8305194,$$

$$\rho_2 = 0.7839769, \qquad \gamma_2 = 2.2558142.$$

We see from Theorem 5.7.3 that asymptotically, the number of permutations avoiding $\underline{123}$ is larger than the number of permutations avoiding $\underline{132}$, which is consistent with Proposition 5.1.10.

From Theorems 5.1.12 and 5.7.2, the following result was obtained in [360], which gives the asymptotic behavior for three equivalence classes modulo trivial bijections.

Theorem 5.7.4. *([360]) Asymptotically, the numbers $s_n(\underline{1342})$, $s_n(\underline{1234})$ and $s_n(\underline{1243})$ are given by*

$$s_n(\underline{1342}) \sim \gamma_1 \cdot (\rho_1)^n \cdot n!,$$

$$s_n(\underline{1234}) \sim \gamma_2 \cdot (\rho_2)^n \cdot n!,$$

$$s_n(\underline{1243}) \sim \gamma_3 \cdot (\rho_3)^n \cdot n!,$$

where $(\rho_1)^{-1}$ is the smallest positive solution of $\int_0^x e^{-t^3/6}dt = 1$, $(\rho_2)^{-1}$ is the smallest positive solution of $\cos x - \sin x + e^{-x} = 0$, and ρ_3 is the solution of a certain equation involving Airy functions. The approximate values are as follows:

$$\rho_1 = 0.954611, \qquad \gamma_1 = 1.8305194,$$

$$\rho_2 = 0.963005, \qquad \gamma_2 = 2.2558142,$$

$$\rho_3 = 0.952891, \qquad \gamma_3 = 1.6043282.$$

We see from Theorem 5.7.4 that asymptotically, the number of permutations avoiding $\underline{1342}$ is larger than the number of permutations avoiding $\underline{1243}$, which is consistent with Proposition 5.1.11.

Aldred et al. [34] found ρ-values characterizing asymptotic behavior for the remaining consecutive patterns of length 4:

- for the pattern $\underline{2413}$, $\rho = 0.957718$;

- for the pattern $\underline{2143}$, $\rho = 0.956174$;

- for the pattern $\underline{1324}$, $\rho = 0.955850$;

- for the pattern $\underline{1423}$, $\rho = 0.954826$.

5.7.2   A spectral approach

Suppose P is a set of consecutive patterns, each of length $m + 1$. Ehrenborg et al. [346] developed a general method to find asymptotics for $\mathcal{S}_n(P)$ using the *spectral theory of integral operators* on $L^2([0,1]^m)$. Kreĭn and Rutman's generalization of the *Perron–Frobenius theory of non-negative matrices* plays a central role here. This approach gives detailed asymptotic expansions and allows for explicit computation of leading terms in many cases. We sketch this method below along with selected relevant results. Our presentation is based on [346]. We refer to [345] by Ehrenborg and Jung for an extension of the spectral method to study the asymptotics related to *weighted consecutive pattern-avoidance*.

Definition 5.7.5. For $x = (x_1, x_2, \ldots, x_n) \in \mathbb{R}^n$ with $x_i \neq x_j$ for $i \neq j$, $\Pi(x)$ denotes the permutation $\pi \in \mathcal{S}_n$ with $\pi_i < \pi_j$ if and only if $x_i < x_j$.

The meaning of $\Pi(x)$ is similar to that of $\mathrm{red}(\sigma)$, taking the reduced form of a permutation σ. The underlying principle that makes the method in [346] work is that one can pick a permutation in $\mathcal{S}_n$ at random with *uniform distribution*, by picking a point $(x_1, x_2, \ldots, x_n)$ in the *unit cube* $[0,1]^n$ and applying the function Π. This method has been used in [347] by Ehrenborg et al. to obtain quadratic inequalities for the *descent set statistics* and in [344] by Ehrenborg and Farjoun to enumerate *alternating* $2 \times n$ *arrays*.

Definition 5.7.6. We define the function χ on the unit cube $[0,1]^{m+1}$ by

$$\chi(x) = \begin{cases} 1, & \text{if } \Pi(x) \notin P, \\ 0, & \text{if } \Pi(x) \in P, \end{cases}$$

and the operator T on $L^2([0,1]^m)$ by

$$(Tf)(x_1, x_2, \ldots, x_m) = \int_0^1 \chi(t, x_1, x_2, \ldots, x_m) \cdot f(t, x_1, x_2, \ldots, x_{m-1})dt.$$

For $n \geq m$, define χ_n on the n-dimensional cube $[0,1]^n$ by

$$\chi_n(x_1, x_2, \ldots, x_n) = \prod_{j=1}^{n-m} \chi(x_j, x_{j+1}, \ldots, x_{m+j}).$$

This allows us to express powers of T. If $k < m$ then

$$\left(T^k f\right)(x) = \int_{[0,1]^k} \chi_{m+k}(t_1, \ldots, t_k, x_1, \ldots, x_m) \cdot f(t_1, \ldots, t_k, x_1, \ldots, x_{m-k})dt_1 \cdots dt_k$$

while if $k \geq m$,

$$\left(T^k f\right)(x) = \int_{[0,1]^k} \chi_{m+k}(t_1, \ldots, t_k, x_1, \ldots, x_m) \cdot f(t_1, \ldots, t_m)dt_1 \cdots dt_k.$$

Definition 5.7.7. The *standard inner product* on $L^2([0,1]^m)$, is defined by

$$(f, g) = \int_{[0,1]^m} f(x) \cdot \overline{g(x)}dx,$$

and hence the L^2 norm is given by $\|f\| = \sqrt{(f, f)}$.

The following proposition is straightforward to prove.

Proposition 5.7.8. *([346]) The number of permutations in $\mathcal{S}_n(P)$, for $n \geq m$, is given by*

$$s_n(P) = n! \cdot \left(T^{n-m}(\mathbf{1}), \mathbf{1}\right),$$

where $\mathbf{1}$ denotes the constant function 1 on $[0,1]^m$. Thus the probability of selecting a P-avoiding permutation at random among all n-permutations is given by $(T^{n-m}(\mathbf{1}), \mathbf{1})$.

Definition 5.7.9. If A is a *bounded linear operator* from a Hilbert space to itself, the *resolvent set* of A is the set of all $z \in \mathbb{C}$ such that $(zI - A)^{-1}$ is also a bounded operator. The complement in $\mathbb{C}$ of the resolvent set is the *spectrum* of A, denoted $\sigma(A)$, and the spectral radius of A is given by

$$r(A) = \sup \{|\lambda| : \lambda \in \sigma(A)\}.$$

The *peripheral spectrum* of A is the intersection of $\sigma(A)$ and the circle of radius $r(A)$ in the complex plane.

The peripheral spectrum of the operator T consists of at least one real eigenvalue at $r(A)$, although this need not be the only eigenvalue in the peripheral spectrum.

Definition 5.7.10. The *adjoint* of an operator A is the operator A^* that satisfies $(f, A^*(g)) = (A(f), g)$.

The asymptotic behavior of powers of a bounded linear operator is determined by its spectrum. Using spectral theory, the following theorem is obtained in [346].

Theorem 5.7.11. *([346]) Let P be a set of forbidden length $m + 1$ consecutive patterns. Then the nonzero spectrum of the associated operator T consists of discrete eigenvalues of finite multiplicity which may accumulate only at 0. Furthermore, let r be a positive real number such that there is no eigenvalue of T with modulus r and let $\lambda_1, \lambda_2, \ldots, \lambda_k$ be the eigenvalues of T with modulus greater than r. Assume that $\lambda_1, \lambda_2, \ldots, \lambda_k$ are simple eigenvalues, with associated eigenfunctions φ_i and that the adjoint operator T^* has eigenfunctions ψ_i corresponding to the eigenvalues λ_i. Then we have the expansion*

$$(5.12) \qquad \frac{s_n(P)}{n!} = \left(T^{n-m}(\mathbf{1}), \mathbf{1}\right) = \sum_{i=1}^{k} \frac{(\varphi_i, \mathbf{1}) \cdot (\mathbf{1}, \overline{\psi_i})}{(\varphi_i, \psi_i)} \cdot \lambda_j^{n-m} + O(r^n).$$

Moreover, when the operator T has a positive spectral radius, that is, $r(T) > 0$ then the spectral radius is an eigenvalue of the operator T.

Remark 5.7.12. When T has spectral radius 0, Theorem 5.7.11 is not useful.

Definition 5.7.13. If (X, μ) is a measure space and $f \in L^2(X, \mu)$ is a real-valued function, we say that $f > 0$ if $f(x) > 0$ for almost every $x \in X$, and $f \geq 0$ if $f(x) \geq 0$ for almost every x. A bounded operator A on $L^2(X, \mu)$ is *positivity improving* if for any $f \geq 0$ different from 0 there is an integer k (possibly depending on f) such that $A^k f > 0$.

For certain sets of forbidden consecutive patterns, we can show that there is a unique eigenvalue with largest modulus and that this eigenvalue is simple and positive, thus obtaining the leading term of the asymptotics of $s_n(P)/n!$.

Theorem 5.7.14. *([346]) If the operator A is positivity improving then its largest eigenvalue is real, positive and simple.*

A sufficient condition for T to be positivity improving can be formulated in combinatorial terms as follows.

Definition 5.7.15. Let P be a set of patterns as above that we consider in this definition to be a subset of $(m+1)$-permutations (slightly abusing the notation, underlines are omitted). Let G_P be the directed graph with vertex set $\mathcal{S}_m$ and a directed edge from π to σ if there is a permutation $\tau = \tau_1 \tau_2 \cdots \tau_{m+1} \in \mathcal{S}_{m+1} \backslash P$ with $\mathrm{red}(\tau_1 \tau_2 \cdots \tau_m) = \pi$ and $\mathrm{red}(\tau_2 \tau_3 \cdots \tau_{m+1}) = \sigma$.

Example 5.7.16. Suppose that the set P of forbidden consecutive patterns of length 4 consists of all patterns beginning with the letter 4 and of all patterns whose first two letters, as well as last two letters, form an ascent. That is,

$$P = \{\underline{1234}, \underline{1324}, \underline{1423}, \underline{2314}, \underline{2413}, \underline{3412}, \underline{4123}, \underline{4132}, \underline{4213}, \underline{4231}, \underline{4312}, \underline{4321}\}.$$

Then the graph G_P is in Figure 5.6.

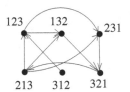

Figure 5.6: The graph G_P for P defined in Example 5.7.16.

Remark 5.7.17. The graph $G_\emptyset$ is known as the *graph of overlapping permutations*, and it is an analogue of the *de Bruijn graph* on words (see Definition 8.6.4) for permutations.

Definition 5.7.18. A directed graph G is *strongly connected* if any vertex of G is connected to any other vertex by a directed path.

Example 5.7.19. The graph in Figure 5.6 is not strongly connected as there is no edge coming in to the node 312, so we cannot travel from, say, the node 123 to the node 312. However, if, for example, the pattern $\underline{4312}$ were to be excluded from the set P in Example 5.7.16 resulting in the appearance of an edge from the node 321 to the node 312, then the graph would become strongly connected.

Theorem 5.7.20. *([346]) Suppose that the graph G_P is strongly connected and that the monotone patterns $\underline{12\cdots(m+1)}$ and $\underline{(m+1)m\cdots 1}$ are not in P. Then the operator T is positivity improving.*

As a corollary to Theorem 5.7.20, Ehrenborg et al. [346] settled a conjecture of Warlimont [792]. We record this result in the following theorem, which, in the case of length 3 consecutive patterns and an arbitrary increasing pattern, was proved in [792].

Theorem 5.7.21. *([346]) For any consecutive pattern p, there exist constants $\gamma > 0$ and $w < 1$ such that*

$$\frac{s_n(p)}{n!} \sim \gamma w^n.$$

As a matter of fact, we can characterize the spectrum of T in terms of another graph associated with P.

Definition 5.7.22. Let Δ_π denote the set of points $x = (x_1, x_2, \ldots, x_m) \in (0,1)^m$ with $x_i \neq x_j$ for $i \neq j$ and $\Pi(x) = \pi$. The graph H_P has vertex set $\bigcup_{\pi \in S_m} \Delta_\pi$ and directed edges from $(x_1, x_2, \ldots, x_m)$ to $(x_2, x_3, \ldots, x_{m+1})$ if $x_1 \neq x_{m+1}$ and $\Pi(x_1, x_2, \ldots, x_{m+1}) \notin P$.

It is shown in [346] that if the graph H_P is strongly connected then there is a non-trivial upper bound on the length of the directed path connecting any two vertices. The *period* of a strongly connected graph G is defined as follows.

Definition 5.7.23. Fix a vertex v of G and, for any non-negative integer k, let X_k be the set of all vertices in G that can be reached from v in exactly k steps. The set Q of all k with $v \in X_k$ is a semigroup and it generates a subgroup $d\mathbb{Z}$ of $\mathbb{Z}$. The integer d is the *period* of the graph G.

Note that if G is strongly connected, then G has period d for some positive integer d.

Definition 5.7.24. A graph is called *ergodic* if it is strongly connected and has period 1.

Theorem 5.7.25. *([346]) Suppose that P is a set of forbidden patterns and that the graph H_P is strongly connected with period d. Then the operator T has positive spectral radius $r(T)$ and T has a simple eigenvalue $\lambda = r(T)$ with strictly positive eigenfunction. Moreover, the spectrum of T is invariant under multiplication by $\exp(2\pi i/d)$.*

In the case when the period is 1, one has the following result.

Theorem 5.7.26. *([346]) Suppose that P is a set of forbidden patterns such that the graph H_P is ergodic. Then the operator T is positivity improving. That is, the operator has a unique largest eigenvalue which is simple, real and positive, and the associated eigenfunction is positive.*

Remark 5.7.27. When P is empty, we have $s_n(P) = n!$ for all $n \geq 0$. In this case, as is pointed out in [346], the associated operator only has one non-zero eigenvalue, namely 1, with eigenfunction and adjoint eigenfunction **1**. This is the only case we know of where the number of eigenvalues is finite.

For certain patterns, we can compute the spectrum and eigenfunctions of T and obtain sharp asymptotic formulas for $s_n(P)/n!$.

Theorem 5.7.28. *([346]) If $P = \{\underline{123}\}$, the operator T has a trivial kernel and spectrum given by $\{\lambda_k\}_{k \in \mathbb{Z}}$ where*

$$\lambda_k = \frac{\sqrt{3}}{2\pi \left(k + \frac{1}{3}\right)}.$$

Furthermore, all the eigenvalues are simple. The eigenfunctions of T and the adjoint operator T^* are

(5.13)
$$\frac{\alpha_n(\underline{123})}{n!} = \exp\left(\frac{1}{2\lambda_0}\right) \cdot \lambda_0^{n+1} + O\left(|\lambda_{-1}|^n\right)$$

where $\lambda_0 = r(T)$ and λ_{-1} is the eigenvalue with next largest modulus. For more terms in the asymptotic expansion see [346].

For certain prohibited sets of patterns, $\rho(T) = 0$. One such set is discussed in [346]. Namely, let $P = \{\underline{132}, \underline{231}\}$. A P-avoiding permutation has no peaks (see Table A.1 for definition) and it is easy to see that $s_n(P) = 2^{n-1}$. It is straightforward to verify that the operator T has no non-zero eigenvalues. Also, note that the graph G_P is not strongly connected.

Several other cases of prohibited patterns of length 3 and the corresponding asymptotic results are discussed in [346] by Ehrenborg et al. We conclude this section by stating two asymptotic multi-avoidance results in [34] by Aldred et al.

Theorem 5.7.29. *([34]) We have*

$$\lim_{n \to \infty} \sqrt[n]{\frac{s_n(\underline{312}, \underline{132})}{n!}} = 0.601730727943943;$$

$$\lim_{n \to \infty} \sqrt[n]{\frac{s_n(\underline{312}, \underline{231})}{n!}} = 0.676388228094035.$$

5.8 Multi-avoidance of consecutive patterns

We have already seen multi-avoidance of consecutive patterns above, for example, in Section 5.7 on asymptotic enumeration. In this section we collect enumerative results on simultaneous avoidance of consecutive patterns in permutations. In Table 5.3 we see the *double factorial* $n!!$ which is defined by $0!! = 1$, and, for $n > 0$,

$$n!! = \left\{ \begin{array}{ll} n \cdot (n-2) \cdots 3 \cdot 1, & \text{if } n \text{ is odd,} \\ n \cdot (n-2) \cdots 4 \cdot 2, & \text{if } n \text{ is even.} \end{array} \right.$$

Avoiding $\underline{123}$ and $\underline{321}$ simultaneously gives alternating and reverse alternating permutations. Thus using a classical result [271] by Comtet (mentioned in Remark 1.0.22) we get the following theorem.

Theorem 5.8.1. *([271]) The e.g.f. for $s_n(\underline{123}, \underline{321})$ is given by*

$$2(\tan x + \sec x) - x - 1.$$

The following theorem deals with so-called *valleyless* permutations.

Theorem 5.8.2. *([505, 360]) We have $s_n(\underline{213}, \underline{312}) = 2^{n-1}$. The corresponding e.g.f. is given by*

$$\frac{1}{1 - \tanh x}.$$

Remark 5.8.3. In Theorem 3.5.9 and Remark 3.5.10, a generalization of Theorem 5.8.2 is discussed, where instead of avoiding the patterns $\underline{213}$ and $\underline{312}$ at the same time, one considers the distribution of the pattern function $\underline{213}+\underline{312}$, equivalent to the distribution of valleys in permutations. Note that by applying the complement operation, we can consider peaks instead of valleys; see Subsection 3.5.2.

We collect results on avoidance of two length 3 consecutive patterns in Table 5.2 and of three length 3 consecutive patterns in Table 5.3, where, unless dealing with e.g.f.s, pattern-avoiding n-permutations are enumerated. These results provide a complete classification for multi-avoidance of two or three consecutive length 3 patterns (the trivial bijections can be applied to obtain results for missing pairs and triples of patterns). We refer to [505] for classification of the cases of avoidance of more than three consecutive patterns. In many cases, the number of permutations avoiding four or more consecutive patterns is a constant in $\{0, 1, 2\}$, while more interesting cases are recorded in the following theorem.

Theorem 5.8.4. *([505]) We have*

$$s_n(\underline{132}, \underline{213}, \underline{312}, \underline{321}) = n - 1$$

Patterns	Enumeration	Ref.
<u>132</u>, <u>231</u>	2^{n-1}	[505]
<u>123</u>, <u>132</u>	e.g.f. $e^{x+\frac{x^2}{2}}$	[251]
<u>123</u>, <u>321</u>	e.g.f. $2(\tan x + \sec x) - x - 1$	[271]
<u>123</u>, <u>231</u>	for $n \geq 3$, $a(n) = a(n-1) + a(n;1) + \cdots + a(n;n-1)$, for $1 \leq i \leq n$, $a(n;i) = \displaystyle\sum_{j=1}^{i-1} a(n-1;j) + \sum_{j=i}^{n-2}(n-1-j)a(n-2;j)$	[516]
<u>132</u>, <u>213</u>	$b(n) = \displaystyle\sum_{i,j=1}^{n} b(n;i,j)$ with $b(n;i,i) = 0$ for $n, i \geq 1$; $$b(n;i,j) = \sum_{k=1}^{i-1} b(n-1;j,k) \text{ if } i > j;$$ $$b(n;i,j) = \sum_{k=1}^{i-1} b(n-1;j-1,k) + \sum_{k=j}^{n-1} b(n-1;j-1,k) \text{ if } i < j;$$ $$b(2;1,2) = b(2;2,1) = 1$$	[516]
<u>213</u>, <u>231</u>	$c(n) = \displaystyle\sum_{i,j=1}^{n} c(n;i,j)$ with $c(n;i,i) = 0$ for $n, i \geq 1$; $$c(n;i,j) = \sum_{k=1}^{j-1} c(n-1;j,k) + \sum_{k=i}^{n-1} c(n-1;j,k) \text{ if } i > j;$$ $$c(n;i,j) = \sum_{k=1}^{i-1} c(n-1;j-1,k) + \sum_{k=j}^{n-1} c(n-1;j-1,k) \text{ if } i < j;$$ $$c(2;2,1) = c(2;2,1) = 1$$	[516]

Table 5.2: Multi-avoidance of two length 3 consecutive patterns in permutations.

Patterns	Enumeration	Ref.
123, 132, 213	$\binom{n}{\lfloor n/2 \rfloor}$	[505]
123, 132, 231	n	[505]
123, 213, 231	$a(n) = \sum_i \binom{n-i-1}{i} a(n-2i-1) + ((n+1) \bmod 2)$ $a(0) = 1,\ a(1) = 1$	[505]
123, 321, 132	$(n-1)!! + (n-2)!!$	[505]
123, 231, 312	e.g.f. $1 + x(\sec(x) + \tan(x))$	[517]
132, 213, 231	$1 + 2^{n-2}$	[505]

Table 5.3: Multi-avoidance of three length 3 consecutive patterns in permutations.

and

$$s_n(\underline{123}, \underline{321}, \underline{132}, \underline{213}) = \begin{cases} 2C_k, & \text{if } n = 2k+1, \\ C_k + C_{k-1}, & \text{if } n = 2k, \end{cases}$$

where C_n is the n-th Catalan number.

Remark 5.8.5. Alternative recurrence relations for $s_n(\underline{213}, \underline{231})$ and $s_n(\underline{132}, \underline{213})$ are offered in [34] by Aldred et al. Also, note the striking similarity between the e.g.f. for Av($\underline{123}, \underline{231}, \underline{312}$) (see Table 5.3) and the e.g.f. for Av($\underline{123}, \underline{321}$) (see Theorem 5.8.1). Aldred et al. [34] constructed a bijection between the subset of permutations in Av($\underline{123}, \underline{231}, \underline{312}$) which begin with the letter 1 and the subset of permutations in Av($\underline{123}, \underline{321}$) which begin with the letter 1.

5.8.1 Avoidance with prescribed beginning

Kitaev [503] considered enumeration of permutations avoiding a consecutive pattern of length 3 and beginning with a prescribed (consecutive) pattern. A motivation for the study was a connection to trimmed trees discussed in Subsection 3.4.1 (see Table 3.2). The requirement to begin with a particular pattern is equivalent to simultaneously avoiding several consecutive patterns having a left hook. For example, to require that the pattern begins with $\underline{123}$ is equivalent to multi-avoidance of the patterns from the set $\{\underline{132}, \underline{213}, \underline{231}, \underline{312}, \underline{321}\}$. However, it is more convenient to indicate the required beginning pattern rather than to list the beginning patterns to avoid.

Definition 5.8.6. Let $E_p^q(x)$ be the e.g.f. for the number of permutations avoiding a pattern p and beginning with a consecutive pattern q. For convenience, we remove the underline and hook from q in $E_p^q(x)$.

Table 5.4 provides enumerative results obtained in [503]. Note that applying the reverse and complement operations, Table 5.4 also contains results on permutations avoiding a length 3 consecutive pattern and ending with an increasing or decreasing pattern. Note that clearly, we have $E_{123}^{12\cdots k}(x) = 0$ for $k \geq 3$. The following theorem provides a description of $E_{132}^{k(k-1)\cdots 1}(x)$ different from that in Table 5.4, in terms of the error function $\text{erf}(x) = \frac{2}{\sqrt{\pi}} \int_0^x e^{-t^2}\, dt$.

Theorem 5.8.7. *([503])* $E_{132}^{k(k-1)\cdots 1}(x)$ *equals*

$$\frac{(k/2 - 1)! 2^{k/2 - 1}}{(k-1)!(1 - \sqrt{\frac{\pi}{2}}\,\text{erf}(x))} \left(1 - e^{-x^2/2} \sum_{i=0}^{k/2-1} \frac{x^{2i}}{2^i i!} \right)$$

if k is even, and

$$\frac{1}{(k-1)!!} \left(-1 + \frac{1}{1 - \sqrt{\frac{\pi}{2}}\,\text{erf}(x)} \left(1 - e^{-x^2/2} \sum_{i=0}^{(k-3)/2} \frac{x^{2i+1}}{(2i+1)!!} \right) \right)$$

if k is odd.

Definition 5.8.8. Let Γ_k^{min} (resp., Γ_k^{max}) denote the set of all length k consecutive patterns such that the least (resp., greatest) letter of the pattern is the rightmost letter.

Example 5.8.9. $\Gamma_3^{min} = \{\underline{231}, \underline{321}\}$ and $\Gamma_4^{max} = \{\underline{1234}, \underline{1324}, \underline{2134}, \underline{2314}, \underline{3124}, \underline{3214}\}$.

The following theorem by Kitaev and Mansour [517] generalizes some of the results in Table 5.4.

Theorem 5.8.10. *([517])* *Suppose $p_1, p_2 \in \Gamma_k^{min}$ and omitting the underlines, $p_1 \in S_k(\underline{132})$ and $p_2 \in S_k(\underline{123})$. Thus, the complements $c(p_1), c(p_2) \in \Gamma_k^{max}$ and $c(p_1) \in S_k(\underline{312})$, $c(p_2) \in S_k(\underline{321})$. Then, for $k \geq 2$,*

$$E_{132}^{p_1}(x) = E_{312}^{c(p_1)}(x) = \frac{\int_0^x t^{k-1} e^{-t^2/2}\, dt}{(k-1)!(1 - \int_0^x e^{-t^2/2}\, dt)}$$

and

$$E_{123}^{p_2}(x) = E_{321}^{c(p_2)}(x) = \frac{e^{x/2} \int_0^x e^{-t/2} t^{k-1} \sin(\frac{\sqrt{3}}{2}t + \frac{\pi}{6}))\, dt}{(k-1)! \cos(\frac{\sqrt{3}}{2}x + \frac{\pi}{6})}.$$

Formulas for $E_p^q(x)$ appearing in [503]

$$E_{\underline{123}}^1(x) = \frac{\sqrt{3}}{2} \frac{e^{x/2}}{\cos\left(\frac{\sqrt{3}}{2}x + \frac{\pi}{6}\right)}$$

$$E_{\underline{123}}^{21}(x) = \frac{\sqrt{3}}{2} \tan\left(\frac{\sqrt{3}}{2}x + \frac{\pi}{6}\right) - x - \frac{1}{2}$$

$$E_{\underline{123}}^{k(k-1)\cdots 1}(x) = \frac{e^{x/2}}{(k-1)! \cos\left(\frac{\sqrt{3}}{2}x + \frac{\pi}{6}\right)} \int_0^x e^{-t/2} t^{k-1} \sin\left(\frac{\sqrt{3}}{2}t + \frac{\pi}{3}\right) dt$$

$$E_{\underline{123}}^{12}(x) = \frac{\sqrt{3}}{2} \frac{e^{x/2}}{\cos\left(\frac{\sqrt{3}}{2}x + \frac{\pi}{6}\right)} - \frac{1}{2} - \frac{\sqrt{3}}{2} \tan\left(\frac{\sqrt{3}}{2}x + \frac{\pi}{6}\right)$$

$$E_{\underline{132}}^1(x) = E_{\underline{213}}^1(x) = \frac{1}{1 - \int_0^x e^{-t^2/2}\, dt}$$

$$E_{\underline{132}}^{12}(x) = \frac{e^{-x^2/2}}{1 - \int_0^x e^{-t^2/2}\, dt} - x - 1$$

$$E_{\underline{132}}^{123}(x) = -\frac{1}{2} - x - \frac{x^2}{2} + \frac{\left(1 + \frac{x}{2}\right)e^{-x^2/2} - \frac{1}{2}}{1 - \int_0^x e^{-t^2/2}\, dt}$$

$$E_{\underline{132}}^{12\cdots k}(x) = E_{\underline{132}}^1(x) \int_0^x \int_0^{t_{k-2}} \cdots \int_0^{t_2} \left(e^{-t_1^2/2} - \frac{t_1 + 1}{E_{\underline{132}}^1(t_1)}\right) dt_1 dt_2 \cdots dt_{k-2}$$

$$E_{\underline{132}}^{k(k-1)\cdots 1}(x) = E_{\underline{132}}^1(x) \frac{1}{(k-1)!} \int_0^x t^{k-1} c^{-t^2/2}\, dt$$

$$E_{\underline{213}}^{12\cdots k}(x) = \int_0^x \int_0^t \frac{s^{k-2} e^{T(t) - T(s)}\, ds dt}{(k-2)!(1 - \int_0^t e^{-m^2/2} dm)} \text{ where } T(x) = -x^2/2 + \int_0^x \frac{e^{-t^2/2}}{1 - \int_0^t c^{-s^2/2} ds}\, dt$$

$$E_{\underline{213}}^{k(k-1)\cdots 1}(x) = -\frac{x^{k-1}}{(k-1)!} + \sum_{n=0}^{k-2} \int_0^x \int_0^{t_n} \cdots \int_0^{t_1} \frac{C_{k-n}(t) + \delta_{n,k-2}}{1 - \int_0^t e^{-m^2/2} dm}\, dt dt_1 \cdots dt_n,$$

$$C_k(x) = e^{T(x)} \int_0^x \int_0^{t_{k-2}} \cdots \int_0^{t_1} e^{-T(t)} \left(\frac{e^{-t^2/2}}{1 - \int_0^t e^{-m^2/2} dm} - t - 1\right) dt dt_1 \cdots dt_{k-2}$$

Table 5.4: Avoiding a length 3 consecutive pattern with prescribed beginning.

We conclude this subsection with a topic that is not about our patterns per se, as it involves (monotone) subsequences of variable lengths, but it also deals with permutations with prescribed beginning. It was first derived by Garsia and Goupil [413] as a consequence of *character polynomial calculations*, and then proved combinatorially by Panova [649], that for $n \geq k$,

$$|T_{n,k}| = |\{\pi = \pi_1 \pi_2 \cdots \pi_n \in \mathcal{S}_n \mid \pi_1 < \pi_2 < \cdots < \pi_{n-k}, lis(\pi) = n - k\}|$$

$$= \sum_{r=0}^{k} (-1)^{k-r} \binom{k}{r} \frac{n!}{(n-r)!}.$$

The set $T_{n,k}$ consists of n-permutations π such that their first $n - k$ letters form an increasing factor and the longest increasing subsequence of π has length $n - k$. Moreover, the last formula has a q-analogue also proved algebraically in [413] and combinatorially in [649]: For $n \geq 2k$,

$$\sum_{\pi \in T_{n,k}} q^{\mathrm{maj}(\pi^{-1})} = \sum_{r=0}^{k} (-1)^{k-r} \binom{k}{r} [n]_q \cdots [n - r + 1]_q,$$

where recall that $\mathrm{maj}(\sigma) = \sum_{i|\sigma_i > \sigma_{i+1}} i$ denotes the *major index* of a permutation (see Definition 3.3.4) and $[n]_q = \frac{1-q^n}{1-q}$.

5.8.2   Avoidance with prescribed beginning and ending

To generalize the problem in Subsection 5.8.1, we can consider avoidance of a consecutive pattern, beginning *and* ending with prescribed patterns. This problem is studied in [516] by Kitaev and Mansour. We provide here just one theorem in this direction, Theorem 5.8.12 below, referring to [516] for more results.

Definition 5.8.11. Let p, q and h be consecutive patterns and $E_p^{q,h}(x)$ be the e.g.f. for the number of permutations avoiding p, beginning with q and ending with h. For convenience, we omit the underlines from q and h in $E_p^{q,h}(x)$.

Theorem 5.8.12. *([516]) Let*

$$\Theta_k(x) = \int_0^x \sec(\Psi_6(t)) \left(\sin(\Psi_3(t)) - \frac{\sqrt{3}}{2} e^{-t/2} \right) \left(\Phi_k(t) + \frac{t^{k-1}}{(k-1)!} \right) dt,$$

where

$$\Phi_k(x) = \frac{e^{x/2}}{(k-1)!} \sec(\Psi_6(x)) \int_0^x e^{-t/2} t^{\ell-1} \sin(\Psi_3(t)) \, dt,$$

and $\Psi_k(x) = \frac{\sqrt{3}}{2} x + \frac{\pi}{k}$. We have

- $E_{\underline{123}}^{12\cdots k,12\cdots \ell}(x) =$

$$\begin{cases} 0, & \text{if } k \geq 3 \text{ or } \ell \geq 3, \\ x - \frac{1}{2} - \frac{\sqrt{3}}{2}\tan(\Psi_6(x)) + \\ \sec(\Psi_6(x))\left(\frac{\sqrt{3}}{2}\left(e^{x/2} + e^{-x/2}\right) - \sin(\Psi_3(x))\right), & \text{if } k = 2 \text{ and } \ell = 2, \\ \frac{\sqrt{3}}{2}e^{x/2}\sec(\Psi_6(x)) - 1, & \text{if } k = 1 \text{ and } \ell = 1, \\ \frac{\sqrt{3}}{2}e^{x/2}\sec(\Psi_6(x)) - \frac{1}{2} - \frac{\sqrt{3}}{2}\tan(\Psi_6(x)), & \text{otherwise;} \end{cases}$$

- $E_{\underline{123}}^{12\cdots k,\ell(\ell-1)\cdots 1}(x) = \begin{cases} 0, & \text{if } k \geq 3, \\ \Phi_\ell(x), & \text{if } k = 1, \\ \Theta_\ell(x), & \text{if } k = 2; \end{cases}$

- $E_{\underline{123}}^{k(k-1)\cdots 1,12\cdots \ell}(x) = \begin{cases} 0, & \text{if } \ell \geq 3, \\ \Phi_k(x), & \text{if } \ell = 1, \\ \Theta_k(x), & \text{if } \ell = 2; \end{cases}$

- $E_{\underline{123}}^{k(k-1)\cdots 1,\ell(\ell-1)\cdots 1}(x)$ *is given by*

$$\begin{cases} E_{\underline{123}}^{\ell(\ell-1)\cdots 1}(x), & \text{if } k = 1, \\ E_{\underline{123}}^{k(k-1)\cdots 1}(x), & \text{if } \ell = 1, \\ E_{\underline{123}}^{k(k-1)\cdots 1}(x) - E_{\underline{123}}^{k(k-1)\cdots 1,12}(x), & \text{if } \ell = 2; \end{cases}$$

For $k \geq 2$ and $\ell \geq 3$, $E_{\underline{123}}^{k(k-1)\cdots 1,\ell(\ell-1)\cdots 1}(x)$ satisfies

$$\frac{d}{dx}E_{\underline{123}}^{k(k-1)\cdots 1,\ell(\ell-1)\cdots 1}(x)$$
$$= \left(E_{\underline{123}}^{\ell(\ell-1)\cdots 1}(x) + \frac{x^{\ell-1}}{(\ell-1)!}\right)E_{\underline{123}}^{k(k-1)\cdots 1,21}(x) + E_{\underline{123}}^{(k-1)\cdots 1,\ell(\ell-1)\cdots 1}(x),$$

where $E_{\underline{123}}^{k(k-1)\cdots 1}(x)$ is given in Table 5.4.

5.9 Consecutive patterns and 312-avoiding permutations

Barnabei et al. [92] used Krattenthaler's bijection between the set $\mathcal{S}_n(312)$ and Dyck paths on $2n$ steps to study the joint distribution of a length 3 consecutive pattern and descents (the pattern $\underline{21}$). In this section we provide selected results from [92].

For a consecutive pattern $\tau \neq \underline{312}$, we define

$$A^\tau(x,y,z) = \sum_{n,k,j \geq 0} a_{n,k,j}^\tau x^n y^k z^j$$

where $a_{n,k,j}^{\tau}$ is the number of permutations in $\mathcal{S}_n(312)$ containing k occurrences of the consecutive pattern τ and j descents.

Theorem 5.9.1. *([92])* $a_{n,k,j}^{\underline{321}} = a_{n,k,n-1-j}^{\underline{123}}$, $a_{n,k,j}^{\underline{231}} = a_{n,k,n-1-j}^{\underline{132}}$ *and* $a_{n,k,j}^{\underline{213}} = a_{n,k,n-1-j}^{\underline{213}}$. *Also,* $a_{n,k,j}^{\underline{12\cdots m}} = a_{n,k,n-1-j}^{\underline{m(m-1)\cdots 1}}$. *Moreover, the following tuples of statistics are equidistributed on* $\mathcal{S}_n(312)$:

- $(\underline{321}, \underline{132}, \underline{213}, des)$ *and* $(\underline{123}, \underline{231}, \underline{213}, asc)$,

- $(\underline{321}, \underline{231}, \underline{213}, des)$ *and* $(\underline{123}, \underline{132}, \underline{213}, asc)$,

that is,

$$\sum_{\sigma \in \mathcal{S}_n(312)} x^{\underline{321}(\sigma)} x^{\underline{132}(\sigma)} x^{\underline{213}(\sigma)} t^{des(\sigma)} = \sum_{\sigma \in \mathcal{S}_n(312)} x^{\underline{123}(\sigma)} x^{\underline{231}(\sigma)} x^{\underline{213}(\sigma)} t^{asc(\sigma)}, \text{ and}$$

$$\sum_{\sigma \in \mathcal{S}_n(312)} x^{\underline{321}(\sigma)} x^{\underline{231}(\sigma)} x^{\underline{213}(\sigma)} t^{des(\sigma)} = \sum_{\sigma \in \mathcal{S}_n(312)} x^{\underline{123}(\sigma)} x^{\underline{132}(\sigma)} x^{\underline{213}(\sigma)} t^{asc(\sigma)}.$$

Theorem 5.9.2. *([92])* *The g.f.* $A^{\underline{321}}(x, y, z)$ *satisfies the following functional equation:*

$$xz(y + x - xy)\left(A^{\underline{321}}(x, y, z)\right)^2 + (x - xyz - 1)A^{\underline{321}}(x, y, z) + 1 = 0.$$

As was remarked in [92], the series $A^{\underline{321}}(x, y, z)$ specializes to some well-known generating functions. For example, $zA^{\underline{321}}(x, 1, z)$ is the generating function for the *Narayana numbers* (see Section A.2 for the definition and see the book by Flajolet and Sedgewick [387] for the result), and $A^{\underline{321}}(x, y, 1)$ is the generating function for the distribution of the statistic "triple down step" on Dyck paths. The last distribution was studied in [709] by Sapounakis et al., where the following expression for the coefficients of the series $A^{\underline{321}}(x, y, 1)$ was obtained:

$$\sum_{i=0}^{n-1} a_{n,k,i}^{\underline{321}} = \frac{1}{n+1} \sum_{j=0}^{k} (-1)^{k-j} \binom{n+j}{n} \binom{n+1}{k-j} \sum_{i=j}^{\lfloor \frac{n+j}{2} \rfloor} \binom{n+j+1-k}{i+1} \binom{n-i}{i-j}.$$

We can deduce the following corollary to the last formula.

Theorem 5.9.3. *([92])* $s_n(312, \underline{321}) = M_n$, *where* M_n *is the* n-th *Motzkin number (see Section A.2 for the definition).*

Theorem 5.9.4. *([92])* *The g.f.* $A^{\underline{231}}(x, y, z)$ *satisfies the following functional equation:*

$$xz(1 - x + xy)\left(A^{\underline{231}}(x, y, z)\right)^2 + (x - xz + x^2z - x^2yz - 1)A^{\underline{231}}(x, y, z) + 1 = 0.$$

Barnabei et al. [92] remark that $A^{\underline{231}}(x, y, 1)$ gives the distribution of the statistic "*dudd*" on Dyck paths (where u stands for "up-step" and d for "down-step"). According to [709] by Sapounakis et al., we have

$$\sum_{i=0}^{n-1} a_{n,k,j}^{\underline{231}} = \sum_{j=k}^{\lfloor \frac{n-1}{2} \rfloor} \frac{(-1)^{j-k}}{n-j} \binom{j}{k} \binom{n-j}{j} \binom{2n-3j}{n-j+1}.$$

We can deduce the following corollary to the last formula.

Theorem 5.9.5. *([92])*

$$s_n(312, \underline{231}) = \sum_{j=0}^{\lfloor \frac{n-1}{2} \rfloor} \frac{(-1)^j}{n-j} \binom{n-j}{j} \binom{2n-3j}{n-j+1}.$$

Theorem 5.9.6. *([92]) For every $n \geq 0$, we have*

$$a_{n,k,j}^{213} = a_{n-1,k,j}^{213} + a_{n-1,k,j-1}^{\underline{213}} + \sum_{s=2}^{n-1} \sum_{h=0}^{k} \sum_{i=1}^{j} a_{s-1,h,i-1}^{213} a_{n-h,k-h-1,j-i}^{\underline{213}}.$$

Thus, $A^{\underline{213}}(x, y, z)$ satisfies the functional equation

$$xyz \left(A^{\underline{213}}(x, y, z) \right)^2 + (x + xz - 2xyz - 1)A^{\underline{213}}(x, y, z) + 1 - xz + xyz = 0.$$

It is remarked in [92] that the series $A^{\underline{213}}(x, y, 1)$ gives the distribution of the statistic "*ddu*" on Dyck paths. This distribution was studied in [300] by Deutsch, where the following expression for the coefficients of $A^{\underline{213}}(x, y, 1)$ was obtained:

$$\sum_{i=0}^{n-1} a_{n,k,i}^{\underline{213}} = 2^{n-2k-1} C_k \binom{n-1}{2k},$$

where $C_k = \frac{1}{k+1} \binom{2k}{k}$ is the k-th Catalan number.

We conclude the section with the following result.

Theorem 5.9.7. *([92]) We have $s_n(312, \underline{213}) = 2^{n-1}$, and the number of permutations in $S_n(312, \underline{213})$ with j descents is given by $a_{n,0,j}^{\underline{213}} = \binom{n-1}{j}$.*

5.10   Consecutive patterns in words

This section is based on Subsection 6.3 of the book [461] by Heubach and Mansour.

As a corollary to a much more general theorem by Heubach and Mansour [458] on *compositions* (see Definition 5.12.1 below), one has the following theorem on the joint distribution of the consecutive patterns $\underline{12}$, $\underline{11}$ and $\underline{21}$ on the set of all ℓ-ary words.

Theorem 5.10.1. *([458])*

$$\sum_{w\in[\ell]^*} x^{\text{length}(w)} r^{\underline{12}(w)} y^{\underline{11}(w)} d^{\underline{21}(w)} = 1 - \frac{1 - \left(\frac{1-x(y-r)}{1-x(y-d)}\right)^{\ell}}{r - d\left(\frac{1-x(y-r)}{1-x(y-d)}\right)^{\ell}}.$$

To obtain the distribution of a single pattern, or the joint distribution of a pair of patterns in the set $\{\underline{11}, \underline{12}, \underline{21}\}$, one needs to set the unwanted variable(s) to 1 in Theorem 5.10.1, while in the case of avoiding a pattern p, the variable corresponding to p is set to 0. For example, the g.f. for $W_{n,\ell}(\underline{11})$ is obtained by setting $r = 1$, $y = 0$ and $d = 1$ in Theorem 5.10.1 (see Definition 1.0.29 for the notions of $w_{n,\ell}(P)$ and $W_{n,\ell}(P)$):

$$\frac{1}{1 - (\ell - 1)x} = (1 + x)\sum_{n\geq 0}(\ell - 1)^n x^n,$$

thus giving $w_{n,\ell}(\underline{11}) = \ell(\ell - 1)^{n-1}$. For another example, the g.f. for $W_{n,\ell}(\underline{12})$ is obtained by setting $r = 0$, $y = 1$ and $d = 1$ in Theorem 5.10.1:

$$\frac{1}{(1 - x)^{\ell}} = \sum_{n\geq 0}\binom{n + \ell - 1}{n}x^n,$$

thus giving $w_{n,\ell}(\underline{12}) = \binom{n+\ell-1}{n}$ which is equal to $w_{n,\ell}(12)$.

More results can be extracted from Theorem 5.10.1. For example, by setting $d = 1$ and $r = y$, one obtains the distribution of *weak rises* (factors ab in a word with $a \leq b$).

Table 5.5 provides distributions for all non-equivalent 3-patterns modulo trivial bijections. Generating functions for the cases of avoidance are obtained by setting $q = 0$ in Table 5.5.

Similarly to the case of permutations, a *peak* in a word is an occurrence of a pattern from the set $\{\underline{121}, \underline{132}, \underline{231}\}$. Let

$$W^{\text{peak}}_{[\ell]}(x, q) = \sum_{w\in[\ell]^*} x^{\text{length}(w)} q^{\text{peak}(w)}.$$

Heubach and Mansour [460] proved the following theorem.

p	g.f. $\sum_{n,k\geq 0} w_{n,\ell}^k(p)x^n q^k$	ref.
$\underline{111}$	$\dfrac{1+(1-q)x(1+\ell x)-(1-q)(\ell-1)x^2}{1-(\ell-1+q)x-(\ell-1)(1-q)x^2}$	[208]
$\underline{112}$	$\dfrac{1-q}{1-\frac{1}{x}-q+\frac{1}{x}(1-x^2(1-q))^\ell}$	[208]
$\underline{212}$	$\dfrac{1}{1-x-x\sum_{j=0}^{\ell}\frac{1}{1+jx^2(1-q)}}$	[208]
$\underline{213}$	$\dfrac{1}{1-x-x\sum_{i=0}^{\ell-2}\prod_{j=0}^{i}(1-jx^2(1-q))}$	[209]
$\underline{123}$	$\dfrac{1}{1-\ell x-\sum_{j=3}^{\ell}(-x)^j\binom{\ell}{j}(1-q)^{\lfloor j/2\rfloor}U_{j-3}(q)}$ where $U_0(q)=U_1(q)=1$, $U_{2j}(q)=(1-q)U_{2j-1}(q)-U_{2j-2}(q)$, $U_{2j+1}(q)=U_{2j}(q)-U_{2j-1}(q)$, and $\sum_{j\geq 0}U_j(q)t^j=\frac{1+t+t^2}{1+(1+q)t^2+t^4}$	[209]

Table 5.5: Distribution of consecutive 3-patterns on words over an ℓ-letter alphabet.

Theorem 5.10.2. *([460])*

$$W_{[\ell]}^{\text{peak}}(x,q) = \frac{1+(1+x-xq)(W_{[\ell-1]}^{\text{peak}}-1)}{1-x-x(x+(1-x)q)(W_{[\ell-1]}^{\text{peak}}-1)}$$

for all $\ell \geq 1$, with initial condition $W_{[0]}^{\text{peak}}(x,q)=1$.

The first few generating functions can be computed using, for instance, Mathematica:

$$W_{[1]}^{\text{peak}}(x,q) = \frac{1}{1-x};$$

$$W_{[2]}^{\text{peak}}(x,q) = \frac{1-(q-1)x^2}{1-2x-(q-1)x^2+(q-1)x^3};$$

$$W_{[3]}^{\text{peak}}(x,q) = \frac{-1+3(q-1)x^2-(q-1)^2x^4}{-1+3x+3(q-1)x^2-4(q-1)x^3-(q-1)^2x^4+(q-1)^2x^5}.$$

The following two theorems are generalizations of the first two formulas in Table 5.5. Again, the case of avoidance is obtained by setting $q=0$ in the theorems.

Theorem 5.10.3. *([209]) Let $p = \underline{11\cdots 1}$ be of length m. Then*

$$\sum_{w\in[\ell]^*} x^{\text{length}(w)}q^{p(w)} = \frac{1+(1-q)x\sum_{j=0}^{m-2}(\ell x)^j-(1-y)(\ell-1)\sum_{d=2}^{m-1}x^d\sum_{j=0}^{m-1-d}(\ell x)^j}{1-(\ell-1+y)x-(\ell-1)(1-y)(1-x^{m-2}\frac{x^2}{1-x})}.$$

Theorem 5.10.4. *([208]) Let $p = \underline{11 \cdots 12}$ be of length m. Then*

$$\sum_{w \in [\ell]^*} x^{\text{length}(w)} q^{p(w)} = \frac{1-q}{1 - x^{2-m} - q + x^{2-m}(1 - x^{m-1}(1-q))^k)}.$$

Remark 5.10.5. Clearly, because of the trivial bijections, Theorem 5.10.4 gives distributions of the patterns $\underline{22 \cdots 21}$, $\underline{211 \cdots 1}$, and $\underline{122 \cdots 2}$. In contrast with this observation, the patterns $\underline{11 \cdots 12}$ and $\underline{22 \cdots 21}$ are *not* Wilf-equivalent in compositions [461] (for the definition of a composition, see Definition 5.12.1 below).

5.11   Non-overlapping consecutive patterns

Definition 5.11.1. Two occurrences of a consecutive pattern τ *overlap* in a permutation or word σ if they contain the same letters. Let $\tau\text{-nlap}(\sigma)$ be the *maximum number of non-overlapping occurrences* of τ in σ.

Example 5.11.2. The two occurrences of $\underline{132}$ in 3254761 overlap and therefore $\underline{132}$-nlap(3254761) = 1. For $\tau = \underline{231}$, $\tau\text{-nlap}(452673918) = 2$, while $\tau(452673918) = 3$. Finally, $\underline{121}$-nlap(32526634322) = $\underline{121}$(32526634322) = 2.

Kitaev [504, 509] used partially ordered patterns to obtain the following theorem connecting the distribution of the maximum number of non-overlapping occurrences of consecutive patterns with avoidance of consecutive patterns.

Theorem 5.11.3. *([504, 509]) Let τ be a consecutive pattern and $A_\tau(x) = \sum_{n \geq 0} s_n(\tau) \dfrac{x^n}{n!}$ be the e.g.f. for the number of permutations avoiding τ. Then*

$$\sum_{n \geq 0} \frac{x^n}{n!} \sum_{\sigma \in \mathcal{S}_n} y^{\tau\text{-nlap}(\sigma)} = \frac{A_\tau(x)}{1 - y + y(1-x)A_\tau(x)}.$$

Therefore, whenever we know the e.g.f. for avoidance of a consecutive pattern, we know the e.g.f. for the maximum number of non-overlapping occurrences of the pattern.

Example 5.11.4. Since for each $n \geq 0$ there is only one n-permutation without descents, namely, the increasing permutation, the e.g.f. for the number of permutations without descents is

$$A(x) = e^x = 1 + x + \frac{x^2}{2!} + \frac{x^3}{3!} + \cdots.$$

Thus the distribution of the maximum number of non-overlapping descents, by Theorem 5.11.3 is given by the following e.g.f.

$$\frac{e^x}{1 - y + y(1 - x)e^x}.$$

Using Theorems 5.1.9 and 5.11.3, the distribution of the maximum number of non-overlapping occurrences of the pattern $\underline{132}$ is given by

$$\frac{1}{1 - yx + (y - 1)\int_0^x e^{-t^2/2}dt}.$$

Mendes and Remmel [629] used the symmetric functions approach discussed in Section 5.3 not only to prove differently Theorem 5.11.3, but also to generalize it (see Theorem 5.11.6 below).

Definition 5.11.5. For a set P of consecutive patterns and a permutation σ, let P-nlap(σ) be the *maximum number of non-overlapping occurrences* of patterns from P in σ. Thus, instead of a single pattern as in Definition 5.11.1, now several patterns can affect the maximum number of non-overlapping occurrences.

Theorem 5.11.6. *([629]) Let P be a set of consecutive patterns of the same length and $A_P(x) = \sum_{n\geq 0} s_n(P)\dfrac{x^n}{n!}$ be the e.g.f. for the number of permutations avoiding every pattern in P. Then*

$$\sum_{n\geq 0} \frac{x^n}{n!} \sum_{\sigma\in S_n} y^{P\text{-nlap}(\sigma)} = \frac{A_P(x)}{1 - y + y(1 - x)A_P(x)}.$$

A q-analogue to Theorem 5.11.3, where q registers the number of inverses in permutations, is obtained in [511] by Kitaev. The notion of $[n]_q!$ is in Definition 3.3.2.

Theorem 5.11.7. *([511]) Let τ be a consecutive pattern and*

$$A_\tau(x, q) = \sum_{\pi\in\text{Av}(\tau)} q^{\text{inv}(\pi)} \frac{x^{|\pi|}}{[|\pi|]_q!}.$$

Then

$$\sum_{\pi\in\text{Av}(\tau)} q^{\text{inv}(\pi)}y^{\tau\text{-nlap}(\pi)}\frac{x^{|\pi|}}{[|\pi|]_q!} = \frac{A_\tau(x, q)}{1 - y + y(1 - x)A_\tau(x, q)}.$$

Definition 5.11.8. The *q-shifted factorial* of an integer $n \geq 0$ is

$$(t; q)_n = \prod_{k=0}^{n-1} (1 - tq^k).$$

Also,

$$\cos_q(x) = \sum_{n \geq 0} (-1)^n \frac{x^{2n}}{(q; q)_{2n}}, \quad \sin_q(x) = \sum_{n \geq 0} (-1)^n \frac{x^{2n+1}}{(q; q)_{2n+1}}, \quad \text{and} \quad \tan_q(x) = \frac{\sin_q(x)}{\cos_q(x)}.$$

A different expression for $\sum_{\pi \in \mathrm{Av}(\tau)} q^{\mathrm{inv}(\pi)} y^{\tau - \mathrm{nlap}(\pi)} \frac{x^{|\pi|}}{[\|\pi\|]_q!}$ in Theorem 5.11.7 is obtained by Mendes and Remmel [629] using the symmetric functions approach. Moreover, this result is generalized in [629] to a q-analogue to Theorem 5.11.6. The following theorem provides an example of a result in this direction.

Theorem 5.11.9. *([672]) Let $P = \{\underline{132}, \underline{231}\}$ be the set of patterns related to peaks in permutations. Then we have the following formula for non-overlapping occurrences of peaks in permutations:*

$$\sum_{n \geq 0} \sum_{\sigma \in S_n} \frac{q^{\mathrm{inv}(\sigma)} y^{P - \mathrm{nlap}(\sigma)} x^n}{(q; q)_n} = \left(1 - \frac{yx}{1 - q} + \sqrt{-1}(1 - y)\tan_q(x\sqrt{-1})\right)^{-1}.$$

Kitaev and Mansour [514] extended Theorem 5.11.3 to the case of words.

Theorem 5.11.10. *([514]) Let τ be a consecutive pattern. Then for all $\ell \geq 1$,*

$$\sum_{n \geq 0} \sum_{w \in [\ell]^n} y^{\tau - \mathrm{nlap}(w)} x^n = \frac{A_{\tau; \ell}(x)}{1 - y + (1 - kx)A_{\tau; \ell}(x, q)},$$

where $A_{\tau; \ell}(x) = \sum_{n \geq 0} w_{n, \ell}(\tau) x^n$.

Using the avoidance results in Section 5.10 (in particular, Table 5.5), we obtain the following examples of applications of Theorem 5.11.10, which appear in [514].

Example 5.11.11. On $[\ell]^*$, the distribution of non-overlapping occurrences of

- descents (the pattern $\underline{21}$) is given by

$$\frac{1}{(1 - x)^\ell + y(1 - \ell x - (1 - x)^\ell)};$$

- the pattern $\underline{122}$ is given by

$$\frac{x}{(1 - x^2)^\ell + x - 1 + y(1 - \ell x^2 - (1 - x^2)^\ell)};$$

- the pattern $\underline{212}$ is given by

$$\frac{1}{1 - x \sum_{j=0}^{\ell-1} \frac{1}{1 + jx^2} + xy \left(\sum_{j=0}^{\ell-1} \frac{1}{1 + jx^2} - \ell \right)}.$$

5.12   A hierarchy of consecutive pattern enumeration problems

Definition 5.12.1. Let $K_n = \{w = w_1 w_2 \cdots w_n \mid w_1, w_2, \ldots, w_n \text{ are positive integers}\}$. An element $w \in K_n$ for which $\operatorname{sum}(w) = w_1 + w_2 + \cdots + w_n = m$ is said to be a *composition of m into n parts*.

Example 5.12.2. 233144 is a composition of 17 into 6 parts.

Since each composition is a word, patterns in compositions can be defined in exactly the same way as patterns in words are defined. See the book [461] by Heubach and Mansour for a comprehensive overview of known results on patterns in compositions.

Definition 5.12.3. A *column-convex polyomino, CCP*, is constructed by successively gluing a finite sequence of columns, each consisting of a finite number of unit square cells, together in the xy-plane so that

- the lower left vertex of the leftmost column has coordinates $(0, 0)$,

- each pair of adjacent columns share an edge of positive integer length, and

- all cell vertices have integer coordinates.

For a CCP Q, we write $Q = Q_1 Q_2 \cdots Q_{\operatorname{col}(Q)}$ where Q_i is the i-th column and $\operatorname{col}(Q)$ is the number of columns in Q.

Example 5.12.4. A column-convex polyomino Q with $\operatorname{col}(Q) = 6$ is in Figure 5.7.

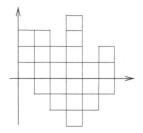

Figure 5.7: A column-convex polyomino.

The enumeration of CCPs by various statistics has been widely studied (see, e.g. [788] by Viennot for a survey). However, the study of consecutive (or *ridge*) patterns on CCPs, of interest here, was essentially initiated in [672] by Rawlings and Tiefenbruck. The simplest ridge patterns are formed between two adjacent columns. For example, for a CCP Q, we say that an *upper ascent* occurs at index i if the top cell in Q_i is lower than the top cell in Q_{i+1}. A *lower ascent* is defined similarly along the lower ridge. In Figure 5.7, Q has upper ascents at indices 3 and 5, and lower ascents at indices 4 and 5.

Definition 5.12.5. Let $\mathcal{PS}$, $\mathcal{PC}$, $\mathcal{PCCP}$ and $\mathcal{PW}$ denote respectively the sets of consecutive pattern enumeration problems on permutations, compositions, column-convex polyominoes and words.

Rawlings and Tiefenbruck [672] showed that there is an intimate relation between the sets in Definition 5.12.5, namely, that

$$\mathcal{PS} \subset \mathcal{PC} \subset \mathcal{PCCP} \subset \mathcal{PW},$$

whose significance is that it allows powerful methods from the larger problem sets to be applied to the smaller problem sets. For example, it is shown in [672] how certain results on words, as well as Bousquet-Mélou's adaptation of Templerley's method for enumerating CCPs (see [672] for references), can be used to count permutations by consecutive patterns.

In Subsection 5.12.1 we will discuss the relation $\mathcal{PS} \subset \mathcal{PC}$ that follows from *Fédou's insertion-shift bijection*. In our presentation we follow [672].

5.12.1   Fédou's insertion shift bijection

For $\sigma \in \mathcal{S}_n$ and $1 \leq i \leq n$, we let

$$inv_i(\sigma) = |\{j \mid i < j \leq n, \sigma_i > \sigma_j\}|.$$

Also, let $\Lambda_n = \{w \in K_n \mid w_1 \leq w_2 \leq \cdots \leq w_n\}$. The inverse of Fédou's [375] insertion-shift bijection $\nabla_n : S_n \times \Lambda_n \to K_n$ is given by the rule $\nabla_n(\sigma, \lambda) = w$ where $w_i = inv_i(\sigma) + \lambda_{\sigma_i}$. For example,

$$\nabla_6(256143, 224444) = 377254.$$

Two key properties here are as follows:

- If $\nabla_n(\sigma, \lambda) = w$, then $\mathrm{inv}(\sigma) + \mathrm{sum}(\lambda) = \mathrm{sum}(w)$;

- ∇_n roughly transfers the overall shape and consecutive patterns of σ to the corresponding shape and consecutive patterns of w. The explanation for ∇_n's preservation of overall shape is in the fact that, if $\nabla_n(\sigma, \lambda) = w$ and $1 \leq i < m \leq n$, then

$$\sigma_i < \sigma_m \text{ if and only if } w_i \leq w_m + |\{j \mid i < j < m, \ \sigma_i > \sigma_j\}|.$$

In particular, ∇_n transfers a peak, that is, a factor $\sigma_k\sigma_{k+1}\sigma_{k+2}$ with $\sigma_k < \sigma_{k+1} > \sigma_{k+2}$, to the factor $w_kw_{k+1}w_{k+2}$ with $w_k \leq w_{k+1} > w_{k+2}$.

We now define the notion of a consecutive pattern in compositions differently from the definition in the book [461] by Heubach and Mansour. The language of the patterns in our definition is included in the language of the usual consecutive patterns considered in [461]. Instead of defining consecutive patterns directly on compositions, it is convenient to define them through ∇_n. For $p \in S_m$, the factor $w_kw_{k+1}\cdots w_{k+m-1}$ is said to be an occurrence of the consecutive pattern p in w provided the corresponding factor $\sigma_k\sigma_{k+1}\cdots\sigma_{k+m-1}$ is an occurrence of the consecutive pattern p in the unique permutation σ satisfying $w = \nabla_n(\sigma, \lambda)$. To be consistent with our notation, we also underline p. For example, from our new definition, w_kw_{k+1} is an occurrence of the pattern $\underline{12}$ in a composition w if $w_k \leq w_{k+1}$ (instead of $w_k < w_{k+1}$ as we would expect). From the perspective of compositions, distinguishing between the cases $w_k < w_{k+1}$ and $w_k = w_{k+1}$, corresponding to *ascents* and *levels*, respectively, is often of interest. So there are problems in $\mathcal{PC}$ that have no analog in $\mathcal{PS}$. However, one has $\mathcal{PS} \subset \mathcal{PC}$.

We conclude this subsection with the following theorem connecting occurrences of consecutive patterns in permutations and compositions through Fédou's bijection. See Definition 5.11.7 for the notion of $(t; q)_n$. Also, see [672] for the discussion of three immediate applications of Theorem 5.12.6.

Theorem 5.12.6. *([672]) Suppose P is a set of consecutive patterns and $B_n \subseteq S_n$. Then*

$$\sum_{n \geq 0}\sum_{\sigma \in B_n}\left(\prod_{p \in P} y^{p(\sigma)}\right)\frac{q^{\mathrm{inv}(\sigma)}z^n}{(q;q)_n} = \sum_{n \geq 0}\sum_{w \in \nabla_n(B_n, \Lambda_n)}\left(\prod_{p \in P} y_p^{p(w)}\right)q^{\mathrm{sum}(w)}\left(\frac{z}{q}\right)^n.$$

Moreover, the above equality remains true if instead of occurrences of some or all
$p \in P$ we are interested in non-overlapping occurrences (for a pattern p, such a
change must be made on both sides of the equality).

5.13   More results on consecutive patterns

Even though in this chapter we have already seen many approaches and results
related to consecutive patterns, we did not exhaust all of them! For example, see
[793] by Warlimont for alternative derivations of some of the results on consecutive
patterns mentioned above. Also, see [38] by Amigó et al. for consecutive patterns
appearing in the context of *shift systems*, which are universal models in *information
theory*, (discrete-time) *dynamical systems* and *stochastic processes*. However, this
final section is dedicated to [34] by Aldred et al., where, in particular, an alternative
treatment for the case of $\underline{12 \cdots k}$-avoiding permutations is offered.

Definition 5.13.1. Let $\lambda_1, \lambda_2, \ldots, \lambda_{k-1}$ be the roots of

$$\lambda^{k-1} + \lambda^{k-2} + \cdots + 1 = 0,$$

which are the non-trivial k-th roots of 1.

Definition 5.13.2. Let Δ_i denote the value of the determinant of the matrix

$$\begin{bmatrix} 1 & 1 & 1 & \cdots & 1 \\ \lambda_1 & \lambda_2 & \lambda_3 & \cdots & \lambda_{k-1} \\ \lambda_1^2 & \lambda_2^2 & \lambda_3^2 & \cdots & \lambda_{k-1}^2 \\ \vdots & \vdots & \vdots & \ddots & \vdots \\ \lambda_1^{k-3} & \lambda_2^{k-3} & \lambda_3^{k-3} & \cdots & \lambda_{k-1}^{k-3} \end{bmatrix}$$

when the i-th column is removed, and the result is multiplied by a sign ($+1$ if i is
odd, -1 if i is even).

Theorem 5.13.3. *([34]) The e.g.f. for $\mathrm{Av}(\underline{12 \cdots k})$ is given by*

$$\frac{\lambda_1^{-2}\Delta_1 + \lambda_2^{-2}\Delta_2 + \cdots + \lambda_{k-1}^{-2}\Delta_{k-1}}{\lambda_1^{-2}\exp(\lambda_1 x)\Delta_1 + \lambda_2^{-2}\exp(\lambda_2 x)\Delta_2 + \cdots + \lambda_{k-1}^{-2}\exp(\lambda_{k-1}x)\Delta_{k-1}}.$$

Two instances of using the formula in Theorem 5.13.3 are provided in [34]:

- The e.g.f. for $\mathrm{Av}(\underline{123456})$ is

$$\frac{3}{\exp(x/2)\cos(\sqrt{3}x/2 + \pi/3) + \sqrt{3}\exp(-x/2)\cos(\sqrt{3}x/2 + \pi/6) + \exp(-x)};$$

- The e.g.f. for Av($\underline{12345678}$) is

$$\frac{4}{\exp(-x)+\cos(x)-\sin(x)+\cos(z)(2\cosh(z)-\sqrt{2}\sinh(z))-\sqrt{2}\cosh(z)\sin(z)},$$

where $z=\sqrt{2}x/2$.

We conclude the section by giving some families of Wilf-equivalent sets appearing in [34] as a consequence of a more general result on equidistribution. Review the notions of the $\oplus$ and $\ominus$ operations on permutations in Definition 2.2.32.

Definition 5.13.4. A pair of permutations (τ,σ) is said to have the *separation property* if τ avoids the consecutive pattern $\underline{1\ominus\sigma}$ and σ avoids the consecutive pattern $\underline{\tau\oplus1}$. In particular, any pair of permutations of equal length has the separation property.

Example 5.13.5. The pair $(231,32415)$ does not have the separation property since $\underline{231\oplus1}=\underline{2314}$ occurs in 32415. On the other hand, the pair $(231,54321)$ has the separation property.

Definition 5.13.6. Let $\alpha\beta$ be any permutation and let $\Pi(\alpha,\beta,k)$ be the set of all permutations $\alpha\gamma\beta$ where the length of γ is k and the letters of γ are greater than the letters of $\alpha\beta$.

Example 5.13.7.

$$\Pi(12,543,3)=\{12678543,12687543,12768543,12786543,12867543,12876543\}.$$

Theorem 5.13.8. *([34]) Suppose that $\alpha\beta$ is a permutation such that $(\mathrm{red}(\alpha),\mathrm{red}(\beta))$ has the separation property. Let*

$$\{\alpha\gamma_1\beta,\alpha\gamma_2\beta,\ldots,\alpha\gamma_t\beta\}\subseteq\Pi(\alpha,\beta,k)$$

and let $m_1,m_2,\ldots,m_t$ be any non-negative integers. Then the number of permutations of length n that, for $1\leq i\leq t$, contain $\underline{\alpha\gamma_i\beta}$ exactly m_i times depends only on $\alpha,\beta,k,t,m_1,m_2,\ldots,m_t$ and not on the permutations γ_i themselves.

Corollary 5.13.9. *([34]) Suppose that $\alpha\beta$ is a permutation such that $(\mathrm{red}(\alpha),\mathrm{red}(\beta))$ has the separation property and suppose that k and t are fixed. Then all t-element subsets of $\Pi(\alpha,\beta,k)$ viewed as consecutive patterns are Wilf-equivalent.*

Example 5.13.10. Using Corollary 5.13.9,

$$\mathcal{S}_n(\underline{3157624},\underline{3165724})=\mathcal{S}_n(\underline{3176524},\underline{3175624})$$

where $\alpha=31$ and $\beta=24$.

Compare Corollary 5.13.9 with Theorem 5.4.23, both dealing with Wilf-equivalence for consecutive pattern-avoidance.

Chapter 6

Classical patterns and POPs

The idea of this chapter is to provide a catalogue of known, mostly enumerative results, on classical and partially ordered patterns in permutations and words. We do not state any proofs, which can be found following the references we provide. When stating a result on (sets of) patterns, we normally pick just one representative from the corresponding symmetry class (the equivalence class modulo trivial bijections). At the same time (strongly) Wilf-equivalent classes of patterns not equivalent modulo trivial bijections (not from the same symmetry class) are normally stated separately. Recall from Definition 3.3.11 that a k-pattern is a pattern of length k (involving k letters).

6.1 Classical patterns in permutations

We have already seen in previous chapters many results, including enumerative ones, on classical patterns. More results on classical patterns can be found in Chapter 8 (e.g., some algorithmic aspects, packing density, and questions on generation of pattern-avoiding sets). However, we would like to collect all of these, and many more results in one place to provide relatively easy access to known enumerative formulas and several other results on classical patterns.

6.1.1 Patterns of length at most 4

We have already stated in Remark 3.1.19 that for the pattern 12 (equivalently, for 21) the distribution on $\mathcal{S}_n$ is given by the coefficients of

$$(1 + q)(1 + q + q^2) \cdots (1 + q + q^2 + \cdots + q^{n-1}).$$

Patterns	Enumeration	Reference
123 132	C_n	MacMahon [578]
123, 132	2^{n-1}	Simion and Schmidt [721]
123, 231	$1 + \binom{n}{2}$	Simion and Schmidt [721]
123, 321	0 for $n \geq 5$	Simion and Schmidt [721]
123, 132, 213	F_{n+1}	Simion and Schmidt [721]
123, 132, 231	n	Simion and Schmidt [721]

Table 6.1: Avoidance of classical 3-patterns in $\mathcal{S}_n$.

As was also mentioned before, MacMahon [578] was the first one to prove that $s_n(123) = C_n$, the n-th Catalan number, while it is easy to show that $s_n(231) = C_n$ (see Proposition 2.1.3). Simion and Schmidt [721] found $s_n(P)$ for any $P \subseteq \mathcal{S}_3$. Leaving aside trivial cases of avoiding more than three 3-patterns simultaneously, which essentially have a constant number (0 or 2) of n-permutations avoiding them, we state the other results in Table 6.1, where F_n is the n-th Fibonacci number.

It follows from Theorem 4.0.3 by Backelin et al. (see [66]) that $1243 \sim 2143$. Applying the composition of reverse and complement to the same theorem, we see that $1234 \sim 1243$. Moreover, Stankova shows in [734] that $1342 \sim 2413$ and in [735] that $1234 \sim 3214$, which completed classification of 4-patterns. These facts are summarized in Table 6.2 where references to the enumerative results by Gessel [417] and Bóna [127] are provided. An alternative derivation of $s_n(1234)$ is given in [158] by Bousquet-Mélou, where the following formula is proved:

$$
\begin{aligned}
s_n(1234) &= \sum_{k=1}^{n} \binom{2k-2}{k-1}\binom{n+2}{k}\binom{n}{k} \frac{2nk - 3k^2 + 4k - n}{n(n+1)(n+2)} \\
&= \frac{1}{(n+1)^2(n+2)} \sum_{k=0}^{n} \binom{2k}{k}\binom{n+1}{k+1}\binom{n+2}{k+1}.
\end{aligned}
$$

Table 6.2 also addresses P-recursiveness issues related to the numbers (see Definition 1.7.5). Note that it is an open problem to find an explicit formula for $s_n(1324)$ or for the corresponding generating function, though a recurrence relation was obtained by Marinov and Radoičić [624] who studied the generating tree for such permutations.

Theorem 6.1.1. *([624]) $s_n(1324) = g(\langle 1 \rangle, n)$, where g is given by the following recursive formula:*

$$g(\langle a_1 a_2 \cdots a_m \rangle, n) = \begin{cases} \sum_{i=1}^{m} a_i & \text{if } n = 1, \\ \sum_{i=1}^{m} g(f(\langle a_1 \ldots a_m \rangle), n-1) & \text{if } n > 1 \end{cases}$$

and $f(\langle a_1 a_2 \cdots a_m \rangle, i) = \langle b_1 b_2 \cdots b_{a_i} \rangle$, where

$$b_j = \begin{cases} a_i + 1 & \text{if } j = 1 \\ \min(i+1, a_j) & \text{if } 2 \le j \le i \\ a_{j-1} + 1 & \text{if } i < j \le a_i. \end{cases}$$

Using Theorem 6.1.1, Marinov and Radoičić [624] obtained the first 20 values of $s_n(1324)$ that begins as follows (starting from $n = 1$):

$$1, 2, 6, 23, 103, 513, 2762, 15793, 94776, 591950, 3824112, \ldots .$$

Bóna [126] shows that for $n \ge 7$,

$$s_n(1342) < s_n(1234) < s_n(1324).$$

Stankova [736] gave a different proof of this result by ordering $\mathcal{S}_3$ up to the stronger *shape-Wilf-order* (not to be defined here), which led to a generalization of the result of Bóna to arbitrary length patterns:

$$s_n((k+2)(k+1)(k+3)\tau) < s_n((k+1)(k+2)(k+3)\tau) < s_n((k+3)(k+1)(k+2)\tau)$$

for any k-pattern τ, $k \ge 1$ and $n \ge 2k + 5$.

Atkinson et al. [54] noted an interesting phenomenon mentioned in Subsection 2.1.1 that $s_n(1342) = s_n(B)$ where B is an infinite (irreducible) set of classical patterns:

$$B = \{2(2m-1)416385 \cdots (2m)(2m-3) \mid m = 2, 3, \ldots\}.$$

Symmetric permutations

Among other problems on *symmetric permutations*, Egge [337] considered the permutations in $\mathcal{S}_n(P)$ that are invariant under the composition of reverse and complement operations for certain sets of patterns P. Many enumerative results are obtained in [337], some of which are listed below, where $s^{rc}(P)$ is the number of permutations in $\mathcal{S}_n(P)$ invariant under the composition. As usual, C_n and F_n are the n-th Catalan and Fibonacci numbers, respectively.

Pattern	Enumeration	Ref.	P-recursive
1234 1243 2143 3214	$$2\sum_{k=0}^{n}\binom{2k}{k}\binom{n}{k}^2\frac{3k^2+2k+1-n-2kn}{(k+1)^2(k+2)(n-k+1)}$$	[417]	yes [814]
1342 2413	$$(-1)^{n-1}\frac{7n^2-3n-2}{2}+$$ $$3\sum_{i=2}^{n}\frac{(2i-4)!}{i!(i-2)!}\binom{n-i+2}{2}(-1)^{n-i}2^{i+1}$$ g.f. $\dfrac{32x}{-8x^2+20x+1-(1-8x)^{3/2}}$	[127]	yes [127]
1324	no nice formula; recurrence is in Theorem 6.1.1		open

Table 6.2: Classification of avoidance of classical 4-patterns in $\mathcal{S}_n$. The patterns from different groups are not Wilf-equivalent.

- $s_{2n}^{rc}(132)=s_{2n}^{rc}(132,213)=s_{2n+1}^{rc}(132)=s_{2n+1}^{rc}(132,213)=2^n$ for $n\ge 0$;

- $s_{2n}^{rc}(132,123,231)=s_{2n}^{rc}(132,321,231)=1$ for $n\ge 3$;

- $s_{2n}^{rc}(132,123)=F_{n+1}$ and $s_{2n+1}^{rc}(132,123)=F_n$ for $n\ge 0$;

- $s_{2n}^{rc}(123)=\binom{2n}{n}$, $s_{2n+1}^{rc}(123)=C_n$ for $n\ge 0$;

- $s_{2n}^{rc}(123,2143)=s_{2n+3}^{rc}=F_{2n}$ for $n\ge 0$.

Avoidance of a classical 3-pattern τ and a classical 4-pattern σ

Note that if τ occurs in σ then $s_n(\tau)=s_n(\tau,\sigma)=C_n$, the n-th Catalan number. By the following well-known theorem of Erdős and Szekeres [363], $s_n(123,4321)=0$ for $n\ge 7$.

Theorem 6.1.2. ([363]) Any sequence of $m\ell+1$ real numbers has either an increasing subsequence of length $m+1$ or a decreasing subsequence of length $\ell+1$.

Stanley [116] shows that

$$s_n(123,3412)=2^{n+1}-\binom{n+1}{3}-2n-1,$$

Patterns	Enumeration
123, 4321	0 for $n \geq 7$
123, 3421	$\binom{n}{4} + 2\binom{n}{3} + n$
132, 4321	$\binom{n}{4} + \binom{n+1}{4} + \binom{n}{2} + 1$
123, 4231	$\binom{n}{5} + 2\binom{n}{4} + \binom{n}{3} + \binom{n}{2} + 1$
123, 3241	$3 \cdot 2^{n-1} - \binom{n+1}{2} - 1$
123, 3412	$2^{n+1} - \binom{n+1}{3} - 2n - 1$
132, 4231	$1 + (n-1)2^{n-2}$
132, 3421	
132, 3214	g.f. $\frac{(1-x)^3}{1-4x+5x^2-3x^3}$

Table 6.3: Avoidance of a classical 3-pattern and a classical 4-pattern. The results are mostly due to West [802].

and Guibert, in his master's thesis, shows that

$$s_n(132, 4231) = 1 + (n-1)2^{n-2}.$$

All other cases were solved by West [802] who used *generating trees* to enumerate the objects. Many of the cases involve Fibonacci numbers, and we state them separately in the following theorem. The cases not involving Fibonacci or Catalan numbers are summarized in Table 6.3.

Theorem 6.1.3. *([802]) The following pairs of patterns are Wilf-equivalent and for any such pair $\{\tau, \sigma\}$, $s_n(\tau, \sigma) = F_{2n}$, the $(2n)$-th Fibonacci number:*

$$\{123, 2143\}, \{123, 2413\}, \{132, 2314\}, \{132, 2341\}, \{312, 2314\},$$
$$\{312, 3241\}, \{312, 3214\}, \{123, 3214\}, \{312, 4321\}, \{312, 3421\},$$
$$\{132, 3241\}, \{132, 3412\}, \{312, 1432\}, \{312, 1342\}.$$

Finally, Tenner [765] shows that the permutations with *Boolean principal order ideals* are exactly $\mathrm{Av}(321, 3412)$ leading some people to call this class the class of *Boolean permutations*. We refer to [765] for details.

Avoidance of a classical 3-pattern and a classical 5-pattern

Results on 3-5 pairs of patterns can be found in [566] by Lipson, in [592] by Mansour, in [611] by Mansour and Stankova, in [617, 616, 618] by Mansour and Vainshtein, and in [779] by Vatter, which gives a complete classification of Wilf-equivalences among these pairs of patterns. There are seven Wilf-equivalent classes of 3-5 pairs containing at least two pairs that are non-trivially Wilf-equivalent. Of the seven classes, the largest is the class of pairs Π which contains a total of 29 pairs (including some that are Wilf-equivalent by symmetry); furthermore, we have $s_n(\Pi) = (3^{n-1} + 1)/2$.

Avoidance of at least two classical 4-patterns

There are 56 non-equivalent (modulo trivial bijections) classes of patterns in this case. By a theorem of Erdős and Szekeres [363] (see Theorem 6.1.2), $s_n(1234, 4321) = 0$ for $n \geq 10$. See Subsections 2.2.5 and 2.2.6 and Table 2.2 for ten classes of permutations avoiding two 4-patterns that are counted by the Schröder numbers. In particular, Subsection 2.2.5 contains results on the set $Av(2413, 3142)$ of separable permutations, and in Subsection 2.2.6 we can find several enumerative results on Schröder permutations appearing in [338] by Egge and Mansour and in [679] by Reifegerste (when extra restrictions, beyond avoiding 1243 and 2143 simultaneously, are imposed). Also, see Theorems 2.3.5 and 2.3.6 for enumeration of five classes of Wilf-equivalent classical patterns equinumerous with the class of smooth permutations $Av(1324, 2143)$.

The following result is obtained by Kremer and Shiu [547] using *finite transition matrices*.

Theorem 6.1.4. *([547]) We have*

$$s_n(1234, 3214) = s_n(4123, 3214) = s_n(2341, 2143) = s_n(1234, 2143) = \frac{4^{n-1} + 2}{3}.$$

Definition 6.1.5. A *site* for a permutation $\pi = a_1 a_2 \cdots a_n$ is the same as a "position" in Definition 3.3.13. Namely, a site is a position lying between two consecutive elements a_i and a_{i+1} for $1 \leq i \leq n - 1$, or to the left of a_1, or to the right of a_n. For a set of prohibited patterns P, a site of a permutation $\pi \in \mathcal{S}_n(P)$ is *active* if the insertion of $n + 1$ in that site gives a permutation in $\mathcal{S}_{n+1}(P)$; otherwise, it is said to be *inactive*.

Example 6.1.6. For $P = \{132, 1234\}$ and the permutation $\pi = 123$, only the site 0 is active since $4123 \in \mathcal{S}_4(132, 1234)$ while 1423, 1243 and 1234 are not in $\mathcal{S}_4(132, 1234)$.

Barcucci et al. [84] found generating functions for $Av(321)$ and $Av(4231, 4132)$ according to the length of a permutation, number of *active sites*, and number of inversions. However, we do not state these results here, instead we refer the interested reader to [84].

A permutation avoiding the pattern 123 is the union of two decreasing subsequences, and these permutations are called *merge permutations*. Chapter 4 deals much with such permutations. The permutations in the definition below have attracted some interest in the literature.

Definition 6.1.7. A permutation is *skew-merged* if it is the union of an increasing subsequence with a decreasing subsequence.

Example 6.1.8. The permutation 2143 is not skew-merged, while the permutation 7413256 is skew-merged (it is the union of 1256 and 743).

Stankova [734] proved that Av(2143, 3412) is exactly the class of skew-merged permutations, while Atkinson 6.1.9 enumerated these permutations (see the following theorem). In [172, 498] the authors considered the more general problem of partitioning a permutation into a given numbers of increasing and decreasing subsequences.

Theorem 6.1.9. *([45]) We have*

$$s_n(2143, 3412) = \binom{2n}{n} - \sum_{m=0}^{n-1} 2^{n-m-1} \binom{2m}{m}$$

and the corresponding g.f. is

$$\sum_{n \geq 0} s_n(2143, 3412) x^n = \frac{1 - 3x}{(1 - 2x)\sqrt{1 - 4x}}.$$

Moreover, for $n \to \infty$,

$$\frac{s_n(2143, 3412)}{\binom{2n}{n}} \to \frac{1}{2}.$$

Zeilberger [816] introduced a wide-reaching method of *enumeration schemes* which was extended by Vatter [781]. The idea here is to provide an algorithm whose input is a set of (classical) patterns P, and whose output can be read as a recurrence relation for $s_n(P)$. The Maple package WILF created by Doron Zeilberger and available at http://www.math.rutgers.edu/~zeilberg/ found an enumeration scheme *of depth* 4 for $s_n(1234, 1324)$. Additionally, WILF obtained an enumeration scheme of depth 5 for $s_n(1234, 1243)$ and one of depth 7 for $s_n(1234, 43215)$, but was unable to obtain enumeration schemes of depth ≤ 7 for any of the yet unsolved pairs of patterns of length 4.

In the electronic appendix to [547], Kremer and Shiu gave finite transition matrices for seven classes of permutations avoiding a pair $(u, v) \in \mathcal{S}_4 \times \mathcal{S}_4$ where u contains the subsequence 123 and v contains the subsequence 321.

Vatter [782] used a practical algorithm computing the accepting automaton for the *insertion encoding* of a permutation class to enumerate two more classes of patterns: $\mathrm{Av}(4321, 1324)$ and $\mathrm{Av}(4321, 3142)$. Analyzing the block structure of permutations, Le [556] enumerated $\mathrm{Av}(1342, 2143)$ and proved that $s_n(1342, 2143) = s_n(3142, 2341)$. Finally, Atkinson et al. [57] proved the following theorem linking the class $\mathrm{Av}(2341, 4123)$ to *(3+1)-free posets*; note that $s_n(2341, 4123) = s_n(1432, 3214)$ by applying the reverse operation r.

Theorem 6.1.10. *([57]) The g.f. f for $\mathrm{Av}(1432, 3214)$ has the form $f = 1/(1 - g)$ where*

$$g = \frac{1 - 2x - \sqrt{1 - 4x}}{2x} -$$

$$\frac{(1 - 13x + 74x^2 - 247x^3 + 539x^4 - 805x^5 + 834x^6 - 595x^7 + 283x^8 - 80x^9 + 8x^{10})x^2}{(1 - x)^7(1 - 2x)(1 - 6x + 12x^2 - 9x^3 + x^4)}.$$

Leaving aside the class $\mathrm{Av}(1234, 4321)$ having the following generating function

$$1 + x + 2x^2 + 6x^3 + 22x^4 + 86x^5 + 306x^6 + 882x^7 + 1764x^8 + 1764x^9$$

and the class $\mathrm{Av}(1432, 3214)$ enumerated in Theorem 6.1.10, we summarize known results for the other pairs of 4-patterns in Table 6.4, and in Table 6.5 we provide pairs of classical 4-patterns whose enumeration is unknown (the numerical data is provided by Julian West for [547] by Kremer and Shiu). Based on the data, we can conjecture the following Wilf-equivalence: $s_n(1243, 2134) = s_n(1342, 3124)$, which was confirmed by Le [556], thus completing the classification of Wilf-equivalent classes for pairs of 4-patterns.

Mansour [600] enumerated $\mathrm{Av}(1324, 2134, 1243)$, which he called the set of *maximal-poset permutations*. These permutations correspond to posets, studied by Tenner [764], that have unique maximal element.

Theorem 6.1.11. *([600]) We have*

$$\sum_{n \geq 0} s_n(1324, 2134, 1243)x^n = \frac{2 - C(x)}{2 - x - C(x)},$$

where $C(x) = \frac{1 - \sqrt{1 - 4x}}{2x}$ is the g.f. for the Catalan numbers C_ns. Moreover,

$$s_n(1324, 2134, 1243) = P_{5n} - 4P_{5(n-1)} + 2P_{5(n-2)} - \sum_{j=0}^{n-2} C_j P_{5(n-2-j)},$$

where P_n is the n-th Padovan number (see Subsection A.2.1 for the definition).

Patterns	Generating function	Ref.
2143, 3412	$\frac{1-3x}{(1-2x)\sqrt{1-4x}}$	[45]
1234, 3214		
4123, 3214	$\frac{x(1-3x)}{(x-1)(4x-1)}$	[547]
2341, 2143		
1234, 2143		
1324, 2143		
1342, 2431		[130]
1342, 3241	$\frac{1-5x+3x^2+x^2\sqrt{1-4x}}{1-6x+8x^2-4x^3}$	
1342, 2314		[734]
1324, 2413		
2413, 3142		
1234, 2134		
1324, 2314		[419]
1342, 2341		
3124, 3214	$\frac{1-x-\sqrt{1-6x+x^2}}{2x}$	[545]
3142, 3214		
3412, 3421		[801]
1324, 2134		
3124, 2314		
2134, 3124		
1243, 3241	$1 + \frac{x(1-9x+31x^2-49x^3+37x^4-14x^5+2x^6)}{(1-x)(1-4x+2x^2)(1-3x+x^2)^2}$	[547]
1243, 3421	$1 + \frac{x(1-9x+34x^2-64x^3+64x^4-28x^5+4x^6)}{(x-1)(2x-1)^5}$	[547]
1423, 3214	$\frac{1-7x+16x^2-13x^3+3x^4}{1-8x+22x^2-25x^3+10x^4-2x^5}$	[547]
1423, 3241	$\frac{1-6x+11x^2-8x^3}{1-7x+16x^2-16x^3+4x^4}$	[547]
1243, 3214	$\frac{1-6x+12x^2-13x^3+5x^4-x^5}{1-7x+17x^2-22x^3+13x^4-4x^5}$	[547]
1234, 3241	$1 + \frac{x(1-11x+54x^2-151x^3+268x^4-313x^5+234x^6-108x^7+29x^8-4x^9)}{(1-3x+x^2)(2x-1)^2(x-1)^6}$	[547]
1234, 3421	$1 + \frac{x(1-7x+24x^2-44x^3+62x^4-39x^5+32x^6-19x^7+4x^8)}{(1-x)^9}$	[547]
1234, 2413	$\frac{(1-x)(1-3x)^2}{(1-2x)^2(1-4x+x^2)}$	[782]
1234, 4231	$\frac{1-11x+56x^2-172x^3+357x^4-519x^5+554x^6-413x^7+217x^8-83x^9+20x^{10}-2x^{11}}{(1-x)^{12}}$	[782]
3142, 2341		
1342, 2143	$\frac{1-\sqrt{1-8x+16x^2-8x^3}}{4x(1-x)}$	[556]

Table 6.4: Generating functions for all known cases, but the classes $Av(1234, 4321)$ and $Av(1432, 3214)$, of avoidance of non-equivalent modulo trivial bijections pairs of 4-patterns.

P	$s_n(P)$ for $n = 4, 5, 6, 7, 8, 9, 10, 11$
1234, 3412	22, 86, 333, 1235, 4339, 14443, 45770, 138988
1243, 4231	22, 86, 335, 1266, 4598, 16016, 53579, 172663
1324, 3412	22, 86, 335, 1271, 4680, 16766, 58656, 201106
1324, 4231	22, 86, 336, 1282, 4758, 17234, 61242, 214594
1243, 3412	22, 86, 337, 1295, 4854, 17760, 63594, 223488
1324, 2341	22, 87, 352, 1428, 5768, 23156, 92416, 367007
1342, 4123	22, 87, 352, 1434, 5861, 24019, 98677, 406291
1243, 2134	22, 87, 354, 1459, 6056, 25252, 105632, 442916
1342, 3124	the same as the previous one (proved by Le [556])
1243, 2431	22, 88, 363, 1507, 6241, 25721, 105485, 430767
1324, 2431	22, 88, 363, 1508, 6255, 25842, 106327, 435965
1243, 2341	22, 88, 365, 1540, 6568, 28269, 122752, 537708
1342, 3412	22, 88, 366, 1556, 6720, 29396, 129996, 580276
1243, 2413	22, 88, 367, 1568, 6810, 29943, 132958, 595227
1243, 3124	22, 88, 367, 1571, 6861, 30468, 137229, 625573
1234, 2341	22, 89, 376, 1611, 6901, 29375, 123996, 518971
1342, 2413	22, 89, 379, 1664, 7460, 33977, 156727, 730619
1324, 1432	22, 89, 380, 1677, 7566, 34676, 160808, 752608
1234, 1342	22, 89, 380, 1678, 7584, 34875, 162560, 766124
1432, 2143	22, 89, 381, 1696, 7781, 36572, 175277, 853410
1243, 1432	22, 89, 382, 1711, 7922, 37663, 182936, 904302
2143, 2413	22, 90, 395, 1823, 8741, 43193, 218704, 1129944

Table 6.5: Open enumeration problems of avoiding pairs of classical 4-patterns.

Albert et al. [18] show that

$$s_n(2431, 4231, 4321) = \sum_{k=0}^{n} \binom{n}{k} f_{n-k},$$

where $\{f_n\}_{n\geq 0}$ is the sequence of *Fine numbers* (see [726, A000957] or [305]), and that

$$\sum_{n\geq 0} s_n(3241, 3421, 4321)x^n = \frac{3 - 13x + 2x^2 + (5x - 1)\sqrt{1 - 4x}}{2(1 - 4x - x^2)}.$$

Green and Losonczy [430] introduced *freely braided permutations*, which are the class Av(3421, 4231, 4312, 4321). Mansour [595] enumerated these permutations, which we record in the following theorem.

Theorem 6.1.12. *([595]) We have*

$$\sum_{n\geq 0} s_n(3421, 4231, 4312, 4321)x^n = \frac{1 - 3x - 2x^2 + (1 + x)\sqrt{1 - 4x}}{1 - 4x - x^2 + (1 - x^2)\sqrt{1 - 4x}}.$$

Barcucci et al. [79] showed that the class Av(4132, 4231, 4312, 4321) (considered in [434] by Guibert) is in one-to-one correspondence with the *forests of binary trees* and with the *directed animals on the triangular lattice* not to be discussed here.

Finally, Albert et al. [30] make use of the *insertion encoding* to obtain

$$s_n(3124, 4123, 3142, 4132) = \binom{2n - 2}{n - 1}.$$

6.1.2 Containing occurrences of patterns

Noonan [643] proved that the number of n-permutations containing the pattern 123 exactly once is

$$s_n^1(123) = \frac{3}{n}\binom{2n}{n - 3},$$

while Fulmek [407] showed that the number of n-permutations containing the pattern 123 exactly twice is

$$s_n^2(123) = \frac{59n^2 + 117n + 100}{2n(2n - 1)(n + 5)}\binom{2n}{n - 3}.$$

The last formula was originally conjectured by Noonan and Zeilberger [644] who suggested a general approach to the problem, in particular, giving another proof for the formula for $s_n^1(123)$. Moreover, Noonan and Zeilberger [644] conjectured that the number of n-permutations containing the pattern 132 exactly once is

$$s_n^1(132) = \binom{2n-3}{n-3}$$

which was proved by Bóna in [131]; the corresponding generating function is

$$\frac{1}{2}\left(x - 1 + \frac{1 - 3x}{\sqrt{1 - 4x}}\right).$$

A general conjecture of Noonan and Zeilberger [644], believed by many people to be false, states that the number of n-permutations containing a given classical pattern p exactly r times is P-recursive (see Definition 1.7.5 and a discussion after that) in n for any r and p. However, Bóna [128] proved that the conjecture is true for the pattern $p = 132$. Beyond P-recursiveness, Bóna [128] proved that the generating function for $s_n^r(132)$ is

$$P_r(x) + Q_r(x)(1 - 4x)^{-r+1/2},$$

where $P_r(x)$ and $Q_r(x)$ are polynomials and $1 - 4x$ does not divide $Q_r(x)$. That is, the generating function for $s_n^r(132)$ is a 2-variable rational function, with variables x and $\sqrt{1 - 4x}$.

Obtaining explicit enumerations, or a description of corresponding permutations, for $s_n^r(p)$, where $r > 0$, are challenging problems. Mansour and Vainshtein [594] suggested an approach for $p = 132$ which allows us to get an explicit expression for $s_n^r(132)$ for any given r. More precisely, they show that finding the generating function for $s_n^r(132)$ can be reduced to a routine check of all permutations in $\mathcal{S}_{2r}$. As a corollary to their study, Mansour and Vainshtein [594] obtain that

$$s_n^2(132) = \frac{n^3 + 17n^2 - 80n + 80}{2n(n-1)}\binom{2n-6}{n-2}$$

with the corresponding generating function

$$\frac{1}{2}\left(x^2 + 3x - 2 + \frac{2x^4 - 4x^3 + 29x^2 - 15x + 2}{(1 - 4x)^{3/2}}\right).$$

In what follows, $C_n^k = \frac{k}{2n+k}\binom{2n+k}{n}$. Callan [213] presented a method, illustrated by several examples, to find an explicit enumeration of (restricted) permutations containing a given 3-pattern. In particular, he shows that for $r \leq 4$, the numbers $s_n^r(123)$ are expressed in terms of *ballot numbers*:

- $s_n^1(123) = C_{n-3}^6$ (originally proved by Noonan [643]);

- $s_n^2(123) = 3C_{n-4}^8 + C_{n-6}^{11}$;

- $s_n^3(123) = 7C_{n-5}^{10} + 6C_{n-7}^{13} + C_{n-9}^{16}$;

- $s_n^4(123) = 13C_{n-6}^{12} + 19C_{n-8}^{15} + 9C_{n-10}^{18} + C_{n-12}^{21} + 4C_{n-7}^{14} + 5C_{n-5}^{10} + C_{n-4}^6 - 2C_{n-5}^8$.

Also, for $r \leq 4$, Callan [213] showed that the number of 123-avoiding n-permutations containing exactly r occurrences of the pattern 132 is expressed in terms of linear combination of powers of 2:

- $r = 0$: 2^{n-1} (originally proved by Simion and Schmidt [721]);

- $r = 1$: $\binom{n-2}{1}2^{n-3} = (n-2)2^{n-3}$ (originally proved by Robertson [694]);

- $r = 2$: $\binom{n-3}{1}2^{n-4} + \binom{n-3}{2}2^{n-5}$;

- $r = 3$: $\binom{n-3}{1}2^{n-4} + \binom{n-3}{2}2^{n-5} + \binom{n-4}{3}2^{n-7}$;

- $r = 4$: $2\binom{n-4}{1}2^{n-5} + 3\binom{n-4}{2}2^{n-6} + \binom{n-4}{3}2^{n-7} + \binom{n-5}{3}2^{n-8} + \binom{n-5}{4}2^{n-9}$.

Robertson [694] proved that the number of n-permutations which contain exactly one 123-pattern and exactly one 132-pattern is given by $(n-3)(n-4)2^{n-5}$, for $n \geq 5$. Extending this result, Robertson [695] studied $\mathcal{S}_n(R;T)$, the set of n-permutations which avoid all patterns in the set R and contain each pattern in the multiset T exactly once. More precisely, $\mathcal{S}_n(\alpha;\beta)$ and $\mathcal{S}_n(\emptyset;\{\alpha,\beta\})$, for all $\alpha \neq \beta \in \mathcal{S}_3$, are enumerated. Let $s_n(R;T) = |\mathcal{S}_n(R;T)|$. We collect the enumerative results from [695] in Table 6.6 referring to the original source for a description of pairs of patterns for each equivalence class. It is stated above in this subsection, that Mansour and Vainshtein found $s_n(\emptyset;\{132,132\}) = s_n^2(132)$ and Callan [213] found $s_n(\emptyset;\{123,123\}) = s_n^2(123)$ thus finishing the study of $\mathcal{S}_n(\emptyset;\{\alpha,\beta\})$ for all $\alpha, \beta \in \mathcal{S}_3$.

Daly [288] continued the study in this direction proving the following theorem on avoidance of a 3-pattern and containment of a 4-pattern and vice versa.

Theorem 6.1.13. *([288]) We have*

$$s_n(3412;321) = s_{n+1}(321;3412) =$$

$$\sum_{i=1}^{n-2} F_{2i}F_{2(n-i-1)} = \frac{2(2n-5)F_{2n-6} + (7n-16)F_{2n-5}}{5},$$

where F_n is the n-th Fibonacci number. The corresponding g.f. is

$$\frac{x^3}{(1 - 3x + x^2)^2}.$$

$\mathcal{S}_n(R;T)$	$s_n(R;T)$
$\mathcal{S}_n(123;321)$	0 for $n \geq 6$
$\mathcal{S}_n(123;132)$	$(n-2)2^{n-3}$ for $n \geq 3$
$\mathcal{S}_n(123;231)$	$2n-5$ for $n \geq 3$
$\mathcal{S}_n(132;213)$	$n2^{n-5}$ for $n \geq 4$
$\mathcal{S}_n(132;231)$	2^{n-3} for $n \geq 3$
$\mathcal{S}_n(\emptyset;\{123;321\})$	0 for $n \geq 6$
$\mathcal{S}_n(\emptyset;\{123;231\})$	$2n-5$ for $n \geq 5$
$\mathcal{S}_n(\emptyset;\{123;132\})$	$\binom{n-3}{2}2^{n-4}$ for $n \geq 5$
$\mathcal{S}_n(\emptyset;\{132;213\})$	$(n^2+21n-28)2^{n-9}$ for $n \geq 7$
$\mathcal{S}_n(\emptyset;\{123;231\})$	2^{n-3} for $n \geq 4$

Table 6.6: Avoiding 3-patterns and containing other 3-patterns once.

In Subsection 6.1.5 we will see other results of more general type when one 3-pattern is forbidden and the distribution of another pattern (or even more than one pattern) is considered.

Definition 6.1.14. For a number n and a fixed pattern p, the sequence

$$(s_n^k(p))_{k \geq 0} = (s_n^0(p), s_n^1(p), s_n^2(p), \ldots)$$

is called the *frequency sequence* of the pattern p for n. Clearly this sequence consists entirely of zeros if n is less than the length of p and we call these sequences *trivial* and all others *nontrivial*.

The pattern $p = 12$ (equivalently, $p = 21$) is the only pattern for which the frequency sequence is well understood (see [744] by Stanley).

Definition 6.1.15. An integer c is called an *internal zero* of the sequence $(s_n^k(p))_{k \geq 0}$ if $s_n^c(p) = 0$, but there exist c_1 and c_2 with $c_1 < c < c_2$ such that $s_n^{c_1}(p) \neq 0$ and $s_n^{c_2}(p) \neq 0$.

Bóna et al. [150] investigated when the sequence $(s_n^k(p))_{k \geq 0}$ has internal zeros. It turns out that if p is a monotone pattern then, except for $p = 12$ or 21, the nontrivial sequences always have internal zeros. For the pattern $p = 1(\ell+1)\ell \cdots 2$ there are infinitely many sequences which contain internal zeros and when $\ell = 2$ there are also infinitely many which do not. In the latter case, the only possible places for internal zeros are the next-to-last or the second-to-last positions. By trivial bijections, this completely determines the existence of internal zeros for all classical 3-patterns.

classes	$\mathcal{S}_1$	$\mathcal{S}_2$	$\mathcal{S}_3$	$\mathcal{S}_4$	$\mathcal{S}_5$	$\mathcal{S}_6$	$\mathcal{S}_7$
symmetry	1	1	2	7	23	115	694
Wilf	1	1	1	3	16	91	595

Table 6.7: The number of symmetry classes and Wilf-classes on k-patterns for $k \leq 7$.

6.1.3 Larger patterns

Babson and West [65] proved the cases $k = 2$ and $k = 3$ in a more general Theorem 4.0.3, thus obtaining Wilf-classification of 5-patterns. A table for Wilf-equivalent classes on 5-patterns can be found in [737].

Theorem 4.0.5 proved by Stankova and West in [737] established the only missing Wilf-equivalence on 6-patterns, namely $546213 \sim 465213$, and it completed classification of Wilf-equivalence on 7-patterns. Table 6.7 (taken from [737]) shows the numbers of *symmetry classes* (equivalence classes obtained by applying trivial bijections) and Wilf-classes (Wilf-equivalent classes) on k-patterns for $k \leq 7$.

See Theorems 4.0.3, 4.0.5 and 4.0.7 for general results on Wilf-equivalence for classical patterns.

Gessel [417] gave an algebraic proof of the following theorem which was later proved differently by Bousquet-Mélou in [159].

Theorem 6.1.16. *([417]) The* Bessel generating function *of* $\mathrm{Av}(12 \cdots (k+1))$ *is*

$$\sum_{\pi \in \mathrm{Av}(12\cdots(k+1))} \frac{x^{2|\pi|}}{|\pi|!^2} = \det(\mathbf{I}_{i-j})_{1 \leq i,j \leq k},$$

where

$$\mathbf{I}_i = \mathbf{I}_{-i} = \sum_{n > \max(0,-i)} \frac{x^{2n+i}}{n!(n+i)!} = \sum_{n \geq 0} \frac{x^{2n+i}}{n!(n+i)!} = \sum_{n \geq 0} \frac{x^{2n-i}}{n!(n-i)!}.$$

(For the last two equalities we need to interpret factorials as Gamma functions, in particular, $1/i! = 1/\Gamma(i+1) = 0$ if $i < 0$.)

The class $\mathrm{Av}(12 \cdots (k+1))$ is closely related to *Young tableaux*. By *RSK-correspondence* (see Subsection 4.1.1) we can associate with any permutation $\pi \in \mathrm{Av}(12 \cdots (k+1))$ a pair of tableaux of the same shape, whose first row has length at most k, and conversely. If $y_k(n)$ is the number of such tableaux on n cells, and if $Y_k(x)$ is the e.g.f. of $\{y_k(n)\}_{n \geq 0}$, then Wilf [804] showed that

$$\sum_{n \geq 0} s_n(12 \cdots (k+1)) \frac{x^{2n}}{(n!)^2} = Y_k(x)Y_k(-x) \ (k = 2, 4, 6, \ldots).$$

Rains [664] showed that the $(2n)$-th *moment* of the trace of a random k-dimensional *unitary matrix* is equal to $s_n(12 \cdots (k+1))$. This correspondence was then generalized in the same paper both to other moments for the *unitary group*, and to the moments of the trace of a random *orthogonal* or *symplectic matrix*. In each case, the moments count objects (*colored permutations, signed permutations,* or *fixed-point-free involutions*) with restricted increasing subsequence length.

Using elementary methods, Bóna [143] shows that

$$s_n(12 \cdots k) \le (k-1)^{2n}.$$

Let p, q, r, s be non-negative integers and let $\mathcal{S}(p, q, r, s)$ be the set of permutations avoiding all of the $(p+q)!(r+s)!$ patterns of length $p + q + r + s = m$ which have the form $\alpha\beta\gamma$ where $|\alpha| = r$, $|\gamma| = s$ and β is any arrangement of $\{1, 2, \ldots, p\} \cup \{m - q + 1, m - q + 2, \ldots, m\}$. Atkinson [47] derived the following recurrence relation to enumerate the permutations of $\mathcal{S}(p, q, r, s)$.

Theorem 6.1.17. *([47]) Let s_n be the number of permutations of length n in $\mathcal{S}(p, q, r, s)$. Let $u = p + q$ and $v = r + s$. Then, for all $n \ge \max(u, v)$,*

$$s_n = uv s_{n-1} - 2!\binom{u}{2}\binom{v}{2}s_{n-2} + 3!\binom{u}{3}\binom{v}{3}s_{n-3} - 4!\binom{u}{4}\binom{v}{4}s_{n-4} + \cdots .$$

The method of proof of Theorem 6.1.17 in [47] shows that $\mathcal{S}(p, q, r, s) = \mathcal{S}(p, q, 1, 0)\mathcal{S}(1, 0, r, s)$ in the sense of *permutation composition* discussed in [19] by Albert et al.

6.1.4   Asymptotics

Avoidance of any classical 3-pattern is given by the Catalan numbers, and thus, asymptotically, for any $p \in \mathcal{S}_3$,

$$s_n(p) \sim \frac{4^n}{n^{3/2}\sqrt{\pi}}.$$

Bóna [129] shows that $s_n(1324) < 36^n$ and that $s_n(1423) \le s_n(1234) < 9^n$ which by Table 6.2 gives asymptotic upper bounds for any 4-pattern. Moreover, the following theorems hold.

Theorem 6.1.18. *([129]) Asymptotically, $s_n(1k23 \cdots (k-1)) < s_n(12 \cdots k)$.*

Theorem 6.1.19. *([674]) For all n, asymptotically*

$$s_n(12\cdots k) \sim \lambda_k \frac{(k-1)^{2n}}{n^{(k^2-2k)/2}}.$$

Here

$$\lambda_k = \gamma_k^2 \int_{x_1\geq} \int_{x_2\geq} \cdots \int_{\geq x_k} [D(x_1, x_2, \ldots, x_k) \cdot e^{-(k/2)x^2}]^2 \, dx_1 dx_2 \cdots dx_k,$$

where $D(x_1, x_2, \ldots, x_k) = \prod_{i<j}(x_i - x_j)$ *and* $\gamma_k = (1/\sqrt{2\pi})^{k-1} k^{k^2/2}$.

The celebrated former *Stanley-Wilf conjecture* is recorded in the following theorem.

Theorem 6.1.20. *([623]) For any pattern $p \in S_k$, the limit $\lim_{n\to\infty} (s_n(p))^{\frac{1}{n}}$ exists and is finite.*

Definition 6.1.21. The limit $L(p) = \lim_{n\to\infty} (s_n(p))^{\frac{1}{n}}$ in Theorem 6.1.20 is called a *Stanley-Wilf limit*, and the sequence $(s_n(p))^{\frac{1}{n}}$ is called a *Stanley-Wilf sequence*.

Since Theorem 6.1.20 was a major breakthrough in the theory of permutation patterns, we provide here a brief historical overview concerning this theorem. The theorem is true for $k = 3$ (see the beginning of this subsection), and for almost all cases of $k = 4$, the theorem follows from results by Bóna [129, 127]. By results of Regev [674] and Gessel [417], for any $k \geq 1$, Theorem 6.1.20 holds for $p = 12\cdots k$. Bóna [132] proved the theorem for *layered patterns* (see Definition 2.1.11). Alon and Friedgut [37] proved a weaker statement of Theorem 6.1.20 that there exists a constant $c = c(p)$ such that $s_n(p) \leq c^{n\gamma^*(n)}$, where γ^* is an extremely slow growing function related to the Ackermann hierarchy. The proof translates this problem into a problem on Davenport-Schinzel sequences [531] related to words that avoid patterns of equalities.

A more detailed overview on the history of attempts to prove Theorem 6.1.20 can be found in [805]. In particular, the following result by Arratia [41] is mentioned there:

$$\lim_{n\to\infty} (s_n(p))^{\frac{1}{n}} = \sup_n (s_n(p))^{\frac{1}{n}}.$$

Theorem 6.1.20 was finally proved by Marcus and Tardos [623]. To provide some details from [623] we define the notion of avoidance of a matrix by another matrix.

Definition 6.1.22. Let A and P be 0-1 matrices (their entries are 0s and 1s). We say that A contains the $m \times \ell$ matrix $P = (p_{i,j})$ if there exists an $m \times \ell$ submatrix $B = (b_{i,j})$ of A with $b_{i,j} = 1$ if and only if $p_{i,j} = 1$. Otherwise, we say that A avoids P.

For a 0-1 matrix, Füredi and Hajnal [409] defined $f(n, P)$ to be the maximum number of 1s in an $n \times n$ 0-1 matrix avoiding P, and they conjectured that for all permutation matrices P we have $f(n, P) = O(n)$. By a result of Klazar [535], the truth of the (former) conjecture of Füredi and Hajnal implies Theorem 6.1.20. Finally, Füredi and Hajnal's conjecture was settled by Marcus and Tardos [623] which proved Theorem 6.1.20.

Stanley-Wilf limits for a single pattern

It is a natural question to ask what the limit $L(p)$ of a Stanley-Wilf sequence is for various patterns p. Based on already mentioned results, we can see that $L(p) = 4$ for $p \in S_3$; $L(12 \cdots k) = (k-1)^2$ and $L(q) = 8$ when q is equivalent modulo trivial bijections to the pattern 1342. Bóna [140] proved that $L(12453) = 9 + 4\sqrt{2}$ showing that $L(p)$ can be irrational. This result was generalized as follows.

Theorem 6.1.23. *([140]) Let $k \geq 4$, then*

$$L(12 \cdots (k-3)(k-1)k(k-2)) = (k - 4 + 2\sqrt{2})^2.$$

Another result obtained in [140] is the fact that the number of permutations avoiding a layered pattern (see Definition 2.1.11) is asymptotically larger than the number of permutations avoiding a monotone pattern of the same length. In [129, 138] Bóna provided bounds $9 \leq L(1324) \leq 288$ for the only unknown enumeration case of avoidance of a classical 4-pattern (see Theorem 6.1.1 for a rather complicated recurrence relation for $s_n(1324)$). Arratia [41] conjectured that $L(p) \leq (k-1)^2$ for all patterns $p \in S_k$. This conjecture was disproved by Albert et al. [29] who established using the transfer matrix method that $L(4231) \geq 9.47$ thus showing that $\mathrm{Av}(4231)$ has the largest Stanley-Wilf limit among all classes of permutations avoiding a single 4-pattern.

An unpublished argument of Pavel Valtr (see [492] by Kaiser and Klazar) shows that $L(p) \geq (k-1)^2/e^3$ for all $p \in S_k$. As noted by Bóna [142], using techniques from [140] we can prove that

$$L(1324567 \cdots k) = (k - 4 + \sqrt{L(1324)})^2 \geq (k - 4 + \sqrt{9.47})^2.$$

Bóna [142] shows that a Stanley-Wilf limit $L(p)$ can be as low as $32 = L(3124675)$ and as high as $4L(1324) \geq 37.88$.

Stanley-Wilf limits: general case

Instead of permutation classes of the form $\text{Av}(p)$ we can consider classes of the form $\text{Av}(\Pi)$ where Π is a set of patterns. It is not known if the Stanley-Wilf limit of $\text{Av}(\Pi)$ exists for arbitrary sets Π.

Definition 6.1.24. A permutation class $\text{Av}(\Pi)$ is of *polynomial growth* if $s_n(\Pi) \leq An^d$ for some constants A and d.

We refer to [23, 467, 492] for discussions of necessary and sufficient conditions on the cases Π for $\text{Av}(\Pi)$ to have polynomial growth, and several other results.

Kaiser and Klazar [492] showed that permutation classes of growth rate less than 2 satisfy one of the following:

- the number s_n of n-permutations in the class is polynomial for large n, or

- $F_{n,k} \leq s_n \leq n^c F_{n,k}$ holds for integers $c \geq 0$ and $k \geq 2$, where $F_{n,k}$ denotes the *k-generalized Fibonacci number* defined by $F_{n,k} = F_{n-1,k} + F_{n-2,k} + \cdots + F_{n-k,k}$ for $n \geq 2$, $F_{n,k} = 0$ for $n \leq 0$ and $F_{1,k} = 1$.

Thus, all growth rates of permutation classes less than 2 are positive roots of $x^{k+1} - x^k - \cdots - x^2 - 1$ for some k. Vatter [780] showed that there are only countably many permutation classes of growth rate less than $\kappa \approx 2.20557$, the unique real root of $x^3 - 2x^2 - 1$, while there are uncountably many permutation classes of growth rate κ. Moreover, Vatter [780] characterized the possible growth rates between 2 and κ, which are roots of one of the following four families of polynomials:

- $(x^3 - 2x^2 - 1)x^{k+1} - x + 3$,

- $(x^3 - 2x^2 - 1)x^{k+2} - x^2 + 2x + 1$,

- $(x^3 - 2x^2 - 1)x^{k+\ell} + x^\ell + 1$,

- $(x^3 - 2x^2 - 1)x^k + 1$,

where $k, \ell \geq 0$ are integers. The set of these growth rates contains no accumulation points from above, but does contain countably many accumulation points from below accumulating to κ.

Albert and Linton [31] conjectured that there is some λ such that every real number at least λ is the growth rate of a permutation class. Vatter [783] confirmed the conjecture which we record in the following theorem.

Theorem 6.1.25. *([783]) Let λ denote the unique real root of $x^5 - 2x^4 - 2x^2 - 2x - 1$, approximately 2.48187. Every real number greater than λ is the growth rate of a permutation class.*

Some limiting distributions

The *average* length, $\mu(n)$, of the longest increasing subsequence of a permutation π, $lis(\pi)$, was found in [363] by Erdős and Szekeres to be $\geq \sqrt{n}/2$ and then in [570] by Logan and Shepp to be $\geq 2\sqrt{n}$. The correct order of magnitude was established in [786] by Vershik and Kerov, where it was shown to be $\sim 2\sqrt{n}$. Also, Vershik and Kerov [786] estimated the standard deviation $\sigma(n)$ to be approximately $n^{1/6}$ but this was not proved to be so. Finally, Baik et al. [71] determined the complete limiting distribution function of the *normalized random variable* $(lis(\pi) - \mu(n))/n^{1/6}$. Also, see [646] by Odlyzko and Rains for a relevant paper that presents the results of both *Monte Carlo* and exact computations exploring the finer structure of the distribution of the expected value of *lis* of a random n-permutation.

Rotem [705] used n-noded *binary trees* to study several statistics on $\mathcal{S}_n(231)$, the set of *stack-sortable permutations* discussed in Section 2.1 (there is a one-to-one correspondence between the trees and permutations). In particular, he proved the following three theorems, where C_n is the n-th Catalan number.

Theorem 6.1.26. *([705]) The expected value of the longest decreasing subsequence in a random permutation of $\mathcal{S}_n(231)$ (each permutation chosen with equal probability C_n^{-1}) is asymptotically*

$$\sqrt{\pi n} - 1.5 + \frac{11}{24}\sqrt{\frac{\pi}{n}} + O(n^{-3/2}).$$

Theorem 6.1.27. *([705]) The expected value of the longest increasing subsequence in a random permutation of $\mathcal{S}_n(231)$ is $\frac{1}{2}(n+1)$.*

Theorem 6.1.28. *([705]) The average number of inversions (occurrences of the pattern 21) in a random permutation of $\mathcal{S}_n(231)$ is $\frac{1}{2}\left(\frac{4^n}{C_n} - 3n - 1\right)$.*

Deutsch et al. [302] proved the following three theorems on limiting distributions for pattern-avoiding permutations, in particular, showing that the 132-avoiding case is identical to the distribution of heights of ordered trees.

Theorem 6.1.29. *([302]) In $\mathcal{S}_n(231)$ and in $\mathcal{S}_n(312)$, the length of the longest increasing subsequence has mean $(n+1)/2$, and standard deviation $\sim \sqrt{n}/2$. Moreover, the random variable $(lis(\pi) - (n+1)/2)/(\sqrt{n}/2)$ has asymptotically the standard normal distribution.*

Theorem 6.1.30. *([302]) In $\mathcal{S}_n(132)$ and in $\mathcal{S}_n(213)$, the length of the longest increasing subsequence has mean $\sqrt{\pi n} + O(n^{1/4})$, and standard deviation $c_1\sqrt{n} + $*

$O(n^{1/4})$ $(c_1 = \sqrt{\pi(\pi/3 - 1)} = 0.38506\ldots)$. *Moreover, the normalized random variable*

$$X_n(\pi) := \frac{lis(\pi) - \sqrt{\pi n}}{\sqrt{n}},$$

defined for $\pi \in \mathcal{S}_n(132)$, satisfies

$$\lim_{n \to \infty} Prob(X_n(\pi) \le \theta) = \sum_{t=-\infty}^{\infty} (1 - 2t^2(\theta + \sqrt{\pi})^2)e^{-(\theta + \sqrt{\pi})^2 t^2} \quad (\theta > -\sqrt{\pi}).$$

Theorem 6.1.31. *([302]) For permutations $\pi \in \mathcal{S}_n(321)$, define the random variable*

$$X_n(\pi) := \frac{lis(\pi) - \frac{n}{2}}{\sqrt{n}}.$$

Then we have

$$\lim_{n \to \infty} Prob(X_n(\pi) \le \theta) = \frac{2}{\sqrt{\pi}} \int_0^{4\theta^2} u^{1/2} e^{-u} du = \frac{\Gamma(3/2, 4\theta^2)}{\Gamma(3/2)},$$

where $\Gamma(z)$ is the Gamma function and $\Gamma(z, w)$ is the incomplete Gamma function

$$\Gamma(z, w) = \int_0^w u^{z-1} e^{-u} du.$$

An *alternating subsequence* $\pi_{i_1} \pi_{i_2} \cdots \pi_{i_k}$ in a permutation $\pi = \pi_1 \pi_2 \cdots \pi_n$ satisfies $\pi_{i_1} < \pi_{i_2} > \pi_{i_3} < \pi_{i_4} > \cdots \pi_{i_k}$. Let $al(\pi)$ be the random variable whose value is the longest alternating subsequence of an n-permutation π chosen uniformly at random. Stanley [747] proved that the mean of $al(\pi)$ is $\frac{4n+1}{6}$ for $n \ge 2$, and the variance of $al(\pi)$ is $\frac{8}{45}n - \frac{13}{180}$ for $n \ge 4$. Moreover, Widom [803] showed that the limiting distribution of the normalized random variable $(al(\pi) - 2n/3)/\sqrt{8n/45}$, as $n \to \infty$, is the standard normal distribution.

The main result in [384] by Firro et al. is the following theorem.

Theorem 6.1.32. *([384]) Let $\tau \in \mathcal{S}_3$. In the class of τ-avoiding permutations of length n, the length of the longest alternating subsequence has mean $\mu_\tau \sim n/2$ and variance $\sigma_\tau^2 \sim n/4$. Moreover, the (almost normalized) random variable $X_n^\tau = \frac{al(\pi) - \frac{n}{2}}{\frac{\sqrt{n}}{2}}$, defined for all τ-avoiding n-permutations, converges in distribution to the standard normal distribution as $n \to \infty$. In other words, $al(\pi)$ satisfies the Gaussian Limit Law.*

Albert [12] considered the distribution of the length of the longest subsequence avoiding an arbitrary pattern p in a random n-permutation. The well-studied case discussed above of a longest increasing subsequence corresponds to $p = 21$. Albert [12] shows that there is some constant c_p such that as $n \to \infty$ the mean value of this length is asymptotic to $2\sqrt{c_p n}$ and that the distribution of the length is tightly concentrated around its mean. Also, some connections between c_p and the Stanley-Wilf limit of the class of p-avoiding permutations are given in [12].

6.1.5   Avoiding a 3-pattern with extra restrictions

We have already seen many results related to this topic in Subsection 6.1.1 (for example, see Table 6.3). Restricting attention to permutations avoiding a 3-pattern and studying various properties of such permutations is a very popular direction of research that contains many interesting results, some of which we state in this subsection. Many other results on this – in connection with different subsets of permutations – are stated in Subsections 6.1.7, 6.1.8, 6.1.9 and 6.1.10 below. A large variety of results was possible to obtain due to the relatively simple structure of the permutations in question. For example, 132-avoiding permutations are those that have each letter to the left of the largest letter greater than any letter to the right of it. For this section, review the definition of the r-th Chebyshev polynomials of the second kind in Definition B.2.1.

The main result of [247] by Chow and West can be formulated as follows.

Theorem 6.1.33. *([247]) We have*

$$s_n(123, (k-1)(k-2)\cdots 1k) = s_n(213, 12\cdots k) = s_n(213, k12\cdots(k-1))$$

and the corresponding generating function is

$$R_k(x) = \frac{U_{k-1}\left(\frac{1}{2\sqrt{x}}\right)}{\sqrt{x}U_k\left(\frac{1}{2\sqrt{x}}\right)},$$

where $U_r(\cos(\theta)) = \sin((r+1)\theta)/\sin(\theta)$ is the Chebyshev polynomial of the second kind. The asymptote for these numbers is $4\cos^2(\pi/(r+2))$.

Theorem 6.1.33 was proved by Krattenthaler [543] differently through Dyck paths. Mansour and Vainshtein [615] generalized Theorem 6.1.33 for the reverse complement of the second of the three pairs of patterns there as follows.

Theorem 6.1.34. *([615]) For any $k \geq 1$, the generating function $F_r(x; k)$ for the number $s_n^r(132; 12\cdots k)$ of 132-avoiding n-permutations with r occurrences of the* *pattern* $12 \cdots k$ *is given by*

$$F_r(x; k) = \frac{x^{\frac{r-1}{2}} U_{k-1}^{r-1}\left(\frac{1}{2\sqrt{x}}\right)}{U_k^{r+1}\left(\frac{1}{2\sqrt{x}}\right)}, \quad 1 \le r \le k,$$

$$F_0(x; k) = \frac{U_{k-1}\left(\frac{1}{2\sqrt{x}}\right)}{\sqrt{x} U_k\left(\frac{1}{2\sqrt{x}}\right)}.$$

Mansour and Vainshtein [616] provided another generalization of Theorem 6.1.33. To state it, we need the notion of a *p-layered pattern* which is a layered pattern (see Definition 2.1.11) with p layers.

Theorem 6.1.35. *([616]) Let* $F_{123,\tau}(x) = \sum_{n \ge 0} s_n(123, \tau) x^n$, *where* $\tau \in \mathcal{S}_k$ *is an arbitrary p-layered pattern* $(k \ge p)$.

- *If* $p = 1$, *then* $F_{123,\tau}(x)$ *is a polynomial of degree* $2k - 2$;

- *If* $p - 2$, *then* $F_{123,\prime}(x) = R_k(x)$;

- *If* $p \ge 3$, *then* $F_{123,\tau}(x) = \frac{1 - \sqrt{1-4x}}{2x}$, *the g.f. for the Catalan numbers.*

Jani and Rieper [479] found the distribution of the pattern $12 \cdots k$ over 132-avoiding permutations.

Theorem 6.1.36. *([479]) The generating function that enumerates 132-avoiding permutations by number of increasing patterns of length* k *is*

$$\sum_{r,n \ge 0} s_n^r(132; 12 \cdots k) x^n q^r = \cfrac{1}{1 - \cfrac{N_1}{1 - \cfrac{N_2}{1 - \cfrac{N_3}{1 - \cfrac{N_4}{\ddots}}}}}$$

in which the ℓ-th numerator N_ℓ is $xq^{\binom{\ell-1}{k-1}}$.

The case $k = 3$ in Theorem 6.1.36 was considered by Robertson et al. [698]. However, Bränden et al. [170] generalized Theorem 6.1.36 by considering joint distribution of increasing patterns of all lengths. To state this result, we need the following notations. For $k \ge 1$, we denote by e_{k-1} the pattern $12 \cdots k$. Thus $e_0(\pi)$ is the length $|\pi|$ of π, and $e_1(\pi)$ counts the number of non-inversions in π. We also define $e_{-1}(\pi) = 1$ for all permutations π, which is equivalent to declaring that all permutations have exactly one increasing subsequence of length 0.

Theorem 6.1.37. *([170]) The following continued fraction expansion holds:*

$$\sum_{\pi \in \mathrm{Av}(132)} \prod_{k \geq 0} x_k^{e_k(\pi)} = \cfrac{1}{1 - \cfrac{x_0^{\binom{0}{0}}}{1 - \cfrac{x_0^{\binom{1}{0}} x_1^{\binom{1}{1}}}{1 - \cfrac{x_0^{\binom{2}{0}} x_1^{\binom{2}{1}} x_2^{\binom{2}{2}}}{1 - \cfrac{x_0^{\binom{3}{0}} x_1^{\binom{3}{1}} x_2^{\binom{3}{2}} x_3^{\binom{3}{3}}}{\ddots}}}}}$$

in which the $(n+1)$-th numerator is $\prod_{k=0}^{n} x_k^{\binom{n}{k}}$.

Several applications of Theorem 6.1.37 are discussed in [170]. For example, the total number of increasing subsequences in a permutation is counted by $e_0 + e_1 + \cdots$, which, using Theorem 6.1.37, is given by

$$\sum_{\pi \in \mathrm{Av}(132)} x^{e_0(\pi) + e_1(\pi) + \cdots} = \cfrac{1}{1 - \cfrac{xt}{1 - \cfrac{x^2 t}{1 - \cfrac{x^4 t}{1 - \cfrac{x^8 t}{\ddots}}}}}.$$

As an application of results on *extended pattern-avoidance* introduced by Eriksson and Linusson [368], Linusson [565] proved a sequence of refinements on the enumeration of permutations avoiding 3-patterns, which we record in the following theorem.

Theorem 6.1.38. *([565]) We have*

1. $|\{\pi = \pi_1 \pi_2 \cdots \pi_n \in \mathcal{S}_n(123) \mid \pi_s = n, \ \pi_n = t\}| =$

   $$|\{\pi \in \mathcal{S}_n(123) \mid \pi_{n+1-s} = 1, \ \pi_1 = n+1-t\}| = C_n(s,t).$$

2. $|\{\pi = \pi_1 \pi_2 \cdots \pi_n \in \mathcal{S}_n(123) \mid \pi_s = n, \ \pi_1 = n+1-t\}|$

   $$= \begin{cases} C_{n-1}(t+s-3), & \text{if } 2 \leq s, t \leq n, \\ C_{n-1}, & \text{if } s = t = 1, \\ 0, & \text{otherwise.} \end{cases}$$

3. $|\{\pi = \pi_1\pi_2\cdots\pi_n \in \mathcal{S}_n(132) \mid \pi_s = n, \ \pi_1 = t\}|$

$$= \begin{cases} C_{n-s}C_{s-1}(n-t), & \text{if } 2 \le s \le n-t+1, \\ C_{n-1}, & \text{if } s = 1, t = n, \\ 0, & \text{otherwise.} \end{cases}$$

4. $|\{\pi = \pi_1\pi_2\cdots\pi_n \in \mathcal{S}_n(132) \mid \pi_s = n, \ \pi_n = t\}|$

$$= \begin{cases} C_{s-1}C_{n-s-t}C_{t-1}, & \text{if } s+t \le n, \\ C_{n-1}, & \text{if } s = t = n, \\ 0, & \text{otherwise.} \end{cases}$$

Here $C_n = \frac{1}{n+1}\binom{2n}{n}$ and $C_n(t) = \binom{2n-t-1}{n-t} - \binom{2n-t-1}{n-t-1}$ are the Catalan and ballot numbers, respectively. Also,

$$C_n(s,t) = \binom{2n-s-t-2}{n-t-1} - \binom{2n-s-t-2}{n-s-t-1},$$

$C_n(s,n) = C_n(n,t) = 0$, and $C_n(n,n) = 1$.

Remark 6.1.39. The ballot numbers defined in Theorem 6.1.38 can also be defined as

$$C_n(t) = |\{\pi = \pi_1\pi_2\cdots\pi_n \in \mathcal{S}_n(213) \mid \pi_t = n\}|,$$

while $C_n(s,t)$ can be defined as

$$C_n(s,t) = |\{\pi = \pi_1\pi_2\cdots\pi_n \in \mathcal{S}_n(213) \mid \pi_s = n, \pi_n = t\}|.$$

132-avoiding 2-stack sortable permutations

2-stack sortable permutations are defined in Subsection 2.1.1. Egge and Mansour [339] enumerated 2-stack sortable permutations which avoid (or contain exactly once) the pattern 132 and which avoid (or contain exactly once) an arbitrary pattern τ. In most cases the number of such permutations is given by a simple formula involving Fibonacci or Pell numbers (the n-th Pell number is $P_n = 2P_{n-1} + P_{n-2}$ with $P_0 = 0$ and $P_1 = 1$). Egge and Mansour [339] gave a bijection between the set of 132-avoiding 2-stack sortable n-permutations and the set of tilings of a $1 \times (n-1)$ rectangle with tiles of size 1×1 and 1×2, where each 1×1 tile can be red or blue. Thus, the number of such permutations is given by Pell numbers. Below we state three theorems as examples of results in [339], where $G_P(x)$ is the g.f. for the number of 132-avoiding 2-stack sortable permutations avoiding each pattern from the set P.

Theorem 6.1.40. *([339]) We have $G_1(x) = 1$, $G_{12}(x) = \frac{1}{1-x}$, and for $k \geq 3$,*

$$G_{12\cdots k}(x) = \frac{(1-x)\sum_{i=0}^{k-3}(1-x-x^2)^{i+1}x^{k-i-3} + x^{k-2}}{(1-x)(1-x-x^2)^{k-2}}.$$

For example, as a corollary to Theorem 6.1.40, we have that for $n \geq 1$, the number of 132-avoiding 2-stack sortable n-permutations that avoid the pattern 1234 is given by

$$1 - \frac{8}{5}F_{n+1} + \frac{n+1}{5}(F_{n+3} + F_{n+1}),$$

where F_n is the n-th Fibonacci number.

Theorem 6.1.41. *([339]) We have $G_{21}(x) = \frac{1}{1-x}$,*

$$G_{312}(x) = \frac{(1-x-x^2)(1-x) + x(1+x)}{(1-x)^2},$$

and, for all $k \geq 4$, $G_{k12\cdots(k-1)}(x) =$

$$\frac{(1-x)(1-x-x^2)^{k-2} + (1-x)^2\sum_{i=0}^{k-4}(1-x-x^2)^{i+1}x^{k-i-3} + x^{k-2}(1+x)}{(1-x)^2(1-x-x^2)^{k-3}}.$$

For example, as a corollary to Theorem 6.1.42, we have that the number of 132-avoiding 2-stack sortable n-permutations that avoid the pattern 4123 is given by $F_{n+4} - 2n - 2$ for $n \geq 0$.

Theorem 6.1.42. *([339]) Let $TS(132)$ denote the set of all 132-avoiding 2-stack sortable permutations. Then*

$$\sum_{\pi \in TS(132)} \prod_{k\geq 1} x_k^{12\cdots k(\pi)} = 1 + \sum_{n\geq 1} \frac{\prod_{i\geq 1} x_i^{\binom{n}{i}}}{\prod_{m=1}^n \left(1 - \prod_{i\geq 1} x_i^{\binom{m-1}{i-1}} - \prod_{i\geq 1} x_i^{2\binom{m-1}{i-1}} x_{i+1}^{2\binom{m-1}{i-1}}\right)}.$$

Remark 6.1.43. Note that by Theorem 2.1.4, $TS(132) = \mathrm{Av}(132, 2341, 3\bar{5}241)$.

Avoidance of at least two 3-patterns

Vatter [777] obtained the following theorem.

Theorem 6.1.44. *([777]) Let Q be any set of patterns that contains two elements of $\mathcal{S}_3$. Then the generating function for $s_n(Q)$ is rational.*

From Table 6.1, we see that $s_n(123, 132, 213) = F_{n+1}$, which was generalized by Egge [333]:

$$s_n(123, 132, (k-1)(k-2)\cdots 1k) = s_n(132, 213, 12\cdots k) = F_{n+1,k-1},$$

where $F_{n,k}$ is the n-th k-generalized Fibonacci number defined in Subsection 6.1.4. Juarna and Vajanovszki [489, 491] provided a bijective proof of the last formula by giving a bijection between the set of length $(n-1)$ binary strings with no $(k-1)$ consecutive 1s and the set $\mathcal{S}_n(132, 213, 12\cdots k)$, which generalizes the bijection given by Simion and Schmidt [721] between the set of length $(n-1)$ binary strings with no two consecutive 1s and the set $\mathcal{S}_n(123, 132, 213)$. Also, Egge [333] shows that

$$s_n(132, 213, 23\cdots k1) = \sum_{i=1}^{n} F_{i,k-2}.$$

Vatter [777] shows that

$$s_n(132, 213, 23\cdots k1, 12\cdots(j+1)) = \sum_{i=1}^{j} F_{i,k-2},$$

and

$$s_n(132, 213, 231, 12\cdots(k+1)) = \min\{n, k\}.$$

Many other results of similar type can be found in [341] by Egge and Mansour, who used generating functions and bijective techniques to give several sets of pattern-avoiding permutations that can be enumerated in terms of Fibonacci or k-generalized Fibonacci numbers. Another relevant paper here is [78] by Barcucci et al. whose idea (coming from [82]) can be described as follows.

Begin with the fact that $s_n(123, 132, 213) = F_n$ and $s_n(123) = C_n$. The patterns 132 and 213 can be seen as particular cases of more general patterns: respectively, $q_k = 1(k+1)k(k-1)\cdots 2$ and $r_k = k(k-1)(k-2)\cdots 21(k+1)$. When $k \to \infty$, the patterns q_k and r_k increase their lengths, and in the limit they can be disregarded in the enumeration of the permutations in $\mathcal{S}_n(123, q_k, r_k)$, since for each $n \geq 0$, any n-permutation for sure does not contain a pattern of infinite length. In other words, starting from $Av(123, 132, 213)$, the case $k = 2$ (involving Fibonacci numbers), for each $k > 2$ we have a class of pattern-avoiding permutations that for increasing k brings us to the class $Av(123)$ enumerated by Catalan numbers, which can be seen as a sort of "continuity" between Fibonacci and Catalan numbers. In [78], the transformation from Fibonacci to Catalan numbers is achieved in two stages: first, only the pattern 132 is eventually eliminated bringing us to the class $Av(123, 213)$ enumerated by $\{2^{n-1}\}_{n\geq 1}$, then the pattern 213 is increased in order to obtain $Av(123)$. Each class of permutations involved in the transformation from Fibonacci to Catalan numbers is enumerated in [78]. In particular, it is shown in [78] that

$$s_n(123, 2143, 3214) = P_n,$$

the n-th Pell number defined with slightly different initial conditions than above: $P_0 = P_1 = 1$, $P_2 = 2$, and $P_n = 2P_{n-1} + P_{n-2}$.

Mansour [588] found the generating functions for $s_n(Q, q)$ for all q and sets $Q \subset S_3$ with $|Q| \geq 2$. An example of the results obtained in [588] is the following theorem.

Theorem 6.1.45. *([588]) Let $p \in S_k(123, 231)$. Then*

- *there exist m and r with $2 \leq m \leq k+1$ and $1 \leq r \leq m-2$ such that*

$$p = (k(k-1) \cdots mr(r-1) \cdots 1(m-1)(m-2) \cdots (r+1));$$

- *for all $n \geq k$,*

$$s_n(123, 231, p) = (k-2)n - \frac{k(k-3)}{2}.$$

For instance, Theorem 6.1.45 gives

$$s_n(123, 231, 4312) = s_n(123, 231, 1432) = s_n(123, 231, 2143)$$

$$= s_n(123, 231, 3214) = 2n - 2.$$

Further, Mansour [596] computed generating functions for the number of permutations that avoid at least two patterns of length three and contain another pattern (of any length) exactly once. An example of the results in [596] is the following theorem, where $C_p(x)$ is the generating function for the number of permutations that avoid the patterns 132 and 213 and contain one occurrence of p.

Theorem 6.1.46. *([596]) Let $p \in S_k(132, 213)$. Then*

- *there exists $k+1 = r_0 > r_1 > \cdots > r_m = 1$ such that*

$$p = r_1(r_1 + 1) \cdots kr_2(r_2 + 1) \cdots (r_1 - 1) \cdots r_m(r_m + 1) \cdots (r_{m-1} - 1);$$

- *for all $0 \leq r \leq k-1$,*

$$C_{(r+1)(r+2)\cdots kp'}(x) = \frac{x^{k-r}(1-x)}{1 - 2x + x^{k-r}} C_{p'}(x),$$

where $p' \neq \varepsilon$, and

$$C_{12\cdots k}(x) = \frac{x^k(1-x)^2}{(1 - 2x + x^k)^2}.$$

Statistics on 3-pattern-avoiding permutations

The following theorem provides the distribution of the statistic $lmin$, left-to-right minima, on 123-avoiding permutations. Several statistics on 321-avoiding permutations were studied by Adin and Roichman in [10] and on 132-avoiding permutations by Mansour in [599], not to be discussed in this book.

Theorem 6.1.47. ([137]) Let $1 \leq m \leq n$. Then the number of 123-avoiding n-permutations having exactly m left-to-right minima is

$$\frac{1}{n}\binom{n}{m}\binom{n}{m+1},$$

a Narayana number (see Subsection A.2.1 to learn more on these numbers).

Define $F_n^k(P)$ to be the set of n-permutations with exactly k fixed points that avoid each pattern from a set P. Robertson et al. [697] investigated $F_n^0(\tau)$, the τ-avoiding derangements, for all $\tau \in \mathcal{S}_3$. In particular, it is shown in [697] that the 321-avoiding derangements are enumerated by the *Fine numbers*, which count *ordered rooted trees* with root of even degree. The first few values of Fine's sequence are $0, 1, 2, 6, 18, 57, 186, 622, \ldots$; the n-th Fine number is

$$\sum_{i=0}^{n}(-1)^i \frac{i+1}{n+1}\binom{2n-i}{n}.$$

Also, Robertson et al. [697] showed that $F_n^k(132) = F_n^k(213) = F_n^k(321)$ and $F_n^k(231) = F_n^k(312)$ for all $0 \leq k \leq n$.

Further, Mansour and Robertson [608] enumerated $F_n^k(P)$ for all $P \subseteq \mathcal{S}_3$ with $|P| \geq 2$ and $0 \leq k \leq n$. We give here just two examples of results in [608].

Theorem 6.1.48. ([608]) For $n \geq 1$, $|F_n^{n-1}(132, 213)| = 0$, $|F_n^n(132, 213)| = 1$ and

$$|F_{2n+i}^0(132, 213)| = \begin{cases} \frac{5 \cdot 4^{n-1}-2}{3}, & \text{if } i = 0, \\ \frac{2(4^n-1)}{3}, & \text{if } i = 1, \end{cases}$$

$$|F_{2n+i}^{2k}(132, 213)| = \begin{cases} 4^{n-k-1}, & \text{if } i = 0 \text{ and } 1 \leq k \leq n-1, \\ 0, & \text{if } i = 1 \text{ and } 1 \leq k \leq n-1, \end{cases}$$

$$|F_{2n+i}^{2k+1}(132, 213)| = \begin{cases} 0, & \text{if } i = 0 \text{ and } 0 \leq k \leq n-1, \\ 4^{n-k-1}, & \text{if } i = 1 \text{ and } 0 \leq k \leq n-1. \end{cases}$$

Theorem 6.1.49. *([608]) For $n \geq 3$, we have*

$$|F_n^0(132, 231)| = \tfrac{1}{3}(2^{n-1} + (-1)^n),$$

$$|F_n^k(132, 231)| = \tfrac{2}{3}(2^{n-k-1} + (-1)^{n-k}), \; for \; 1 \leq k \leq n - 2,$$

$|F_n^{n-1}(132, 231)| = 0$ and $|F_n^n(132, 231)| = 1.$

Elizalde [354] studied the distribution of the statistics "number of fixed points" and "number of excedances" in permutations avoiding subsets of 3-patterns, which is a more general set up compared to the work of Mansour and Robertson [608]. All the cases of simultaneous avoidance of more than one pattern are enumerated, and some cases are generalized to patterns of arbitrary length, while for avoidance of one single pattern only partial results are given. Among many enumerative results, we state only the following five theorems. Notice that setting $x = 0$ in these theorems gives us pattern-avoidance results on derangements.

Theorem 6.1.50. *([354, 358]) The g.f. for 321-avoiding permutations with respect to fixed points, excedances and descents is*

$$\sum_{n \geq 0} \sum_{\pi \in \mathcal{S}_n(321)} x^{\mathrm{fp}(\pi)} q^{\mathrm{exc}(\pi)} p^{\mathrm{des}(\pi)} z^n =$$

$$\frac{2}{1 + (1 + q - 2x)z + \sqrt{1 - 2(1 + q)z + ((1 + q)^2 - 4qp)z^2}}.$$

Similarly,

$$1 + \sum_{n \geq 1} \sum_{\pi \in \mathcal{S}_n(132)} x^{\mathrm{fp}(\pi)} q^{\mathrm{exc}(\pi)} p^{\mathrm{des}(\pi)+1} z^n = 1 + \sum_{n \geq 1} \sum_{\pi \in \mathcal{S}_n(213)} x^{\mathrm{fp}(\pi)} q^{\mathrm{exc}(\pi)} p^{\mathrm{des}(\pi)+1} z^n$$

$$= \frac{2(1 + xz(p - 1))}{1 + (1 + q - 2x)z - qz^2(p - 1)^2 + \sqrt{f_1(q, z)}},$$

where $f_1(q, z) = 1 - 2(1 + q)z + [(1 - q)^2 - 2q(p - 1)(p + 3)]z^2 - 2q(1 + q)(p - 1)^2 z^3 + q^2(p - 1)^4 z^4$.

Theorem 6.1.51. *([354]) We have*

$$\sum_{n \geq 0} \sum_{\pi \in \mathcal{S}_n(123,132)} x^{\mathrm{fp}(\pi)} q^{\mathrm{exc}(\pi)} z^n = \sum_{n \geq 0} \sum_{\pi \in \mathcal{S}_n(123,213)} x^{\mathrm{fp}(\pi)} q^{\mathrm{exc}(\pi)} p^{\mathrm{des}(\pi)} z^n =$$

$$\frac{1 + xz + (x^2 - 4q)z^2 + (-3xq + q + q^2)z^3 + (xq + xq^2 - 3x^2q + 3q^2)z^4}{(1 - qz^2)(1 - 4qz^2)}.$$

Theorem 6.1.52. *([354]) We have*

$$\sum_{n\geq 0}\sum_{\pi\in\mathcal{S}_n(132,213)} x^{\text{fp}(\pi)} q^{\text{exc}(\pi)} z^n =$$

$$\frac{1-(1+q)z-2qz^2+4q(1+q)z^3-(xq^2+xq+5q^2)z^4+2xq^2z^5}{(1-z)(1-xz)(1-qz)(1-4qz^2)}.$$

Theorem 6.1.53. *([354]) We have*

$$\sum_{n\geq 0}\sum_{k\geq 0}\sum_{\pi\in\mathcal{S}_n(132,(k+1)k\cdots1)} x^{\text{fp}(\pi)} q^{\text{exc}(\pi)} z^n p^k =$$

$$\frac{2(1+xz(p-1))}{(1-p)[1+(1+q-2x)z-qz2(p-1)2+\sqrt{f_1(q,z)}]},$$

where $f_1(q,z) = 1-2(1+q)z+[(1-q)^2-2q(p-1)(p+3)]z^2-2q(1+q)(p-1)^2z^3+q^2(p-1)^4z^4$.

Theorem 6.1.54. *([354]) We have*

$$\sum_{n\geq 0}\sum_{\pi\in\mathcal{S}_n(132,213)} x^{\text{fp}(\pi)} q^{\text{exc}(\pi)} z^n =$$

$$\frac{1+xz+(x^2-q)z^2+(-xq+q^2+q)z^3-x^2qz^4}{(1+qz^2)(1-3qz^2+q^2z^4)}.$$

Among many enumerative results in [606] by Mansour et al., the following three theorems are proved.

Theorem 6.1.55. *([606]) The g.f. for* Av(321) *with respect to the number of fixed points and the number of right-to-left minima is given by*

$$\sum_{n\geq 0}\sum_{\pi\in\mathcal{S}_n(321)} x^{\text{fp}(\pi)} q^{\text{rmin}(\pi)} z^n = \frac{2}{1+z(1+q-2qx)+\sqrt{(1-z(1+q))^2-4z^2q}}.$$

Theorem 6.1.56. *([606]) For* $k > 0$, *let*

$$B_k(x,p,q) = \sum_{n\geq 0}\sum_{\pi\in\mathcal{S}_n(321,23\cdots(k+1)1)} x^n p^{\text{fp}(\pi)} q^{\text{rmin}(\pi)}.$$

Then

$$B_k(x,p,q) = \frac{1}{1+x(1-pq)-xB_{k-1}(x,1,q)}.$$

with $B_0(x, p, q) = 1$. Thus, B_k can be expressed as a continued fraction of the form

$$B_k(x, p, q) = \cfrac{1}{1 + x(1 - pq) - \cfrac{x}{\ddots \cfrac{}{1 + x(1 - q) - \cfrac{x}{1 + x(1 - q) - x}}}}$$

where the fraction has k levels, or in terms of Chebyshev polynomials of the second kind, as

$$B_k(x, p, q) = \frac{U_{k-1}(t) - \sqrt{x}U_{k-2}(t)}{\sqrt{x}[U_k(t) - \sqrt{x}(1 - q(1-p))U_{k-1}(t) - xq(1-p)U_{k-2}(t)]},$$

where $t = 1 + \frac{x(1-q)}{2\sqrt{x}}$.

Theorem 6.1.57. *([606]) We have*

$$\sum_{n \geq 0} \sum_{\pi \in \mathcal{S}_n(231)} x^n y^{\mathrm{inv}(\pi)} z^{\mathrm{rmin}(\pi)} q^{\mathrm{asc}(\pi)} =$$

$$1 - \frac{1}{q} + \cfrac{\frac{1}{q}}{1 + x(1 - zq) - \cfrac{x}{1 + yx(1 - zq) - \cfrac{yx}{1 + y^2 x(1 - zq) - \cfrac{y^2 x}{\ddots}}}}.$$

Bandlow and Killpatrick [74] defined an *area* statistic on Dyck paths (not to be defined here) and gave a bijection from Dyck paths to 312-avoiding permutations that sends the area statistic on Dyck paths to the inv statistic on 312-avoiding permutations. In addition, occurrences of the patterns $k(k-1)\cdots 1$, $k \geq 3$, on 312-avoiding permutations are classified in terms of Dyck paths. We refer to [322] by Dukes and Reifegerste for other relevant results.

The *diagram* of a permutation π is obtained from the permutation matrix of $\pi = \pi_1 \pi_2 \cdots \pi_n$ by shading, for every square (i, π_i), all squares to the east in row i and squares to the south in column π_i. Fulton [408] introduced the *essential set* of a permutation π as the set of southeast corners of the diagram of π, which together with a *rank function* was used as a powerful tool in Fulton's algebraic treatment of *Schubert polynomials* and *degeneracy loci*. Eriksson and Linusson [368] showed that the average size of the essential set of a permutation in $\mathcal{S}_n$ is

$$\frac{\binom{n-1}{3} + 6\binom{n}{2}}{6n} \sim \frac{1}{36}n^2,$$

while the average size of the essential set of a 321-avoiding permutation in $\mathcal{S}_n$ is

$$\frac{4^{n-2}}{C_n} \sim \frac{\sqrt{\pi}}{16} n^{3/2},$$

where C_n is the n-th Catalan number.

Reduced decompositions

For an n-permutation π and $i \in \{1, 2, \ldots, n-1\}$, the map s_i transposes i and $i+1$, and fixes all other elements in π. The symmetric group $\mathcal{S}_n$ is the *Coxeter group of type A_{n-1}*, and it is generated by the adjacent transpositions $\{s_i \mid i = 1, 2, \ldots, n-1\}$. The adjacent transpositions satisfy the *Coxeter relations*:

- $s_i^2 = 1$ for all i;

- (*the short braid relation*) $s_i s_j = s_j s_i$, if $|i - j| > 1$;

- (*the long braid relation*) $s_i s_{i+1} s_i = s_{i+1} s_i s_{i+1}$ for $1 \le i \le n - 2$.

For a permutation $\pi = \pi_1 \pi_2 \cdots \pi_n$, $s_i \pi$ interchanges the positions of values i and $i + 1$ in π, while πs_i interchanges the values in positions i and $i + 1$ in π; that is, $\pi s_i = \pi_1 \pi_2 \cdots \pi_{i-1} \pi_{i+1} \pi_i \pi_{i+2} \pi_{i+3} \cdots \pi_n$. Because $\mathcal{S}_n$ is generated by adjacent transpositions, any $\pi \in \mathcal{S}_n$ can be written as $\pi = s_{i_1} s_{i_2} \cdots s_{i_\ell}$ for some $\{i_1, i_2, \ldots, i_\ell\}$. The least such ℓ is called the *length* of π, denoted $\ell(\pi)$. It is well-known that for a permutation π, $\mathrm{inv}(\pi) = \ell(\pi)$, and that is why the permutation $n(n-1) \cdots 1$ is called the *longest element* in $\mathcal{S}_n$.

Definition 6.1.58. For a permutation π with $\ell(\pi) = \ell$, a word $i_1 i_2 \cdots i_\ell$ such that $\pi = s_{i_1} s_{i_2} \cdots s_{i_\ell}$ is a *reduced decomposition* of, or a *reduced word* for π. Let $R(\pi)$ denote the set of all reduced decompositions of π.

Definition 6.1.59. Let $\mathbf{i} = i_1 i_2 \cdots i_\ell$ be a reduced decomposition of $\pi = \pi_1 \pi_2 \cdots \pi_n$. For $M \in \mathbb{N}$, the *shift* of $\mathbf{i}$ by M is

$$\mathbf{i}^M := (i_1 + M) \cdots (i_\ell + M) \in R(12 \cdots M (\pi_1 + M)(\pi_2 + M) \cdots (\pi_n + M)).$$

Billey et al. [116] showed that 321-avoiding permutations are exactly those permutations where the subsequence $i(i \pm 1)i$ never occurs in a reduced decomposition. Reiner [687] showed that the number of $i(i \pm 1)i$ occurrences in a reduced decomposition of the longest element in the symmetric group, which has the maximal number of occurrences of 321, is equal to the number of such reduced decompositions. Stanley [742] had previously shown that this is the number of *standard Young tableaux of a staircase shape*. Among other results, Tenner [764] generalized the result of Billey et al. via a new characterization of vexillary (that is, 2143-avoiding) permutations, which is recorded in the following theorem.

Theorem 6.1.60. *([764]) The permutation π is vexillary if and only if for every permutation σ containing an occurrence of π there exists a reduced decomposition $\mathbf{j} \in R(\sigma)$ containing some shift of an element $\mathbf{i} \in R(\pi)$ as a factor.*

321-k-gon-avoiding permutations

Definition 6.1.61. A 321-k-gon-avoiding permutation π avoids 321 and the following four patterns:

$$k(k+2)(k+3)\cdots(2k-1)1(2k)23\cdots(k-1)(k+1),$$
$$k(k+2)(k+3)\cdots(2k-1)(2k)12\cdots(k-1)(k+1),$$
$$(k+1)(k+2)(k+3)\cdots(2k-1)1(2k)23\cdots k,$$
$$(k+1)(k+2)(k+3)\cdots(2k-1)(2k)123\cdots k.$$

We let $f_k(n)$ denote the number of 321-k-gon-avoiding permutations in $\mathcal{S}_n$ and $f_k(x) = \sum_{n\geq 0} f_k(n)x^n$.

Remark 6.1.62. Note that 321-4-gon-avoiding permutations, also called 321-hexagon-avoiding permutations, are discussed in Subsection 2.3.2 in connection with the Kazhdan-Lusztig polynomials.

Stankova and West [738] gave an exact enumeration of the 321-k-gon-avoiding permutations in terms of linear recurrences with constant coefficients for the cases $k = 2, 3, 4$.

Theorem 6.1.63. *([738]) For $k = 2, 3, 4$, and, respectively, $n \geq 3, 5, 6$, the sequences $f_k(n)$ satisfy the recursive relations:*

$$f_4(n) = 6f_4(n-1) - 11f_4(n-2) + 9f_4(n-3) - 4f_4(n-4) - 4f_4(n-5) + f_4(n-6);$$
$$f_3(n) = 4f_3(n-1) - 4f_3(n-2) + 3f_3(n-3) + f_3(n-4) - f_3(n-5);$$
$$f_2(n) = 3f_2(n-1) - 3f_2(n-2) + f_2(n-3) = (n-1)^2 + 1.$$

Mansour and Stankova [611] extended these results by proving the following theorem.

Theorem 6.1.64. *([611]) For $k \geq 3$,*

$$f_k(x) = \frac{(1 + 2x^2 + x^3)U_{2k-3}\left(\frac{1}{2\sqrt{x}}\right) - \sqrt{x}(1+x)U_{2k-4}\left(\frac{1}{2\sqrt{x}}\right)}{\sqrt{x}\left[(1 + 2x^2 + x^3)U_{2k-2}\left(\frac{1}{2\sqrt{x}}\right) - \sqrt{x}(1+x)U_{2k-3}\left(\frac{1}{2\sqrt{x}}\right)\right]},$$

where $U_n(x)$ is the n-th Chebyshev polynomial of the second kind.

6.1.6   More results on classical patterns

In this subsection we collect (in no particular order) miscellaneous facts on classical patterns, normally providing very little detail on a particular result/research direction, and thus inviting the reader to consult the corresponding original sources.

Remark 6.1.65. Recall that a *pattern class* is a set of permutations closed under pattern containment or, equivalently, defined by certain subsequence avoidance conditions. The study of pattern classes is an interesting and well-developed area of research with many (enumerative) applications to (usual) pattern-avoidance theory. However, in this book, we discuss pattern classes and associated problems in a neither comprehensive nor systematic way, just from several perspectives. One should consult, for example, [53, 55, 13, 18, 46, 466, 467, 182, 58, 186, 181, 635, 634, 350, 784, 797] and references therein for a proper overview of known results in this direction.

We refer to Section 8.1 for the notion of the *basis* of a permutation class and *inflation*. New permutation classes can be formed using constructions involving one or more old classes: see [46] by Atkinson, [58] by Atkinson and Stitt, and [634] by Murphy. The *wreath product* of two sets of permutations X and Y (not necessarily permutation classes) is the set $X \wr Y$ of all permutations which can be expressed as an inflation of a permutation in X by permutations in Y, that is, the set of permutations of the form $\pi[\alpha_1, \alpha_2, \ldots, \alpha_n]$ with $\pi \in X$ and $\alpha_1, \alpha_2, \ldots, \alpha_n \in Y$. The wreath product of two permutation classes is again a permutation class. Moreover, as is shown in [58], the wreath product is associative: $(A \wr B) \wr C = A \wr (B \wr C)$.

Atkinson and Stitt [58] proved that for any finitely based class X, the wreath product $X \wr \mathrm{Av}(21)$ is also finitely based, and that $\mathrm{Av}(21) \wr \mathrm{Av}(321654)$ is not finitely based. Brignall [181] generalized both of these results by observing the connection to "pin sequences" introduced in [186] by Brignall et al. Skipping the definition of pin sequences, we state the following theorem from [181].

Theorem 6.1.66. *([181]) For any finitely based class Y not admitting arbitrarily long pin sequences, the wreath product $X \wr Y$ is finitely based for all finitely based classes X.*

Additionally, Brignall [181] indicated a general construction for basis elements in the case where $X \wr Y$ is not finitely based.

Elder [350] studied pattern-avoidance from a *formal language theory* point of view. He proved that a variant of the *insertion encoding* of Albert et al. [30] for any class of pattern-avoiding permutations is *context-sensitive*, and every finitely based class of permutations bijects to a context-sensitive language.

The *containment order* introduced in Definition 2.2.38 is a partial order (for two permutations, σ and τ, $\sigma \leq \tau$ if and only if σ occurs in τ as a pattern). Spielman and Bóna [730] proved that this partial order contains an infinite antichain, that is, that there exists an infinite set of permutations no one of which contains another as a pattern.

Barcucci et al. [83] studied pattern-avoidance from the following perspective. Let $F^k = \{\sigma(k+1)(k+2) \mid \sigma \in \mathcal{S}_k\}$. For $k = 1$, $F^1 = \{123\}$ and $\mathrm{Av}(F^1)$ is counted by the Catalan numbers C_n, while $F^2 = \{1234, 2134\}$ and $\mathrm{Av}(F^2)$ is counted by the Schröder numbers. For $k > 2$, $C_n \leq s_n(F^k) \leq n!$, and $\mathrm{Av}(F^k)$ is enumerated in [83] registering two statistics.

Theorem 6.1.67. *([83]) The generating function of permutations in $\mathrm{Av}(F^k)$ according to the length (variable x), the number of left-to-right minima (variable y), and non-inversions (variable q) is given by*

$$x^k y f_k(x,y,q) \prod_{i=1}^{k-1}(y + q[i]_q) + \sum_{j=2}^{k} x^{j-1} y \prod_{i=1}^{j-2}(y + q[i]_q), \quad k \geq 1,$$

where

$$f_k(x,y,q) = \frac{\sum_{n \geq 0} \frac{(-1)^n x^n q^{kn + \binom{n+1}{2}}}{(q)_n (x(y + q[k-1]_q))_{n+1}}}{\sum_{n \geq 0} \frac{(-1)^n x^n q^{kn + \binom{n+1}{2}}}{(q)_n (x(y + q[k-1]_q))_n}}.$$

Here $[i]_q = \frac{1-q^i}{1-q}$ and $(a)_n = \prod_{i=0}^{n-1}(1 - aq^i)$.

Erdős and Szekeres [363] showed that any permutation of length $n \geq k^2 + 1$ contains a monotone subsequence of length $k + 1$ (see Theorem 6.1.2). It is conjectured by Myers [640] that any n-permutation, for $n \geq k^2 + 1$, has at least

$$M_k(n) = (n \mod k)\binom{\lceil \frac{n}{k} \rceil}{k+1} + (k - (n \mod k))\binom{\lfloor \frac{n}{k} \rfloor}{k+1} \approx \frac{1}{k^k}\binom{n}{k+1}$$

monotone subsequences of length $k+1$ (this number is achieved on a simple example). In other words, the conjecture is that, for $n \geq k^2 + 1$, the value of the function $(12 \cdots (k+1) + (k+1)k \cdots 1)(\pi)$ is at least $M_k(n)$.

The case $k = 2$ is proved in [640] with a complete characterization of the extremal permutations. Moreover, for $k > 2$ and $n \geq k(2k-1)$, Myers [640] characterized the permutations containing the minimum number of monotone subsequences of length $k+1$ subject to the additional constraint that either all such subsequences are occurrences of the pattern $12 \cdots (k+1)$ or all of them are occurrences of the pattern $(k+1)k \cdots 1$. It is shown in [640] that there are $2\binom{k}{n \mod k} C_k^{2k-2}$ such extremal permutations, where C_k is the k-th Catalan number. It is further conjectured by Myers [640], based on computer experiments, that permutations with a minimum number of monotone $(k + 1)$-subsequences must have all such subsequences in the same direction (all ascending or all descending) if $n \geq k(2k - 1)$, except for the case of $k = 3$ and $n = 16$.

6.1.7 Classical patterns and involutions

Definition 6.1.68. An n-permutation π is an *involution* if, in the usual group-theoretic sense, the square of π is the identity permutation, that is, if $\pi(\pi(i)) = i$ for all i, $1 \leq i \leq n$; equivalently, $\pi = \pi^{-1}$. Let I_n denote the set of all involutions – called *n-involutions* – in $\mathcal{S}_n$, and let $I_n(P)$ denote the set of n-involutions avoiding each pattern from a set of patterns P; $I(P) = \cup_{n \geq 0} I_n(P)$. We also let $i_n(P) = |I_n(P)|$. Moreover, $I_n^r(P; \tau)$ denotes the set of n-involutions avoiding P and containing r occurrences of τ. Finally, $i_n^r(P; \tau) = |I_n^r(P; \tau)|$ and

$$I_{P;\tau}^r(x) = \sum_{n=0}^{\infty} i_n^r(P; \tau)x^n; \quad I_{P;\tau}(x, y) = \sum_{n=0}^{\infty} \sum_{r=0}^{\infty} i_n^r(P; \tau)x^n y^r.$$

Example 6.1.69. 52361478 is an involution, while 23145 is not.

The permutation matrix corresponding to an involution is symmetric with respect to the line $y = x$, which is useful in dealing with involutions. For example, the permutation matrix for the involution 52361478 is in Figure 6.1. If one assumes, for instance, that an involution avoids the pattern 321, then its permutation matrix has two increasing symmetric subsequences.

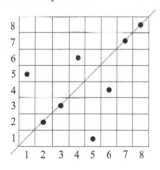

Figure 6.1: The permutation matrix corresponding to the involution 52361478.

Definition 6.1.70. Patterns p_1 and p_2 are *I-Wilf-equivalent*, denoted $p_1 \sim_I p_2$, if $i_n(p_1) = i_n(p_2)$ for all $n \geq 0$.

We begin with selected results by Egge and Mansour [340]. Review Definition 2.1.11 for the notion of a layered pattern. Also, the definition of k-generalized Fibonacci numbers $F_{n,k}$ is in Section 6.1.4. Note that for $k \geq 1$,

$$\sum_{n=0}^{\infty} F_{n,k} x^n = \frac{x}{1 - x - x^2 - \cdots - x^k}.$$

Proposition 6.1.71. *([340]) For all $n \geq 0$, the sets $I_n(231)$, $I_n(312)$, and $\mathcal{S}_n(231, 312)$ are all equal to the set of layered n-permutations, and thus*

$$I_n(231, p) = I_n(312, p) = \mathcal{S}_n(231, 312, p) = \mathcal{S}_n(132, 213, r(p))$$

for any pattern p ($r(p)$ is the reverse of p).

Theorem 6.1.72. *([340]) For $k \geq 1$,*

$$I_{231;k(k-1)\cdots 1}(x, y) = \frac{1}{1 - \sum_{j \geq 1} x^j y^{\binom{j}{k}}}.$$

Theorem 6.1.73. *([340]) For $0 \leq r \leq k$ and $n \geq 0$,*

$$i_n^r(231; k(k-1)\cdots 1) = \sum_{s_0, s_1, \ldots, s_r} \prod_{i=0}^{r} F_{s_i+1, k-1},$$

where the sum is over all sequences s_0, $s_1, \ldots, s_r$ of nonnegative integers such that $\sum_{i=0}^{r} s_i = n - kr$. Moreover,

$$I_{231,k(k-1)\cdots 1}^r(x) = \frac{x^{kr}}{(1 - x - \cdots - x^{k-1})^{r+1}}.$$

Theorem 6.1.74. *([340]) Let p be a layered pattern with m layers of lengths ℓ_1, $\ell_2, \ldots, \ell_m$. Set $k_0 = \ell_1 - 1$, $k_m = \ell_m - 1$ and $k_i = \min(\ell_i - 1, \ell_{i+1} - 1)$ for $1 \leq i \leq m-1$. Then*

$$i_n^1(231; p) = \sum_{s_0, s_1, \ldots, s_m} \prod_{i=0}^{m} F_{s_i+1, k_i},$$

where the sum is over all sequences of nonnegative integers such that $\sum_{i=0}^{m} s_i = n - \sum_{i=1}^{m} \ell_i$. Moreover,

$$I_{231;p}^1(x) = x^{\sum_{i=1}^{m} \ell_i} \prod_{i=0}^{m} \frac{1}{1 - x - x^2 - \cdots - x^{k_i}}.$$

A number of results in [340] are on involutions containing one occurrence of the pattern 231. We present selected results on that.

Theorem 6.1.75. *([340]) Fix $k \geq 4$. Then for all $n \geq 0$, the number of n-involutions containing exactly one occurrence of the pattern 231 and avoiding the pattern $k(k-1)\cdots 1$ is given by*

$$\sum_{i=0}^{n-4} F_{i+1,k-1} F_{n-i-3,k-1}.$$

Moreover, the corresponding generating function is given by

$$\frac{x^4}{(1 - x - \cdots - x^{k-1})^2}.$$

The last theorem can be generalized.

Theorem 6.1.76. *([340]) For all $k \geq 4$, the generating function for the number of involutions containing exactly one occurrence of the pattern 231 and counting occurrences of the pattern $k(k-1)\cdots 1$ (variable y) is given by*

$$\frac{x^4}{\left(1 - \sum_{j \geq 1} x^j y^{\binom{j}{k}}\right)^2}.$$

Many enumerative results, not to be stated here, on avoidance of the pattern 132 and containment of it exactly once, subject to extra restrictions, are given by Guibert and Mansour [437]. In particular, it is shown in [437] that the number of n-involutions containing 132 exactly once which have k fixed points, $1 \leq k \leq n$, is the *ballot number*

$$\binom{n-2}{\frac{n+k}{2}-1} - \binom{n-2}{\frac{n+k}{2}}.$$

Research on involutions with a specified number of fixed points was conducted in [304] by Deutsch et al. In particular, it is shown in [304] that, for all $0 \leq k \leq n$, the number of involutions with k fixed points is the same in the following sets:

- $I_n(132)$, $I_n(213)$ and $I_n(321)$;

- $I_n(231)$ and $I_n(312)$;

- $I_n^1(\emptyset; 132)$ and $I_n^1(\emptyset; 213)$;

- $I_n^1(\emptyset; 231)$ and $I_n^1(\emptyset; 312)$.

Moreover, several enumerative results are presented in [304] on involutions with k fixed points that either avoid or contain exactly once a classical 3-pattern. In the following two theorems we provide examples of these results.

Theorem 6.1.77. *([304]) Let $p \in \{132, 213, 321\}$. For $0 \le k \le n$, the number of n-involutions with k fixed points that avoid the pattern p is given by*

$$\frac{k+1}{n+1} \binom{n+1}{\frac{n-k}{2}}$$

for $n + k$ even, and 0 otherwise.

Theorem 6.1.78. *([304]) For $n \ge 3$, $0 \le k \le n$, the number of n-involutions with k fixed points that contain exactly one occurrence of 321 is given by*

$$\frac{k(k+3)}{n+1} \binom{n+1}{\frac{n-k}{2}-1}.$$

Elizalde [354] presented results on avoidance of more than one 3-pattern on involutions when fixed points are registered.

Theorem 6.1.79. *([354]) For a set of patterns P, let $G_P(x, z) = \sum\limits_{n \ge 0} \sum\limits_{\pi \in I_n(P)} x^{\mathrm{fp}(\pi)} z^n$.*

- $G_{123,132}(x, z) = G_{123,213}(x, z) = \frac{1 + xz + (x^2 - 1)z^2}{1 - 2z^2}$;

- $G_{231,321}(x, z) = G_{312,321}(x, z) = \frac{1}{1 - xz - z^2}$;

- $G_{132,213}(x, z) = \frac{1 - z^2}{(1 - xz)(1 - 2z^2)}$;

- $G_{231,312}(x, z) = \frac{1 - z^2}{1 - xz - 2z^2}$;

- $G_P(x, z) = \frac{1 + xz^3}{(1 - xz)(1 - z^2)}$ for $P \in \{\{132, 231\}, \{213, 231\}, \{132, 312\}, \{213, 312\}\}$;

- $G_{132,321}(x, z) = G_{213,321}(x, z) = \frac{1}{(1 - xz)(1 - z^2)}$;

- $G_{123,231}(x, z) = G_{123,312}(x, z) = \frac{1 + xz + (x^2 - 1)z^2 + xz^3 + z^4}{(1 - z^2)^2}$;

- $G_{123,321}(x, z) = 1 + xz + (x^2 + 1)z^2 + 2xz^3 + 2z^4$.

Guibert and Mansour [438] studied certain statistics on 132-avoiding involutions. In particular, they obtained the following result.

Theorem 6.1.80. *([438]) The number of 132-avoiding n-involutions having r ascents is given by*

$$\binom{\lfloor n/2 \rfloor}{\lfloor (r+1)/2 \rfloor} \binom{\lfloor (n-1)/2 \rfloor}{\lfloor r/2 \rfloor}.$$

Even and *odd* involutions are also studied in [438] (see Subsection 6.1.10 for a discussion on patterns in even/odd permutations). For example, it is shown in [438] that the number of even and odd n-involutions containing exactly one occurrence of the pattern 132, for $n \geq 5$ and $n \geq 3$, respectively, is given by

$$\binom{n-3}{1+2\lfloor (n-5)/4 \rfloor} \quad \text{and} \quad \binom{n-3}{2\lfloor (n-3)/4 \rfloor}.$$

In what follows one needs the notion of a *simple permutation* and *inflation* discussed in Section 8.1 below (see Definitions 8.1.1 and 8.1.4). Compare the following proposition with Propositions 6.1.103 and 8.1.7.

Proposition 6.1.81. *([187]) A permutation $\pi = 12[\alpha_1, \alpha_2]$ is an involution if and only if α_1 and α_2 are involutions.*

Proposition 6.1.82. *([187]) The involutions that are inflations of 21 are precisely those of the form:*

- $21[\alpha_1, \alpha_2]$, *where neither α_1 nor α_2 are skew indecomposable and $\alpha_1 = \alpha_2^{-1}$, and*

- $321[\alpha_1, \alpha_2, \alpha_3]$, *where α_1 and α_3 are skew indecomposable, $\alpha_1 = \alpha_3^{-1}$, and α_2 is an involution.*

Proposition 6.1.83. *([187]) Let $\pi = \sigma[\alpha_1, \alpha_2, \ldots, \alpha_m]$ be the inflation of a simple permutation $\sigma = \sigma_1 \sigma_2 \cdots \sigma_m \neq 21$. If π is an involution, then σ is an involution, and the following equalities hold: $\alpha_i = \alpha_{\sigma_i^{-1}}^{-1} = \alpha_{\sigma_i}^{-1}$, for $i = 1, 2, \ldots, m$. In other words, every transposition in σ must be inflated with two permutations one the inverse of the other.*

Manara and Cippo [582] studied involutions belonging to $I(321)$, the class of 321-avoiding involutions. In particular, they calculated several generating functions. Simple 321-avoiding involutions are shown in [582] to be counted by *Riordan numbers*, and involutions are linked to *Motzkin* and *Dyck paths*. Motzkin and Dyck paths are defined in Subsection A.2.2, while the Riordan number R_n (sequence A005043 in [726]) is the number of Riordan paths of length n where a Riordan path is a Motzkin path containing no horizontal steps at ground level (x-axis). These numbers begin

$$1, 1, 3, 6, 15, 36, 91, 232, 603, 1585, 4213, \ldots.$$

We state below selected results in [582]. We need the following definition.

Definition 6.1.84. A permutation is of *type* 12 (resp., 21) if it is an inflation of 12 (resp., 21).

Example 6.1.85. 21354 and 45123 are examples of type 12 and 21 permutations, respectively.

Proposition 6.1.86. *([582]) If $\pi \in I(321)$ is of type 21, then*

$$\pi = 21[12\cdots m, 12\cdots m] = (m+1)(m+2)\cdots(2m)12\cdots m \in I_{2m}(321).$$

Proposition 6.1.87. *([582]) Let $n > 2$. If $\sigma \in I_n(321)$ is a simple involution, then n is even. If π is an inflation of a simple involution $\sigma \in I_n(321)$ then the length of π is even.*

Proposition 6.1.88. *([582]) The set $I(321)$ has infinitely many simple involutions.*

Theorem 6.1.89. *([582]) The g.f. for $I(321)$ is given by*

$$\frac{1 - 4x^2 - \sqrt{1 - 4x^2}}{2x(2x-1)},$$

and thus $i_n(321) = \binom{n}{\lfloor n/2 \rfloor}$.

Theorem 6.1.90. *([582]) The g.f. for the number of involutions in $I(321)$ of type 12 is given by*

$$\frac{1 + x - 4x^2 - 4x^3 - \sqrt{1 + 2x - 7x^2 - 12x^3 + 16x^4 + 16x^5 - 16x^6}}{2x(2x-1)}.$$

4-patterns and involutions

Jaggard [477] completed classification of avoidance of a 4-pattern on involutions. The corresponding I-Wilf-classes are as follows:

$$\{1324\}, \{1234, 2143, 3412, 4321, 1243, 2134, 1432, 3214\}, \{2431, 4132, 3241, 4213\},$$

$$\{4231\}, \{1342, 1423, 2314, 3124\}, \{2341, 4123\}, \{3421, 4312\}, \{2413, 3142\}.$$

Egge [334] connected generating functions for various subsets of $I(3412)$ with continued fractions and Chebyshev polynomials of the second kind, and gave a recursive formula for computing them. For example, the following two theorems dealing with 3412- and $(3412, (2k)(2k-1)\cdots 1)$-avoiding involutions, respectively, are proved in [334].

Theorem 6.1.91. *([334]) We have*

$$\sum_{\pi \in I(3412)} \prod_{k \geq 1} x_k^{k(k-1)\cdots 1(\pi)} = \cfrac{1}{1 - x_1 - \cfrac{x_1^2 x_2}{1 - x_1 x_2^2 x_3 - \cfrac{x_1^2 x_2^5 x_3^4 x_4}{1 - x_1 x_2^4 x_3^6 x_4^4 x_5 - \cdots}}},$$

where the n-th numerator is $\prod_{i=1}^{2n} x_i^{\binom{2n-2}{i-1}+\binom{2n-1}{i-1}}$ *and the n-th denominator is* $1 - \prod_{i=1}^{2n+1} x_i^{\binom{2n}{i-1}}$.

Theorem 6.1.92. *([334]) For all* $k \geq 1$,

$$\sum_{\pi \in I(3412,(2k)(2k-1)\cdots 1)} x^{|\pi|} = \frac{U_{k-1}\left(\frac{1-x}{2x}\right)}{xU_k\left(\frac{1-x}{2x}\right)}$$

and

$$\sum_{\pi \in I(3412,(2k-1)(2k-2)\cdots 1)} x^{|\pi|} = \frac{U_{k-1}\left(\frac{1-x}{2x}\right) + U_{k-2}\left(\frac{1-x}{2x}\right)}{x\left(U_k\left(\frac{1-x}{2x}\right) + U_{k-1}\left(\frac{1-x}{2x}\right)\right)},$$

where $U_n(x)$ *is the n-th Chebyshev polynomial of the second kind.*

Egge and Mansour [342] studied involutions that either avoid the pattern 3412 or contain it exactly once, subject to extra restrictions which are avoidance, or containment exactly once, of another pattern. In many cases the obtained results are expressed in terms of Chebyshev polynomials of the second kind. We provide several examples of the enumerative results in [342]. In what follows,

$$V_n(x,y) = U_n\left(\frac{1-x}{2x\sqrt{y}}\right).$$

Theorem 6.1.93. *([342]) For all* $k \geq 1$, *we have the following generating functions for the number of* $(3412, (2k)(2k-1)\cdots 1)$- *and* $(3412, (2k-1)(2k-2)\cdots 1)$-*avoiding involutions, respectively, with 2-cycles registered (variable y):*

$$\frac{V_{k-1}(x,y)}{x\sqrt{y}V_k(x,y)} \quad and \quad \frac{\sqrt{y}V_{k-1}(x,y) + V_{k-2}(x,y)}{x\sqrt{y}(\sqrt{y}V_k(x,y) + V_{k-1}(x,y))}.$$

Theorem 6.1.94. *([342]) The g.f. for those involutions that contain exactly one occurrence of* 3412 *with 2-cycles registered (variable y) is given by*

$$\frac{(1+x)(4 - 12x + 12x^2 - 12x^2y - 4x^3 + 12x^3y)}{8x^2y(1-x^2y)\sqrt{(x-1)^2 - 4x^2y}} - \frac{(1+x)(4 - 8x + 4x^2 - 4x^2y)}{8x^2y(1-x^2y)}.$$

Theorem 6.1.95. *([342]) For all* $k \geq 1$, *we have the following generating functions for the number of* $(2k)(2k-1)\cdots 1$- *and* $(2k-1)(2k-2)\cdots 1$-*avoiding involutions, respectively, which contain exactly one occurrence of* 3412; *variable y registers 2-cycles in involutions:*

$$\frac{\sum_{j=0}^{k-2} w_{2j+1}(x,y)V_j^2(x,y)}{(1-x^2y)V_k^2(x,y)};$$

$$\frac{\sum_{j=0}^{k-2} w_{2j}(x,y)(\sqrt{y}V_j(x,y) + V_{j-1}(x,y))^2}{(1-x^2y)(\sqrt{y}V_k(x,y) + V_{k-1}(x,y))^2},$$

where $w_{2k}(x,y) = 1 + x - x^{2k+1}y^k - x^{2k+2}y^{k+1}$ and $w_{2k-1}(x,y) = 1 + x - x^{2k}y^k - x^{2k+1}y^k$.

As a corollary to Theorem 6.1.95, the number of 321-avoiding n-involutions with exactly one occurrence of 3412 is

$$\frac{n-3}{5}F_{n-1} + \frac{n-1}{5}F_{n-3},$$

where F_n is the n-th Fibonacci number, and the number of 4321-avoiding n-involutions with exactly one occurrence of 3412 is

$$2^{n-5}(3n - 10).$$

Vexillary involutions

The following result was first conjectured by Guibert [434] and proved in a bijective manner by Guibert et al. [440]. Bousquet-Mélou [158] provided an alternative proof of the result and refined it.

Theorem 6.1.96. *([440, 158]) The number of vexillary (2143-avoiding) involutions in $\mathcal{S}_n$ is the n-th Motzkin number:*

$$i_n(2143) = \sum_{k=0}^{\lfloor n/2 \rfloor} \frac{1}{k+1}\binom{2k}{k}\binom{n}{2k}.$$

Combining results in [434, 440] we have

$$i_n(1234) = i_n(3412) = i_n(4321) = i_n(2143) = i_n(1243) = M_n.$$

It was also conjectured by Guibert [434] that $i_n(1432) = M_n$.

Theorem 6.1.97. *([440, 158]) Let $Z = Z(t)$ be the unique power series in t satisfying*

$$Z = t(1 + Z + Z^2).$$

Then the generating function $G(t; u, v)$ that counts vexillary involutions by the length, the position of the first (leftmost) descent and the number of active sites is algebraic of degree 2:

$$G(t; u, v) = \frac{uv(1 - tv(1+Z) + t^2uv^2Z)}{(1 - tuv)(1 - tuvZ)(1 - tv(1+Z))}.$$

In particular, $G(t; 1, 1) = Z/t$, and the Lagrange inversion formula *gives Theorem 6.1.96.*

Prior to [440], Guibert and Pergola [439] found the number of vexillary involutions π such that $\pi = r.c(\pi)$, which is recorded in the next theorem.

Theorem 6.1.98. *The number of vexillary involutions on $\mathcal{S}_{2n}$ which are fixed under the composition of the reverse and complement operations is*

$$\sum_{i=0}^{n} \frac{n!}{(n-i)! \lfloor i/2 \rfloor! \lceil i/2 \rceil!}.$$

The corresponding generating function is

$$\frac{\sqrt{(1+x)/(1-3x)} - 1}{2}.$$

Counting patterns in involutions

The following two theorems are by Guibert and Mansour [437] and by Egge and Mansour [340], respectively.

Theorem 6.1.99. *([437]) The number of involutions in $\mathcal{S}_n$ which contain exactly one occurrence of 132 is given by $\binom{n-2}{\lfloor (n-3)/2 \rfloor}$ for all $n \geq 3$.*

Theorem 6.1.100. *([340]) The number of involutions in $\mathcal{S}_n$ which contain exactly one occurrence of 231 is given by $(n-1)2^{n-6}$ for all $n \geq 5$.*

Mansour et al. [622] presented an algorithm that computes the generating function $I_r^{231}(x)$ for the number of involutions with r occurrences of the pattern 231 for any $r \geq 0$. To get the result for a given r, the algorithm performs certain routine checks for all involutions of length at most $2r + 2$. Moreover, the algorithm was implemented in C, and yielded explicit enumerative results for $0 \leq r \leq 7$. We state below the formulas for $r = 2, 3$ referring to [622] for the remaining cases.

Theorem 6.1.101. *([622]) For $r = 2, 3$,*

$$I_r^{231}(x) = \frac{(1-x)^2}{(1-2x)^{r+1}} Q_r(x),$$

where

$$Q_2(x) = x^4(1 - 3x^2 - 2x^3 + x^4 - x^5) \text{ and}$$
$$Q_3(x) = x^5(4 - 14x + 8x^2 + 11x^3 - 6x^4 - 2x^5 + 2x^6 + 5x^7 - 2x^8 + x^9).$$

The corresponding numbers of n-involutions, for $n \geq 4r + 1$, are $2^n Q_r(n)$ where

$$Q_2(n) = \tfrac{1}{2^{12}}(n^2 + 137n - 234) \text{ and}$$
$$Q_3(n) = \tfrac{1}{3 \cdot 2^{24}}(n^3 + 414n^2 + 12227n - 30762).$$

McKay et al. [625] show that the probability that an involution in I_n contains an occurrence of any fixed classical pattern $\tau \in \mathcal{S}_k$ is $1/k! + o(1)$ as $n \to \infty$. This result was sharpened by Jaggard [478] who proved the following theorem.

Theorem 6.1.102. ([478]) For a fixed pattern $\tau = \tau_1\tau_2\cdots\tau_k \in \mathcal{S}_k$ and $n \geq k$, the number of n-involutions containing an occurrence of τ equals

$$\sum_j \binom{n-k}{k-j} |I_{n-2k+j}|$$

where the sum is taken over $j = 0$ and those $j \in [k]$ such that the reduced form of $\tau_1\tau_2\cdots\tau_j$ is an involution in $\mathcal{S}_j$.

As a result of his studies, Jaggard [478] gives an interesting characterization of layered patterns (see Definition 2.1.11) in terms of involutions.

Proposition 6.1.103. ([478]) A permutation $\pi = \pi_1\pi_2\cdots\pi_n \in \mathcal{S}_n$ is a layered permutation if and only if the reduced form of $\pi_1\pi_2\cdots\pi_i$ is an involution for every $i \in [n]$.

Involutions invariant under reverse complement

Egge [337] studied pattern-avoiding involutions that are invariant under the composition of the reverse and complement operations. Below we list results on that from [337], where $i_n^{rc}(P)$ denotes the number of involutions in $I_n(P)$ that are invariant under the reverse complement.

- $i_{2n}^{rc}(132) = i_{2n}^{rc}(132, 213) = i_{2n+1}^{rc}(132) = i_{2n+1}^{rc}(132, 213) = 2^n$ for $n \geq 0$;

- $i_n^{rc}(132, 231) = i_{2n}^{rc}(132, 321) = 2$ for $n \geq 2$;

- $i_{2n}^{rc}(132, 123) = i_{2n+3}^{rc}(132, 123) = F_{n+1}$, the $(n+1)$-th Fibonacci number;

- $i_{2n+1}^{rc}(132, 321) = i_n^{rc}(132, 123, 231) = i_n^{rc}(132, 321, 231) = 1$ for $n \geq 3$;

- $i_{2n+1}^{rc}(123) = \binom{n}{\lfloor \frac{n}{2} \rfloor}$ and $i_{2n}^{rc}(123) = 2^n$ for $n \geq 1$.

More results on involutions

Jaggard [477] proved that the following two pairs of patterns are I-Wilf-equivalent for any permutations τ on $\{3, 4, \ldots, k\}$ and $\{4, 5, \ldots, k\}$, respectively:

$$12\tau \sim_I 21\tau, \quad \text{and} \quad 123\tau \sim_I 321\tau.$$

Jaggard [477] also conjectured that the following pairs of patterns are I-Wilf-equivalent:

- $12 \cdots k\tau \sim_I k(k-1) \cdots 1\tau$ for any $k \geq 1$, where τ is any permutation on $\{k+1, k+2, \ldots, m\}$;

- $12345 \sim_I 45312$ (or equivalently, $54321 \sim_I 45312$);

- $123456 \sim_I 456123 \sim_I 564312$ (or equivalently, $654321 \sim_I 456123$).

The first conjecture was settled in [162] by Bousquet-Mélou and Steingrímsson, while the last two conjectures were settled in [317] by Dukes et al., completing the I-Wilf classification for patterns of lengths 5, 6 and 7.

Regev [674] provided an asymptotic formula for $i_n(12 \cdots k)$ (recorded in the following theorem) and also showed that $i_n(1234) = M_n$, the n-th Motzkin number (see Subection A.2.1 for the definition).

Theorem 6.1.104. *([674]) We have*

$$i_n(12 \cdots k) \sim k^n \left(\frac{k}{n}\right)^{k(k-1)/4} \frac{1}{k!} \Gamma\left(\frac{3}{2}\right)^{-k} \prod_{i=1}^{k} \Gamma\left(1 + \frac{j}{2}\right).$$

Gessel [417] gave the e.g.f. for $I_n(k(k-1) \cdots 1)$ (equivalently, for $I_n(12 \cdots k)$):

$$\sum_{\pi \in I((k+1)k \cdots 1)} \frac{x^{|\pi|}}{|\pi|!} = \begin{cases} e^x \det(\mathbf{I}_{i-j} - \mathbf{I}_{i+j})_{1 \leq i,j \leq \ell}, & \text{if } m = 2\ell + 1; \\ \det(\mathbf{I}_{i-j} + \mathbf{I}_{i+j-1})_{1 \leq i,j \leq \ell}, & \text{if } m = 2\ell, \end{cases}$$

where $\mathbf{I}_i$ is defined as in Theorem 6.1.16. Bousquet-Mélou [159] provided another proof of the last formula as well as an alternative proof of the following refinement, which can be derived from known results in the literature (see [159] for details).

Theorem 6.1.105. *([159]) If $k = 2\ell + 1$, the e.g.f. of involutions in $I_n((k+1)k \cdots 1)$ that have r fixed points is*

$$\sum_{\pi \in I((k+1)k \cdots 1), fp(\pi)=r} \frac{x^{|\pi|}}{|\pi|!} = \frac{x^r}{r!} \det(\mathbf{I}_{i-j} - \mathbf{I}_{i+j})_{1 \leq i,j \leq \ell},$$

or equivalently,

$$\sum_{\pi \in I((k+1)k \cdots 1)} \frac{x^{|\pi|}}{|\pi|!} q^{\text{fp}(\pi)} = e^{qx} \det(\mathbf{I}_{i-j} - \mathbf{I}_{i+j})_{1 \leq i,j \leq \ell}.$$

If $k = 2\ell$, then

$$\sum_{\pi \in I((k+1)k\cdots 1), fp(\pi)=r} \frac{x^{|\pi|}}{|\pi|!} = \det \left(\begin{array}{c} (\mathbf{I}_{r+\ell-j} - \mathbf{I}_{r+\ell+j})_{1 \le j \le \ell} \\ (\mathbf{I}_{i+j-1} - \mathbf{I}_{i-j-1})_{2 \le i \le \ell, 1 \le j \le \ell} \end{array} \right),$$

where the first row of the determinant is described separately from the other $\ell - 1$ rows.

Gouyou-Beauchamps gave bijective proofs of exact formulas for $i_n(12345)$ and $i_n(123456)$. The following theorem was first obtained via a bijective proof by Gouyou-Beauchamps [424], and then by Gessel [417] using symmetric functions. Bousquet-Mélou [158] presented an alternative proof.

Theorem 6.1.106. *([424, 417, 158]) The number of 12345-avoiding involutions in $\mathcal{S}_n$ is*

$$i_n(12345) = C_{\lceil n/2 \rceil} C_{1 + \lfloor n/2 \rfloor},$$

where C_n is the n-th Catalan number.

Theorem 6.1.107. *([424]) The number of 123456-avoiding involutions in $\mathcal{S}_n$ is*

$$i_n(123456) = \sum_{i=0}^{\lfloor n/2 \rfloor} \frac{3! n! (2i+2)!}{(n-2i)! i! (i+1)! (i+2)! (i+3)!}.$$

Bousquet-Mélou [158] refined Theorem 6.1.106.

Hultman and Vorwerk [468] show that the principal order ideal of an element w in the *Bruhat order* on involutions in a symmetric group is a *Boolean lattice* if and only if w avoids the classical patterns 4321, 45312 and 456123. An involution is called *Boolean* if its principle order ideal in the Bruhat order on involutions is isomorphic to a *Boolean lattice*.

Theorem 6.1.108. *([468]) The g.f. for the number of Boolean involutions with respect to the inversion number and the excedance number is given by*

$$\sum_{\sigma} x^{\text{length}(\sigma)} y^{\text{inv}(\sigma)} z^{\text{exc}(\sigma)} = \frac{x^2 yz + x - x^2 y^2 - x^3 y^3 z}{1 - x - x^2 yz - xy^2 + x^2 y^2 - x^2 y^3 z + x^3 y^3 z}.$$

We let $\mathcal{A}_k = \{\pi_1 \pi_2 \cdots \pi_k \in \mathcal{S}_k \mid \pi_1 = 1, \pi_2 = 2\}$ and $J_n(p) = \sum_{i=0}^{n} I_{n;i} p^i$, where $I_{n;i}$ is the number of involutions in I_n with i fixed points. For example, $J_3(p) = 3p + p^3$. We can see that $J_0(p) = 1$, $J_1(p) = p$ and, for $n \ge 2$,

$$J_n(p) = p J_{n-1}(p) + (n-1) J_{n-2}(p).$$

The e.g.f. for the sequence $\{J_n(p)\}_{n \ge 0}$ is given by $e^{px + x^2/2}$. The main result of [292] by Deng et al. is the following theorem.

Theorem 6.1.109. *([292]) Let $k \geq 2$. The g.f. for the number of $\mathcal{A}_k$-avoiding involutions with fixed points registered is given by*

$$
\sum_{n \geq 0} \sum_{\pi \in I_n(\mathcal{A}_k)} x^n p^{\mathrm{fp}(\pi)} = \sum_{j=0}^{k-3} J_j(p)x^j - \frac{x^{k-3}}{2} \left(p + (p(k-3)x^2 - 2x - p)u_0(x) \right) J_{k-2}(p)
$$

$$
- \frac{x^{k-4}}{2} \left(x + p - (x^3(k-3) - px^2(k-1) + x + p)u_0(x) \right),
$$

where $u_0(x) = 1/\sqrt{1 - 2(k-1)x^2 + (k-3)^2 x^4}$.

See Subsection A.2.2 for the notion of a Schröder path. A Schröder path of length $2n$ is called *symmetric* if it is symmetric with respect to the line $x = n$. Deng et al. [292] found a bijection between symmetric Schröder paths of length $2n$ and the set $I_{n+1}(1234, 1243)$ of $(1234, 1243)$-avoiding $(n + 1)$-involutions. Statistics such as rmax (the number of right-to-left maxima) and fp (the number of fixed points) on involutions correspond to the number of steps in the symmetric Schröder path of a particular type.

Barnabei et al. [90] proved that the descent statistic on involutions is not *log-concave*, while Dukes [315] also considered certain permutation statistics on involutions, not to be discussed in this book. Baik and Rains [72] considered the asymptotics of the distribution of the lengths of the longest monotone subsequences of random involutions. Finally, Walsh [791] and Poneti and Vajnovszki [655] studied generation of involutions, to be discussed briefly in Subsection 8.2.2.

6.1.8   Classical patterns and alternating permutations

Review the notion of an alternating permutation in Definition 1.0.18. For a survey on alternating permutations see [749] by Stanley.

Definition 6.1.110. We let A_n denote the set of alternating permutations of length n. $A_n(P)$ is the set of those permutations in A_n that avoid each pattern from a set of patterns P. Also, $A_P(x) = \sum_{n \geq 0} A_n(P)x^n$.

It is well-known that

$$
\sum_{n=0}^{\infty} |A_{2n}| \frac{x^{2n}}{(2n)!} = \sec(x) \quad \text{and} \quad \sum_{n=0}^{\infty} |A_{2n+1}| \frac{x^{2n+1}}{(2n + 1)!} = \tan(x).
$$

Avoidance of any 3-pattern is given by the Catalan numbers (see [464, 748]). For example, the subsets of 132-avoiding alternating permutations in $\mathcal{S}_6$ and $\mathcal{S}_7$ are as follows:

$$
\{342516, 452316, 453612, 562314, 563412\} \text{ and}
$$

$$\{4536271, 5634271, 5647231, 6734251, 6745231\},$$

while the subsets of 321-avoiding alternating permutations in $\mathcal{S}_4$ and $\mathcal{S}_5$ are as follows:

$$\{1324, 2314, 1423, 2413, 3412\} \text{ and } \{13254, 23154, 14253, 24153, 34152\}.$$

Lewis [558] constructed a bijection between $\mathcal{S}_n(132)$ and $A_{2n+1}(132)$ which restricts to a bijection between $\mathcal{S}_n(132, p_1, p_2, \ldots, p_k)$ and $A_{2n+1}(132, q_1, q_2, \ldots, q_k)$, where $\{p_1, p_2, \ldots, p_k\} = P$ is any set of 132-avoiding patterns and $\{q_1, q_2, \ldots, q_k\}$ is a certain set of patterns related to P.

Lewis [560] used bijections between permutations and *standard Young tableaux* of certain shapes to prove the following theorem.

Theorem 6.1.111. *([560]) We have*

$$|A_{2n}(1234)| = \frac{2(3n)!}{n!(n+1)!(n+2)!} \quad and \quad |A_{2n+1}(1234)| = \frac{16(3n)!}{(n-1)!(n+1)!(n+3)!}.$$

Later, using the generating trees approach, Lewis [559] obtained an enumeration of 2143-avoiding alternating permutations.

Theorem 6.1.112. *([559]) We have*

$$|A_{2n}(2143)| = |A_{2n}(1234)| = \frac{2(3n)!}{n!(n+1)!(n+2)!} \quad and$$

$$|A_{2n+1}(2143)| = \frac{2(3n+3)!}{n!(n+1)!(n+2)!(2n+1)(2n+2)(2n+3)}.$$

Based on numerical evidence, Lewis [559] posed several conjectures on 4-patterns with respect to alternating permutations that we collect in the following statement.

Conjecture 6.1.113. ([559]) We have

- $|A_{2n}(p)| = |A_{2n}(1234)|(= |A_{2n}(2143)|)$ for all $n \geq 1$ and every

$$p \in \{1243, 2134, 1432, 3214, 2341, 4123, 3421, 4312\};$$

- $|A_{2n}(3142)| = |A_{2n}(3241)| = |A_{2n}(4132)|$ and $|A_{2n}(2413)| = |A_{2n}(1423)| = |A_{2n}(2314)|$ for all $n \geq 1$;

- $|A_{2n+1}(p)| = |A_{2n+1}(1234)|$ for all $n \geq 0$ and every $p \in \{2134, 4312, 3214, 4123\}$;

- $|A_{2n+1}(p)| = |A_{2n+1}(2143)|$ for all $n \geq 0$ and every $p \in \{1243, 3421, 1432, 2341\}$.

We now consider selected results by Mansour [591] who studied alternating permutations that avoid or contain exactly one occurrence of the pattern 132 and also avoid or contain exactly once another pattern. In what follows $U_n(x)$ is the n-th Chebyshev polynomial of the second kind and $R_k(x)$ is defined in Theorem 6.1.33.

Theorem 6.1.114. *([591]) The number of alternating permutations in A_n that contain exactly one occurrence of the pattern 132 is given by*

$$\binom{n-1}{(n-3)/2} + \binom{n-1}{(n-4)/2},$$

where $\binom{a}{b}$ is assumed to be 0 whenever $a < b$, $b < 0$, or b is a non-integer number.

Theorem 6.1.115. *([591]) For all $k \geq 2$,*

$$A_{132,12\cdots k}(x) = \frac{(1+x)U_{k-2}\left(\frac{1}{2x}\right)}{xU_{k-1}\left(\frac{1}{2x}\right)} = (1+x)R_{k-1}(x^2).$$

In particular, $|A_n(132, 12345)| = F_{2\lfloor n/2 \rfloor - 1}$, the $(2\lfloor n/2 \rfloor - 1)$-th Fibonacci number.

Theorem 6.1.116. *([591]) For all $k \geq 3$, we have*

$$A_{132,23\cdots k1}(x) = \frac{(1+x)U_{k-3}^2\left(\frac{1}{2x}\right) - x}{U_{k-2}^2\left(\frac{1}{2x}\right)}.$$

In particular,

$$|A_{2n}(132, 234561)| = \frac{7}{10}nL_{2n} - \frac{1}{10}(15n - 4)F_{2n} \quad and \quad |A_{2n+1}(132, 234561)| = F_{2n-1},$$

where L_n is the n-th Lucas number defined by $L_0 = 2$, $L_1 = 1$, and $L_n = L_{n-1} + L_{n-2}$ for all $n \geq 2$.

Theorem 6.1.117. *([591]) For all $k \geq 3$, the g.f. for the number of alternating permutations that avoid the pattern 132 and contain the pattern $12\cdots k$ exactly once is*

$$\frac{1+x}{U_{k-1}^2\left(\frac{1}{2x}\right)}.$$

6.1.9   Classical patterns and Dumont permutations

Definition 6.1.118. A *Dumont permutation of the first kind* is a permutation $\pi = \pi_1 \pi_2 \cdots \pi_{2n} \in S_{2n}$ where each even letter is a descent and each odd letter is an ascent or ends the permutation. In other words, for every $i = 1, 2, \ldots, 2n$,

- π_i is even implies $i < 2n$ and $\pi_i > \pi_{i+1}$, and

- π_i is odd implies $\pi_i < \pi_{i+1}$ or $i = 2n$.

A *Dumont permutation of the second kind* is a permutation $\pi \in S_{2n}$ where all entries at even positions are deficiencies and all entries at odd positions are fixed points or excedances. In other words, for every $i = 1, 2, \ldots, n$, we have $\pi_{2i} < 2i$ and $\pi_{2i-1} \geq 2i - 1$.

Definition 6.1.119. We denote the set of Dumont permutations of the first (resp., second) kind of length $2n$ by $\mathcal{D}_{2n}^1$ (resp., $\mathcal{D}_{2n}^2$). For $i = 1, 2$, let $\mathcal{D}_{2n}^i(P)$ denote the set of those permutations in $\mathcal{D}_{2n}^i$ that avoid each pattern from a set of patterns P.

Example 6.1.120. We have

$$\mathcal{D}_2^1 = \mathcal{D}_2^2 = \{21\}, \quad \mathcal{D}_4^1 = \{2143, 3421, 4213\}, \quad \mathcal{D}_4^2 = \{2143, 3142, 4132\}.$$

Remark 6.1.121. Dumont permutations of odd length can be defined similarly to those of even length. Then $\mathcal{D}_{2n+1}^1$ and $\mathcal{D}_{2n+1}^2$ are obtained simply by adjoining the letter $2n+1$ to the end of each permutation in $\mathcal{D}_{2n}^1$ and $\mathcal{D}_{2n}^2$, respectively. Obviously, $|\mathcal{D}_{2n+1}^1| = |\mathcal{D}_{2n}^1|$ and $|\mathcal{D}_{2n+1}^2| = |\mathcal{D}_{2n}^2|$.

Dumont [328] showed that

$$|\mathcal{D}_{2n}^1| = |\mathcal{D}_{2n}^2| = G_{2n+2} = 2(1 - 2^{2n+2})B_{2n+2},$$

where G_n is the n-th *Genocchi number*, a multiple of the *Bernoulli number* B_n. The exponential generating functions for the *unsigned* and *signed* Genocchi numbers are given by

$$\sum_{n=1}^{\infty} G_{2n} \frac{x^{2n}}{(2n)!} = x \tan\left(\frac{x}{2}\right), \quad \sum_{n=1}^{\infty} (-1)^n G_{2n} \frac{x^{2n}}{(2n)!} = \frac{2x}{e^x + 1} - x = -x \tanh\left(\frac{x}{2}\right).$$

Studies of pattern-avoiding Dumont permutations are conducted in [196] by Burstein, in [598] by Mansour, and in [198] by Burstein et al. We summarize most of the (un)known enumerative results on classical 3- and 4-pattern-avoidance on Dumont permutations in Tables 6.8 and 6.9, where C_n is the n-th Catalan number and $C(x)$ is the g.f. for these numbers; s_{n+1} is the $(n+1)$-th small Schröder number given by $s_1 = 1$,

$$s_{n+1} = -s_n + 2\sum_{k=1}^{n} s_k s_{n+1-k}$$

for $n \geq 1$; $a_{2m} = \frac{1}{2m+1}\binom{3m}{m}$, $a_{2m+1} = \frac{1}{2m+1}\binom{3m+1}{m+1}$, and

$$C(2;n) = \sum_{m=0}^{n-1} \frac{n-m}{n}\binom{n-1+m}{m}2^m.$$

Mansour [598] studied $\mathcal{D}_n^1(132)$ (he observed that $|D_n^2(132)| = 0$ for $n \geq 4$). A number of general theorems are proved in [598] for permutations in $\mathcal{D}_n^1(132)$ avoiding, or containing exactly one occurrence of, another pattern. We state here selected results, where $F_k(x)$ and $G_k(x)$ satisfy the same recurrence relation:

$$Q_k(x) = 1 + \frac{x^2 Q_{k-1}(x)}{1 - x^2 Q_{k-2}(x)},$$

with $F_0(x) = 0$ and $F_1(x) = G_0(x) = G_1(x) = 1$. For example, $F_2(x) = 1 + x^2$, $F_3(x) = \frac{1+x^4}{1-x^2}$, $G_2(x) = \frac{1}{1-x^2}$, and $G_3(x) = \frac{1-x^2+x^4}{(1-x^2)^2}$.

Theorem 6.1.122. *([598]) The g.f.s for $\mathcal{D}_n^1(132, 12\cdots k)$ and $\mathcal{D}_n^1(132, 2134\cdots k)$ are, respectiively,*

$$\mathcal{D}_{132,12\cdots k}^1(x) = F_k(x) + x F_{k-1}(x) \text{ and } \mathcal{D}_{132,2134\cdots k}^1(x) = G_{k-1}(x) + x G_{k-2}(x).$$

Theorem 6.1.123. *([598]) The g.f. for the number of 132-avoiding Dumont permutations of the first kind with k descents is given by $C_k(1 + x)x^{2k+1}$ for all $k \geq 0$. Here C_k is the k-th Catalan number.*

Theorem 6.1.124. *([598]) Let*

$$A_k(x) = \frac{x^2}{1 - x^2 F_{k-2}(x)} A_{k-1}(x) + \frac{x^4 F_{k-1}(x)}{(1 - x^2 F_{k-2}(x))^2} A_{k-2}(x)$$

for all $k \geq 2$, where $A_1(x) = 0$ and $A_2(x) = x^4$. Then for $k \geq 2$, the g.f. for the number of 132-avoiding Dumont permutations of the first kind containing exactly one occurrence of the pattern $12\cdots k$ is $A_k(x) + x A_{k-1}(x)$.

More results on $\mathcal{D}_{2n}^1(132)$, as well as on $\mathcal{D}_{2n}^1(231)$ and $\mathcal{D}_{2n}^2(331)$ can be found in [198] by Burstein et al. Moreover, Burstein et al. [198] studied statistics in connection with pattern-avoiding permutations in $\mathcal{D}_{2n}^2$. For example, the following result was obtained there.

p	$|\mathcal{D}_{2n}^1(p)|$	Reference
123	Open	
132	C_n	[598]
213	C_{n-1}	[196]
231	C_n	[598]
312	C_n	[598]
321	1	[196]
2143	Open	
3421	Open	
4213	Open	
132, 123	$1 + (-1)^n$	[598]
132, 213	g.f. $\frac{1+x-x^3}{1-x}$	[598]
132, 1234	g.f. $\frac{1+2x+x^2+2x^6+x^7+x^8}{(1+x)(1-x^2-x^4)}$	[598]
132, 2134	g.f. $\frac{1+x-x^2-x^3+x^4}{(1-x^2)^2}$	[598]
231, 4213	1	[196]
1342, 1423	s_{n+1}	[196]
2341, 2413	s_{n+1}	[196]
1342, 2413	s_{n+1}	[196]
2341, 1423	$b_n = 3b_{n-1} + 2b_{n-2}, \ b_0 = b_1 = 1, \ b_2 = 3$	[196]
1342, 4213	2^{n-1}	[196]
2413, 3142	$C(2; n)$	[196]
1423, 4132	g.f. $\frac{1-3x-(1+x)\sqrt{1-4x}}{1-3x-2x^2-(1+x)\sqrt{1-4x}}$	[196]
1423, 3142 2413, 4132	g.f. $\frac{1+xC(x)-\sqrt{1-xC(x)-5x}}{2x(1+C(x))}$	[196]

Table 6.8: Avoidance of classical patterns of lengths 3 and 4 on Dumont permutations of the first kind.

p	$|\mathcal{D}^2_{2n}(p)|$	Reference
123	Open	
132	0	Obvious
213	0	Obvious
231	2^{n-1}	[196]
312	1	[196]
321	C_n	[598]
3142	C_n	[196]
4132	C_n	[198]
2143	$a_n a_{n+1}$	[198]

Table 6.9: Avoidance of classical patterns of lengths 3 and 4 on Dumont permutations of the second kind.

Theorem 6.1.125. ([198]) Let $A(q, t, x) = \sum_{n \geq 0} \sum_{\pi \in \mathcal{D}^2_{2n}(3142)} q^{\mathrm{fp}(\pi)} t^{\#2\text{-}\mathrm{cyc}(\pi)} x^n$ be the g.f. for 3142-avoiding Dumont permutations of the second kind with respect to the number of fixed points and the number of 2-cycles. Then

$$A(q, t, x) = \frac{1 + x(q - t) - \sqrt{1 - 2x(q + t) + x^2((q + t)^2 - 4q)}}{2xq(1 + x(1 - t))}.$$

6.1.10   Classical patterns and even/odd permutations

Definition 6.1.126. The *signature* of a permutation π is $\mathrm{sign}(\pi) = (-1)^{\mathrm{inv}(\pi)}$. We say that π is an *even permutation* (resp., *odd permutation*) if $\mathrm{sign}(\pi) = 1$ (resp., $\mathrm{sign}(\pi) = -1$). We denote by E_n (resp., O_n) the set of all even (resp., odd) permutations in S_n. Also, $e_n(P) = |E_n(P)|$ and $o_n(P) = |O_n(P)|$, where $E_n(P)$ (resp., $O_n(P)$) is the number of even (resp., odd) n-permutations avoiding each pattern from a set P.

Example 6.1.127. $312 \in E_3$ and $25314 \in O_5$.

It is a well-known fact that $|E_n| = |O_n| = \frac{n!}{2}$. Simion and Schmidt [721] proved that the g.f. for $E_n(132)$ is

$$\frac{1}{2} \left(\frac{1 - \sqrt{1 - 4x}}{2x} + 1 + \frac{1 - \sqrt{1 - 4x^2}}{2x} \right),$$

while the g.f. for $O_n(132)$ is

$$\frac{1}{2} \left(\frac{1 - \sqrt{1 - 4x}}{2x} - 1 - \frac{1 - \sqrt{1 - 4x^2}}{2x} \right).$$

Moreover, Simion and Schmidt [721] showed that the number of even 321-avoiding n-permutations is equal to the number of odd ones if n is even, and exceeds it by the $\left(\frac{n-1}{2}\right)$-th Catalan number otherwise. Reifegerste [682] presented an involution that proves a refinement of this sign-balance property respecting the length of the longest increasing subsequence of the permutation.

Definition 6.1.128. Patterns p_1 and p_2 are *E-Wilf-equivalent*, denoted $p_1 \sim_E p_2$, if $e_n(p_1) = e_n(p_2)$ for all $n \geq 0$. Similarly, we can define *O-Wilf-equiivalence* $\sim_O$.

It follows from Simion and Schmidt's work [721] that $123 \sim_E 231 \sim_E 312$ and $132 \sim_E 213 \sim_E 321$. However, it is not true in general that $\sigma \sim_E \tau$ whenever $\sigma \sim \tau$ and σ and τ have the same parity, which is demonstrated by two even patterns $1234 \sim 4321$ having $e_6(1234) = 258$ and $e_6(4321) = 255$. In either case, Baxter and Jaggard [97] showed that whenever k is odd, $k(k-1)\cdots 1P \sim_E (k-1)(k-2)\cdots 1kP$, where P is any permutation on $\{k+1, k+2, \ldots, \ell\}$, which should be compared with the fact that $k(k-1)\cdots 1P \sim (k-1)(k-2)\cdots 1kP$ (see the discussion after Remark 4.0.4). The result of Baxter and Jaggard allows us to classify all patterns $\sigma \in \mathcal{S}_4$ according to even-Wilf-equivalence. However, since even-Wilf-equivalence is not preserved by all the symmetries of the square (e.g., $1234 \nsim_E 4321$), this result does not complete the classification of $\sigma \in \mathcal{S}_5$ according to even-Wilf-equivalence.

Mansour [594] studied the number of even (or odd) n-permutations containing exactly $r \geq 0$ occurrences of the pattern 132. Similarly to [594] dealing with counting the pattern 132 in $\mathcal{S}_n$ and [622] counting the pattern 231 in involutions, he shows that finding this number for a given r can be done by a routing check on all $(2r)$-permutations. Explicit results are given in [594] for $r \leq 6$. Below we state the cases of $r = 1, 2$.

- The g.f. for the number of even permutations containing the pattern 132 exactly once is given by

$$-\frac{1}{2}(1 - 2x - x^2) + \frac{1 - 3x}{4\sqrt{1 - 4x}} + \frac{1 - 3x - 4x^2 + 4x^3}{4\sqrt{1 - 4x^2}};$$

- The g.f. for the number of odd permutations containing the pattern 132 exactly once is given by

$$-\frac{1}{2}(x + x^2) + \frac{1 - 3x}{4\sqrt{1 - 4x}} - \frac{1 - 3x - 4x^2 + 4x^3}{4\sqrt{1 - 4x^2}};$$

- The g.f. for the number of even permutations containing the pattern 132 exactly twice is given by

$$\frac{1}{2}x(x^3 + 3x^2 - 4x - 1) + \frac{1}{4}(2x^4 - 4x^3 + 29x^2 - 15x + 2)(1 - 4x)^{-3/2}$$
$$-\frac{1}{4}(16x^7 - 48x^6 - 76x^5 + 64x^4 + 36x^3 - 21x^2 - 5x + 2)(1 - 4x^2)^{-3/2};$$

- The g.f. for the number of odd permutations containing the pattern 132 exactly twice is given by

$$-\tfrac{1}{2}(x^4 + 3x^3 - 5x^2 - 4x + 2)$$
$$+\tfrac{1}{4}(2x^4 - 4x^3 + 29x^2 - 15x + 2)(1 - 4x)^{-3/2}$$
$$+\tfrac{1}{4}(16x^7 - 48x^6 - 76x^5 + 64x^4 + 36x^3 - 21x^2 - 5x + 2)(1 - 4x^2)^{-3/2}.$$

Moreover, Mansour [602] studied the generating functions for 132-avoiding even (odd) permutations that additionally avoid, or contain exactly once or twice, another classical pattern of length k of certain general form. In several interesting cases, the generating functions depend only on k and can be expressed in terms of Chebyshev polynomials of the second kind. Among many enumerative results in [602], it was shown that the g.f. for the permutations in $E_n(132)$ (or, in $O_n(132)$) that contain the pattern $12 \cdots (2k + 1)$ exactly twice is given by

$$\frac{1}{2}\left(\frac{\sqrt{x}U_{k-1}\left(\frac{1}{2\sqrt{x}}\right)}{U_k^3\left(\frac{1}{2\sqrt{x}}\right)} + \frac{x^2\left(U_{k-1}\left(\frac{1}{2x}\right) - U_k\left(\frac{1}{2x}\right)\right)}{U_k^3\left(\frac{1}{2x}\right)}\right),$$

where $U_n(x)$ is the n-th Chebyshev polynomial of the second kind.

Affine permutations

Elements of the *affine symmetric group*, $\tilde{S}_n$, can be expressed as an infinite sequence of integers, and it is still natural to ask if a (classical) pattern occurs in the sequence. We can then ask how many $w \in \tilde{S}_n$ avoid a given pattern. More formally, $\tilde{S}_n$ is the set of all bijections $w : \mathbb{Z} \to \mathbb{Z}$ with $w(i + n) = w(i) + n$ for all $i \in \mathbb{Z}$ and

$$\sum_{i=1}^{n} w(i) = \binom{n + 1}{2}.$$

The elements of $\tilde{S}_n$ are called *affine permutations*.

Crites [286] proved the following theorems.

Theorem 6.1.129. *([286]) Let $p \in S_m$. For any $n \geq 2$ there exist only finitely many $w \in \tilde{S}_n$ that avoid p if and only if p avoids the pattern 321.*

Theorem 6.1.130. *([286]) For $p \in S_m$, let $|\tilde{S}_n(p)|$ be the number of $w \in \tilde{S}_n$ avoiding p and let $F_p(x) = \sum_{n=2}^{\infty} |\tilde{S}_n(p)|x^n$. Then*

$$F_{123}(x) = 0, \quad F_{132}(x) = F_{213}(x) = \sum_{n=2}^{\infty} x^n, \quad and$$

$$F_{231}(x) = F_{312}(x) = \sum_{n=2}^{\infty} \binom{2n-1}{n} x^n.$$

By Theorem 6.1.129, $|\tilde{\mathcal{S}}_n(321)| = |\tilde{\mathcal{S}}_n(p)| = \infty$ for all n and any

$$p \in \{1432, 2431, 3214, 3241, 3421, 4132, 4213, 4231, 4312, 4321\},$$

and the corresponding generating functions are not defined. Regarding some other 4-patterns, the following results are obtained in [286].

Theorem 6.1.131. *([286]) We have $F_{1234}(x) = 0$,*

$$F_{1243}(x) = F_{1324}(x) = F_{2134}(x) = F_{2143}(x) = \sum_{n=2}^{\infty} x^n,$$

$$F_{1342}(x) = F_{1423}(x) = F_{2314}(x) = F_{3124}(x) = \sum_{n=2}^{\infty} \binom{2n-1}{n} x^n.$$

Finally, Crites [286] posed the following conjecture.

Conjecture 6.1.132. ([286]) The following equalities hold:

$$|\tilde{\mathcal{S}}_n(3142)| = |\tilde{\mathcal{S}}_n(2413)| = \sum_{k=0}^{n-1} \frac{n-k}{n} \binom{n-1+k}{k} 2^k,$$

$$|\tilde{\mathcal{S}}_n(3412)| = |\tilde{\mathcal{S}}_n(4123)| = |\tilde{\mathcal{S}}_n(2341)| = \frac{1}{3} \sum_{k=0}^{n} \binom{n}{k}^2 \binom{2k}{k}.$$

6.2 Classical patterns in words

The book [461] by Heubach and Mansour contains more information and details on classical patterns in words than we include in our presentation here.

6.2.1 Wilf-equivalence

Review Definition 1.0.30 for the notion of Wilf-equivalence on words and for the corresponding notation.

The classical patterns 11 and 12 are not Wilf-equivalent, which is not difficult to see. Thus there are two Wilf-equivalence classes for 2-patterns on words.

Burstein [195] shows analytically that $123 \sim 132$, while Burstein and Mansour [208] proved that $112 \sim 121$. This allows us to obtain the Wilf-equivalence classes for 3-patterns:

$$\{111\}, \{112, 121\}, \{123, 132\}.$$

Above, as usual, in showing the classes we did not include patterns equivalent modulo trivial bijections. To proceed, we need the following definitions.

Definition 6.2.1. The *content* of a word w is the unordered multiset of the letters appearing in w. Also, for a word w, we denote by w^{+i} the word obtained by adding i to each letter of w.

Example 6.2.2. The content of the word 22413231 is the multiset $\{1, 1, 2, 2, 2, 3, 3, 4\}$. $(321143)^{+2} = 543365$.

Definition 6.2.3. We say that two patterns p_1 and p_2 are *st-Wilf-equivalent*, denoted by $p_1 \sim_{st} p_2$, if for an ℓ-letter alphabet, there is a bijection f from $W_{n,\ell}(p_1)$ to $W_{n,\ell}(p_2)$ that preserves the content of every word.

Remark 6.2.4. The relation $p_1 \sim_{st} p_2$ introduced in Definition 6.2.3 is known in the literature as *strong Wilf-equivalence*. However, in this book, strong Wilf-equivalence means equidistribution, as defined in Definition 1.0.30. Clearly, $p_1 \sim_{st} p_2$ implies $p_1 \sim p_2$.

The fact that $112 \sim 121$ also follows from the following, more general theorem by Jelínek and Mansour [482], which was proved by considering certain 0-1-*fillings of Ferrers diagrams* not to be discussed here.

Theorem 6.2.5. *([482]) For $i, j \geq 0$ and any word p, $2^i 12^j p^{+2} \sim_{st} 12^{i+j} p^{+2}$.*

Another general result was obtained by Krattenthaler [544] who also used 0-1-fillings of Ferrers diagrams. The following theorem is an analogue for words of Theorem 4.0.3 on permutations.

Theorem 6.2.6. *([544]) For $k \geq 1$ and any word w, $12 \cdots k w^{+k} \sim_{st} k(k-1) \cdots 1 w^{+k}$.*

The following theorem was proved using an idea coming from the context of pattern-avoiding set partitions.

Theorem 6.2.7. *([482]) For any $t \geq 3$, all patterns that consist of a single 1, a single 3 and $t - 2$ letters 2 are st-Wilf-equivalent.*

Using the theorems stated above, we can obtain a classification of Wilf-equivalences of patterns of length at most 6. In Table 6.10 we state the classes of equivalences

$$\{1123, 1132\}, \{1112, 1121\}, \{1223, 1232, 1322, 2132\}, \{1234, 1243, 1432, 2143\}$$

$$\{12435, 13254\}, \{12443, 21143\}, \{12134, 12143\}$$
$$\{11123, 11132\}, \{12534, 21534\}, \{12453, 21453\}$$

$$\{11234, 11243, 11432\}, \{11223, 11232, 11322\}, \{11112, 11121, 11211\}$$

$$\{12223, 12232, 12322, 13222, 21232, 21322\}$$
$$\{12345, 12354, 12543, 15432, 21354, 21543\}$$
$$\{12234, 12243, 12343, 12433, 21243, 21433\}$$

Table 6.10: Wilf-equivalence classes for patterns of lengths 4 and 5.

for patterns of lengths 4 and 5, referring the interested reader to [461] for the length 6 case.

Another example of Wilf-equivalence is the following result by Mansour [601]: For $k \geq 1$,

$$w_{n,\ell}(132, 12 \cdots k) = w_{n,\ell}(132, k12 \cdots (k-1)).$$

6.2.2 Enumeration

Burstein [195] proved the following theorem, where C_n is the n-th Catalan number.

Theorem 6.2.8. *([195]) We have*

$$w_{n,\ell}(123) = w_{n,\ell}(132) = 2^{n-2(\ell-2)} \sum_{i=0}^{\ell-2} a_{\ell-2,i} \binom{n+2i}{n},$$

where

$$a_{\ell,i} = \sum_{j=i}^{\ell} C_j \binom{2(\ell-j)}{\ell-j}.$$

The corresponding generating function is

$$\sum_{\ell,n\geq 0} w_{n,\ell}(123) y^\ell \frac{x^n}{n!} = 1 + \frac{y}{2x(1-x)} \left(1 - \sqrt{\frac{(1-2x)^2 - y}{1-y}} \right).$$

See [461] for three different proofs of Theorem 6.2.8, namely, by the *Noonan-Zeilberger algorithm*, by the *block decomposition method*, and by the *scanning-element algorithm*.

Theorem 6.2.9. *([208]) The e.g.f. for the number of ℓ-ary words of length n that avoid the pattern 111 is given by*

$$\sum_{n \geq 0} w_{n,\ell}(111) \frac{x^n}{n!} = \left(1 + x + \frac{1}{2}x^2\right)^\ell,$$

and therefore the number of ℓ-ary words of length n that avoid 111 is

$$w_{n,\ell}(111) = n! \sum_{i=0}^{\ell} \frac{\binom{\ell}{i}\binom{i}{n-i}}{2^{n-i}}.$$

The following theorem in [208] by Burstein and Mansour involves an interesting connection to the (unsigned) Stirling numbers of the first kind (see Subsection A.2.1).

Theorem 6.2.10. *([208]) We have*

$$w_{n,\ell}(112) = \sum_{i=0}^{\ell} \binom{n+\ell-i-1}{n} \begin{bmatrix} n \\ n-i \end{bmatrix}$$

and

$$\sum_{\ell,n \geq 0} w_{n,\ell}(112) y^\ell \frac{x^n}{n!} = \frac{1}{1-y} \left(\frac{1-y}{1-y-xy}\right)^{1/y},$$

where $\begin{bmatrix} n \\ i \end{bmatrix}$ *is the unsigned Stirling number of the first kind.*

The following theorem on words by Mansour [601], is in the direction of Subsection 6.1.5 dealing with permutations.

Theorem 6.2.11. *([601]) Let* $A_k(x,y) = \sum_{n,\ell \geq 0} w_{n,\ell}(132, 12 \cdots k) x^n y^\ell$. *For* $k \geq 3$,

$$A_k(x,y) = 1 + \cfrac{y}{(1-x)(1-y) + \cfrac{xy(1-x)(1-y)}{x(1-y) + \cfrac{y}{1 - A_{k-1}(x,y)}}}$$

with $A_2(x,y) = \frac{1-x}{1-x-y}$. *Thus,* $A_k(x,y) - 1$ *can be expressed as a continued fraction with k levels:*

$$\cfrac{y}{(1-x)(1-y) - \cfrac{xy(1-x)(1-y)}{(1-2x)(1-y) - \cfrac{xy(1-x)(1-y)}{(1-2x)(1-y) - \cfrac{\cdots}{(1-2x)(1-y) - xy}}}},$$

or in terms of Chebyshev polynomials of the second kind, as

$$A_k(x,y) = 1 + \sqrt{xy^3(1-x)(1-y)}\frac{\alpha U_{k-2}(t) + \beta U_{k-3}(t)}{\alpha U_{k-3}(t) + \beta U_{k-4}(t)},$$

where $\alpha = (1 - 2x - y)$, $\beta = \frac{xy^2(1-x-y)}{\sqrt{xy^3(1-x)(1-y)}}$ *and* $t = \frac{2x-1}{2}\sqrt{\frac{1-y}{xy(1-x)}}$.

As an example of applications of Theorem 6.2.11, Mansour [601] provides the following generating function for $w_{n,\ell}(132, 1234)$:

$$\frac{(1-x)(1-2x)^2 - (1 - 3x + 4x^2 - 3x^3)y}{(1-x)((1-2x)^2 - (2-3x)(1-x)y + (1-x)y^2)}.$$

Another result in [601] is the generating function for $w_{n,\ell}(132, 212)$:

$$\frac{(1-x)^2 - (1-2x)y - \sqrt{(1-x)^4 - 2(1-x)^2y + (1 - 4x^2 + 4x^3)y^2}}{2xy(1-y)}.$$

There are many more enumerative results of similar type in [601]; in particular, the generating function for $w_{n,\ell}(132, 212, 12\cdots k)$ is presented there. More results of this kind can be found in [208] by Burstein and Mansour.

We end this subsection on enumeration by stating two theorems by Firro and Mansour [383] on containing a 3-pattern exactly once.

Theorem 6.2.12. *([383]) The g.f. for the number of ℓ-ary words of length n that contain the pattern 123 exactly once is given by*

$$\frac{y[2x^3(x-1) + x^2(9 - 10x + 2x^2)(1-y) - x(3-2x)(1-y)^2 + (1-y)^3]}{2x^3(1-x)^2(1-y)}$$

$$-\frac{y(1-x-y)((1-x)(1-2x) - y)}{2x^3(1-x)^2}\sqrt{\frac{(1-2x)^2 - y}{1-y}}.$$

Theorem 6.2.13. *([383]) The g.f. for the number of ℓ-ary words of length n that contain the pattern 132 exactly once is given by*

$$\frac{y^2\left[x + y - 1 - ((1-2x)(1-x) - y)\left(\frac{(1-2x)^2 - y}{1-y}\right)^{-\frac{1}{2}}\right]}{2(1-x)^2(1-y)^2}.$$

6.2.3 An asymptotic result

Regev [675] found asymptotics for the number of words avoiding the pattern $12 \cdots k$.

Theorem 6.2.14. *([675]) For all $\ell \geq k$ and $n \to \infty$, we have*

$$w_{n,\ell}(12 \cdots (k+1)) \simeq C_{\ell,k} n^{k(\ell-k)} k^n$$

where

$$C_{\ell,k}^{-1} = k^{k(\ell-k)} \prod_{i=1}^{k} \prod_{j=1}^{\ell-k} (i+j-1).$$

An alternative, simple proof of Theorem 6.2.14 that uses the transfer matrix method is given in [171] by Bränden and Mansour.

6.2.4 Longest alternating subsequences

Similarly to the case of permutations, we define an *alternating subsequence* $w_{i_1} w_{i_2} \cdots w_{i_k}$ in a word $w = w_1 w_2 \cdots w_n$ to be such that $w_{i_1} < w_{i_2} > w_{i_3} < w_{i_4} > \cdots w_{i_k}$. Let $al(w)$ be the length of the longest alternating subsequence in w.

Mansour [604] gave the following generating function for the number of ℓ-ary words w of length n having $al(w) = m$:

$$\sum_{n \geq 1} \sum_{m \geq 1} \sum_{w \in [\ell]^n, al(w)=m} q^m x^n = \frac{q(1-x-u+xuq)(u^\ell - 1)}{1-x-u+xuq^2 - q(1-x)(1-u)u^\ell},$$

where $u = \frac{2-2x+x^2-x^2q^2+\sqrt{(2-2x+x^2-x^2q^2)^2-4(1-x)^2}}{2(1-x)}$. As a consequence of this formula, Mansour [604] found the mean

$$\mu(n,k) = \frac{1}{\ell^n} \sum_{w \in [\ell]^n} al(w) = \frac{(\ell+1)\ell^3 + \ell^{n+1}((\ell-2)(\ell+1) + 2n(2\ell-1)(\ell-1))}{6(\ell-1)\ell^{n+2}}$$

and the variance

$$\sigma^2(n,k) = \frac{1}{\ell^n} \sum_{w \in [\ell]^n} (al(w) - \mu(n,k))^2 = \frac{(\ell+1)(-5(\ell+1)\ell^{5-2n} + 2\ell^{3-n}a_1 + a_2)}{180\ell^3(\ell-1)^2},$$

where

$$a_1 = -2\ell^4 + 9\ell^3 + 3\ell^2 - 16\ell + 10 + (\ell-1)(\ell^3 - 5\ell^2 - 14\ell + 10)n,$$
$$a_2 = -\ell(\ell-2)(13\ell^2 - 13\ell + 2) + 4\ell(\ell-1)(4\ell-3)(2\ell-1)n.$$

In [603], Mansour studied ℓ-ary words avoiding any pattern $p \in \mathcal{S}_3$ with respect to the statistic $al(w)$. We give here just two results in this direction.

Theorem 6.2.15. *([603]) The generating function for 312-avoiding words by length (variable x), number of letters in the alphabet (variable y), and the longest alternating subsequence (variable q) is given by*

$$1 + \frac{y(1 - y + x(q - 1))}{2x(1 - x)(1 - y)(1 - y + x(q^2 - 1))} \times$$

$$\left(1 - y + x^2(q^2 - 1) - \sqrt{((1 - x)^2 + x^2q^2 - y)^2 - 4x^2(1 - x)^2q^2}\right).$$

Theorem 6.2.16. *([603]) The generating function for 132-avoiding words by length (variable x), number of letters in the alphabet (variable y), and the longest alternating subsequence (variable q) is given by*

$$\frac{1 - x + xqy}{(1 - x)(1 - y)} + \frac{yq(1 - y + x(q - 1))}{2x(1 - x)(1 - y + x(q^2 - 1))^2} \times$$

$$\left(1 - y - 2x - \frac{1 + y}{1 - y}(q^2 - 1)x^2 - \sqrt{((1 - x)^2 + x^2q^2 - y)^2 - 4x^2(1 - x)^2q^2}\right).$$

6.3 Partially ordered patterns in permutations

Many results on POPs are presented in Section 3.5. In this section we collect several of the remaining known results on POPs. We refer to [511] by Kitaev for several open problems on POPs, the first one of which on $(2m + 1)$-*alternating n-permutations* was solved by Rawlings and Tiefenbruck in [672].

It is easy to see that $\mathrm{Av}(12) = \mathrm{Av}(\underline{12})$ and, as shown by Claesson [251], $\mathrm{Av}(213) = \mathrm{Av}(2\underline{13})$. As a corollary to a classification theorem on POPs (not to be stated here), Hardarson [448] showed that the two examples above are the only cases, modulo trivial bijections, of vincular patterns P and Q, $P \neq Q$, such that $\mathrm{Av}(P) = \mathrm{Av}(Q)$, while, in contrast, there exist infinitely many such pairs for partially ordered patterns. For example, assuming a and b are incomparable and both are larger than 1, one has

$$\mathrm{Av}(a1b) = \mathrm{Av}(\underline{a1}b) = \mathrm{Av}(a\underline{1b}),$$

while assuming $1 < 2, 2'$ and $1' < 2, 2'$ are the only relations on the set $\{1, 1', 2, 2'\}$, we have

$$\mathrm{Av}(12\ 1'2') = \mathrm{Av}(12\underline{1'2'}) = \mathrm{Av}(12\underline{1'}2') = \mathrm{Av}(1\underline{21'}2').$$

Let σ_i, $i = 0, 1, \ldots, k$, be consecutive – possibly empty – patterns. Assuming the σ_is are built on incomparable alphabets, we call the pattern $p = \sigma_0\sigma_1 \cdots \sigma_k$ a *multi-pattern*. Notice that assuming at least two of the σ_is are non-empty, p is *not* a consecutive pattern. For example, if $p = \underline{2'1'3'}\,\underline{2''1''}\,\underline{acdb}$, where $1' < 2' < 3'$, $1'' < 2''$ and $a < b < c < d$ are the only relations among p's letters, then p is a multi-pattern.

For σ_is as above, the pattern $q = \sigma_0 a_1 \sigma_1 a_2 \cdots \sigma_{k-1} a_k \sigma_k$ where $a_1 a_2 \cdots a_k \in \mathcal{S}_k$ and any a_i is greater than every letter in σ_j, for $j = 0, 1, \ldots, k$, is called a *shuffle pattern*.

The motivations of Kitaev [509] for introducing the multi-patterns were, respectively, dealing with non-overlapping occurrences of consecutive patterns in permutations and a generalization of a result of Claesson to be discussed below. As is pointed out in [509], to avoid a multi-pattern p, or a shuffle pattern q, where $|\sigma_i| = \ell_i$ for $i = 0, 1, \ldots, k$ is the same as to avoid

$$\prod_{i=1}^{k}\binom{\ell_0 + \ell_1 + \cdots + \ell_i}{\ell_i} = \binom{\ell_0 + \ell_1}{\ell_1}\binom{\ell_0 + \ell_1 + \ell_2}{\ell_2}\cdots\binom{\ell_0 + \ell_1 + \cdots + \ell_k}{\ell_k}$$

certain (ordinary) vincular patterns.

6.3.1 Multi-patterns

As is shown in [509], permuting σ_is in a multi-pattern and then replacing each σ_i with a permutation obtained by applying a trivial bijection (possibly the identity), one gets a multi-pattern that is Wilf-equivalent to the original one.

The following theorem expresses the e.g.f. for avoidance of a multi-pattern $p = \sigma_1 \sigma_2 \cdots \sigma_k$ in terms of the e.g.f.s for the $s_n(\sigma_i)$s, $i = 1, 2, \ldots, k$.

Theorem 6.3.1. *([509]) Let $p = \sigma_1 \sigma_2 \cdots \sigma_k$ be a multi-pattern and let $A_i(x)$ be the e.g.f. for $s_n(\sigma_i)$, $i = 1, 2, \ldots, k$. Then the e.g.f. $B(x)$ for $s_n(p)$ is*

$$B(x) = \sum_{i=1}^{k} A_i(x) \prod_{j=1}^{i-1}((x-1)A_j(x) + 1).$$

For example, applying Theorem 6.3.1, if $\sigma_i \in \{\underline{12}, \underline{21}\}$ then the e.g.f. for $\mathrm{Av}(\sigma_1 \sigma_2 \cdots \sigma_k)$ is easily seen [509] to be given by

$$\frac{1 - (1 + (x-1)e^x)^k}{1 - x}.$$

6.3.2 Shuffle patterns

Let us begin by considering a shuffle pattern that, in fact, is an ordinary generalized pattern. This pattern is $p = \sigma k$, where σ is an arbitrary consecutive pattern built on letters $1, 2, \ldots, k - 1$. So the last letter of p is greater than any other letter.

Theorem 6.3.2. *([509],[355]) Let $p = \sigma k$ as described above, and let $A(x)$ (resp., $B(x)$) be the e.g.f. for $s_n(\sigma)$ (resp., $s_n(p)$). Then*

$$B(x) = e^{\int_0^x A(t)dt}.$$

We provide two examples of applying Theorem 6.3.2.

Example 6.3.3. Let $p = 12$. Here $\sigma = 1$, so $A(x) = 1$ since $s_n(1) = 0$ for all $n \geq 1$ and $s_0(1) = 1$. So $B(x) = e^x$. This corresponds to the fact that for each $n \geq 1$ there is exactly one permutation that avoids the pattern p, namely $n(n-1)\cdots 1$.

Example 6.3.4. Suppose $p = 1\underline{23}$. Here $\sigma = \underline{12}$, so $A(x) = e^x$, since there is exactly one permutation that avoids the pattern σ. So

$$B(x) = \sum_{n \geq 0} \frac{B_n}{n!} x^n = e^{e^x - 1}.$$

This result agrees well with [251, Proposition 2] claiming that for all $n \geq 1$, $s_n(p)$ is the n-th *Bell number* (see Subsection A.2.1 for the definition).

Consider now the problem of simultaneous avoidance of two shuffle patterns, $1\sigma 2$ and $2\sigma 1$ (σ is a consecutive pattern), which is equivalent to avoiding the single POP $p = k_1 \sigma k_2$, where k_1 and k_2 are incomparable, and they are larger than any letter in σ. Similarly to the fact in Theorem 6.3.2 that knowing $A(x)$, the e.g.f. for the avoidance of σ, we obtain the e.g.f. $B(x)$ for the avoidance of σk, we have that knowing $B(x)$ gives us the e.g.f. $C(x)$ for avoidance of $k_1 \sigma k_2$ as was shown by Hardarson [448].

Theorem 6.3.5. *([448]) Let the POP $p = k_1 \sigma k_2$ be as described above, and let $B(x)$ (resp., $C(x)$) be the e.g.f. for $s_n(\sigma k)$ (resp., $s_n(p)$). Then*

$$C(x) = 1 + \int_0^x B^2(t)dt \quad and \quad s_{n+1}(p) = \sum_{i=0}^n \binom{n}{i} s_i(\sigma k) s_{n-i}(\sigma k).$$

Thus, by Theorem 6.3.2, for $A(x)$ defined there,

$$C(x) = 1 + \int_0^x e^{2\int_0^t A(y)dy} dt.$$

In fact, in Theorem 6.3.5, the pattern σ is allowed to be any consecutive POP, not just a regular consecutive pattern. In particular, we can use Theorem 6.3.5 together with the fact that the e.g.f. for $\mathcal{S}_n(\underline{312}, \underline{312})$ is $\frac{e^{2x}+1}{2}$ (see Table 5.2 for a Wilf-equivalent pair of patterns $\{\underline{132}, \underline{231}\}$) to obtain the e.g.f. for $\mathcal{S}_{n+1}(xy1zw)$ in Theorem 3.5.53 ($\mathcal{S}_n(\underline{312}, \underline{312}) = \mathcal{S}_n(\underline{y1z})$):

$$1 + \int_0^x \exp\left(\frac{e^{2t} + 2t - 1}{2}\right) dt.$$

Theorem 6.3.6. *([509]) Let p be a shuffle pattern of the form σkτ for some consecutive patterns σ and τ. Thus k is the largest letter of the pattern, and each letter of σ is incomparable with any letter of τ. Let $A(x)$, $B(x)$ and $C(x)$ be the e.g.f.s for the number of permutations that avoid σ, τ and p, respectively. Then $C(x)$ is the solution of the differential equation*

$$C'(x) = (A(x) + B(x))C(x) - A(x)B(x),$$

with initial condition $C(0) = 1$.

It follows from Theorem 6.3.6 that applying any trivial bijections to σ and τ, and possibly swapping them, gives a pattern that is Wilf-equivalent to the original pattern p.

6.4 Partially ordered patterns in words

POPs in words have not been much studied. Below we state selected results from [514] by Kitaev and Mansour, which extends [509] by Kitaev (on POPs in permutations) to the case of words.

The following theorem is an extension of Theorem 6.3.1 to the case of words. Note that in what follows, σ_i is allowed to have repeated letters.

Theorem 6.4.1. *([514]) Let $p = \sigma_1\sigma_2\cdots\sigma_k$ be a multi-pattern and let $A_i(x;\ell) = \sum_{n\geq 0} w_{n,\ell}(\sigma_i)x^n$, $i = 1, 2, \ldots, k$. Then the g.f. $B(x;\ell)$ for $w_{n,\ell}(p)$ is*

$$B(x;\ell) = \sum_{i=1}^{k} A_i(x;\ell) \prod_{j=1}^{i-1}((\ell x - 1)A_j(x;\ell) + 1).$$

For example, applying Theorem 6.4.1, with $\sigma_i \in \{\underline{12}, \underline{21}\}$, the g.f. for the avoidance of $\sigma_1\sigma_2\cdots\sigma_k$ on $[\ell]^*$ is given by

$$\frac{1 - \left(1 + \frac{\ell x - 1}{(1-x)^\ell}\right)^k}{1 - \ell x}.$$

Another result on multi-patterns in [514] is the following formula, where 1 is incomparable to $1' < 2'$:

$$w_{n,\ell}(\underline{1'2'}) = \ell\binom{n+\ell-2}{n-1}.$$

Suppose that τ and τ' are two equal consecutive patterns built on incomparable alphabets, m is the largest letter comparable with all letters in both alphabets, and p is the shuffle pattern $\tau m \tau'$.

Theorem 6.4.2. *([514]) For the shuffle pattern p defined above and $\ell \geq m$ (ℓ is the number of letters in the alphabet),*

$$A_p(x; \ell) = \frac{1}{(1 - xA_\tau(x; \ell - 1))^2} \left(A_p(x; \ell - 1) - xA_p^2(x; \ell - 1)\right).$$

As an example of an application of Theorem 6.4.2, we can derive the g.f. for $W_{n,\ell}(1'21'')$, where $1'$ and $1''$ are incomparable and $1', 1'' < 2$. Here $\tau = 1$, so $A_\tau(x; \ell) = 1$ for all $\ell \geq 1$. Hence

$$A_{1'21''}(x; \ell) = \frac{A_{1'21''}(x; \ell - 1) - x}{(1 - x)^2},$$

which together with $A_{1'21''}(x; 1) = \frac{1}{1-x}$ (for any n, only the word $11 \cdots 1$ of length n avoids $1'21''$) gives

$$A_{1'21''}(x; \ell) = \frac{1}{(1 - x)^{2\ell - 1}} - \sum_{i=1}^{\ell-1} \frac{x}{(1 - x)^{2i}}.$$

Chapter 7

VPs, BVPs and BPs

In this chapter we collect known enumerative results and/or references to them related to vincular patterns in permutations and words, and to bivincular patterns and barred patterns in permutations. As is typical for this book, no proofs are given, but they can easily be found following the references.

7.1 Vincular patterns in permutations

There is a recent survey [751] on vincular patterns by Steingrímsson. Several parts of this section follow the survey. Note that we have already seen many results on VPs in Chapters 2, 3 and 5. In particular, in Subsection 2.2.4, we saw that $Av(24\underline{1}3, 3\underline{1}42)$ is exactly the set of Baxter permutations; in Table 3.2 we linked VPs to other combinatorial objects; in Subsection 3.1.3 we discussed the relevance of the pattern $2\underline{1}3$ to PASEP, a statistical mechanics model; in Subsection 2.2.3 we connected the class $Av(2413, 3142)$ to 2-stack sortable permutations and rooted non-separable planar maps, as well as to barred pattern-avoidance. We refer to Chapter 5 for results on consecutive patterns and to Section 3.3 for the original motivation for study of vincular patterns in [64] by Babson and Steingrímsson. This section contains a comprehensive survey of enumerative results on VPs, except for consecutive patterns.

Remark 7.1.1. In the case of classical patterns, patterns that are each other's inverses belong to the same *symmetry class* (defined by trivial bijections) which is not the case for vincular patterns, since the inverse of a vincular pattern is not defined (as opposed to the inverse of a bivincular pattern). Thus, in the case of VPs, a symmetry class is defined by complement, reverse, and their composition, which is a group of order 4 on the set of VPs. For example, the VPs $2\underline{31}$, $2\underline{13}$, $\underline{13}2$

and $\underline{312}$ form an entire symmetry class.

Regev and Roichman [677] linked vincular patterns with q-*Bell numbers* $B_q(n)$ defined by $B_q(n) = \sum_{k=0}^{n} q^k S(n, k)$, where $S(n, k)$ are the *Stirling numbers of the second kind* (see Subsection A.2.1). Note that $B_1(n)$ is the usual n-th Bell number. Such a way to define q-Bell numbers generalizes the classical formula

$$B_1(n) = \frac{1}{e} \sum_{r \geq 0} \frac{r^n}{r!}$$

as follows:

$$B_q(n) = \frac{1}{e^q} \sum_{r \geq 0} \frac{q^r r^n}{r!}.$$

The connection made by Regev and Roichman [677] states that

$$s_{n+q-1}(P) = (q-1)! B_q(n),$$

where P is the set of $q!$ vincular patterns $\{p_1 p_2 \cdots p_q \underline{(q+2)(q+1)} \mid p_1 p_2 \cdots p_q \in \mathcal{S}_q\}$.

Stating results on VPs, it is convenient to subdivide the patterns into groups based on their type defined as follows.

Definition 7.1.2. A vincular pattern $p = p_1 p_2 \cdots p_k$ is of *type* $(i_1, i_2, \ldots, i_m)$ if the first (leftmost) i_1 letters of it are underlined, then the following i_2 letters of p are underlined, etc, and there is no common line under p_{i_1} and p_{i_1+1}, $p_{i_1+i_2}$ and $p_{i_1+i_2+1}$, and so on. We assume that if $i_j = 1$ then there is no line under the letter $p_{i_1+i_2+\cdots+i_{j-1}+1}$.

Vincular patterns have shown up implicitly in the literature in several places before they were introduced in [64]. For example, Simion and Stanton [722] essentially studied the patterns $\underline{231}$, $\underline{213}$, $\underline{132}$ and $\underline{312}$ in connection with a set of *orthogonal polynomials* generalizing the *Laguerre polynomials*. Also, as was already mentioned, many permutation statistics can be described in terms of VPs, for example, the *valleys* $(\underline{213} + \underline{312})$, the *peaks* $(\underline{231} + \underline{132})$, the *double ascents* $(\underline{123})$ and *double descents* $(\underline{321})$, and more generally, *increasing* and *decreasing runs* of length k $(\underline{12 \cdots k}$ and $\underline{k(k-1) \cdots 1}$, respectively).

Example 7.1.3. The patterns $\underline{123}$ and $\underline{231}$ are of type $(2, 1)$, while the pattern $\underline{24}13\underline{5}$ is of type $(3, 2)$.

7.1.1   Avoidance of a vincular 3- or 4-pattern

As was already mentioned in Section 3.4, the first systematic study of vincular patterns of length 3 was done by Claesson in [251]. See Table 3.2 for a collection of enumerative results related to vincular 3-patterns, as well as for their connections to other combinatorial objects. In particular, three cases of avoidance of two VPs of length 3 studied in [251] are provided in Table 3.2 that involve Motzkin numbers M_ns (see Subsection A.2.1 for the definition), Bessel numbers B_n^*s (see Subsection 3.4.1 for the definition), and the number of involutions i_n.

It is worth mentioning that the fact that Av($1\underline{23}$) is counted by the Bell numbers shows that the former Stanley-Wilf conjecture (see Theorem 6.1.20) does not hold for some VPs (see Subsection 7.1.3 for known asymptotic results for VPs). Moreover, the same fact shows that the conjecture of Noonan and Zeilberger [644] mentioned in Subsection 6.1.2 is also false for vincular patterns, namely, the number of permutations avoiding a vincular pattern is not necessarily polynomially recursive.

Claesson and Mansour [265] extended results in [251] by presenting a complete solution for the number of permutations avoiding a pair of VPs of type (1,2) or (2,1). The enumeration results can be summarized as follows:

# pairs	2	2	4	34	8	2	4	4	4	2
$s_n(p,q)$	$0, \ n > 5$	$2(n-1)$	$1 + \binom{n}{2}$	2^{n-1}	M_n	a_n	b_n	i_n	C_n	B_n^*

where C_n is the n-th Catalan number, $b_{n+2} = b_{n+1} + \sum_{k=0}^{n} \binom{n}{k} b_k$, $b_0 = b_1 = 1$ and

$$\sum_{n>0} a_n x^n = \frac{1}{1 - x - x^2 \sum_{n \geq 0} B_n^* x^n}.$$

Claesson and Mansour [265] also conjectured enumerative results for avoidance of any set of three or more VPs of type (1,2) or (2,1). These conjectures were proved for sets of size 3 by Bernini et al. [105], and for sizes 4, 5 and 6 by Bernini and Pergola [109]. In these enumerations, one meets the Fibonacci and Motzkin numbers, (central) binomial coefficients, powers of 2, and other rather simple answers.

Avoidance of vincular patterns of length 4

As is stated in [751], there are 48 symmetry classes of VPs of length 4, and computer experiments show that there are at least 24 Wilf-equivalent classes (their exact number is unknown). For vincular non-classical patterns enumeration results are known for seven Wilf classes (out of at least 24) which are as follows.

- Elizalde and Noy [360] gave the e.g.f.s for the number of occurrences of a consecutive 4-pattern for three of the seven Wilf classes for consecutive patterns, namely the classes with representatives $\underline{1234}$, $\underline{1243}$ and $\underline{1342}$ (see Theorem 5.1.12).

- Kitaev [509] and Elizalde [355], based on results by Elizalde and Noy [360] on consecutive 3-patterns, enumerated two Wilf classes with representatives $\underline{1234}$ and $\underline{1324}$ (see Theorem 6.3.2). The corresponding generating functions for distribution (avoidance is obtained by setting $u = 0$) for these cases are

$$\exp\left(\frac{\sqrt{3}}{2}\int_0^x \frac{e^{t/2}dt}{\cos\left(\frac{\sqrt{3}}{2}t + \frac{\pi}{6}\right)}\right) \quad \text{and} \quad \exp\left(\int_0^x \frac{dt}{1 - \int_0^x e^{-u^2/2}du}\right).$$

- Callan [214] proved that $s_n(\overline{35}241) = s_n(\underline{3}142)$ and in [215] he gave two recursive formulas for Av($\underline{3}142$). One of the recursions for $a_n = s_n(\underline{3}142)$ is as follows: $a_0 = c_1 = 1$ and

  1. $a_n = \sum_{i=0}^{n-1} a_i c_{n-i}$ for $n \geq 1$;
  2. $c_n = \sum_{i=0}^{n-1} i a_{n-1,i}$ for $n \geq 2$;
  3. $a_{n,k} = \begin{cases} \sum_{i=0}^{k-1} a_i \sum_{j=k-i}^{n-1-i} a_{n-1-i,j} & \text{for } 1 \leq k \leq n-1; \\ a_{n-1} & \text{for } k = n. \end{cases}$

  Note that the barred pattern $\overline{35}241$ is involved in a description of the class Av($2341, \overline{35}241$) of 2-stack sortable permutations (see Subsection 2.1.1).

- Callan [219] showed that $s_n(\underline{1234}) = \sum_{k=1}^n u(n,k)$ where for $1 \leq k \leq n$, $u(n,k)$ is defined by the recurrence

$$u(n,k) = u(n-1, k-1) + k \sum_{j=k}^{n-1} u(n-1, j)$$

with initial conditions $u(0,0) = 1$ and $u(n,0) = 0$ for $n \geq 1$. The result of Callan was an improvement of a complicated recurrence for $s_n(\underline{1234})$ posted at [726, A113227].

Remark 7.1.4. Actually, a complicated recursion for counting Av($\underline{1234}$) is available at [726, A113226]. However, we do not count it as a case in which enumeration is known.

The following theorem from [42] by Asinowski et al. shows a connection between the set Av($2\underline{413}, 3\underline{142}$) of Baxter permutations (see Subsection 2.2.4) and the class Av($2\underline{143}, 3\underline{412}$); the latter class is enumerated there. Moreover, the class Av($2\underline{143}, 3\underline{412}$) was shown in [42] to be in bijection with *pairs of orders* induced by neighborhood relations between *segments* of a *floorplan partition*. A floorplan partition is a certain tiling of a rectangle by other rectangles. There are natural ways to order floorplan partitions' elements (rectangles and segments). Pairs of ordered *rectangles* of a floorplan partition were studied in [4] by Ackerman et al., where a natural bijection between these pairs and the Baxter permutations was obtained.

Theorem 7.1.5. *([42]) The generating function for* $\mathrm{Av}(2\underline{14}3, 3\underline{41}2)$ *is given by*

$$\sum_{n \geq 1} x^{n-1}(1-x)^n b_n$$

where

$$b_n = s_n(2\underline{41}3, 3\underline{14}2) = \sum_{m=0}^{n} \frac{2}{n(n+1)^2} \binom{n+1}{m}\binom{n+1}{m+1}\binom{n+1}{m+2}$$

is the number of Baxter n-permutations. Therefore

$$s_n(2\underline{14}3, 3\underline{41}2) = \sum_{i=0}^{\lfloor (n+1)/2 \rfloor} (-1)^i \binom{n+1-i}{i} b_{n+1-i}.$$

As particular cases of more general results, Elizalde [356] obtained the following theorem.

Theorem 7.1.6. *([356]) We have*

$$\sum_{n \geq 0} s_n(1\underline{23}, 3\underline{12}, 3\underline{421})x^n = -1 + \sum_{k \geq 0} \frac{(1+kx)x^{2k}}{(1-(k+1)x)\prod_{j=1}^{k-1}(1-jx)};$$

$$\sum_{n \geq 0} s_n(1\underline{23}, 3\underline{421})x^n = \sum_{k \geq 0} \frac{(1+kx)x^{k+1}}{(1+x)^k(1-kx)(1-(k+1)x)}.$$

7.1.2 Containing occurrences of vincular patterns

We have already seen several results on permutations containing occurrences of a VP. Remark 3.1.19 discusses the distribution of the consecutive 2-patterns, while Theorems 3.1.17 and 3.1.20 deal with the distribution of the pattern $\underline{231}$ with descents registered; see Theorem 7.1.9 below for a relevant result. Elizalde and Noy [360] obtained the e.g.f. for the distribution of $\underline{1342}$, and the differential equations for the e.g.f.s for $\underline{1234}$ and $\underline{1243}$ (see Theorem 5.1.12). Also, Table 3.2 contains the result by Gudmundsson [433] stating that

$$|\{\pi \in \mathcal{S}_n \mid 123(\pi) = 1, \ 1\underline{23}(\pi) = 1\}| = \frac{5}{n+2}\binom{2n-2}{n+1}.$$

With respect to being equidistributed, there are three different classes of vincular patterns of type (1,2) or (2,1). In two of these cases, Claesson and Mansour [264] considered permutations with exactly one occurrence of the corresponding patterns, while they gave the entire distribution in terms of continued fractions for the third class. In the last case, explicit formulas for the number of permutations containing exactly r occurrences of the corresponding patterns are obtained for $r = 1, 2, 3$. We state selected results from [264] below.

Theorem 7.1.7. *([264]) Let B_n be the n-th Bell number (see Subsection A.2.1 for the definition). Then*

$$s_{n+2}^1(1\underline{23}) = 2s_{n+1}^1(1\underline{23}) + \sum_{k=0}^{n-1} \binom{n}{k} [s_{k+1}^1(1\underline{23}) + B_{k+1}]$$

for $n \geq -1$, with the initial condition $s_0^1(1\underline{23}) = 0$. Moreover,

$$s_{n+1}^1(1\underline{32}) = s_n^1(1\underline{32}) + \sum_{k=1}^{n-1} \left[\binom{n}{k} s_k^1(1\underline{32}) + \binom{n-1}{k-1} B_k \right],$$

where $n \geq 0$, with the initial condition $s_0^1(1\underline{32}) = 0$.

Theorem 7.1.8. *([264]) We have*

$$\sum_{n\geq 0} s_n^1(1\underline{23})x^n = \sum_{n\geq 1} \frac{x}{1-nx} \sum_{k\geq 0} \frac{kx^{k+n}}{(1-x)(1-2x)\cdots(1-(k+n)x)}.$$

Based on a result by Clarke et al. [269], the following theorem is obtained in [264].

Theorem 7.1.9. *([264]) The following Stieltjes continued fraction expansion holds*

$$\sum_{\pi \in \mathcal{S}_\infty} x^{1+12(\pi)} y^{21(\pi)} p^{2\underline{31}(\pi)} q^{3\underline{12}(\pi)} t^{|\pi|} = \cfrac{1}{1 - \cfrac{x[1]_{p,q}t}{1 - \cfrac{y[1]_{p,q}t}{1 - \cfrac{x[2]_{p,q}t}{1 - \cfrac{y[2]_{p,q}t}{\ddots}}}}}$$

where $[n]_{p,q} = q^{n-1} + pq^{n-2} + \cdots + p^{n-2}q + p^{n-1}$. *In particular,*

$$\sum_{\pi \in \mathcal{S}_\infty} p^{213(\pi)} = \cfrac{1}{1 - \cfrac{[1]_p t}{1 - \cfrac{[1]_p t}{1 - \cfrac{[2]_p t}{1 - \cfrac{[2]_p t}{\ddots}}}}}$$

where $[n]_p = 1 + p + \cdots + p^{n-1}$.

Theorem 7.1.10. *([264]) We have*

$$s_n^1(2\underline{13}) = \binom{2n}{n-3}, \quad s_n^2(2\underline{13}) = \frac{n(n-3)}{2(n+4)}\binom{2n}{n-3}, \quad s_n^3(2\underline{13}) = \frac{1}{3}\binom{n+2}{2}\binom{2n}{n-5}.$$

7.1.3 Asymptotic results for vincular patterns

This subsection is based entirely on [355] by Elizalde. We present here selected theorems from [355]. Asymptotic results for consecutive patterns can be found in Section 5.7.

Using [251] by Claesson, and the asymptotic behavior of the Catalan and Bell numbers, we have the following facts.

- For $p \in \{2\underline{13}, 2\underline{31}, 3\underline{12}, 1\underline{32}\}$,

$$s_n(p) \sim \frac{4^n}{\sqrt{\pi n}};$$

- For $q \in \{\underline{123}, 3\underline{21}, \underline{321}, \underline{123}, 1\underline{32}, 2\underline{31}, 3\underline{12}, 2\underline{13}\}$,

$$s_n(q) \sim \frac{1}{\sqrt{n}}\lambda(n)^{n+1/2}e^{\lambda(n)-n-1},$$

where $\lambda(n)$ is defined by $\lambda(n)\ln(\lambda(n)) = n$. Another useful description of the asymptotic behavior of $s_n(q)$ is the following:

$$\frac{\ln(s_n(q))}{n} = \ln(n) - \ln(\ln(n)) + O\left(\frac{\ln(\ln(n))}{\ln(n)}\right).$$

Proposition 7.1.11. *([355]) Let p be a VP having three consecutive underlined letters. Then there exist constants $0 < c, d < 1$ such that, for all $n \geq k$,*

$$c^n n! < s_n(p) < d^n n!.$$

Based on Theorem 6.3.2 the following statement can be proved.

Proposition 7.1.12. *([355]) For a consecutive $(k-1)$-pattern σ,*

$$\lim_{n \to \infty} \left(\frac{s_n(\sigma k)}{n!} \right)^{1/n} = \lim_{n \to \infty} \left(\frac{s_n(\sigma)}{n!} \right)^{1/n}.$$

Given two formal power series $F(x) = \sum_{n \geq 0} f_n x^n$ and $G(x) = \sum_{n \geq 0} g_n x^n$, we use the notation $F(x) < G(x)$ to indicate that $f_n < g_n$ for all n.

Theorem 7.1.13. *([355]) For $k \geq 1$, let*

$$h_k = 1 + \frac{1}{2} + \cdots + \frac{1}{k},$$

$$b_k(x) = \sum_{i=0}^{k} \binom{k}{i}^2 [x + 2(h_{k-i-h_i})] e^{ix},$$

$$c_k(x) = \frac{e^{(k+1)x}}{k+1} - \sum_{i=0}^{k} \binom{k}{i} \binom{k+1}{i} \left[x + 2(h_{k-i} - h_i) + \frac{1}{k+1-i} \right] e^{ix},$$

$$S(x) = \sum_{k \geq 1} b_k(x) + \sum_{k \geq 1} c_k(x).$$

Then

$$e^{S(x)} < \sum_{n \geq 0} s_n(\underline{1234}) \frac{x^n}{n!} < e^{S(x)+e^x+x-1}.$$

Proposition 7.1.14. *([355]) We have*

$$\frac{1}{2} \int_0^x e^{2e^y - 2} dy - \frac{x}{2} < \sum_{n \geq 0} s_n(1\underline{234}) \frac{x^n}{n!} < \sum_{n \geq 0} \frac{C_n(e^x - 1)^n}{n!},$$

where C_n is the n-th Catalan number.

As a corollary to Proposition 7.1.14, Elizalde [355] obtained

$$\lim_{n \to \infty} \left(\frac{s_n(1\underline{234})}{n!} \right)^{1/n} = 0.$$

For a consecutive pattern $\sigma = \sigma_1\sigma_2\cdots\sigma_{k-2}$, let $1\sigma k$ denote the vincular pattern $1(\sigma_1+1)(\sigma_2+1)\cdots(\sigma_{k-2}+1)k$. Lower and upper bounds are given in [355] for $s_n(1\sigma k)$ in terms of formal power series (not to be stated here). As a corollary to this result, Elizalde [355] obtained

$$\lim_{n\to\infty}\left(\frac{s_n(1\sigma k)}{n!}\right)^{1/n} = \lim_{n\to\infty}\left(\frac{s_n(\sigma)}{n!}\right)^{1/n}.$$

7.1.4 VPs and a classical 3-pattern

Elizalde [356] proved the following theorem.

Theorem 7.1.15. *([356]) The g.f. for* $\mathrm{Av}(213, 23\underline{41}, 32\underline{41})$ *where u marks the value of the rightmost letter is*

$$\sum_{n\geq 1}\sum_{\pi\in\mathcal{S}_n(213,23\underline{41},32\underline{41})} u^{last(\pi)}x^n = \frac{1 - x - 2xu - \sqrt{1 - 2x - 3x^2}}{2x\left(1 + u + \frac{1}{u}\right) - 2}.$$

Generating functions for the following sets were obtained in [356] when two certain statistics are taken into account: $\mathrm{Av}(213, \underline{3421})$, $\mathrm{Av}(213, 12\underline{34})$ and $\mathrm{Av}(213, \underline{1234})$. We do not state these enumerative results here.

Mansour [589] studied generating functions for the number of n-permutations avoiding the pattern 132 (or containing 132 exactly once) and another VP τ (or containing τ exactly once). Many enumerative results are given in [589] out of which we state here four theorems.

Theorem 7.1.16. *([589]) The g.f. for* $\mathrm{Av}(213, \underline{1234})$ *is*

$$\frac{1 - 2x - x^2 - \sqrt{1 - 4x + 2x^2 + x^4}}{2x^2}.$$

Theorem 7.1.17. *([589]) Let $\tau \in \{12, \underline{12}, 21, \underline{21}\}$. For any $k \geq 2$, we have*

$$\sum_{n\geq 0} s_n(132, \tau34\cdots k)x^n = \frac{U_{k-1}\left(\frac{1}{2\sqrt{x}}\right)}{\sqrt{x}U_k\left(\frac{1}{2\sqrt{x}}\right)},$$

where $U_n(x)$ is the n-th Chebyshev polynomial of the second kind.

Theorem 7.1.18. *([589]) The g.f. for the number of permutations that avoid the pattern 132 and contain the VP $\underline{1234}\cdots k$ exactly once is given by*

$$\frac{1}{(1 - x)U_k^2\left(\frac{1}{2\sqrt{x}}\right)}.$$

Theorem 7.1.19. *([589]) For any $k \geq 2$, the g.f. for the number of permutations containing exactly one occurrence of the pattern 132 and avoiding the VP $\underline{1234\cdots k}$ is given by*

$$\frac{x}{U_k^2\left(\frac{1}{2\sqrt{x}}\right)} \sum_{j=1}^{k-2} U_j^2\left(\frac{1}{2\sqrt{x}}\right).$$

Further, Mansour [587, 593] studied distributions of the VPs $\underline{1234\cdots k}, \underline{2134\cdots k}$, $\underline{12\cdots k}$ and $\underline{k(k-1)\cdots 1}$ on Av(132). Let $F_p(x,y) = \sum_{r\geq 0} F_{p;r}y^r$, $F_{p;r}(x) = \sum_{n\geq 0} f_{p;r}(n)x^n$, and $f_{p;r}(n)$ be the number of 132-avoiding n-permutations containing exactly r occurrences of a pattern p. As particular cases of general theorems, the following enumerative results on consecutive 3-patterns were obtained.

Theorem 7.1.20. *([587]) We have*

$$F_{\underline{123}}(x,y) = F_{\underline{321}}(x,y) = \frac{1 + xy - x - \sqrt{1 - 2x - 3x^2 - xy(2 - 2x - xy)}}{2x(x + y - xy)};$$

$$F_{\underline{231}}(x,y) = \frac{1 - 2x + 2xy - \sqrt{1 - 4x + 4x^2 - 4x^2 y}}{2xy};$$

$$F_{\underline{213}}(x,y) = F_{\underline{312}}(x,y) = \frac{1 - x^2 + x^2 y - \sqrt{1 + 2x^2 - 2x^2 y + x^4 - 2x^4 y + x^4 y^2 - 4x}}{2x(1 + xy - x)}.$$

The generating functions for 123-avoiding permutations with $r = 0, 1, 2, 3$ occurrences of the pattern $\underline{123}$ are given in [587]. We state the $r = 0, 1, 2$ cases in the following theorem.

Theorem 7.1.21. *([587]) We have*

$$F_{\underline{123};0}(x) = \frac{1 - x - \sqrt{1 - 2x - 3x^2}}{2x^2};$$
$$F_{\underline{123};1}(x) = \frac{x-1}{2x} + \frac{1 - 2x - x^2}{2x\sqrt{1 - 2x - 3x^2}};$$
$$F_{\underline{123};2}(x) = \frac{x^4}{(1 - 2x - 3x^2)^{3/2}}.$$

The following theorem generalizes Theorem 7.1.17 and is of similar nature to Theorem 6.1.37.

Theorem 7.1.22. *([593]) The g.f. $\sum_{\pi\in\mathrm{Av}(132)} x_1^{|\pi|} \prod_{k\geq 2} x_k^{\underline{1234\cdots k}(\pi)}$ is given by the following continued fraction:*

$$\cfrac{1}{1 - x_1 + x_1 x_2^{\binom{0}{0}} - \cfrac{x_1 x_2^{\binom{0}{0}}}{1 - x_1 + x_1 x_2^{\binom{1}{0}} x_3^{\binom{1}{1}} - \cfrac{x_1 x_2^{\binom{1}{0}} x_3^{\binom{1}{1}}}{1 - x_1 + x_1 x_2^{\binom{2}{0}} x_3^{\binom{2}{1}} x_4^{\binom{2}{2}} - \cfrac{x_1 x_2^{\binom{2}{0}} x_3^{\binom{2}{1}} x_4^{\binom{2}{2}}}{\ddots}}}},$$

where the $(n+1)$-th numerator is $x_1 \prod_{i=0}^{n} x_{i+2}^{\binom{n}{i}}$.

The g.f. $\sum_{\pi \in Av(132)} x_1^{|\pi|} \prod_{k \geq 2} x_k^{2134 \cdots k(\pi)}$ is also expressed in [593] in the form of a continued fraction. [587]

Motzkin permutations

Elizalde and Mansour [359] studied $Av(132, \underline{123})$ which they call the class of *Motzkin permutations* because, as is shown bijectively in [359], $s_n(132, \underline{123}) = M_n$, the n-th Motzkin number (see Subsection A.2.1 for the definition). Many enumerative results on Motzkin permutations are given in [359], some of which we state below.

Theorem 7.1.23. *([359]) The g.f. for Motzkin permutations with respect to the length of the longest decreasing subsequence (variable v) and the number of ascents (variable y) is*

$$\sum_{n \geq 0} \sum_{\pi \in Av(132, \underline{123})} v^{lds(\pi)} y^{asc(\pi)} x^n = \frac{1 - vx - \sqrt{1 - 2vx + (v^2 - 4vy)x^2}}{2vyx^2}$$

$$= \sum_{n \geq 0} \sum_{m \geq 0} \frac{1}{n+1} \binom{2n}{n} \binom{m+2n}{2n} x^{m+2n} v^{m+n} y^n.$$

Theorem 7.1.24. *([359]) The g.f. $\sum_{n \geq 0} \sum_{\pi \in Av(132, \underline{123})} \prod_{i \geq 1} x_i^{12 \cdots i(\pi)}$ is given by the following continued fraction:*

$$\cfrac{1}{1 - x_1 - \cfrac{x_1^2 x_2}{1 - x_1 x_2 - \cfrac{x_1^2 x_2^3 x_3}{1 - x_1 x_2^2 x_3 - \cfrac{x_1^2 x_2^5 x_3^4 x_4}{\ddots}}}},$$

where the n-th numerator is $\prod_{i=1}^{n+1} x_i^{\binom{n}{i-1} + \binom{n-1}{i-1}}$ and the n-th denominator is $\prod_{i=1}^{n} x_i^{\binom{n-1}{i-1}}$.

Theorem 7.1.25. *([359]) Fix $k \geq 2$. The g.f. for the number of Motzkin permutations which contain the pattern $12 \cdots k$ exactly r times is given by*

$$\frac{\left(U_{k-2} \left(\frac{1-x}{2x} \right) - x U_{k-3} \left(\frac{1-x}{2x} \right) \right)^{r-1}}{U_{k-1}^{r+1} \left(\frac{1-x}{2x} \right)},$$

for all $r = 1, 2, \ldots, k$.

Theorem 7.1.26. *([359]) The g.f.* $\sum_{n\geq 0}\sum_{\pi \in Av(132,1\underline{23})} x^n q^{\mathrm{lmax}(\pi)}$ *where the left-to-right maxima (see Definition A.1) are registered by the variable q, is given by the following continued fraction:*

$$\cfrac{1}{1-xq-\cfrac{x^2q}{1-x-\cfrac{x^2}{1-x-\cfrac{x^2}{\ddots}}}}.$$

7.1.5   More results on vincular patterns

Note that results in Chapter 5 combined with Theorem 6.3.2 give expressions for the e.g.f.s for the following patterns:

$$\underline{12\cdots k}(k+1), \underline{k(k-1)\cdots 1}(k+1), \underline{12\cdots a\tau(a+1)}(k+1), \underline{(a+1)\tau a(a-1)\cdots 1}(k+1),$$

$$\underline{k(k-1)\cdots(k+1-a)\tau'(k-a)}(k+1), \underline{(k-a)\tau'(k+1-a)(k+2-a)\cdots k}(k+1),$$

where k and a are positive integers with $a \leq k-2$, τ is any permutation of $\{a+2, a+3, \ldots, k\}$ and τ' is any permutation of $\{1, 2, \ldots, k-a-1\}$.

Elizalde [355] proved the following result.

Proposition 7.1.27. *([355]) If σ and τ are two consecutive k-patterns and $\sigma \sim \tau$ (σ and τ are Wilf-equivalent) then $\sigma(k+1)(k+2) \sim \tau(k+1)(k+2)$. Also, $\sigma(k+1)(k+2) \sim \sigma(k+2)(k+1)$.*

Bernini et al. [103] used the *ECO* method together with a graphical representation of permutations to study avoidance of any VP of type (1,2) or (2,1) when the statistic "first/last letter" (head/last in our terminology) is registered. Examples of results obtained are as follows:

$$|\{\pi_1 \cdots \pi_n \in \mathcal{S}_n(2\underline{13}) \mid \pi_n = k\}| = |\{\pi_1 \cdots \pi_n \in \mathcal{S}_n(\underline{31}2) \mid \pi_1 = k\}| = \frac{k}{n}\binom{2n-k-1}{n-1};$$
$$|\{\pi_1 \cdots \pi_n \in \mathcal{S}_n(2\underline{31}) \mid \pi_n = k\}| = |\{\pi_1 \cdots \pi_n \in \mathcal{S}_n(\underline{13}2) \mid \pi_1 = k\}| = \frac{n-k+1}{n}\binom{n+k-2}{n-1}.$$

In Subsections 5.8.1 and 5.8.2 we considered the problem of counting permutations that avoid a consecutive 3-pattern, and begin and/or end with a prescribed pattern. Kitaev and Mansour [517, 516] obtained enumerations for the cases of avoiding a type (2,1) or (1,2), or a classical 3-pattern, and beginning and/or ending with increasing or decreasing patterns, or the patterns $(k-1)(k-2)\cdots 1k$ or $23\cdots k1$. We state here two theorems as examples of the results in [517, 516]. Recall that $E_p^q(x)$ denotes the e.g.f. for the number of permutations avoiding a pattern p and beginning with a (consecutive) pattern q. We also let $G_p^{q,h}(x)$ denote the g.f. for the number of permutations avoiding p, beginning with q and ending with h.

Theorem 7.1.28. *([517]) We have*

$$E_{\underline{132}}^{(k-1)(k-2)\cdots 1k}(x) = \begin{cases} e^{e^x} \int_0^x e^{-e^t} \sum_{n \geq k-1} \frac{t^n}{n!} dt, & \text{if } k \geq 2, \\ e^{e^x-1}, & \text{if } k = 1. \end{cases}$$

Theorem 7.1.29. *([517]) Let* $C(x) = \frac{1-\sqrt{1-4x}}{2x}$. *Then*

$$G_{\underline{132}}^{12\cdots k, 12\cdots \ell}(x) = x^{k+\ell-1} C^{\ell+1}(x) + \frac{x^m - x^{k+\ell-1}}{1-x}.$$

7.1.6 VPs and alternating permutations

The reader may wish to review the notion of an alternating permutation in Definition 1.0.18 and the notation in Definition 6.1.110.

Han et al. [447] defined the following q-analogue of the Euler numbers counting alternating permutations.

Definition 7.1.30. The q-*tangent* numbers $E_{2n+1}(q)$ are defined by

$$\sum_{n=0}^{\infty} E_{2n+1}(q)x^n = \cfrac{1}{1 - \cfrac{[1]_q[2]_q x}{1 - \cfrac{[2]_q[3]_q x}{1 - \cfrac{[3]_q[4]_q x}{\ddots}}}}$$

where $[n]_q = \frac{1-q^n}{1-q}$, and the q-*secant numbers* $E_{2n}(q)$ are defined by

$$\sum_{n=0}^{\infty} E_{2n}(q)x^n = \cfrac{1}{1 - \cfrac{[1]_q^2 x}{1 - \cfrac{[2]_q^2 x}{1 - \cfrac{[3]_q^2 x}{\ddots}}}}.$$

Example 7.1.31. The initial values for $E_n(q)$ are $E_0(q) = E_1(q) = E_2(q) = 1$, $E_3(q) = 1 + q$, $E_4(q) = 2 + 2q + q^2$ and $E_5(q) = 2 + 5q + 5q^2 + 3q^3 + q^4$.

The following closed formulas for $E_n(q)$ were obtained by Josuat-Vergés [488]:

$$E_{2n+1}(q) = \frac{1}{(1-q)^{2n+1}} \sum_{k=0}^{n} \left(\binom{2n+1}{n-k} - \binom{2n+1}{n-k-1} \right) \sum_{i=0}^{2k+1} (-1)^{i+k} q^{i(2k+2-i)};$$

$$E_{2n}(q) = \frac{1}{(1-q)^{2n}} \sum_{k=0}^{n} \left(\binom{2n}{n-k} - \binom{2n}{n-k-1} \right) \sum_{i=0}^{2k} (-1)^{i+k} q^{i(2k-i)+k}.$$

Chebikin [240] proved combinatorially that

$$\sum_{\sigma \in A_n} q^{\underline{312}(\sigma)} = E_n(q).$$

We continue with selected results by Mansour [591].

Theorem 7.1.32. ([591]) For $\tau \in \{12, 21, \underline{12}, \underline{21}\}$, $A_{132,\tau34\cdots k}(x)$ is the same and is given by Theorem 6.1.115. Moreover, the same generating function counts the number of 132-avoiding alternating permutations that contain exactly one occurrence of $\tau34\cdots k$ with one exception for $\tau = 21$ when the generating function is 0 for $k \geq 3$.

Theorem 7.1.33. ([591]) Letting $A_{132;\tau}^r(x)$ denote the generating function for 132-avoiding alternating permutations that contain exactly r occurrences of a pattern τ, we have that for $r \geq 0$,

- $A_{132;\underline{123}}^r(x) = A_{132;\underline{321}}^r(x) = 0$;

- $A_{132;\underline{213}}^r(x) = A_{132;\underline{312}}^r(x) = \displaystyle\sum_{n \geq r+1} \frac{r+1}{n(n-r)} \binom{n}{r+1}^2 (x^{2n} + x^{2n+1})$;

- $A_{132;\underline{231}}^r(x) = \frac{1}{r+1} \binom{2r}{r} (x^{2r+2} + x^{2r+1})$ for $r \geq 1$.

7.1.7   VPs and Dumont permutations

Note that all necessary definitions are contained in Subsection 6.1.9. Mansour [598] proved the following theorems.

Theorem 7.1.34. ([598]) The g.f.s for $\mathcal{D}_n^1(132, \underline{1234}\cdots k)$ and $\mathcal{D}_n^1(132, \underline{2134}\cdots k)$ are, respectively,

$$\mathcal{D}_{132,\underline{1234}\cdots k}^1(x) = F_k(x) + xF_{k-1}(x) \text{ and } \mathcal{D}_{132,\underline{2134}\cdots k}^1(x) = G_{k-1}(x) + xG_{k-2}(x).$$

Remark 7.1.35. Note that by Theorems 6.1.122 and 7.1.34,

$$|\mathcal{D}_n^1(132, 12\cdots k)| = |\mathcal{D}_n^1(132, \underline{123}\cdots k)|, \quad |\mathcal{D}_n^1(132, 213\cdots k)| = |\mathcal{D}_n^1(132, \underline{213}\cdots k)|.$$

However, no bijective proofs of these facts are known.

Theorem 7.1.36. *([598]) Let*

$$A_k(x) = \frac{x^2}{1 - x^2 F_{k-2}(x)} A_{k-1}(x) + \frac{x^4 F_{k-1}(x)}{(1 - x^2 F_{k-2}(x))^2} A_{k-2}(x)$$

for all $k \geq 4$, where $A_1(x) = x^2$ and $A_2(x) = 2x^4$. Then for all $k \geq 2$, the g.f. for the number of 132-avoiding Dumont permutations of the first kind containing exactly one occurrence of the pattern $\underline{123}4\cdots k$ is $A_k(x) + xA_{k-1}(x)$.

A similar statement to Theorem 7.1.36 is proved in [598] for avoidance on $\mathcal{D}_n^1$ of the pattern 132 and containing the pattern $\underline{213}4\cdots k$ exactly once. These results were extended to obtaining explicit formulas for 132-avoiding Dumont permutations containing more than 1 occurrence of the pattern $\underline{123}4\cdots k$ (or, $\underline{213}4\cdots k$).

7.2 Vincular patterns in words

Vincular patterns in words have not been much studied. The case of consecutive patterns in words is considered in Section 5.10. In this section we collect the remaining known enumerative results.

Burstein and Mansour [210] obtained the following theorems.

Theorem 7.2.1. *([210]) For $\ell \geq 1$, we have*

$$\sum_{n\geq 0} w_{n,\ell}(\underline{111})x^n = (-1)^{\ell-1} \prod_{i=1}^{\ell} A_i(x) + \sum_{j=1}^{\ell} \left((-1)^{\ell-j} B_j(x) \prod_{i=j+1}^{\ell} A_i(x) \right),$$

where $A_j(x) = \frac{jx^2}{1-(j-1)x}$ and $B_j(x) = \frac{1+x}{1-(j-1)x}$. Also,

$$\sum_{n\geq 0} w_{n,\ell}(\underline{112})x^n = \prod_{j=0}^{\ell-1} \frac{1-(j-1)x}{1-(j+x)x}.$$

In what follows, let $F_p(x; \ell) = \sum_{n\geq 0} w_{n,\ell}(p)x^n$.

Theorem 7.2.2. *([210]) For $\ell \geq 0$,*

$$F_{\underline{212}}(x;\ell) = F_{\underline{211}}(x;\ell) = 1 + \sum_{d=0}^{\ell-1}\left(x^{d+1}F_{\underline{211}}(x;\ell-d)\sum_{i=d}^{\ell-1}(1-x)^{i-1}\binom{i}{d}\right).$$

The following generating functions are corollaries to Theorem 7.2.2:

$$\begin{aligned}
F_{\underline{212}}(x;0) &= F_{\underline{211}}(x;0) = 1;\\
F_{\underline{212}}(x;1) &= F_{\underline{211}}(x;1) = \tfrac{1}{1-x};\\
F_{\underline{212}}(x;2) &= F_{\underline{211}}(x;2) = \tfrac{1}{(1-x)^2} + \tfrac{x^2}{(1-x)^3};\\
F_{\underline{212}}(x;3) &= F_{\underline{211}}(x;3) = \tfrac{1-3x+6x^2-5x^3+3x^4-x^5}{(1-x)^6}.
\end{aligned}$$

Theorem 7.2.3. *([210]) For $\ell \geq 1$,*

$$\sum_{n\geq0} w_{n,\ell}(\underline{123})x^n = \sum_{n\geq0} w_{n,\ell}(\underline{213})x^n = \prod_{j=0}^{\ell-1}\frac{1}{1-\frac{x}{(1-x)^j}}.$$

Burstein and Mansour [210] provided a recurrence for calculating $F_{\underline{132}}(x;\ell) = F_{\underline{123}}(x;\ell)$ (see also [461, Proposition 6.39]). Base on this result, they derived the following particular cases:

$$\begin{aligned}
F_{\underline{132}}(x;1) &= \tfrac{1}{1-x};\\
F_{\underline{132}}(x;2) &= \tfrac{1}{1-2x};\\
F_{\underline{132}}(x;3) &= \tfrac{(1-x)^2}{(1-2x)(1-3x+x^2)};\\
F_{\underline{132}}(x;4) &= \tfrac{1-4x+6x^2-3x^3}{(1-3x)(1-2x)(1-3x+x^2)}.
\end{aligned}$$

Bernini et al. [106] enumerated several classes of words simultaneously avoiding two type $(1,2)$ VPs of length 3. They proved the following theorems where $(a)_b = a(a-1)\cdots(a-b+1)$ is the usual falling factorial.

Theorem 7.2.4. *([106]) We have*

$$w_{n,\ell}(\underline{112},\underline{221}) = \begin{cases} \sum_{k=0}^{n-1} k(\ell)_k + (\ell)_n, & \text{if } n \leq \ell;\\ \sum_{k=0}^{\ell-1} k(\ell)_k, & \text{if } n > \ell. \end{cases}$$

Theorem 7.2.5. *([106]) We have*

$$w_{n,\ell}(\underline{121},\underline{212}) = \sum_{k=0}^{n} k\binom{\ell}{k}(n-1)_{k-1}.$$

Theorem 7.2.6. *([106]) We have*

$$\sum_{n\geq 0} w_{n,\ell}(1\underline{11},1\underline{12})x^n = \sum_{k=0}^{\ell} \binom{\ell}{k}(1+x)^k C_k(x),$$

where $C_0(x) = 1$ and, for $k \geq 1$, $C_k(x) = \sum_{j=1}^{k}\binom{k}{j}x^j C_{k-1}(x)$.

Theorem 7.2.7. *([106]) We have*

$$\sum_{n\geq 0} w_{n,\ell}(1\underline{11},1\underline{21})x^n = \sum_{k=0}^{\ell} \binom{\ell}{k}x(1+x)^k \left(\prod_{i=0}^{k-2}\left((1+x)^{k-i}-1\right)\right).$$

Theorem 7.2.8. *([106]) We have*

$$\sum_{n> 0} w_{n,\ell}(1\underline{11},1\underline{22})x^n = \sum_{k=0}^{\ell} \binom{\ell}{k}\frac{x^k(x+k)_k}{\prod_{i=1}^{k-1}(1-ix)}.$$

Theorem 7.2.9. *([106]) We have*

$$\sum_{n\geq 0} w_{n,\ell}(2\underline{11},1\underline{22})x^n = \sum_{k=0}^{\ell} \frac{(\ell)_k x^{k-1}}{(1-x)\left(\prod_{i=1}^{k-1}(1-ix)\right)}.$$

The unsolved problems in avoidance of two type (1,2) VPs of length 3, up to Wilf-equivalence, are to enumerate the following sets:

$$W_{n,\ell}(1\underline{12},1\underline{21}) \quad W_{n,\ell}(1\underline{12},1\underline{22})$$
$$W_{n,\ell}(1\underline{12},2\underline{11}) \quad W_{n,\ell}(1\underline{12},2\underline{12})$$
$$W_{n,\ell}(1\underline{21},1\underline{22}) \quad W_{n,\ell}(1\underline{21},2\underline{11}).$$

7.3 Bivincular patterns in permutations

There are not so many results in the literature on bivincular patterns. As was mentioned in Section 1.8, we can provide the following references: [255, 263, 318, 321, 143, 145, 161, 527, 528, 557, 636, 771, 772, 811]. Many of these results are already discussed in Sections 2.3 and 3.2.1. In particular, the generating function for $s_n(\overline{\underline{123}}_{\underline{231}})$ is given in Theorem 3.2.66, and a more general enumeration result on the pattern $\overline{\underline{123}}_{\underline{231}}$ is presented in Theorem 3.2.68. In this section we provide some of the remaining results.

Theorem 7.3.1. *([474, 475]) For all positive integers n, and any $k \le n$, $s_n(\overline{\underline{12\cdots k}})$ is equal to the coefficient of x^n in*

$$\sum_{m \ge 0} m! x^m \left(\frac{1 - x^{k-1}}{1 - x^k} \right)^m .$$

Thus, for $k \le n < 2k$, the number of permutations of length $k + r$ containing at least one occurrence of the pattern $\overline{\underline{12\cdots k}}$ is $r!(r^2 + r + 1)$.

The avoidance of bivincular patterns where the top row is overlined and the bottom row is underlined (see Theorem 7.3.1 for such a pattern) is called "very tight pattern-avoidance" by Bóna in [143, 145]. Avoidance of such patterns is also studied by Myers [636]. Bóna [143] described a set P of length k bivincular patterns (with the top row overlined and the bottom row underlined) such that for $p \in P$, $s_n(p) \le s_n(\overline{\underline{12\cdots k}})$. Moreover, using a result in [145], it is shown in [143] that the set P consists of almost all such bivincular patterns (of length k). One should compare this result with the fact that $s_n(p)$ is neither minimal nor maximal for the pattern $p = 12 \cdots k$ compared to the other classical pattern of length k, while $s_n(p)$ is believed to be maximal for the pattern $\underline{12 \cdots k}$ compared to the other consecutive patterns of length k.

Bóna [143] also studies the *limiting distribution* of the BVP $\overline{\underline{12}}$.

Theorem 7.3.2. *([143]) Let Z_n be the random variable that counts occurrences of $\overline{\underline{12}}$ in a randomly selected n-permutation. Then Z_n converges to a Poisson distribution with parameter $\lambda = 1$.*

7.4   Barred patterns in permutations

If needed, review the definition of barred patterns (BPs) in Section 1.2.

We have already seen BPs in various contexts in this book. It is recorded in Theorem 2.1.4 that $\mathcal{S}_n(2341, 3\overline{5}241)$ is the set of 2-stack sortable permutations; the corresponding cardinality is $s_n(2341, 3\overline{5}241) = \frac{2(3n)!}{(2n+1)!(n+1)!}$. The same numbers count other sets related to the avoidance of BPs that can be found in Table 2.19: $\mathrm{Av}(2413, 41\overline{3}52) = \mathrm{Av}(2413, 3\underline{1}42)$ (non-separable permutations), $\mathrm{Av}(2413, 45\overline{3}12)$, $\mathrm{Av}(2413, 21\overline{3}54)$, $\mathrm{Av}(2413, 5\overline{1}324)$, $\mathrm{Av}(2413, \overline{4}2315)$, $\mathrm{Av}(2314, \overline{4}2513)$, $\mathrm{Av}(3241, \overline{2}4153)$ and $\mathrm{Av}(3214, \overline{2}4135)$. See Theorem 2.1.5 and the formula right above it, for counting

2-stack sortable permutations when right-to-left maxima and descents, respectively, are taken into account.

In several cases (but not always!), avoidance of barred patterns can be described in the language of vincular patterns. For example, $\mathrm{Av}(1\bar{4}23) = \mathrm{Av}(\underline{123})$ (see [663] by Pudwell), $\mathrm{Av}(3\bar{5}241) = \mathrm{Av}(\underline{31}42)$ (see [214] by Callan; enumeration of this class is discussed in Subsection 7.1.1), $\mathrm{Av}(25\bar{3}14, 41\bar{3}52) = \mathrm{Av}(24\underline{13}, 3\underline{142})$ (Baxter permutations; see Subsection 2.2.4, in particular, for two enumeration formulas), and, as mentioned in the previous paragraph, $\mathrm{Av}(2413, 41\bar{3}52) = \mathrm{Av}(2413, 3\underline{142})$. Theorem 2.3.7 says that the Schubert variety X_π is factorial if and only if $\pi \in \mathrm{Av}(1324, 21\bar{3}54) = \mathrm{Av}(1324, 2\underline{143})$. Remark 2.3.9 says that the last set of permutations is exactly the class of forest-like permutations studied in [160] by Bousquet-Mélou and Butler; see Theorem 2.3.10 for enumeration of these permutations. By Remark 2.2.18, we also have

$$s_n(24\underline{13}) = s_n(3\underline{41}2) = s_n(21\bar{3}54).$$

Another already mentioned result on BPs is given by Theorem 3.2.77 where $\mathrm{Av}(3\bar{1}52\bar{4})$ is enumerated which settled a conjecture of Pudwell. The last set is in one-to-one correspondence with (2+2)- and N-avoiding posets (see Subsection 3.2.6).

Barcucci et al. [79] linked the class $\mathrm{Av}(321, 4\bar{1}523)$ with *single-source directed animals* and with *forests of* 1-2 *trees*. It is remarked in [79] that

$$\sum_{\pi \in \mathrm{Av}(321, 4\bar{1}523)} x^{\mathrm{length}(\pi)} q^{\mathrm{inv}(\pi)} = \frac{x(J_1(x,q) + J_0(x,q))}{J_0(x,q)(1-xq) - xq(1+q)J_1(x,q)},$$

where

$$J_1(x,q) = \sum_{n \geq 0} \frac{(-1)^n x^{n+1} q^{3n}}{(x,q)_{n+1}(q,q)_n}, \quad J_0(x,q) = \frac{(-1)^n x^n q^{3n}}{(x,q)_n (q,q)_n}$$

and $(a,q)_n = \prod_{k=0}^{n-1}(1 - aq^k)$. Also, Barcucci et al. [84] described a bijection between the set $\mathcal{S}_n(321, 3\overline{1}42)$ and *steep parallelogram polyominoes* having half-perimeter equal to $n + 2$. The permutations in $\mathrm{Av}(321, 3\overline{1}42)$ are enumerated in [84] with respect to three statistics and length, in particular showing that

$$s_n(321, 3\overline{1}42) = M_n, \text{ the } n\text{-th Motzkin number.}$$

A characterization of $(321, 3\overline{1}42)$-avoiding permutations in terms of their *canonical reduced decompositions* (see Subsection 4.1.9) is obtained in [243] by Chen et al., which allows one construct a bijection between $\mathrm{Av}(321, 3\overline{1}42)$ and Motzkin paths. This bijection sends descents of permutations to up-steps in the corresponding paths. Similar results are obtained in [243] for $\mathrm{Av}(321, 4\overline{1}32)$. Also, in [243], the inversion

numbers of the permutations in question are linked to the areas of the corresponding
Motzkin paths. Finally, Chen et al. [242] established a correspondence between
Riordan paths and $(321, 3\bar{1}42)$-avoiding *derangements*. Riordan paths are Motzkin
paths without horizontal steps on the x-axis. Riordan paths are counted by the
Riordan numbers

$$\frac{1}{n+1} \sum_{k=1}^{n-1} \binom{n+1}{k} \binom{n-k-1}{k-1}.$$

Generalizing considerations of $Av(321, 3\bar{1}42)$ and $Av(321, 4\bar{1}523)$, Barcucci et
al. [82] studied "discrete continuity" between the Motzkin and Catalan numbers
similarly to the continuity between Fibonacci and Catalan numbers considered by
Barcucci et al. [78], which is discussed in Subsection 6.1.5. More precisely, Barcucci
et al. [82] studied the sequence of classes $Av(321, (k+2)\bar{1}(k+3)23\cdots(k+1))$. The
case $k = 1$ gives the Motzkin numbers, while the case $k = \infty$ gives the Catalan
numbers; each $j > 1$ gives a counting sequence that lies between the Motzkin and
Catalan numbers. These classes of permutations are enumerated in [82] according to
several parameters. See [606] by Mansour et al. for some relevant studies. Moreover,
research in the same vein is conducted in [549] by Labelle et al., where barred
patterns restricted permutations are used to go from $4\bar{1}32$-avoiding permutations
counted by the Bell numbers to all permutations counted by $n!$. We omit details of
that paper, also dealing with certain q-analogues, just mentioning that it is shown
there that the permutations in $\mathcal{S}_n(4\bar{1}32)$ with m right-to-left minima are counted by
the *Stirling numbers of the second kind* (see Subsection A.2.1 for the definition).

A rather unusual barred pattern is considered in [322] by Dukes and Reifegerste.
This is $1\bar{4}3\bar{4}2$ which denotes an occurrence of the pattern 132 which is part of neither
an occurrence of the pattern 1432 nor the pattern 1342. The following proposition
was proved in [322] where the *left border number* of a permutation π, lbsum(π), is
the sum of the initial positions of the occurrences of all the consecutive patterns
$\underline{(k+1)\sigma k}$ with $\sigma \in \mathcal{S}_{k-1}$ arbitrary.

Proposition 7.4.1. *([322]) We have* lbsum$(\pi) = 21(\pi) + 1\bar{4}3\bar{4}2(\pi)$.

It was shown in [322] that the distribution of *lbsum*, $F_n(x) = \sum_{\pi \in \mathcal{S}_n} x^{\text{lbsum}(\pi)}$,
satisfies $F_0(x) = 1$ and

$$F_n(x) = \sum_{k=1}^{n} \binom{n-1}{k-1} F_{k-1}(x) F_{n-k}(x) x^{k(n-k)}.$$

7.4.1 A systematic enumeration for BPs avoidance

Pudwell [662, 663] gave the first comprehensive collection of enumeration results for permutations that avoid barred patterns of length ≤ 4. We state the results obtained/cited in [662, 663] for BPs having at least one bar. The following general lemmas are of help while classifying enumerations of permutations avoiding a single BP.

Lemma 7.4.2. *Let $\bar{p} \in \overline{\mathcal{S}}_k$ be such that every letter of $\bar{p}$ is barred. Then $\mathcal{S}_n(\bar{p})$ is the set of permutations that contain p (the pattern obtained by removing all bars).*

Lemma 7.4.3. *Let $\bar{p} \in \overline{\mathcal{S}}_k$ be such that the letter p_i is barred and either (i) p_{i+1} is unbarred with $p_{i+1} = p_i \pm 1$, or (ii) p_{i-1} is unbarred with $p_{i-1} = p_i \pm 1$. Then $s_n(\bar{p}) = s_n(p')$ where recall from Definition 1.2.3 that p' is the pattern obtained from $\bar{p}$ by removing all barred letters and taking the reduced form of the remaining permutation.*

Lemma 7.4.4. *Suppose that $\bar{p} \in \overline{\mathcal{S}}_k$ with only one unbarred letter. Then $s_n(\bar{p}) = 0$ for all $n \geq 1$.*

We do not state in this book considerations by Pudwell [662, 663] on length 5 BPs (the classification is incomplete) and some computational data related to them. Moreover, we refer to [662, 663] to learn about *prefix enumeration schemes* to find recurrences counting permutations that avoid a BP of length > 4 or a set of BPs.

Avoiding barred patterns of length 1 or 2

The following results are easy to see:

$$s_n(\bar{1}) = \begin{cases} 0, & \text{if } n = 0, \\ n!, & \text{if } n \geq 1; \end{cases}$$

$$s_n(\overline{12}) = s_n(\overline{21}) = n! - 1;$$

$$s_n(\bar{1}2) = s_n(2\bar{1}) = s_n(1\bar{2}) = s_n(\bar{2}1) = s_n(1) = \begin{cases} 1, & \text{if } n = 0, \\ 0, & \text{if } n \geq 1. \end{cases}$$

Avoiding barred patterns of length 3

For any 3-pattern p having all letters barred, by Lemma 7.4.2 we have $s_n(p) = n! - \frac{1}{n+1}\binom{2n}{n}$. By Lemma 7.4.4 it only remains to consider the case of patterns with one bar. Further, using Lemma 7.4.3 and trivial Wilf-equivalences, we obtain:

$$s_n(\overline{1}23) = s_n(32\overline{1}) = s_n(12\overline{3}) = s_n(\overline{3}21) = s_n(1\overline{2}3) = s_n(3\overline{2}1) = s_n(3\overline{1}2) = 1;$$
$$s_n(1\overline{3}2) = s_n(2\overline{3}1) = s_n(13\overline{2}) = s_n(\overline{2}31) = s_n(31\overline{2}) = s_n(\overline{2}13) = s_n(2\overline{1}3) = 1;$$
$$s_n(\overline{1}32) = s_n(23\overline{1}) = s_n(21\overline{3}) = s_n(\overline{3}12) = (n-1)!.$$

Avoiding barred patterns of length 4

Lemma 7.4.2 together with Table 6.2 answer the enumeration question associated with avoidance of a BP of length 4 with all letter barred. In the case of two bars we can either apply Lemma 7.4.3 to reduce the counting problem to avoidance of a classical 2-pattern (giving the trivial answer 1), or we have the following cases for all $n \geq 0$:

$$s_n(\overline{12}43) = s_n(\overline{13}2\overline{4}) = (n-2)!.$$

The case of barred patterns with precisely one bar was first comprehensively studied by Callan [215] and later by Pudwell [663] who provided similar results to Callan's arguments using slightly modified notation. The permutations avoiding the BPs of length 4 with one bar that are not mentioned in the following three theorems are counted by the Catalan numbers, which is an easy application of Lemma 7.4.3 (there are 64 such patterns).

Theorem 7.4.5. *([215, 663]) $s_n(p) = B_n$, the n-th Bell number (see Subsection A.2.1) if p is from the following set*

$$\{1\overline{4}23, 134\overline{2}, 23\overline{1}4, \overline{2}431, \overline{3}124, 32\overline{4}1, 4\overline{1}32, 421\overline{3},$$
$$\overline{2}413, 2\overline{4}13, 24\overline{1}3, 241\overline{3}, \overline{3}142, 3\overline{1}42, 31\overline{4}2, 314\overline{2}\}.$$

Theorem 7.4.6. *([215, 663]) For $p \in \{\overline{1}432, 234\overline{1}, 321\overline{4}, \overline{4}123\}$,*

$$s_n(p) = (n-1)! + \sum_{j=2}^{n} \frac{(n-2)!}{(j-2)!} + \sum_{i=3}^{n} \sum_{j=2}^{n-i+2} \sum_{\ell=j+i-2}^{n} \frac{(n-i)!(\ell-j-1)!}{(\ell-i)!(i-3)!}.$$

Theorem 7.4.7. *([215, 663]) For p in*

$$\{\overline{1}423, \overline{1}342, 231\overline{4}, 243\overline{1}, 312\overline{4}, 324\overline{1}, \overline{4}132, \overline{4}213, \overline{1}324, 132\overline{4}, 423\overline{1}, \overline{4}231\},$$

we have $s_n(p) = \sum_{i=1}^{n}(n-i)!s_{i-1}(p)$. Thus, from [726, A051295], the g.f. $G(x)$ for Av(p) satisfies:

$$G(x) = (1-x)G^2(x) - x^2 G'(x).$$

Also,

$$G(x) = \cfrac{1}{1 - \cfrac{x}{1 - \cfrac{x}{1 - \cfrac{x}{1 - \cfrac{2x}{1 - \cfrac{2x}{1 - \cfrac{3x}{1 - \cfrac{3x}{1 - \cdots}}}}}}}}.$$

Initial values for $s_n(p)$ are $1, 1, 2, 5, 15, 54, 235, 1237, 7790, \ldots$.

7.4.2 Barred vincular patterns

Elizalde [356] considered a mix of the notions of barred and vincular patterns. More precisely, he considered the pattern $\overline{2}31$ and certain of its refinements to be discussed below. From our definitions, a permutation π avoids the pattern $\overline{2}31$ if every descent in π (an occurrence of the VP $\underline{21}$) is part of an occurrence of the VP $2\underline{31}$.

Proposition 7.4.8. *([356]) $s_n(213, \overline{2}\underline{31}) = M_{n-1}$, the $(n-1)$-th Motzkin number.*

A refinement of the pattern $\overline{2}31$ considered by Elizalde [356] is the pattern $\overline{2}^{\circ}31$. We say that a permutation π avoids the pattern $\overline{2}^{\circ}31$ if every descent in π is the "31" part of an odd number of occurrences of $\overline{2}\underline{31}$, that is, for any index i such that $\pi_i > \pi_{i+1}$, the number of indices $j < i$ such that $\pi_i > \pi_j > \pi_{i+1}$ is odd.

Proposition 7.4.9. *([356]) We have*

$$s_n(213, \overline{2}^{\circ}\underline{31}) = \begin{cases} \frac{1}{2k+1}\binom{3k}{k}, & \text{if } n = 2k, \\ \frac{1}{2k+1}\binom{3k+1}{k+1}, & \text{if } n = 2k+1. \end{cases}$$

Elizalde [356] mentioned several combinatorial objects equinumerous with the set $\mathcal{S}_n(213, \overline{2}^{\circ}\underline{31})$, for example, *symmetric ternary trees* with $3n$ edges, *symmetric diagonally convex directed polyominoes* of area n, and lattice paths from $(0,0)$ to $(n, \lfloor n/2 \rfloor)$ with steps $E = (1,0)$ and $N = (0,1)$ that never go above the line $y = x/2$. Also, the number of 2143-avoiding Dumont $2n$-permutations of the second kind is given by the product $s_n(213, \overline{2}^{\circ}\underline{31})s_{n+1}(213, \overline{2}^{\circ}\underline{31})$.

Changing "odd" to "even" in the definition of an occurrence of the pattern $\overline{2}^{\circ}31$, we can define an occurrence of the pattern $\overline{2}^{e}31$.

Proposition 7.4.10. *([356]) We have*

$$s_n(213, \overline{2}^e\underline{31}) = \frac{1}{n} \sum_{k=0}^{\lfloor n/2 \rfloor} \left[2\binom{n}{2k}\binom{n-k}{k-1} + \frac{n}{n-k}\binom{n}{2k+1}\binom{n-k}{k} \right].$$

Prior to the work of Elizalde [356], Burstein and Lankham [205] linked barred vincular patterns avoidance to *Patience Sorting*, a combinatorial algorithm that can be viewed as an iterated, non-recursive form of the *Schensted Insertion Algorithm* (see the survey paper [33] by Aldous and Diaconis). The term "Patience Sorting" was introduced in 1962 by Mallows [579, 580] while studying a card sorting algorithm invented by Ross. The algorithm works as follows.

Given a shuffled deck of cards $\pi = \pi_1 \pi_2 \cdots \pi_n$ that we think of as a permutation in $\mathcal{S}_n$, we do the following:

Step 1. Use a "patience sorting procedure" to form the subsequences (called *piles*) r_1, $r_2, \ldots, r_m$ of π as follows:

- Place the first card π_1 from the deck into a pile r_1 by itself.
- For each remaining card π_i ($i = 2, 3, \ldots, n$), consider the cards $d_1, d_2, \ldots, d_k$ atop the piles $r_1, r_2, \ldots, r_k$ that have already been formed.
  * If $\pi_i > \max\{d_1, d_2, \ldots, d_k\}$, then put π_i into a new right-most pile r_{k+1} by itself.
  * Otherwise, find the left-most card d_j that is larger than π_i and put the card π_i atop pile r_j.

Step 2. Gather the cards up one at a time from these piles in ascending order.

Denote by $R(\pi) = \{r_1, r_2, \ldots, r_m\}$ the *pile configuration* associated with the permutation π. Also, given any pile configuration R, we can form its *reverse patience word* $RPW(R)$ by listing the piles in R "from bottom to top, left to right" (that is, by reversing the so-called "far-eastern reading"). See [205, 207] for examples of formation of $R(\pi)$ and $RPW(R)$.

From our definitions, a permutation π avoids the pattern $3\overline{1}\underline{42}$ if it contains occurrences of the pattern $\underline{231}$ only as parts of occurrences of the pattern $31\underline{42}$. Burstein and Lankham [205] gave the following characterization of reverse patience words.

Theorem 7.4.11. *([205]) We have*

- *the set of permutations $\mathcal{S}_n(3\overline{1}\underline{42})$ is exactly the set $RPW(R(\mathcal{S}_n))$ of reverse patience words obtainable from $\mathcal{S}_n$;*

- $s_n(3\overline{1}\underline{42}) = B_n$, the n-th Bell number.

Further, Burstein and Lankham [207] proved the following theorem.

Theorem 7.4.12. *([207]) A pile configuration R has a unique preimage $\pi \in \mathcal{S}_n$ under Patience Sorting if and only if $\pi \in \mathcal{S}_n(3\overline{1}\underline{42}, 3\overline{1}\underline{24})$. Moreover, if*

$$f(n,k) = |\{\pi = \pi_1\pi_2\cdots\pi_n \in \mathcal{S}_n(3\overline{1}\underline{42}, 3\overline{1}\underline{24}) \mid \pi_1 = k\}|,$$

then $s_n(3\overline{1}\underline{42}, 3\overline{1}\underline{24}) = \sum_{k=1}^{n} f(n,k)$, and we have the following recurrence for $f(n,k)$ subject to the initial conditions $s_0(3\overline{1}\underline{42}, 3\overline{1}\underline{24}) = f(0,0) = 1$:

$$
\begin{array}{ll}
f(n,0) = 0, & \text{for } n \geq 1; \\
f(n,1) = f(n,n) = s_{n-1}(3\overline{1}\underline{42}, 3\overline{1}\underline{24}), & \text{for } n \geq 1; \\
f(n,2) = 0, & \text{for } n \geq 3; \\
f(n,k) = f(n,k-1) + f(n-1,k-1) + f(n-2,k-2), & \text{for } n \geq 3.
\end{array}
$$

The following facts are recorded in [207]:

$$
\begin{array}{c}
\mathcal{S}_n(31\overline{4}2) = \mathcal{S}_n(31\overline{4}2) = \mathcal{S}_n(31\underline{2}); \\
\mathcal{S}_n(\overline{2}41\underline{3}) = \mathcal{S}_n(\overline{2}413) = \mathcal{S}_n(241\overline{3}) = \mathcal{S}_n(241\overline{3}); \\
s_n(\overline{2}41\underline{3}) = s_n(31\overline{4}\underline{2}) = s_n(3\overline{1}\underline{42}) = B_n.
\end{array}
$$

Finally, yet another description of layered patterns in $\mathcal{S}_n$ (see Definition 2.1.11) is given in [207] by Burstein and Lankham, which is $\mathcal{S}_n(3\overline{1}\underline{42}, \overline{2}41\underline{3})$.

Chapter 8

Miscellaneous on patterns in permutations and words

In this chapter we discuss several topics without a common thread. Section 8.1 deals with so-called *simple permutations* intimately related with enumeration of permutation classes, while in Section 8.2 we consider *pattern matching problems* and *Gray codes* related to pattern-restricted permutations. Section 8.3 discusses packing patterns in permutations and words. Several approaches to tackle pattern problems are presented in Section 8.4. Section 8.5 briefly discusses some basic notions of combinatorics on words, a very important field of science, as well as *crucial/maximal* words and permutations thus providing a bridge between our patterns and combinatorics on words. Sections 8.6 and 8.7 deal with, respectively, *universal cycles* and *simsun permutations*. In Section 8.8 we consider an involution on Catalan objects and use it to prove equidistribution results on pattern-restricted permutations. Finally, in Section 8.9, we briefly discuss permutation avoidance games.

8.1 Simple permutations

If needed, review the notion of a *permutation class* in Definition 2.2.38. Also, recall Definition 1.0.23 which says that an *interval* in a permutation is a factor that contains a set of contiguous values. Intervals of permutations have applications to genetic algorithms and matching gene sequences (see [278] by Corteel et al. and references therein).

Definition 8.1.1. An n-permutation is simple if it only contains intervals of lengths 0, 1, and n.

Example 8.1.2. The permutation 2413 is simple, while 6243715 is not as it contains the non-trivial interval 243.

Brignall [183] wrote a survey of simple permutations. In this section we follow the survey in presenting selected materials on these permutations and we extend it by a few results. We refer to [183] and references therein for more details; in particular, for the discussion of complexity of computing the substitution decomposition, for so-called *pin sequences*, and for *partial well-orders* on permutation classes. We will need the following definition.

Definition 8.1.3. A given permutation class is often described in terms of its *minimal avoidance set*, or *basis*. Formally, the basis B of a permutation class $\mathcal{C}$ is the smallest set for which $\mathcal{C} = \{\pi \mid \pi$ avoids β for all $\beta \in B\}$.

Simple permutations form frames, or "building blocks" upon which all other permutations are constructed by means of the *substitution decomposition* (also called the *modular decomposition*, *disjunctive decomposition* or *X-join* in other contexts).

Definition 8.1.4. Given a permutation $\sigma = \sigma_1\sigma_2\cdots\sigma_m$ and nonempty permutations $\alpha_1, \alpha_2, \ldots, \alpha_m$, the *inflation* of σ by $\alpha_1, \alpha_2, \ldots, \alpha_m$, denoted $\sigma[\alpha_1, \alpha_2, \ldots, \alpha_m]$, is the permutation obtained by replacing the entry σ_i by an interval that is order-isomorphic to α_i for $1 \leq i \leq m$. Conversely, a *deflation* of π is any expression of π as an inflation $\pi = \sigma[\pi_1, \pi_2, \ldots, \pi_m]$, and σ is called a *quotient* of π.

Example 8.1.5. $3142[12, 12, 312, 1] = 45128673$.

Albert and Atkinson [22] proved the following two propositions.

Proposition 8.1.6. *([22]) Every permutation may be written as the inflation of a unique simple permutation. Moreover, if π can be written as $\sigma[\alpha_1, \alpha_2, \ldots, \alpha_m]$ where σ is simple and $m \geq 4$, then the α_is are unique.*

Proposition 8.1.7. *([22]) If π is an inflation of 12, then there is a unique sum indecomposable α_1 such that $\pi = 12[\alpha_1, \alpha_2]$ for some α_2 which is itself unique. The same holds with 12 replaced by 21 and "sum" replaced by "skew".*

Remark 8.1.8. Note that $12[\alpha_1, \alpha_2] = \alpha_1 \oplus \alpha_2$ and $21[\alpha_1, \alpha_2] = \alpha_1 \ominus \alpha_2$ in the notation of Definition 2.2.32.

Definition 8.1.9. The *substitution decomposition tree* for a permutation is obtained by recursively decomposing until we are left only with inflations of simple permutations by singletons.

Example 8.1.10. The permutation 657341289 is decomposed as

$$
\begin{aligned}
657341289 &= 2314[21, 1, 3412, 12] \\
&= 2314[21[1, 1], 1, 21[12, 12], 12[1, 1]] \\
&= 2314[21[1, 1], 1, 21[12[1, 1], 12[1, 1]], 12[1, 1]]
\end{aligned}
$$

and its substitution tree is given in Figure 8.1.

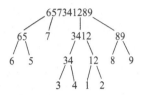

Figure 8.1: The substitution tree corresponding to the permutation 657341289.

8.1.1 Enumeration of simple permutations

Albert et al. [25] obtained the following result in a straightforward way.

Theorem 8.1.11. *([25]) The number, s_n, of simple permutations for $n \geq 4$ is given by*

$$
s_n = -Com_n + (-1)^{n+1} \cdot 2,
$$

where Com_n is the coefficient of x^n in the functional inverse of $f(x) = \sum_{n=1}^{\infty} n! x^n$ (sequence A059372 of [726]). The sequence $\{s_n\}_{n=1}^{\infty}$ begins

$$
1, 2, 0, 2, 6, 46, 338, 2926, 28146, \ldots.
$$

Theorem 8.1.12. *([25]) The number of simple permutations of length n is asymptotically given by*

$$
\frac{n!}{e^2} \left(1 - \frac{4}{n} + \frac{2}{n(n-1)} + O(n^{-3}) \right).
$$

8.1.2 Exceptional simple permutations

The following theorem is due to Schmerl and Trotter [714].

Theorem 8.1.13. *([714]) Every simple permutation of length $n \geq 2$ contains a subsequence of length $n - 1$ or $n - 2$ which is order-isomorphic to a simple permutation. In other words, for any simple permutation, by deleting at most two letters one obtains, after taking reduced form, a shorter simple permutation.*

In most cases, a single letter deletion in Theorem 8.1.13 is sufficient, and thus simple permutations requiring deletion of two letters are called *exceptional*. Schmerl and Trotter [714] call exceptional simple permutations *critically indecomposable*, and present a complete characterization in the analogous problem for partially ordered sets.

Definition 8.1.14. A *horizontal alternation* is a permutation in which every odd entry lies to the left of every even entry, or the reverse of such a permutation. A *vertical alternation* is the inverse of a horizontal alternation. Two families of these alternations, the *parallel* and *wedge alternations*, are identified, in which each "side" of the alternation forms a monotone sequence.

Example 8.1.15. In Figure 8.2, the permutations 54637281 and 86421357 are wedge alternations, while the permutation 13572468 is a parallel alternation.

Remark 8.1.16. Any parallel alternation is already simple or very nearly so, while any wedge alternation may be extended to form a simple permutation by placing a single point in one of two places. See [183] for details.

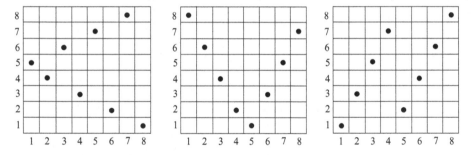

Figure 8.2: The permutations represented by the two permutation matrices to the left are wedge alternations, the permutation matrix to the right gives a parallel alternation.

Albert and Atkinson [22] show that the exceptional simple permutations are precisely the set of parallel alternations.

Theorem 8.1.17. *([22]) The set of exceptional simple permutations is exactly the set of simple parallel alternations, that is, those of the form*

$$246\cdots(2m)135\cdots(2m-1)\quad(m\geq 2)$$

and every symmetry of this permutation.

Brignall and Georgiou [185] characterized the *almost exceptional* simple permutations, where there is a unique point that can be removed that keeps the permutation simple. Such permutations are, roughly, variants on parallel and wedge alternations.

8.1.3 Simple extensions

We can ask how, for an arbitrary permutation π, we may embed π into a simple permutation, and how long this permutation has to be. In the case of an increasing permutation $12 \cdots n$, an additional $\lceil (n + 1)/2 \rceil$ letters are both necessary and sufficient. The following result is due to Brignall et al. [189].

Theorem 8.1.18. *([189]) Every permutation π on n letters has a simple extension with at most $\lceil (n + 1)/2 \rceil$ additional letters.*

8.1.4 Substitution closures

Definition 8.1.19. A class $\mathcal{C}$ of permutations is *substitution-closed* if we have $\sigma[\alpha_1, \alpha_2, \ldots, \alpha_m] \in \mathcal{C}$ for all σ, α_1, $\alpha_2, \ldots, \alpha_m \in \mathcal{C}$. The *substitution-closure*, also called the *wreath closure*, of a set X, $\langle X \rangle$, is defined as the smallest substitution-closed class containing X.

Remark 8.1.20. The concept of the substitution-closure is well-defined since the intersection of substitution-closed classes is substitution-closed, and the set of all permutations is substitution-closed.

Let $\mathrm{Si}(\mathcal{C})$ denote the set of simple permutations in the class $\mathcal{C}$. It is easy to see that $\mathrm{Si}(\mathcal{C}) = \mathrm{Si}(\langle \mathcal{C} \rangle)$ (see, e.g., [183] for an argument). Since substitution-closed classes are uniquely determined by the sets of simple permutations they contain, $\langle \mathcal{C} \rangle$ is the largest class with this property. For instance, the substitution closure of $\mathrm{Av}(132)$ is the largest class whose only simple permutations are 1, 12, and 21, which is precisely the class $\mathrm{Av}(2413, 3142)$ of separable permutations (see Subsection 2.2.5 for a discussion on these permutations).

Albert and Atkinson [22] show that it is rather easy to decide if a permutation class given by a finite basis is substitution-closed, which is stated in the proposition below. A "basis" for a permutation class is defined in Definition 8.1.3.

Proposition 8.1.21. *([22]) A permutation class is substitution-closed if and only if each of its basis elements is simple.*

8.1.5   Algebraic generating functions

Albert and Atkinson [22] obtained the following result by first proving it for substitution-closed classes, and then showing that adding an extra basis restriction does not affect the algebraicity of the generating function.

Theorem 8.1.22. *([22]) A permutation class with only finitely many simple permutations has a readily computable algebraic generating function. Moreover, if a class with finitely many simple permutations avoids a decreasing permutation of some length n, then the class is enumerated by a rational generating function.*

Definition 8.1.23. Calling any set of permutations P a *property*, we say that a permutation π satisfies P if $\pi \in P$. A set $\mathcal{P}$ of properties is said to be *query-complete* if, for every simple permutation σ and property $P \in \mathcal{P}$, one may determine whether $\sigma[\alpha_1, \alpha_2, \ldots, \alpha_m]$ satisfies P simply by knowing which properties of $\mathcal{P}$ each α_i satisfies.

Example 8.1.24. Avoidance of a given pattern is query-complete, since the property $\mathrm{Av}(\beta)$ lies in the query-complete set $\{\mathrm{Av}(\delta) \mid \beta \text{ contains } \delta\}$.

   Brignall et al. [187] proved the following theorem.

Theorem 8.1.25. *([187]) Let $\mathcal{C}$ be a permutation class containing only finitely many simple permutations, $\mathcal{P}$ a finite query-complete set of properties, and $\mathcal{Q} \subseteq \mathcal{P}$. The generating function for the set of permutations in $\mathcal{C}$ satisfying every property in $\mathcal{Q}$ is algebraic over $\mathbb{Q}(x)$.*

Proposition 8.1.26. *([183]) The set of properties consisting of*

- *{alternating permutations},*

- *{permutations beginning with an ascent},*

- *{permutations ending with an ascent}, and*

- *{1},*

is query-complete.

   See [183] for more facts on finite query-complete sets.

8.1.6 Finite basis and three more theorems

Using Theorem 8.1.13, the following result was obtained.

Proposition 8.1.27. *([183]) If the longest simple permutation in C has length k then the basis elements of $\langle C \rangle$ have length at most $k + 2$.*

For example, using Proposition 8.1.27 it can be computed that the substitution-closure of 1, 12, 21, and 2413 is $\mathrm{Av}(3142, 25314, 246135, 362514)$.

The following theorem demonstrates the significance of simple permutations for permutation pattern-avoidance theory.

Theorem 8.1.28. *([634, 22]) Every permutation class containing only finitely many simple permutations is finitely based.*

Let $\mathrm{Av}(\beta_1^{\leq r_1}, \beta_2^{\leq r_2}, \ldots, \beta_k^{\leq r_k})$ denote the set of permutations that have at most r_1 occurrences of β_1, at most r_2 occurrences of β_2, and so on. Clearly any such set is a permutation class. Atkinson [46] showed that the basis elements of the class can have length at most $\max\{(r_i + 1)|\beta_i| \mid i \in [k]\}$. For example, $\mathrm{Av}(132^{\leq 1}) = \mathrm{Av}(1243, 1342, 1423, 1432, 2143, 35142, 354162, 461325, 465132)$.

Brignall et al. [186] obtained the following result.

Theorem 8.1.29. *([186]) If the class $\mathrm{Av}(\beta_1, \beta_2, \ldots, \beta_k)$ contains only finitely many simple permutations then the class $\mathrm{Av}(\beta_1^{\leq r_1}, \beta_2^{\leq r_2}, \ldots, \beta_k^{\leq r_k})$ also contains only finitely many simple permutations for all choices of nonnegative integers $r_1, r_2, \ldots, r_k$.*

From an algorithmic aspect, Brignall et al. [188] proved the following theorem (the proof relies on *pin-permutations*, which are not discussed in this book).

Theorem 8.1.30. *([188]) It is possible to decide if a permutation class given by a finite basis contains infinitely many simple permutations.*

Given a permutation class $\mathrm{Av}(B)$, if the basis B contains only simple permutations, $\mathrm{Av}(B)$ is said to be *wreath-closed*. Bassino et al. [95] reduced the problem of determining if a simple permutation contains a given simple pin-permutation to deciding if a word is a factor of another word, which resulted in an $O(n \ln(n))$ algorithm to determine if a wreath-closed class of permutations contains a finite number of simple permutations. Moreover, Bassino et al. [94] proved the following theorem, which also provides an effective algorithm.

Theorem 8.1.31. *([94]) Deciding if a permutation class $C(\alpha_1, \alpha_2, \ldots, \alpha_k)$ contains a finite number of simple permutations can be done in time $O(n^{3k})$ where n is the size of the largest permutation in the basis.*

8.1.7   Permutation classes with two restrictions

Permutation classes (also called *pattern classes* or *permutation pattern classes*) of the form $\mathrm{Av}(\alpha)$ have a finite number of simple permutations only if $\alpha = 1, 12, 231$ (or a permutation equivalent to one of these by a trivial bijection). Atkinson [48] found a criterion for a pattern class of the form $\mathrm{Av}(\alpha, \beta)$ to have a finite number of simple permutations that we state in the following theorem.

Theorem 8.1.32. *([48])* $\mathrm{Av}(\alpha, \beta)$ *has a finite number of simple permutations if and only if α and β are equivalent by a trivial bijection to one of the following types:*

1. $\alpha = 231$,

2. $\alpha = 2413$, $\beta = 3142$.

3. $\alpha = k(k-1)\cdots 1$, $\beta = (12\cdots a) \oplus \theta \oplus (12 \cdots b)$, *where θ is one of the following permutations: 1, 21, 231.*

By the results of Albert and Atkinson [22], the pattern classes listed in the three cases of Theorem 8.1.32 have algebraic generating functions. As a matter of fact, the generating function for the first type is a rational function $p(x)/q(x)$ and an efficient recursive procedure for calculating it is give in [617] by Mansour and Vainshtein. The growth rate of the first class is then determined as the reciprocal of the smallest root of the polynomial $q(x)$. The simple permutations of these classes are 1, 12, 21.

In the second class of Theorem 8.1.32, the reader may recognize the class of separable permutations discussed in Subsection 2.2.5. The generating function for this class is

$$\frac{1 - x - \sqrt{1 - 6x + x^2}}{2}$$

and thus the corresponding growth rate is $3 + \sqrt{2}$. The simple permutations of this class are also 1, 12, 21.

For the third class of Theorem 8.1.32, if $\theta = 1$ then we deal with

$$\mathrm{Av}(k(k-1)\cdots 1, 12 \cdots (a+b+1))$$

which is finite by the Erdős-Szekeres theorem (see Theorem 6.1.2). In the case $\theta = 21$, we can use [23] by Albert et al. to state that the number of n-permutations in

$$\mathrm{Av}(k(k-1)\cdots 1, (12\cdots a) \oplus 21 \oplus (12\cdots b))$$

is polynomial in n and thus the growth rate here is 1. Atkinson [48] considered the case of $\theta = 231$.

Theorem 8.1.33. *([48]) The growth rate of*

$$\text{Av}(k(k-1)\cdots 1,(12\cdots a)\oplus 231\oplus(12\cdots b))$$

is independent of a, b and hence is equal to the growth rate of $\text{Av}(k(k-1)\cdots 1,231)$.

The growth rate of $\text{Av}(k(k-1)\cdots 1,231)$ can be found by the methods of Mansour and Vainshtein in [617].

8.2 Some algorithmic aspects of patterns

We have already encountered several algorithmic issues in this book, for example, while discussing stack-sortable permutations in Section 2.1, and Kazhdan-Lusztig polynomials in Subsection 2.3.2. In this section, we discuss pattern matching problems, and generation of pattern-restricted permutations using *Gray codes*. We do not provide any background on complexity here, assuming the reader has some basic knowledge.

8.2.1 Pattern matching problems

The *pattern matching problem*, also known as the *pattern involvement problem*, for permutations is to determine whether a given n-permutation π contains a given classical k-pattern p, $k \le n$. Suppose p is fixed and π is the input to the problem. In practice, most pattern containment studies are of this type. A *brute force approach* of simply examining all subsequences of π of length k would have a worst case execution time of $O(n^k)$. Therefore, the problem lies in the complexity class P. Albert et al. [15] developed general algorithms whose worst case complexity is considerably smaller than $O(n^k)$ and in a number of cases further improvements can be made. The study in [15] indicated an unproven observation that the algorithms should never be worse than $O(n^{2+k/2}\log n)$ and in some cases they are considerably better.

Chang and Wang [237] showed how to solve the problem for $p = 12\cdots k$ in time $O(n\log(\log n))$. The permutations 132, 213, 231 and 312 can all be handled in linear time by *stack sorting algorithms*. Albert et al. [15] established that any 4-pattern can be detected in $O(n\log n)$ time. Further, Ibarra [469] presented an $O(kn^4)$ time and $O(kn^3)$ space algorithm for finding an occurrence of p in π or determining that no occurrence exists, given that p is a *separable permutation*, that is, $p \in \mathcal{S}_k(2413,3142)$. Bose et al. [155] used *dynamic programming* to compute the number of occurrences of a separable pattern p in a permutation π in time $O(kn^6)$.

It is mentioned in [155] that rather than just computing the number of occurrences, the algorithm on separable patterns can be augmented to compute an occurrence (if there is one) or to compute a list of all occurrences (in the latter case the running time will have an additional factor of the number of matchings). Yet another algorithm, of complexity $O(n^5 \log n)$, for detection of a separable pattern is presented in [15] by Albert et al. However, Bose et al. [155] proved that the general pattern matching problem (where p is not fixed) is *NP-complete* (*3-SATISFIABILITY* was reduced to this problem).

The *longest common pattern problem for two permutations* is a generalization of the pattern involvement problem, since finding if the longest pattern between permutations σ_1 and σ_2 is equal to σ_1 is equivalent to the pattern matching problem. Based on [155], Bouvel and Rossin [165] gave a polynomial ($O(n^8)$) algorithm for finding a longest common pattern between two permutations of size n given that one is separable. Moreover, an algorithm for general permutations σ_1 and σ_2 is given in [165]. The complexity of the algorithm depends on the length of the longest *simple permutation* involved in one of the permutations (see Section 8.1 for a discussion on these permutations), and in general, the algorithm does not run in polynomial time. However, in the special case where σ_1 comes from a permutation class whose simple permutations are of length at most d, the general algorithm runs in $O(\min(n_1, n_2)n_1 n_2^{2d+2})$, where $n_1 = |\sigma_1|$ and $n_2 = |\sigma_2|$.

8.2.2 Gray codes

The problem of generating and exhaustively listing the objects of a given class is important for several areas of science such as *computer science, hardware* and *software testing, biology* and *(bio)chemistry.*

There are many algorithms for generating various types of permutations, for example, for *alternating permutations* [96], *derangements* [11], *involutions* [700], permutations with a given number of *inversions* (occurrences of the pattern 21) [332], permutations with a given number of *excedances* [86], and *Dyck paths* (and thus, through known bijections, for any objects counted by the Catalan numbers, e.g., for 132- or 123-avoiding permutations) [104]. As particular cases of more general studies, Poneti and Vajnovszki [655] obtained generating algorithms for permutations counted by the *Stirling numbers of the first* and *second kinds, even permutations, fixed-point-free involutions* and *derangements (fixed-point-free permutations).*

The idea of so-called *Gray codes* (or *combinatorial Gray codes*) is to list the objects in such a way that two successive objects, encoded by words, have codes that differ as little as possible in a specified sense. In particular, a *Gray code* can be a list of words such that each word differs from its successor by a number of letters which is bounded independently of the length of the word.

Originally, a Gray code was used in a telegraph demonstrated by French engineer Émile Baudot in 1878. However, we say "the Gray code" to refer to the *reflected binary code* introduced by Frank Gray in 1947 to list all binary words of length n. This list can be generated recursively from the list of binary words of length $n-1$ by reflecting it (that is, listing the entries in reverse order), concatenating the original list with the reversed list, prefixing the entries in the original list with 0, and then prefixing the entries in the reflected list with 1. See Figure 8.3 for examples when $n = 1, 2, 3$. Note that a Gray code for binary words of length n corresponds to a *Hamiltonian path* (a path passing through each node exactly once) in the *unit n-cube*. In Figure 8.4 we show the Hamiltonian paths corresponding to the lists produced in Figure 8.3 for $n = 1, 2, 3$.

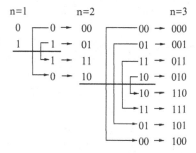

Figure 8.3: The reflected binary codes for $n = 1, 2, 3$.

Much has been discovered and written about the Gray code (see [309] by Doran for a survey) and it was used, for example, in *error corrections in digital communication* and in solving the *Tower of Hanoi puzzle*.

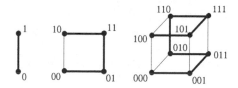

Figure 8.4: The Hamiltonian paths corresponding to the reflected binary codes for $n = 1, 2, 3$.

More relevant to this book is a Gray code for permutations. The problem is to generate n-permutations in such a way that successive permutations differ only

$n = 2$	$n = 3$	$n = 4$			
12	123	1234	1342	4321	2431
21	132	1243	1324	3421	4231
	312	1423	3124	3241	4213
	321	4123	3142	3214	2413
	231	4132	3412	2314	2143
	213	1432	4312	2341	2134

Table 8.1: Generating permutations by adjacent transpositions (for $n = 4$, to be read up-down, left-right).

by the exchange of two letters. Such a Gray code for permutations was shown to be possible in several papers (see [712] by Savage for references), whose disadvantage was that the letters exchanged are not necessarily in adjacent positions. It was shown independently by Johnson [486] and Trotter [768] that it is possible to generate permutations by transpositions even if the two letters exchanged are required to be in adjacent positions. The recursive scheme, illustrated in Table 8.1 inserts into each $(n-1)$-permutation on the list the letter n in each of the n possible positions, moving alternately from right to left, then left to right.

Walsh [791] gave a Gray code for the set of involutions of length n and a Gray code for the set of n-involutions with a given number of 2-cycles. Relative to these Gray codes, [791] provides efficient algorithms determining the immediate successor of a given involution or stating that it is the last one in the list. Moreover, Walsh [791] gave efficient algorithms determining the *rank* (the number of predecessors) of an involution, and determining an involution given its rank. As an example of more general studies in [791], non-recursive algorithms for word lists are given, in which all words with a given suffix form an interval of consecutive words.

Baril and Vajnovszki [88] gave a Gray code and a *constant average time* generating algorithm for *derangements*. In this Gray code, each derangement is transformed into its successor either via one or two transpositions or a rotation of three elements. These results were generalized in [88] to permutations with the number of fixed points bounded between two constants. Baril [85] gave a Gray code for the set of n-permutations with a given number of cycles. In this code, each permutation is transformed into its successor by a product with a cycle of length three, which is optimal. If each permutation is represented by its *transposition array* then the obtained list still remains a Gray code which allowed Baril [85] to construct an algorithm for generating these codes. Also, a Gray code and a generating algorithm for n-permutations with a fixed number of *left-to-right minima* are discussed in [85].

A general technique for generation of Gray codes for a large class of combinatorial families is presented in [108] by Bernini et al. The technique is based on the *ECO method* and produces objects using their encoding via *generating trees*.

Juarna and Vajnovszki [489] used a *loopless algorithm* of Vajnovszki for generating a Gray code for the binary strings without k consecutive 1s counted by the k-generalized Fibonacci numbers, to provide a Gray code for $\mathrm{Av}(132, 213, 12 \cdots k)$ where any two consecutive permutations differ in at most $k - 1$ positions. In particular, the class $\mathrm{Av}(132, 213, 123)$ counted by the Fibonacci numbers can be listed by a Gray code with distance (number of places in which successive permutations differ) 2. Also, Juarna and Vajnovszki [490] presented Gray codes and generating algorithms for three classes of patterns avoiding permutations: $\mathcal{S}_n(123, 132)$, $\mathcal{S}_n(123, 132, k(k - 1) \cdots 1(k + 1))$ and permutations in $\mathcal{S}_n(123, 132)$ having exactly $\binom{n}{2} - k$ inversions.

Generating algorithms and Gray codes for several families of pattern-avoiding permutations appeared in [316] by Dukes et al. Among the families under consideration in [316] are those counted by *Catalan, large Schröder, Pell,* and *even-index Fibonacci numbers* and the *central binomial coefficients*. In particular, [316] provides Gray codes for the set of all n-permutations avoiding any pattern $p \in \mathcal{S}_3$ and a Gray code for $\mathcal{S}_n(1243, 2143)$ (counted by Schröder numbers) of distances 4 and 5, respectively. This should be compared with the algorithm by Bóna [137, Section 8.1.2] for generating $\mathcal{S}_n(231)$ where the successor of the permutation $n(n - 1) \cdots 1(2n + 1)2n(2n - 1) \cdots (n + 2)(n + 1)$ is the permutation $12 \cdots (n - 1)(2n + 1)n(n + 1) \cdots (2n)$; the number of places in which these two permutations differ is linear in n and thus the algorithm does not produce a Gray code.

A general method to construct Gray codes for permutations avoiding a set of patterns is presented in [316] by Dukes et al., which works for a large family of patterns and can be seen to be complementary to the general method in [108] by Bernini et al. However, the method of Dukes et al. has some disadvantages, for example, it gives only a distance-5 Gray code for $\mathrm{Av}(231)$ instead of a more optimal distance-4 code mentioned above, and it does not work for $\mathrm{Av}(1243, 2143)$ (again, mentioned above) since $\{1243, 2143\}$ does not satisfy the required criteria. Below, we list classes of permutations in [316] for which the method of Dukes et al. works (we skip the avoidance of a 3-pattern case):

$$\mathrm{Av}(321, 312), \mathrm{Av}(321, 3412, 4123), \mathrm{Av}(321, 3412), \mathrm{Av}(321, 4123),$$
$$\mathrm{Av}(4321, 4312), \mathrm{Av}(4231, 4132), \mathrm{Av}(4123, 4213),$$
$$\mathrm{Av}(4321, 4231, 4312, 4132), \mathrm{Av}(4231, 4132, 4213, 4123),$$
$$\mathrm{Av}(321, (k + 1)12 \cdots k), \mathrm{Av}(321, 3412, (k + 1)12 \cdots k),$$
$$\mathrm{Av}(P) \text{ for } P = \cup_{\tau \in \mathcal{S}_{k-1}}\{(k + 1)\tau k\}.$$

The work of Dukes et al. [316] was refined and extended by Baril [87] who presented a unified construction of Gray codes. In particular, Gray codes for $\mathcal{S}_n(321)$ and $\mathcal{S}_n(312)$, where two consecutive permutations differ in at most three positions instead of four in [316], are obtained in [87], which is optimal for odd n. Other refinements for the pattern-restricted sets mentioned above have been given in [87] along with a general efficient generating algorithm.

Finally, Baril [86] obtained a Gray code for permutations having one excedance, which induced, through a bijection, a Gray code for the set

$$\mathcal{S}_n(321, 2413, 3412, 21534) \backslash \{12 \cdots n\}.$$

8.3   Packing patterns

In 1992, Herb Wilf introduced the study of pattern containment questions, during his address to the SIAM meeting on Discrete Mathematics. There are three basic problems in this direction.

- Given a set of patterns, which n-permutations contain the greatest number of occurrences of the pattern or patterns from the set, respectively? What is this number?

- What is the length of a shortest permutation containing all k-patterns?

- How many distinct patterns can be contained in a permutation of length n? This question was asked by Wilf at the Conference on Permutation Patterns at Otago, New Zealand, in 2003.

Most of the research done on the questions above is related to classical patterns and permutations. However, some research on this topic is done on words, as well as on packing of vincular patterns. We begin with a brief discussion of the last two questions above, and then give an overview of known results on the first one.

Wilf showed that the n-permutation $1n2(n-1)3(n-2) \cdots \lfloor \frac{n+1}{2} \rfloor$ has asymptotically more than $\left(\frac{1+\sqrt{5}}{2} \right)^n$ different patterns (see [270]). Coleman [270] improved this result with the following theorem.

Theorem 8.3.1. *([270]) For $k \geq 2$, there exists a k^2-permutation containing more than $2^{(k-1)^2}$ distinct patterns.*

The construction used to prove Theorem 8.3.1 is as follows:

$$\pi_k = k(2k) \cdots k^2(k-1)(2k-1) \cdots (k^2-1) \cdots 1(k+1)(2k+1) \cdots (k^2-k+1) \in \mathcal{S}_{k^2}.$$

For example $\pi_3 = 369258147$ and $\pi_4 = 48(12)(16)37(11)(15)26(10)(14)159(13)$.

Let $f(\pi)$ be the number of distinct patterns contained in a permutation π, and let $h(n)$ denote the maximum $f(\pi)$, where the maximum is taken over all permutations of length n. Theorem 8.3.1 says that if n is a perfect square, then

$$f(\pi_n) > 2^{n-2\sqrt{n}+1}$$

and $\sqrt[n]{h(n)} \to 2$ (for all n, not just perfect squares, since $h(n)$ is non-decreasing). However, Albert et al. [28] significantly improved Theorem 8.3.1 by showing that the maximum number of patterns that can occur in a permutation of length n is asymptotically 2^n. More precisely, Albert et al. [28] refined the counting arguments concerning the number of patterns in π_n when n is an even perfect square, and then extended the construction to all other values n to obtain

$$h(n) > 2^n \left(1 - 6\sqrt{n}2^{-\sqrt{n}/2}\right).$$

Also, in [28] an upper bound is discussed for $h(n)$: trivially $h(n) \leq 2^n - 1$, and slightly more precisely, for any n-permutation π,

$$f(\pi) \leq \sum_{k=1}^{n} \min\left(k!, \binom{n}{k}\right),$$

which is asymptotically 2^n.

Finally, Miller [631] found the amount of redundancy due to patterns that are contained multiple times in a given permutation to obtain

$$2^n - O(n^2 2^{n-\sqrt{2n}}) \leq h(n) \leq 2^n - \Theta(n2^{n-\sqrt{2n}}).$$

Let L_k be the length of a shortest permutation containing all patterns of length k (such permutations are called k-*superpatterns*). Eriksson et al. [366] established the bounds

$$e^{-2}k^2 < L_k \leq (2/3 + o(1))k^2,$$

and proved that as $k \to \infty$, there are permutations of length $(1/4 + o(1))k^2$ containing almost all patterns of length k. Remarkably, the permutation π_k defined above in the context of study of $h(n)$, appears in considerations of L_k.

Eriksson et al. [366] conjectured that asymptotically $L_k \sim \frac{k^2}{2}$. Miller [631] gave a simple construction related to the *tilted checkerboard* yielding a large family of k-superpatterns of length $\frac{k(k+1)}{2}$, and thus showing that L_k is at most $\frac{k(k+1)}{2}$.

8.3.1   Pattern densities

Recall that $p(\pi)$ is the number of occurrences of a pattern p in a permutation π. For each $n > 0$, we let

$$g(p, n) := \max_{\pi \in \mathcal{S}_n} p(\pi).$$

If $\pi \in \mathcal{S}_n$ is such that $g(p, n) = p(\pi)$, we say that π is p-*optimal* over $\mathcal{S}_n$.

Clearly, $g(p, n) \leq \binom{n}{k}$ for every $p \in \mathcal{S}_k$. It was conjectured by Wilf that $g(p, n)$ is asymptotically proportional to $\binom{n}{k}$, and the following stronger result was later proved by Galvin (see [805] for the proof).

Proposition 8.3.2. *(Galvin) The sequence* $\left(g(p, n) / \binom{n}{k}\right)_{n \geq k}$ *is non-increasing in* n *and thus* $\lim_{n \to \infty}(g(p, n) / \binom{n}{k})$ *exists.*

Definition 8.3.3. The *packing density* $\rho(p)$ of a k-pattern p is defined to be the limit in Proposition 8.3.2:

$$\rho(p) := \lim_{n \to \infty} \frac{g(p, n)}{\binom{n}{k}}.$$

It is easy to see that for any pattern p, $\rho(p) \leq 1$ and $\rho(12 \cdots k) = 1$ since $12 \cdots n$ is the $(12 \cdots k)$-optimal permutation for $n \geq k$. Kleitman, Galvin and Stromquist, independently, showed that the packing density of the pattern 132 is $2\sqrt{3} - 3$; the proof is recorded in [757] by Stromquist.

Stromquist [757] proved the following theorem (see Definition 2.1.11 for the notion of a layered pattern); the proof of the theorem can also be found in [661] by Price or in [24] by Albert et al.

Theorem 8.3.4. *([757]) Among all k-patterns, the maximum possible packing density is achieved by a layered pattern.*

In his PhD thesis, Price [661] showed that for any pattern $p \in \mathcal{S}_k$,

$$\rho(p) \geq \frac{k!}{k^k} > \frac{1}{e^k}.$$

Moreover, in [661] it is shown that if $\tilde{\kappa}$ is any number greater than a certain constant $\kappa \approx 0.569$ then for all k sufficiently large, there is a pattern $p \in \mathcal{S}_k$ with $\rho(p) < \tilde{\kappa}^k$. However, the main focus of [661] is a study of the packing density of layered patterns. Here we state three theorems from [661], referring to the original source for many more results.

Theorem 8.3.5. *([661]) Let $p = 1(k + 1)k \cdots 2$ be the layered pattern with exactly two layers of length 1 and $k \geq 2$. Then*

$$\rho(p) = k\bar{\alpha}(1 - \bar{\alpha})^{k-1},$$

where $\bar{\alpha}$ is the unique root of the polynomial

$$q(x) = kx^{k+1} - (k + 1)x + 1$$

on the interval $(0, 1)$.

Theorem 8.3.6. *([661]) Let p be the layered pattern with exactly two layers of length a and b, respectively, where $1 < a \leq b$. Then, for each n, an optimal permutation $\pi \in S_n$ to pack p into has only two layers, and*

$$\rho(p) = \binom{a + b}{a} \left(\frac{a}{a + b}\right)^a \left(\frac{b}{a + b}\right)^b.$$

In particular, $\rho(2143) = \frac{3}{8}$.

Theorem 8.3.7. *([661]) Let p be any nontrivial layered pattern, that is, p is not an increasing or decreasing permutation. Then $\rho(p) \leq \rho(132) = 2\sqrt{3} - 3 \approx 0.464$.*

Theorem 8.3.5 is proved by showing that

$$\frac{g(1(k + 1)k \cdots 2, n)}{\binom{n}{k+1}} = \rho(1(k + 1)k \cdots 2) + O\left(\frac{\log n}{n}\right).$$

Hildebrand et al. [462] reproved Theorem 8.3.5 by giving precise bounds on $g(1(k + 1)k \cdots 2, n)$.

Theorem 8.3.8. *([462]) For all $n \leq k \leq 2$,*

$$\rho(1(k+1)k\cdots 2)\frac{(n-k)^{k+1}}{(k+1)!} \leq g(1(k+1)k\cdots 2, n) \leq \rho(1(k+1)k\cdots 2)\frac{(n+\delta_{2,k})^{k+1}}{(k+1)!},$$

where $\delta_{2,k}$ is the Kronecker delta. Theorem 8.3.5 follows by dividing all sides of the inequalities by $\binom{n}{k+1}$ and taking $n \to \infty$.

Several other quantities related to $\rho(1(k + 1)k \cdots 2)$ are bounded in [462].

Albert et al. [24] pointed out the following dichotomy: for some patterns the number of layers in an optimal n-permutation is unbounded as $n \to \infty$, and for other patterns this number is bounded. Examples of both types are given above: Theorem 8.3.5 deals with patterns of the unbounded type, whereas Theorem 8.3.6 deals with patterns of bounded type. Albert et al. [24] generalized these results and showed that the issue is quite sensitive to the presence or absence of layers of size 1. More precisely, they proved the following theorem.

Theorem 8.3.9. *([24]) A pattern p, none of whose layers has size 1, is of the bounded type. Moreover, any pattern which has three layers of sizes a, 1, and b with a, b > 1 is of the bounded type. Finally, if p is a pattern whose first or last layer is of size 1 then p is of the unbounded type.*

The following result is obtained in [24] by Albert et al.

Proposition 8.3.10. *([24]) $\rho(1243) = \frac{3}{8}$.*

Warren [794] generalized the last proposition in two directions by computing the packing densities of layered patterns of type $[1^\alpha, \alpha]$ and $[1, 1, \beta]$, where 1^α abbreviates $\underbrace{1, 1, \ldots, 1}_{\alpha \text{ times}}$. Examples of such patterns are $[1, 1, 1, 3] = 123654$ and $[1, 1, 5] = 1276543$.

Theorem 8.3.11. *([794]) We have*

$$\rho([1^\alpha, \alpha]) = \frac{\binom{2\alpha}{\alpha}}{2^{2\alpha}};$$

$$\rho([1, 1, \beta]) = \binom{\beta + 2}{2} \left(\frac{2}{\beta + 2}\right)^2 \left(\frac{\beta}{\beta + 2}\right)^\beta.$$

Note that the second formula in Theorem 8.3.11 is the same as the formula in Theorem 8.3.6 when $a = 2$ and $b = \beta$. However, the theorems deal with different patterns (a 2-layer one and a 3-layer one, respectively). Warren [795] provided the following generalization of Theorem 8.3.11 again essentially obtaining the formula appearing in Theorem 8.3.6.

Theorem 8.3.12. *([795]) Suppose $\beta \le \alpha \le 2$ and 2β is divisible by α. Then*

$$\rho([1^\alpha, \alpha]) = \binom{\alpha + \beta}{\alpha} \left(\frac{\alpha}{\alpha + \beta}\right)^\alpha \left(\frac{\beta}{\alpha + \beta}\right)^\beta.$$

Albert et al. [24] found lower and upper bounds for packing densities for the non-layered patterns 1342 and 2413 (non-layered patterns are much more difficult to deal with than layered ones). The authors believe that their lower bounds in both cases are closer to the packing density than the upper bounds.

Proposition 8.3.13. *([24]) We have*

$$0.19657 \le \rho(1342) \le \frac{2}{9};$$
$$\frac{51}{511} \le \rho(2413) \le \frac{2}{9}.$$

The lower bound for $\rho(2413)$ was improved to $\approx 0.104250980068974874\ldots$ in [659] by Presutti, who generalized methods for obtaining lower bounds for the packing density of any pattern and demonstrated the methods' usefulness when patterns are non-layered. A further (slight) improvement of the lower bound for $\rho(2413)$, $0.10472422757673209041\ldots$, is obtained in [660] by Presutti and Stromquist, who conjectured that this value is actually equal to the packing density. Presutti and Stromquist [660] also defined the *packing rate* of a permutation with respect to a *measure*, and showed that maximizing the packing rate of a pattern over all measures gives the packing density of the pattern.

Hästö [451] introduced the notion of *simple layered patterns* (not defined here) for which it is very easy to calculate the packing density. In particular, the following theorem is proved in [451].

Theorem 8.3.14. *([451]) Let $p \in \mathcal{S}_k$ be a layered pattern of type $[k_1, k_2, \ldots, k_r]$, that is, the i-th layer is of length k_i, $1 \leq i \leq r$. If $\log_2(r+1) \leq \min\{k_i\}$ then p is simple and*

$$\rho(p) = \frac{k!}{k^k} \prod_{i=1}^{r} \frac{k_i^{k_i}}{k_i!}.$$

Moreover, Hästö [451] found good estimates of the packing densities of the layered patterns p_k of type $[k, 1, k]$ with $k \geq 3$:

$$\binom{2k+1}{k, k, 1} \frac{k^{2k} + (k/2)^k}{(2k+1)^{2k+1}} \leq \rho(p_k) \leq \binom{2k+1}{k, k, 1} \frac{k^{2k} + 2k^k}{(2k+1)^{2k+1}}.$$

8.3.2 Packing sets of patterns

The definitions of $g(p, n)$ and $\rho(p)$ extend naturally if instead of a single pattern p we have a set of patterns P. Here we count the total number of occurrences of patterns from P in a given permutation/word.

Albert et al. [24] discussed $\rho(P)$ for sets of 3-patterns.

Proposition 8.3.15. *([24]) If $P = \{132, 231\}$ then $\rho(P) = \frac{1}{2}$.*

Proposition 8.3.16. *([24]) If $P = \{132, 213\}$ or $P = \{132, 213, 231\}$ or $P = \{132, 213, 231, 312\}$ then $\rho(P) = \frac{3}{4}$.*

Burstein and Hästö [200] proved the following theorem.

Theorem 8.3.17. *([200]) Let $m, n \geq 2$ and let $P(m, n)$ be the set of all permutations whose first layer has length m and whose subsequent layers have total length n. Then* *we have*

$$\rho(P(m,n)) = \binom{m+n-1}{n} \frac{(m-1)^{m-1}n^n}{(m+n-1)^{m+n-1}}.$$

Note that $\rho(P(m,n)) = \rho([m-1,n])$ if $m \geq 3$.

8.3.3 Packing patterns in words

Burstein et al. [201] initiated the study of packing patterns in words. We find it convenient to follow the notation in [201].

Given a word $w \in [k]^n$ and a set of patterns $P \subseteq [\ell]^m$, let $\nu(P,w)$ be the total number of occurrences of patterns in P (P-patterns) in w. Define

$$\mu(P,k,n) = \max\{\nu(P,w)|w \in [k]^n\},$$

$$d(P,w) = \frac{\nu(P,w)}{\binom{n}{m}} \text{ and}$$

$$\delta(P,k,n) = \frac{\mu(P,k,n)}{\binom{n}{m}} = \max\{d(P,w)|w \in [k]^n\},$$

respectively, the maximum number of P-patterns in a word $[k]^n$, the probability that a subsequence of w of length m is an occurrence of a P-pattern, and the maximum such probability over words in $[k]^n$. The goal here is to consider the asymptotic behavior of $\delta(P,k,n)$ as $n \to \infty$ and $k \to \infty$. Barton [93] proved that

$$\delta(P) := \lim_{n\to\infty} \delta(P,n,n) = \lim_{n\to\infty}\lim_{k\to\infty} \delta(P,k,n) = \lim_{k\to\infty}\lim_{n\to\infty} \delta(P,k,n)$$

and $\delta(P)$ is called the packing density of P. The following theorem shows that the packing density on permutations agrees with the packing density on words, and thus all the above results on packing patterns in permutations can be seen as results on packing patterns in words.

Theorem 8.3.18. *([201]) Let $P \subseteq \mathcal{S}_m$ be a set of permutation patterns, then*

$$\rho(P) = \delta(P) = \lim_{n\to\infty} \frac{\max\{\nu(P,w)|w \in \mathcal{S}_n\}}{\binom{n}{m}},$$

that is, the packing density of P on words is equal to that on permutations.

The following results are obtained in [201].

Proposition 8.3.19. *([201]) For $q \geq 2$,*

$$\delta(k(k-1)\cdots 1(k+1)^q) = \binom{k+q}{k} \frac{k^k q^q}{(k+q)^{k+q}}.$$

Symmetry class	111	112	121	123	132
Packing density	1	$2\sqrt{3}-3$	$\frac{2\sqrt{3}-3}{2}$	1	$2\sqrt{3}-3$

Table 8.2: Packing densities of 3-patterns.

Theorem 8.3.20. *([201]) For $p, q, r \geq 1$,*

$$\delta(1^p 2^r 1^q) = \binom{p+q}{p} \frac{p^p q^q}{(p+q)^{p+q}} \delta(1^{p+q} 2^r) = \binom{p+q}{p} \frac{p^p q^q}{(p+q)^{p+q}} \delta([p+q, r]),$$

where $[p+q, r]$ is the layered pattern of length $p+q+r$ with layers of sizes $p+q$ and r. If $r > 1$ then we can use Theorem 8.3.6 to obtain

$$\delta(1^p 2^r 1^q) = \binom{p+q+r}{p, q, r} \frac{p^p q^q r^r}{(p+q+r)^{p+q+r}}.$$

If $r = 1$ then

$$\delta(1^p 2 1^q) = \binom{p+q}{p} p^p q^q (1 - (p+q)\alpha) \alpha^{p+q-1},$$

where $\alpha \in (0, 1)$ is the unique solution of $(1 - sx)^{s+1} = 1 - (s+1)x$ and $s = p + q$.

Among many results related to packing densities, the following theorem is obtained in [93] by Barton.

Theorem 8.3.21. $\delta(1^p 3^q 2^r) = \delta([p, q+r]) \delta([q, r])$. *Using Theorems 8.3.5 and 8.3.6 the packing density can be computed exactly for any pattern of the form $1^p 3^q 2^r$.*

The packing densities of 3-patterns are summarized in Table 8.2.

Definition 8.3.22. Let $n(\ell, m)$ be the length of the shortest superpattern for $[\ell]^m$, that is, the shortest word that contains every pattern of length m on at most ℓ letters.

Clearly, $n(\ell, m) = n(m, m)$ for $m \leq \ell$, therefore we are interested only in the values of $n(\ell, m)$ for $m \geq \ell$. For example, $n(2, 2) = 3$ since 121 contains the patterns 11, 12 and 21, and $n(3, 3) = 7$ since 3123132 contains the patterns 111, 112, 121, 211, 122, 212, 221, 123, 132, 213, 231, 312 and 321; in both cases we can see that no shorter word with the required properties exists.

Proposition 8.3.23. *([201]) For any $m \geq \ell \geq 3$, $n(m, \ell) \leq (m - 2)\ell + 4$.*

A long-standing conjecture is that for $\ell \geq 3$, $n(\ell, \ell) = \ell^2 - 2\ell + 4$ (see [201] for references).

We continue by stating several results for vincular patterns from [201].

Theorem 8.3.24. *([201]) Let $p = 1^{a_1} 2^{a_2} \cdots \ell^{a_\ell} \in [\ell]^m$ ($\ell > 1$) be a consecutive pattern. If there exists a positive integer $j \leq \ell - 2$ such that $a_1 \leq a_{j+1}$, $a_i = a_{i+j}$ ($2 \leq i \leq \ell - j - 1$) and $a_{\ell-j} \geq a_\ell$, then we denote by j_0 the least such j and define $M_p = a_2 + a_3 + \cdots + a_{j_0+1}$. Otherwise, we set $M_p = \max\{a_1, a_\ell\} + a_2 + a_3 + \cdots + a_{\ell-1}$. In either case we have $\delta(p) = 1/M_p$.*

Theorem 8.3.25. *([201]) Let $p = [a_1, a_2, \ldots, a_\ell] \in \mathcal{S}_m$ be any ℓ-layer ($\ell > 1$) consecutive pattern and M_p be as in Theorem 8.3.24. Then $\delta(p) = 1/M_p$.*

Corollary 8.3.26. *([201]) Let $p_1 = \underline{11 \cdots 12} \in [2]^m$ and $p_2 = \underline{1m(m-1) \cdots 2} \in [m]^m$, then $\delta(p_1) = \delta(p_2) = 1/(m-1)$.*

Proposition 8.3.27. *([201]) $\delta(\underline{112}) = \delta(\underline{123}) = \delta(\underline{213}) = 1$.*

Burstein and Hästö [200] proved the following theorem.

Theorem 8.3.28. *([200]) Let S be a set of 3-letter patterns on $[2]^3$.*

1. *If S includes either of the patterns 111 and 222, then $\delta(S) = 1$.*

2. *Otherwise, if S includes either of the sets $\{112, 122\}$ or $\{211, 221\}$, then $\delta(S) = 3/4$.*

3. *Otherwise, if $S = \{112, 121, 211, 212\}$ or $S = \{121, 122, 221, 212\}$, then $\delta(S) = \frac{5}{4}(2\sqrt{3} - 3)$.*

4. *Otherwise, if S includes any of the patterns 112, 122, 211 or 221, then $\delta(S) = 2\sqrt{3} - 3$.*

5. *Otherwise, if S equals $\{121\}$ or $\{212\}$, then $\delta(S) = \frac{1}{4}(2\sqrt{3} - 3)$.*

6. *Otherwise, if $S = \{121, 212\}$, then $\delta(S) \geq 1/4$. This is the only unknown case of packing density of subsets of 3-patterns.*

Also, Burstein and Hästö [200] studied the average *co-occurrence of patterns* and calculated the leading term of the covariance of any two permutation patterns of the same length n (as a polynomial in n) and, in general, of any two patterns of the same length over the same alphabet. However, we do not discuss this in any detail.

8.4   Several approaches to study patterns

In this section we consider several rather generic methods used to tackle problems on patterns. A major method we do not describe in this book is that of *enumeration schemes*, which can be used to completely automate the enumeration of many permutation classes. The method was first introduced by Zeilberger [816], and then extended by Vatter [781] to be applied to many more permutation classes; Pudwell [662, 663] adapted the method for words and for barred patterns. We also do not discuss the *permutation diagrams approach* (e.g., see [680]) here, in particular, *Fulton essential sets* (see [408]). On the other hand, *substitution decomposition*, not considered in this section, is discussed briefly in Section 8.1.

While describing approaches, we follow [515] by Kitaev and Mansour and [781] by Vatter.

8.4.1   Transfer matrices

The main idea of the transfer matrix method can be described as follows (see Theorem 4.7.2 in [743]). Consider a directed multigraph on n vertices $v_1, v_2, \ldots, v_n$, and let A denote its weighted adjacency matrix, that is, a_{ij} is the number of edges directed from v_i to v_j. Then the generating function for the number of walks from v_r to v_s is given by

$$\frac{(-1)^{r+s} \det(I - xA; r, s)}{\det(I - xA)}$$

where I is the identity matrix and $\det(B; r, s)$ is the minor of B with the r-th row and s-th column deleted.

To apply this approach in our context, one constructs a bijection between the permutations in question and walks in an appropriate directed graph. The transfer matrix method seems to have been used for the first time in connection with pattern-restricted permutations by Chow and West in [247] (see Theorem 6.1.33 proved by the method).

8.4.2   Generating trees

Generating trees were introduced by Chung et al. in [250] and were applied to permutation classes enumeration by West in [801, 802].

An n-permutation σ is a *child* of $\pi \in \mathcal{S}_{n-1}$ if σ can be obtained by inserting n into π, which defines a rooted tree T on $\mathcal{S}_\infty$. The subtree of T with nodes $\mathrm{Av}(P)$

is the *pattern-avoidance tree* $T(P)$ of Av(P). For example, the first four levels of $T(123, 231)$ are shown in Figure 8.5.

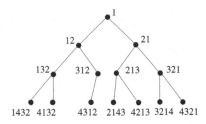

Figure 8.5: The first four levels of the pattern-avoiding tree $T(123, 231)$.

A *generating tree* is a rooted, labeled tree such that the labels of the children of each node are determined by the label of that node. Often the labels of the tree are natural numbers, but sometimes it is inconvenient. One defines a generating tree by providing the label of the *root* (sometimes called the *axiom*) and a set of *succession rules* (also called *inductive steps*). For example, the complete binary tree may be given by

$$\text{Root:} \quad (2)$$
$$\text{Rule:} \quad (2) \rightsquigarrow (2)(2)$$

where "$(2) \rightsquigarrow (2)(2)$" means that every node labeled (2) has two children, each also labeled (2).

In order to enumerate Av(P), one can try to find a generating tree isomorphic (as a rooted tree) to $T(P)$. For example, one can check that $T(132, 231)$ is isomorphic to the complete binary tree and thus to the generating tree given above. A more complicated example can be found in [801] by West where a generating tree isomorphic to $T(1234)$ is described. However, the generating function for Av(1234) from $T(1234)$ was obtained by Bousquet-Mélou [158], which was far from trivial.

Let $T(P; \pi)$ denote the subtree of $T(P)$ that is rooted at π and contains all descendants of π. It is explained by Vatter [781] that $T(P)$ is isomorphic to a finitely labeled generating tree if and only if the set of all principle subtrees $\{T(P; \pi) : \pi \in \text{Av}(P)\}$ contains only finitely many isomorphism classes. When this happens, Av(P) has a rational generating function which may be (routinely) computed using the transfer matrix method (see Subsection 8.4.1). The following theorem characterizes the finitely based permutation classes for which this is possible.

Theorem 8.4.1. *([779]) Let $\mathcal{C}$ be a finitely based permutation class. The pattern-avoidance tree of $\mathcal{C}$ is isomorphic to a finitely labeled generating tree if and only* *if C omits both a child of an increasing permutation and a child of a decreasing permutation.*

8.4.3 ECO method

The *ECO method* (*Enumerating Combinatorial Objects method*) is closely related to the generating trees method (see [80] by Barcucci et al. for a survey) and can be viewed as its extension to other settings. The main idea of the ECO method is as follows. Using an operator that performs a "local expansion" on the objects in question, one gives a recursive construction of the class of objects. Then one introduces certain function equations which, when solved, enumerate the combinatorial objects according to various parameters.

We follow [80] to describe the ECO method. Let $\mathcal{X}$ be a class of combinatorial objects with a *parameter p*, $p : \mathcal{X} \to \mathcal{N}$, and $\mathcal{X}_n = \{x \in \mathcal{X} \mid p(x) = n\}$. An operator μ on $\mathcal{X}$ is a function from $\mathcal{X}_n$ to the power set of $\mathcal{X}_{n+1}$. We say that the operator μ is *recursive* on $\mathcal{X}$ if it satisfies the following conditions:

1. For each element $y \in \mathcal{X}_{n+1}$ there exists $x \in \mathcal{X}_n$ such that $y \in \mu(x)$, and

2. If $x_1, x_2 \in \mathcal{X}_n$ and $x_1 \neq x_2$, then $\mu(x_1) \cup \mu(x_2) = \emptyset$.

Proposition 8.4.2. *([80]) If μ is a recursive operator on $\mathcal{X}$, then $\{\mu(x) \mid x \in \mathcal{X}_n\}$ is a partition of $\mathcal{X}_{n+1}$.*

Therefore, such an operator on $\mathcal{X}$ gives a recursive description of the class $\mathcal{X}$, that is, Proposition 8.4.2 allows us to construct each object $y \in \mathcal{X}_{n+1}$ from an object $x \in \mathcal{X}_n$ and every $y \in \mathcal{X}_{n+1}$ is obtained from only one $x \in \mathcal{X}_n$. In many cases the recursive description is given by generating trees (see Subsection 8.4.2).

The ECO method has proved to be a useful tool in many cases of pattern-restricted permutations. For example, see [80] for applying the approach to deal with $Av(321)$ and $Av(321, 3\bar{1}42)$.

8.4.4 The insertion encoding

The *insertion encoding* was introduced by Albert et al. [30], and is a correspondence between permutation classes and languages which allows the tools of *formal language theory* to be used in tackling permutation class enumeration problems. The main idea here is that every permutation can be generated from the empty permutation by successive insertion of a new maximum element, and to each permutation we can associate a word describing how that permutation evolved. More precisely, at each stage until the desired permutation has been constructed, at least one open slot $\diamond$ exists in the intermediate *configuration*. To proceed to the next configuration one inserts a new maximum element into one of these slots. There are four different types of insertion of a new maximum element n within a slot:

- $\diamond \rightarrow n$ represented by $\mathbf{f}$ (for fill);

- $\diamond \rightarrow n\diamond$ represented by ℓ (for left);

- $\diamond \rightarrow \diamond n$ represented by $\mathbf{r}$ (for right);

- $\diamond \rightarrow \diamond n\diamond$ represented by $\mathbf{m}$ (for middle).

Since each of these operations can be performed on any open slot at any stage, we subscript their symbols with the number of the slot to which they were applied to (read from left to right). For example, the permutation 423615 has the insertion encoding $\mathbf{m}_1\mathbf{m}_1\ell_2\mathbf{f}_1\mathbf{f}_2\mathbf{f}_1$ because its evolution is

$$\diamond \rightarrow \diamond 1\diamond \rightarrow \diamond 2 \diamond 1\diamond \rightarrow \diamond 23 \diamond 1\diamond \rightarrow 423 \diamond 1\diamond \rightarrow 423 \diamond 15 \rightarrow 423615.$$

In the case when the insertion encoding of a permutation class forms a regular language, that class has a rational generating function that can be (routinely) computed. The only obstructions to having a regular insertion encoding are *vertical alternations*, which are $(2n)$-permutations $\pi = \pi_1\pi_2 \cdots \pi_{2n}$ such that either $\pi_1, \pi_3, \ldots, \pi_{2n-1} < \pi_2, \pi_4, \ldots, \pi_{2n}$ or $\pi_1, \pi_3, \ldots, \pi_{2n-1} > \pi_2, \pi_4, \ldots, \pi_{2n}$.

Theorem 8.4.3. *([30]) The insertion encoding of a finitely based class forms a regular language if and only if the class contains only finitely many vertical alternations.*

It is pointed out by Vatter [781] that any class satisfying the hypotheses of Theorem 8.4.1, that is, that omits both a child of an increasing permutation and a child of a decreasing permutation, contains only finitely many vertical alternations, and thus Theorem 8.4.3 applies to every class with a finitely labeled generating tree.

8.4.5   Block decompositions

The *block decomposition* approach was initiated by Mansour and Vainshtein [617] to study the structure of 132-avoiding permutations, and permutations containing a given number of occurrences of the pattern 132 [619]. It is easy to see that if $\pi = \pi_1\pi_2 \cdots \pi_n \in \mathcal{S}_n(132)$ and $\pi_i = n$, then $\pi = \pi' n \pi''$ where π' is a permutation of $\{n - i + 1, n - i + 2, \ldots, n - 1\}$, π'' is a permutation of $\{1, 2, \ldots, n - i\}$, and both π' and π'' avoid the pattern 132. This way of representing π is called the block decomposition of π.

The simple observation above allows one to obtain a general result concerning permutations that avoid the pattern 132 and another pattern $\tau = \tau_1\tau_2\cdots\tau_k \in \mathcal{S}_k(132)$. Let $m_0, m_1, \ldots, m_r$ be the right-to-left maxima (see Definition A.1.1) of τ written from left to right. Then τ can be represented as

$$\tau = \tau^0 m_0 \tau^1 m_1 \cdots \tau^r m_r,$$

where each of τ^is is a possibly empty permutation, and any letter of τ^i is greater than every letter of τ^{i+1}. Define the i-th *prefix* of τ by $P_i = \tau^0 m_0 \tau^1 m_1 \cdots \tau^i m_i$ for $1 \le i \le r$, $P_0 = \tau^0$ and $P_{-1} = \varepsilon$ (the empty word). The i-th *suffix* of τ is defined by $S_i = \tau^i m_i \tau^{i+1} m_{i+1} \cdots \tau^r m_r$ for $0 \le i \le r$, and $S^{r+1} = \varepsilon$.

Theorem 8.4.4. *([617]) For any $\tau \in \mathcal{S}_k(132)$, the g.f. for* $\mathrm{Av}(132, \tau)$, $F_\tau(x)$, *is a rational function satisfying the relation*

$$F_\tau(x) = 1 + x \sum_{j=0}^{r} \left(F_{P_j}(x) - F_{P_{j-1}}(x)\right) F_{S_j}(x).$$

The proof is rather straightforward. Let $\pi = \pi'n\pi''$ be the block decomposition of $\pi \in \mathcal{S}_n(132)$. It is easy to see that π contains τ if and only if there exists i, $0 \le i \le r + 1$, such that π' contains P_{i-1} and π'' contains S_i. Therefore, π avoids τ if and only if there exists i, $0 \le i \le r$, such that π' avoids P_i and contains P_{i-1}, while π'' avoids S_i. We thus get the following relation:

$$f_\tau(n) = \sum_{t=1}^{n} \sum_{j=0}^{r} f_{P_j}^{P_{j-1}}(t-1) f_{S_j}(n-t),$$

where $f_\tau(n) = s_n(132, \tau)$, and $f_\tau^\rho(n)$ is the number of permutations in $\mathcal{S}_n(\tau)$ containing ρ at least once. To obtain the recursion for $F_\tau(x)$ it remains to observe that

$$f_{P_j}^{P_{j-1}}(n) + f_{P_{j-1}}(n) = f_{P_j}(n)$$

for any n and j, and to proceed with the generating functions. The rationality of $F_\tau(x)$ follows easily by induction.

This technique is used in many papers for different structures (for example, see [618, 601] and references therein).

8.5 A bridge to combinatorics on words

Combinatorics on words is a field of mathematics devoted to the systematic study of properties of words. It was originated by Axel Thue [453] at the beginning of the last century. It has connections not only to several branches of mathematics such as *number theory*, *symbolic dynamics*, *algebra*, *probability theory* and *discrete mathematics*, but also to *theoretical computer science* (through the study of *automata* and *formal languages*) and *biology*. There are three books [571, 572, 573] by Lothaire on the subject that give a comprehensive introduction to the field with an overview of the most important results in the area.

However, when asking a combinatorialist on words what a pattern is, one should expect to receive a definition that is different from our notion of a pattern (in words or permutations) considered in this book. As a matter of fact, alphabets under consideration in combinatorics on words are often unordered which makes our definition of a pattern meaningless in that domain. Still, some of the pattern problems studied in combinatorics on words can be naturally formulated in terms of the patterns considered in this book. Lothaire does not do much to find analogues of combinatorics on word problems for our patterns, leaving this as a direction of future research.

The aim of this section is to introduce patterns in combinatorics on words with selected typical results to give the reader a flavor of that area. Then, at the end of this section we discuss *square-free permutations* related to consecutive patterns; study of these permutations was motivated by the classical combinatorics on words notion of squares in words. More details and relevant references for the combinatorics on words results in this section can be found in [572], which we use as the main reference.

8.5.1   Morphisms and square-free words

Definition 8.5.1. A^* and A^+ denote the set of all finite and finite nonempty words over A, respectively.

Definition 8.5.2. Let A and B be two alphabets (possibly $A = B$). A map $\varphi : A^* \to B^*$ is called a *morphism*, if we have $\varphi(uv) = \varphi(u)\varphi(v)$ for any $u, v \in A^*$. A morphism φ can be defined by defining $\varphi(a)$ for each $a \in A$. A particular property of a morphism φ is that $\varphi(\varepsilon) = \varepsilon$. A morphism φ is called *non-erasing* if $\varphi(a) \neq \varepsilon$ for each $a \in A$. A morphism φ is called *uniform* if $|\varphi(a)| = |\varphi(b)|$ for all $a, b \in A$.

The *Thue-Morse sequence* is probably the most famous example of a sequence defined by iterations of a (non-erasing, uniform) morphism. This sequence can be defined by the morphism h such that:

$$h(0) = 01$$
$$h(1) = 10.$$

The first few iterations of h are as follows:

$$0, \ 01, \ 0110, \ 01101001, \ 0110100110010110, \dots.$$

A remarkable property of this sequence is that it does not contain a factor of the form XXx, where X is itself a factor and x is the first letter in X. In the language defined below, we say that the Thue-Morse sequences is 2^+-free (in particular, it is *cube*-free). We can also say that the Thue-Morse sequence is *overlap-free* as it does not contain a factor of the form $axaxa$, where a is a letter and x is a word. Another way to define this sequence is recursive: Let $h^0(0) = 0$ and, for $n \geq 1$, $h^n(0) = h^{n-1}(0)c(h^{n-1}(0))$, where c is the *complement* (switching 0 and 1). Then $h^1(0) = 01$, $h^2(0) = 0110$, $h^3(0) = 01101001$, etc.

Definition 8.5.3. A word is *square-free* if it does not contain two equal factors that are adjacent to each other in the word.

Example 8.5.4. The word 123132 is square-free while 1232311 is not, as it contains squares 2323 and 11.

It is easy to see that the longest square-free word on a 2-letter alphabet, say $\{a, b\}$, is aba, while the longest square-free word on a 1-letter alphabet is of length 1. Axel Thue [766] proved in 1906 that there are infinitely many square-free words on a 3-letter alphabet, and this result was rediscovered many times. An infinite sequence on a 3-letter alphabet avoiding squares is given by iterating the following morphism

$$a \to abc$$
$$b \to ac$$
$$c \to b.$$

The initial values of the sequence are as follows: $abcacbabcbacabcacbacabcb\dots.$

A formula for the number of different square-free words of length n is rather complicated: for example, for a 3-letter alphabet, it is shown [541] by Kolpakov that the number of such words is $3^{cn(1-\epsilon_n)}$ where $1.30173\dots < c < 1.30178\dots$ and $\epsilon_n \to 0$ when $n \to \infty$.

8.5.2 Patterns in combinatorics on words

Definition 8.5.5. A *pattern* (in combinatorics on words) is a word that contains special symbols called *variables*, and the associated (infinite) *pattern language* is obtained by replacing the variables with arbitrary nonempty words, with the condition that two occurrences of the same variable have to be replaced with the same word. Patterns are also allowed to contain constant letters, which, unlike variables, are never replaced with arbitrary words, but this type of pattern is not to be considered in this book.

For an example of a pattern, consider the square $\alpha\alpha$ having associated pattern language $L = \{uu | u \in A^+\}$. It is a classical result (see above) that L is an *avoidable set* of words (there exists an infinite sequence avoiding L) if A has at least three elements, whereas it is an *unavoidable set* of words if A has only one or two elements. We will say that the pattern $\alpha\alpha$ is *3-avoidable* and *2-unavoidable*. More generally, we have the following definition.

Definition 8.5.6. A pattern is *k-avoidable* if and only if it is avoidable on any k-letter alphabet. A pattern which is not k-avoidable is *k-unavoidable*. A pattern which is avoidable on A for some (finite) alphabet A will be called *avoidable*, and a pattern which is unavoidable on A for every A will be called *unavoidable*.

There is the (recursive) *Zimin algorithm* (using so-called *pattern adjacency graphs*, *free sets*, and the concept of *irreducibility*, not to be discussed in this book) to find out if a given pattern is avoidable or not.

Another example of a pattern is $\alpha\beta\alpha$, which represents words of the form uvu, with $u, v \in A^+$ (this pattern is unavoidable whatever the size of the alphabet is — see Theorem 8.5.7). Yet another example of a pattern is $p = \alpha\alpha\beta\beta\alpha$. The pattern language associated with p on the alphabet A is $p(A^+) = \{uuvvu | u, v \in A^+\}$. The word 1011011000111 contains p (through the morphism/substitution $h : \alpha \to 011$, $\beta \to 0$), whereas the word 0000100010111 avoids p (this pattern is 2-avoidable).

8.5.3 Powers and sesquipowers

The simplest class of patterns is the class of *powers* of a single variable, α^n. The first two, $\alpha^0 = \varepsilon$ and $\alpha^1 = \alpha$, are trivially unavoidable as they are encountered by any nonempty word. For $n \geq 2$, $\alpha^2 = \alpha\alpha$ is 3-avoidable, and α^n for $n \geq 3$ is 2-avoidable (the morphism $\mu : a \to abc$, $b \to ac$, $c \to b$ and the Thue-Morse morphism $h : a \to ab$, $b \to ba$ support the respective facts).

The pattern $\alpha\beta$ is obviously unavoidable (any word of length at least 2, regardless of the alphabet, contains an occurrence of $\alpha\beta$).

Theorem 8.5.7. *([572]) The pattern $\alpha\beta\alpha$ is unavoidable. More precisely, if the cardinality of the alphabet A is k, any word of length at least $2k + 1$ contains an occurrence of $\alpha\beta\alpha$.*

Proof. Let $w \in A^*$ be a word of length at least $2k + 1$. Then one of the letters in A, say a, occurs at least three times in w. Write $w = w_0 a w_1 a w_2 a w_3$, and let $h(\alpha) = a$

and $h(\beta) = w_1 a w_2$. Then $h(\alpha\beta\alpha)$ is a factor of w. Note that we cannot improve $2k + 1$ in the statement as for the alphabet $A = \{1, 2, \ldots, k\}$, the word $1122 \cdots kk$ is a word of length $2k$ avoiding $\alpha\beta\alpha$. $\qquad\square$

Theorem 8.5.8. *([572]) Let p be a pattern unavoidable on A, and x a variable that does not occur in p. Then the pattern pxp is unavoidable on A.*

Using the last theorem, we can recursively construct an infinite family of unavoidable patterns. Let α_n, for $n \in \mathbb{N}$, be different variables. Let $Z_0 = \varepsilon$, and for all $n \in \mathbb{N}$, $Z_{n+1} = Z_n \alpha_n Z_n$. The patterns Z_n are called *Zimin words*, or *sesquipowers*. The first four Zimin words are:

$$Z_1 = 1, \ Z_2 = 121, \ Z_3 = 1213121, \ Z_4 = 121312141213121.$$

There are more results on (un)avoidable patterns, for example, those stated in the following three theorems.

Theorem 8.5.9. *([572]) If all variables that occur in a pattern p occur at least twice, then p is avoidable.*

Theorem 8.5.10. *([572]) Let p be a pattern with n variables. If $|p| \geq 2^n$, then p is avoidable.*

Theorem 8.5.11. *([572]) A pattern is unavoidable if and only if it is a factor of some sesquipower Z_n.*

8.5.4 Avoidability on a fixed alphabet

Definition 8.5.12. The *avoidability index $\mu(p)$* of a pattern p is the smallest integer k such that p is k-avoidable, or ∞ if p is unavoidable.

Clearly, $2 \leq \mu(p) \leq \infty$ since no pattern is 1-avoidable. Roughly speaking, the *avoidability index* of a pattern measures how easy it is to avoid this pattern. There is no algorithm to compute avoidability index. Even for very short patterns the value of $\mu(p)$ may be unknown. For instance, it is not known whether $\mu(\alpha\alpha\beta\beta\gamma\gamma)$ is equal to 2 or 3, although there is some experimental evidence that the index is 2.

Apart from the final remark below, we consider only the class of *binary patterns*, that is patterns with at most two different variables, say from the set $E = \{\alpha, \beta\}$.

The fact that $\alpha\alpha$ is a 3-avoidable 2-unavoidable pattern gives us some information about binary patterns. Only a finite number of binary patterns are 3-unavoidable. Indeed, a pattern that contains $\alpha\alpha$ as a factor must be 3-avoidable, and since $\alpha\alpha$ is 2-unavoidable, there are only finitely many binary patterns which do not contain $\alpha\alpha$, namely ε, α, β, $\alpha\beta$, $\beta\alpha$, $\alpha\beta\alpha$ and $\beta\alpha\beta$. All of these are in fact unavoidable, which implies that the avoidability index of a binary pattern can only be 2, 3, or ∞.

The remaining question is to distinguish 3-avoidable from 2-avoidable patterns.

Theorem 8.5.13. *([572]) Binary patterns fall in three categories:*

- *The 7 binary patterns ε, α, β, $\alpha\beta = \beta\alpha$, $\alpha\beta\alpha = \beta\alpha\beta$ are unavoidable (their avoidability index is ∞).*

- *The 22 binary patterns $\alpha\alpha = \beta\beta$, $\alpha\alpha\beta = \beta\beta\alpha$, $\alpha\beta\beta = \beta\alpha\alpha$, $\alpha\alpha\beta\alpha = \beta\beta\alpha\beta$, $\alpha\alpha\beta\beta = \beta\beta\alpha\alpha$, $\alpha\beta\alpha\alpha = \beta\alpha\beta\beta$, $\alpha\beta\alpha\beta = \beta\alpha\beta\alpha$, $\alpha\beta\beta\alpha = \beta\alpha\alpha\beta$, $\alpha\alpha\beta\alpha\alpha = \beta\beta\alpha\beta\beta$, $\alpha\alpha\beta\alpha\beta = \beta\beta\alpha\beta\alpha$, and $\beta\alpha\beta\alpha\alpha = \alpha\beta\alpha\beta\beta$ have avoidability index 3.*

- *All other binary patterns, and in particular all binary patterns of length 6 or more, have avoidability index 2.*

As a final remark, we note the following fact: the pattern $\alpha\alpha$ separates the binary and ternary alphabets meaning that it is 3-avoidable but 2-unavoidable. Similarly, the pattern $\alpha\beta u\theta\gamma v\gamma\alpha w\beta\alpha z\alpha\gamma$ separates 3- and 4-alphabets, that is, it is 4-avoidable but 3-unavoidable. However, no pattern separating larger alphabets is known.

8.5.5   Crucial and maximal words for sets of prohibitions

Even though powers of words were already discussed above, we define them again in parallel with *abelian powers* for an easier comparison.

Definition 8.5.14. A word w over the alphabet $[n]$ contains a *k-th power* if w has a factor of the form $X^k = XX\cdots X$ (k times) for some non-empty word X. A word w contains an *abelian k-th power* if w has a factor of the form $X_1 X_2 \cdots X_k$ where X_i is a permutation of X_1 for $2 \le i \le k$. The cases $k = 2$ and $k = 3$ give us (*abelian*) *squares* and *cubes*, respectively.

Example 8.5.15. The word 13243232323243 over [4] contains the 4-th power $(32)^4 = 32323232$ and it contains the abelian square $43232\,32324$, while the word $123\,312\,213$ is an abelian cube.

Definition 8.5.16. A word w is (*abelian*) *k-power-free* if w avoids (abelian) k-th powers, that is, if w does not contain any (abelian) k-th powers.

Example 8.5.17. The word 1234324 is abelian cube-free, but not abelian square-free since it contains the abelian square 234 324.

Definition 8.5.18. A word w is *crucial* with respect to a given set of *prohibited words* (or simply *prohibitions*) if w avoids the prohibitions, but wx does not avoid the prohibitions for any letter x occurring in w. A *minimal crucial word* is a crucial word of shortest length.

Example 8.5.19. The word $w = 21211$ (of length 5) is crucial with respect to abelian cubes since it is abelian cube-free and the words $w1$ and $w2$ end with the abelian cubes 111 and 21 21 12, respectively. Actually, w is a minimal crucial word over $\{1, 2\}$ with respect to abelian cubes. Indeed, we can easily verify that there do not exist any crucial abelian cube-free words on two letters of length less than 5.

Abelian squares were first introduced by Erdős [364], who asked whether or not there exist words of arbitrary length over a fixed finite alphabet that avoid patterns of the form XX' where X' is a permutation of X (i.e., abelian squares). This question has since been answered in the affirmative in a series of papers from 1968 to 1992 (see [420] for references).

The Zimin words Z_n are defined in Subsection 8.5.3 and the k-*generalized Zimin word* $Z_{n,k} = X_n$ is defined as

$$X_1 = 1^{k-1} = 11 \cdots 1, \ X_n = (X_{n-1}n)^{k-1}X_{n-1} = X_{n-1}nX_{n-1}n \cdots nX_{n-1}$$

where the number of consecutive 1s, as well as the number of ns, is $k - 1$. Thus $Z_n = Z_{n,2}$. It is easy to see (by induction) that each $Z_{n,k}$ avoids (abelian) k-th powers and has length $k^n - 1$. Moreover, it is known that $Z_{n,k}$ is a minimal crucial word avoiding k-th powers.

Crucial abelian square-free words (also called *right maximal abelian square-free words*) of exponential length are given in [287] by Cummings and Mays and in [371] by Evdokimov and Kitaev, and it is shown in [371] that a minimal crucial abelian square-free word over an n-letter alphabet has length $4n - 7$ for $n \geq 3$.

Glen et al. [420] extended the study of crucial abelian k-power-free words to the case of $k > 2$. In particular, they provided a complete solution to the problem of determining the length of a minimal crucial abelian cube-free word (the case $k = 3$).

Theorem 8.5.20. *([420]) A minimal crucial word over $[n]$ with respect to abelian cubes has length $9n - 13$ for $n \geq 5$, and it has length 2, 5, 11, and 20 for $n = 1, 2, 3$, and 4, respectively.*

For $n \geq 4$ and $k \geq 2$, Glen et al. [420] give a construction of length $k^2(n - 1) - k - 1$ of a crucial word over $[n]$ avoiding abelian k-th powers, as well as a trivial lower bound for the shortest crucial word length: $3kn - (4k + 1)$ for $k \geq 4$ and $n \geq 5$. This construction gives the minimal length for $k = 2$ and $k = 3$, and was conjectured to be optimal for any $k \geq 4$ and sufficiently large n. Avgustinovich et al. [60] improved the lower bound by proving that for $n \geq 2k - 1$, the shortest length of a crucial word avoiding abelian k-th powers is at least $k^2n - (2k^3 - 3k^2 + k + 1)$, and thus showing that the conjecture above is true asymptotically (up to a constant term) for growing n.

Definition 8.5.21. A word w over $[n]$ is *maximal* with respect to a given set of prohibitions if w avoids the prohibitions, but xw and wx do not avoid the prohibitions for any letter $x \in [n]$.

Example 8.5.22. The word 323121 is a maximal word of minimal length over $\{1, 2, 3\}$ with respect to abelian squares.

Clearly, the length of a minimal crucial word with respect to a given set of prohibitions is at most the length of a shortest maximal word. Thus, we get a lower bound for the length of a shortest maximal word by obtaining the length of a minimal crucial word.

Korn [542] proved that the length $\ell(n)$ of a shortest maximal abelian square-free word over $[n]$ satisfies $4n - 7 \leq \ell(n) \leq 6n - 10$ for $n \geq 6$, and Bullock [193] refined Korn's methods to show that $6n - 29 \leq \ell(n) \leq 6n - 12$ for $n \geq 8$.

8.5.6   Square-free permutations

Avgustinovich et al. [62] extended the notion of squares in words to the notion of squares in permutations. Also, the notions of crucial and maximal words were transferred naturally to the case of permutations.

Definition 8.5.23. A permutation is *square-free* if it does not contain two consecutive factors of length more than one that are equal in the reduced form (as patterns).

Example 8.5.24. The permutation 246153 contains the square 4615 (in the reduced form the first and the last two letters form the pattern 12), whereas the permutation 246513 is square-free.

Avgustinovich et al. [62] prove that the number of square-free permutations of length n is $n^{n(1-a_n)}$ where $a_n \to 0$ when $n \to \infty$.

Definition 8.5.25. A permutation π of length n is *crucial with respect to squares* if π avoids squares, but any of its extensions to the right (i.e. permutations of length $n + 1$ whose letters, all but the rightmost one, reduce to the pattern π) are not square-free.

Example 8.5.26. The permutations 1234 and 123 are not crucial, as the former one is not square-free, whereas the second one has a square-free extension to the right, e.g., to the permutation 1243 (the first 3 letters in the last permutation reduce to the pattern 123).

As a matter of fact, the shortest crucial permutation with respect to squares is of length 7. Below, we list all the crucial permutations of length 7. The reader can check his/her understanding of the definition by considering any of the permutations below:

2136547, 2137546, 2146537, 2147536, 2156437, 2157436, 2167435,
3146527, 3147526, 3156427, 3157426, 3167425, 3246517, 3247516,
3256417, 3257416, 3267415, 3421675, 3521674, 3621574, 3721564,
4156327, 4157326, 4167325, 4256317, 4257316, 4267315, 4356217,
4357216, 4367215, 4521673, 4531672, 4532671, 4621573, 4631572,
4632571, 4721563, 4731562, 4732561, 5167324, 5267314, 5367214,
5467213, 5621473, 5631472, 5632471, 5641372, 5642371, 5721463,
5731462, 5732461, 5741362, 5742361, 6721453, 6731452, 6732451,
6741352, 6742351, 6751342, 6752341.

Definition 8.5.27. A permutation is called *maximal with respect to squares* if *both* the permutation and its reverse are crucial with respect to squares.

Avgustinovich et al. [62] proved that there exist crucial permutations with respect to squares of any length at least 7, and there exist maximal permutations with respect to squares of odd lengths $8k + 1, 8k + 5, 8k + 7$ for $k \geq 1$. It is an open problem to find out if there exist any maximal permutations with respect to squares of even length or of length $8k + 3$, $k \geq 1$.

8.6 Universal cycles

Even though we assume that the reader is familiar with basic definitions in graph theory, we define the following notions in order to proceed.

Definition 8.6.1. The degree of a vertex v in a graph is the number of edges with which v is incident (i.e., is connected to). A *walk* is an alternating sequence of vertices and edges, with each edge being incident to the vertices immediately preceding and succeeding it in the sequence. A *trail* is a walk with no repeated edges. A *path* is a walk with no repeated vertices. A walk is *closed* if the initial vertex is also the terminal vertex. A *cycle* is a closed trail with at least one edge and with no repeated vertices except that the initial vertex is the terminal vertex. The word *Hamiltonian* in the expression "a Hamiltonian path" or "a Hamiltonian cycle" means a path or a cycle, respectively, visiting *each* vertex in a graph exactly once. Likewise, the word "Eulerian" in "an Eulerian cycle" means visiting *each* edge in a cycle exactly once. A graph is *connected* if, for every pair of vertices, there is a walk whose ends are the given vertices.

Definition 8.6.2. The edges, or *arcs*, in a *directed graph* are arrows pointing to one endpoint or the other. Directed graphs are often called *digraphs*. The number of edges directed into a vertex is called the *in-degree* of the vertex, and the number of edges directed out is called the *out-degree*. A digraph is *balanced* if the in-degree equals the out-degree for each of its vertices.

Definition 8.6.3. Given a graph G, its line graph $L(G)$ is the graph such that

- each vertex of $L(G)$ represents an edge of G; and

- two vertices of $L(G)$ are adjacent if and only if their corresponding edges share a common endpoint (i.e., are adjacent) in G. In the case of a digraph G, there is an arc from a vertex u to a vertex v in $L(G)$ if in G there is a directed path of length 2 first going through the arc corresponding to u and then the arc corresponding to v.

It is not difficult to prove that a graph contains an Eulerian cycle if and only if the graph is connected and the degree of each of its vertices is even. On the other hand, a digraph contains an Eulerian cycle if and only if it is balanced and strongly connected (strong connectivity is defined in Definition 5.7.18).

Definition 8.6.4. An *n-dimensional de Bruijn graph* $\vec{D}_{n,m}$ of m letters is a *directed graph* on m^n vertices labeled by the words of length n over the m-letter alphabet $A = \{0, 1, \dots, m-1\}$. There is an arc from a vertex u to a vertex v if and only if $u = aw$ and $v = wb$ for some word w over A of length $n-1$, and $a, b \in A$, that is, if the end of the word u overlaps with the beginning of v.

Example 8.6.5. See Figure 8.6 for n-dimensional de Bruijn graphs, $n = 1, 2, 3$, for binary words.

It is not hard to see that $\vec{D}_{n,m}$ is strongly connected and balanced, thus there exists an Eulerian cycle in $\vec{D}_{n,m}$. Moreover, we can see that $\vec{D}_{n+1,m} = L(\vec{D}_{n,m})$ and therefore the existence of an Eulerian cycle in $\vec{D}_{n,m}$ gives the existence of a Hamiltonian cycle in $\vec{D}_{n+1,m}$. Following an Eulerian cycle in $\vec{D}_{n,m}$, we can write down a word corresponding to it that will contain each word of length n over the

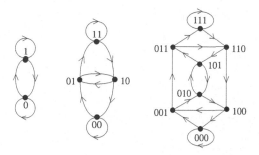

Figure 8.6: $\vec{D}_{1,2}$, $\vec{D}_{2,2}$ and $\vec{D}_{3,2}$.

alphabet A exactly once as a factor. For example, the word 0001011100 is obtained from $\vec{D}_{3,2}$ following the Eulerian cycle

$$000 \to 001 \to 010 \to 101 \to 011 \to 111 \to 110 \to 100,$$

where we skipped recording the last step.

More generally, given a family C of combinatorial objects that can be represented by words of length n, a *universal cycle*, or *U-cycle*, for such a family is a word whose length n factors, read cyclically, represent all the elements of C without repetition. Sometimes, the requirement for factors to be read cyclically is omitted, in which case we can talk about *universal words*. The example above is a universal word for the set of all binary words of length 3, which can be turned into a universal cycle for the same set by removing, for example, the last two 0s: 00010111.

The words containing all words of given length over a given alphabet as factors, whose existence we discussed above, are often called *de Bruijn sequences*, the most well-known universal cycles. Chung et al. [249] extended the study of de Bruijn sequences to other combinatorial structures, such as the set of all permutations, the set of all *partitions*, and the set of all *subsets of a finite set*. In the case of permutations, one needs to modify the notion of a U-cycle since for $n \geq 3$ it is not possible to list all n-permutations in the way we treated words. Instead of looking at a factor of length n, we will look at the reduced form of the factor. Then, our goal is to come up with a word, a U-cycle, such that each n-permutation appears in it exactly once as a consecutive pattern. The existence of such U-cycles for any n was proved in [249] by Chung et al. For example, for $\mathcal{S}_3$ such a U-cycle is 145243. Indeed, reading the reduced forms cyclically starting from the factor 145 we see the following permutations: 123, 231, 312, 132, 321 and 213. Another U-cycle for $\mathcal{S}_3$ is 142342. To list all n-permutations, we need $> n$ different letters. Chung et al. [249] conjectured that $n + 1$ letters are always enough for constructing a U-cycle for $\mathcal{S}_n$

for $n \geq 3$. This conjecture was proved in [485] by Johnson.

The existence of U-cycles for all permutations was proved in [249] by showing that the graph of overlapping permutations considered in Section 5.6 has a Hamiltonian cycle. Similarly, Burstein and Kitaev [203] proved the existence of U-cycles for all *word patterns* (which are the same as *multi-permutations*) by proving that a certain graph contains a Hamiltonian cycle. We note however that a Hamiltonian cycle in a graph associated with a problem cannot necessarily be "lifted" to a U-cycle.

See [472] by Jackson et al. for a recent collection of research problems (and references) on U-cycles.

Albert and West [32] studied U-cycles for permutation classes. Here, as in the case of all permutations, we are interested in finding a word whose factors of length n in the reduced form provide, without repetition, all permutations from a given set $\mathcal{S}_n(P)$. For example, the word 16784325 is a U-cycle for the set $\mathcal{S}_4(132, 312)$, which can be verified by taking each of the 8 factors 1678, 6784, 7843, 8432, 4325, 3251, 2516 and 5167 in the reduced form.

In the case when it is unknown whether a U-cycle exists, we can at least try to see whether the graph of overlapping permutations associated with a pattern-restricted set contains a Hamiltonian cycle, which may or may not bring us to a construction of a U-cycle. Similar to the case of words, the existence of a Hamiltonian cycle is equivalent in the case of pattern-restricted permutations to the existence of an Eulerian cycle (see [32] for an explanation). A set $\mathcal{S}(P)$ is called *pattern cyclic* if the graph of overlapping permutations from $\mathcal{S}_n(P)$ contains an Eulerian cycle. The following proposition is proved in [32], where we say that a set of permutations is *cyclically closed* if, given any permutation in the set, its cyclic shifts are also in the set (e.g., $\{1243, 3124, 4312, 2431\}$ is a cyclically closed set).

Proposition 8.6.6. *([32]) If the set of forbidden patterns P is cyclically closed, and the corresponding graph of overlapping permutations is connected, then $\mathcal{S}_n(P)$ is pattern cyclic.*

The following sets of permutations are shown in [32] to be pattern cyclic for any n:

$$\mathcal{S}_n(132, 213), \ \mathcal{S}_n(132, 312, 123), \ \mathcal{S}_n(132, 312, 1234), \ \mathcal{S}_n(132, 312, 3241, 2314)$$
$$\mathcal{S}_n(132, 213, 3412, 4231), \ \mathcal{S}_n(132, 213, 4321), \ \mathcal{S}_n(132, 213, 123, 3412).$$

The following generalization for the set $\mathcal{S}_n(132, 213, 4321)$ is obtained in [32].

Proposition 8.6.7. *([32]) For all $k \geq 3$, the set $\mathcal{S}_n(132, 213, k(k-1)\cdots 1)$ is pattern cyclic for all n.*

The following sets of permutations are shown in [32] to have U-cycles for any n:

$$\mathcal{S}_n(132, 312), \ \ \mathcal{S}_n(132, 213, 321), \ \ \mathcal{S}_n(123, 3142, 3412).$$

Finally, the associated graphs for the following sets of permutations fail to have Eulerian cycles for any size of permutations, and thus, in these cases there are sizes of permutations for which no U-cycle exists:

$$\mathcal{S}_n(123), \ \ \mathcal{S}_n(132, 213, 123), \ \ \mathcal{S}_n(123, 3142, 2413), \ \ \mathcal{S}_n(123, 3142, 3421, 4312),$$
$$\mathcal{S}_n(123, 2143, 3412), \ \ \mathcal{S}_n(123, 2143, 2413), \ \ \mathcal{S}_n(123, 2143, 3421, 4312),$$
$$\mathcal{S}_n(132, 312, 3241, 2314, 1234), \ \ \mathcal{S}_n(132, 213, 3412, 4231, 1234),$$
$$\mathcal{S}_n(132, 213, 3412, 4231, 4321), \ \ \mathcal{S}_n(132, 213, 1234).$$

8.7 Simsun permutations

Definition 8.7.1. A permutation $\pi = \pi_1 \pi_2 \cdots \pi_n \in \mathcal{S}_n$ is called *simsun* if for all $3 \le k \le n$, the restriction of π to $\{1,2,\ldots,k\}$ (that is, considering the subsequence of π formed by the k smallest letters) has no double descents (that is, belongs to $\mathcal{S}_k(\underline{321})$)). These permutations are named after Rodica *Sim*ion and Sheila *Sundaram* [758]. A permutation π is *double simsun* if both π and π^{-1} are simsun. Let $\mathcal{RS}_n$ and $\mathcal{DRS}_n$ denote, respectively, the sets of simsun and double simsun permutations of length n. Also, $\mathcal{RS}_n(P)$ and $\mathcal{DRS}_n(P)$ denote, respectively, the sets of simsun and double simsun n-permutations avoiding each pattern from a set of patterns P.

Example 8.7.2. The permutation 41325 is simsun, while 32415 is not. Also, the permutation 51324 is simsun but *not* double simsun since its inverse 24351 is not simsun.

Simsun permutations are a variant of *André permutations* studied by Foata and Schützenberger [399], and they are relevant to the enumeration of the *monomials* of the *cd-index* of $\mathcal{S}_n$ (see [455] by Hetyei). Simion and Sundaram proved that $|\mathcal{RS}_n| = E_{n+1}$, the $(n+1)$-th *Euler number* also counting the alternating permutations in $\mathcal{S}_{n+1}$. Using generating functions, Chow and Shiu [246] enumerated simsun permutations by descents.

Deutsch and Elizalde [301] enumerated simsun permutations avoiding a pattern or a set of classical 3-patterns, which involves Catalan numbers C_n, Fibonacci numbers F_n, Motzkin numbers M_n and *secondary structure* numbers S_n related to the *secondary structures of RNA molecules*. See [301] for the definition of the secondary structure numbers whose generating function is

$$\sum_{n \ge 0} S_n x^n = \frac{1 - x - x^2 - \sqrt{1 - 2x - x^2 - 2x^3 + x^4}}{2x^3}.$$

p	$|\mathcal{RS}_n(p)|$	$|\mathcal{DRS}_n(p)|$
123	6 (for $n \geq 4$)	2 (for $n \geq 6$)
132	S_n	S_n
213	M_n	S_n
231	M_n	2^{n-1}
312	2^{n-1}	2^{n-1}
321	C_n	C_n

Table 8.3: The number of simsun and double simsun permutations avoiding a 3-pattern from [301] and [248], respectively.

$\{p_1, p_2\}$	$|\mathcal{RS}_n(p_1, p_2)|$
$\{123, 132\}$	2 (for $n \geq 4$)
$\{123, 213\}$	3 (for $n \geq 3$)
$\{123, 231\}$, $\{123, 312\}$, $\{123, 321\}$	0 (for $n \geq 5$)
$\{132, 213\}$, $\{213, 231\}$, $\{231, 312\}$	F_{n+1}
$\{132, 231\}$, $\{132, 312\}$, $\{213, 312\}$	n
$\{132, 321\}$, $\{213, 321\}$	$\frac{1}{2}(n^2 - n + 2)$
$\{231, 321\}$, $\{312, 321\}$	2^{n-1}

Table 8.4: The number of simsun permutations avoiding two 3-patterns.

Chuang et al. [248] presented a new bijection between simsun permutations and *increasing* 1-2 *trees*, and showed a number of interesting consequences of this bijection in the enumeration of pattern-avoiding simsun and double simsun permutations. Moreover, Chuang et al. [248] enumerated the double simsun permutations that avoid each classical 3-pattern and shown that $|\mathcal{DRS}_n(132, 213)| = |\mathcal{DRS}_n(312, 231)| = F_{n+1}$. We collect results on avoidance of a single pattern on simsun and double simsun permutations in Table 8.3, and on avoidance of two 3-patterns on simsun permutations in Table 8.4. Finally, we note that there are only two cases modulo Wilf-equivalence giving a non-constant number of simsun permutations avoiding more than two 3-patterns, which are $|\mathcal{RS}(231, 312, 321)| = F_{n+1}$ and

$$|\mathcal{RS}(132, 213, 321)| = |\mathcal{RS}(132, 231, 321)| = |\mathcal{RS}(132, 312, 321)|$$

$$= |\mathcal{RS}(213, 231, 321)| = |\mathcal{RS}(213, 312, 321)| = n.$$

8.8 An involution on Catalan objects

This section is based on [261] by Claesson et al.

Consider a triangulation of an n-gon labeled by $1, 2, \ldots, n$, such as the one to the left in the top row in Figure 8.7 (this is a 9-gon). Apply the complement operation to the labels ($i \to n - i + 1$ for $1 \leq i \leq n$) and take the mirror image with respect to the line going through the middle of the segment $(1, n)$ and orthogonal to it (see Figure 8.7 for an example). Clearly we get an involution, which we call h.

Even though h is a simple involution, it has remarkable applications in proving equidistribution results for different combinatorial objects counted by the Catalan numbers, and, in particular, for pattern-avoiding permutations. It is not a coincidence that we used the same name, h to denote the above involution and the involution discussed in Subsection 2.2.2 (in particular, see Theorem 2.2.6). It turns out that the specialization of h on $\beta(1, 0)$-trees to rooted plane trees (discussed immediately after Theorem 2.2.6) can be defined by the involution h on triangulations after applying a standard bijection between the trees and triangulations.

Theorem 2.2.7 is an example of results obtained using h. Claesson et al. [261] used h to describe the equidistribution of (a tuple of) five statistics on Dyck paths, 2-*row rectangular standard Young tableaux*, *binary trees* and *non-crossing set partitions*. More relevant for us is the equidistributions on 231- and 321-avoiding permutations obtained in [261] and stated below. We do not provide a description of h on the mentioned objects, but we give examples of applications of h on the objects in Figure 8.7 (all objects there but the permutations can be obtained from each other by standard bijections, thus giving an induced definition of h). We would like to stress that h can be used to obtain non-trivial equidistribution results on any objects counted by the Catalan numbers (there are many such objects – see [748]) as long as we can translate statistics on the objects mentioned above to the corresponding statistics on other Catalan structures through some bijections.

In the theorem below, for an n-permutation π, the statistic $\mathrm{nonrmin}(\pi) = n + 1 - \mathrm{rmin}(\pi)$, and the statistic $\mathrm{ascs}(\pi)$ is 1 + the number of ascents to the right of maximum i such that $ij(j + 1) \cdots (i - 1)$ is a factor in π; by definition, $\mathrm{ascs}(12 \cdots n) = n$. Definitions of the other statistics can be found in Table A.1.

Theorem 8.8.1. *([261]) On $\mathcal{S}_n(231)$, the following equidistribution holds:*

$$(\mathrm{comp}, \mathrm{rmax}, \mathrm{rmin}, \mathrm{ldr}, \mathrm{ascs}) \sim (\mathrm{rmax}, \mathrm{comp}, \mathrm{nonrmin}, \mathrm{ascs}, \mathrm{ldr}),$$

and on $\mathcal{S}_n(321)$,

$$(\mathrm{comp}, \mathrm{comp}.r, \mathrm{rmin}, \mathrm{lir}, s) \sim (\mathrm{comp}.r, \mathrm{comp}, \mathrm{nonrmin}, s, \mathrm{lir}),$$

where a definition of the statistic s corresponding to lir is yet to be found.

As a corollary to considerations on rooted plane trees, it is shown in [261] that there are no fixed points of h on $\mathcal{S}_n(231)$ or $\mathcal{S}_n(321)$ for even n, and there are Catalan many fixed points for odd n. We can define a sign on permutations to make h be a sign-reversing involution, for instance, inducing it from the following definition of a sign on triangulations.

Suppose T is a triangulation of an n-gon. For $1 \leq i \leq \lfloor \frac{n}{2} \rfloor$, let $\max i$ be the maximum label not equal to n that is connected by an edge to i in T. Similarly, for $\lceil \frac{n}{2} \rceil < i \leq n$, let $\min i$ be the minimum label not equal to 1 that is connected by an edge to i in T.

Let m, $1 \leq m \leq \lfloor \frac{n}{2} \rfloor$, be minimum such that

$$(\max m) - m \neq (n + 1 - m) - \min(n + 1 - m).$$

In other words, m is minimum such that

$$d = n + 1 - (\max m) - \min(n + 1 - m) \neq 0.$$

Then the sign of T is $+,-$, or 0 if $d > 0$, $d < 0$, or such m does not exist, respectively.

For example, for the first triangulation in Figure 8.7, $m = 1$, $d = 9+1-7-7 < 0$ making the triangulation negative, while for the second triangulation $m = 1$, $d = 9 + 1 - 3 - 3 > 0$ making it positive.

The following proposition is straightforward to prove.

Proposition 8.8.2. *([261]) With respect to the definition of sign on triangulations of n-gon, h is a sign-reversing involution.*

8.9 Permutation avoidance games

At Permutation Patterns Conference 2007, Dukes, Parton and West introduced a new combinatorial game, in which players alternate placing 1s (or dots, or rooks) in an initially empty permutation matrix, always playing so as to avoid some specified pattern(s). The player who cannot place a 1 without creating a prohibited pattern, or simply because the permutation matrix is already filled, loses. The authors demonstrated the nature of the game by giving a complete analysis of the case where the avoided pattern is 12, and gave some partial results for the case of the 123 pattern. The slides for the conference presentation are available at
http://www-circa.mcs.st-and.ac.uk/PermutationPatterns2007/talks/west.pdf.

The case of avoidance of the pattern 12 is rather trivial — the first player wins. A rather intriguing behavior occurs when the prohibited pattern is 123: the first player wins for the following sizes of a permutation matrix: 1, 3, 5, 7, 8, 10, while the second player wins for the remaining sizes ≤ 9. The problem can be generalized as follows: Players alternate placing nonattacking rooks on an $m \times n$ board subject to no three rooks forming 123-pattern. A recursive program created by Dukes, Parton and West easily finds winning strategies for all values m, n with $m + n \leq 20$.

Unfortunately, no other mention of games on pattern-restricted permutations seems to appear in the literature. Study of such games is an attractive direction of further research.

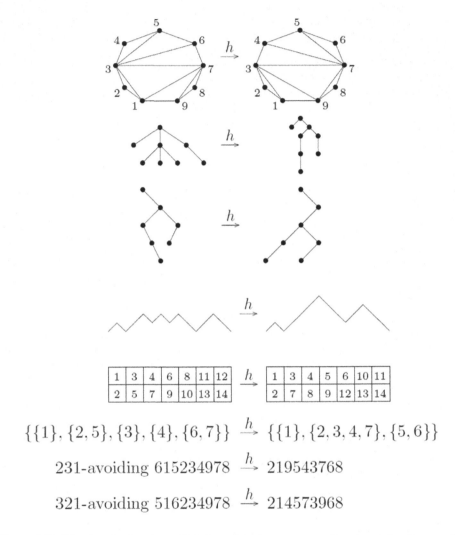

Figure 8.7: The involution h on Catalan objects corresponding to each other under natural bijections

Chapter 9

Extending research on patterns in permutations and words to other domains

In this section we discuss several extensions of research on patterns in permutations and words to other contexts. Among these extensions are a requirement on parity of elements (see Section 9.1), considerations of multisets (see Section 9.3) and signed/colored permutations (see Section 9.6), increasing the number of dimensions (see Subsections 9.8.2 and 9.8.3), and some other directions/generalizations.

Note that unlike the previous chapters, this chapter is neither thorough in the sense of definitions/examples, nor comprehensive in listing the results in question. However, additional information can be obtained by following the references we provide.

9.1   Refining patterns/statistics

The statistic "descent" is refined in [526] by Kitaev and Remmel, by fixing the parity of (exactly) one of the descent's two letters. Let $\mathbb{E} = \{0, 2, 4, \ldots\}$ and $\mathbb{O} = \{1, 3, 5, \ldots\}$ denote the set of even and odd numbers, respectively. Given $\sigma = \sigma_1 \sigma_2 \cdots \sigma_n \in \mathcal{S}_n$, we define the following notions, where $\chi(\sigma_1 \in X)$ is 1 if σ_1 is of type X, and it is 0 otherwise.

- $\overleftarrow{Des}_X(\sigma) = \{i : \sigma_i > \sigma_{i+1} \ \& \ \sigma_i \in X\}$; $\overleftarrow{des}_X(\sigma) = |\overleftarrow{Des}_X(\sigma)|$ for $X \in \{\mathbb{E}, \mathbb{O}\}$;

- $\overrightarrow{Des}_X(\sigma) = \{i : \sigma_i > \sigma_{i+1} \ \& \ \sigma_{i+1} \in X\}$; $\overrightarrow{des}_X(\sigma) = |\overrightarrow{Des}_X(\sigma)|$ for $X \in \{\mathbb{E}, \mathbb{O}\}$;

- $\sum_{\sigma \in S_n} x^{\overleftarrow{\mathrm{des}_{\mathbb{E}}}(\sigma)} = \sum_{k=0}^n R_{k,n} x^k$;

- $\sum_{\sigma \in S_n} x^{\overleftarrow{\mathrm{des}_{\mathbb{O}}}(\sigma)} = \sum_{k=0}^n M_{k,n} x^k$;

- $\sum_{\sigma \in S_n} x^{\overrightarrow{\mathrm{des}_{\mathbb{E}}}(\sigma)} z^{\chi(\sigma_1 \in \mathbb{E})} = \sum_{k=0}^n \sum_{j=0}^1 P_{j,k,n} z^j x^k$;

- $\sum_{\sigma \in S_n} x^{\overrightarrow{\mathrm{des}_{\mathbb{O}}}(\sigma)} z^{\chi(\sigma_1 \in \mathbb{O})} = \sum_{k=0}^n \sum_{j=0}^1 Q_{j,k,n} z^j x^k$.

The following explicit formulas for the distribution of the four new statistics are obtained in [526] using *differential operators*.

Theorem 9.1.1. *([526]) We have*

- $R_{k,2n} = \binom{n}{k}^2 (n!)^2$;

- $R_{k,2n+1} = (k+1)\binom{n}{k+1}^2 (n!)^2 + (2n+1-k)\binom{n}{k}^2 (n!)^2 = \frac{1}{k+1}\binom{n}{k}^2 ((n+1)!)^2$;

- $P_{1,k,2n} = \binom{n-1}{k}\binom{n}{k+1}(n!)^2$;

- $P_{1,k,2n+1} = \frac{(n+1)(n-k)}{k+1}\binom{n}{k}^2 (n!)^2$;

- $P_{0,k,2n} = \binom{n-1}{k}\binom{n}{k}(n!)^2$;

- $P_{0,k,2n+1} = (k+1)\binom{n}{k}\binom{n+1}{k+1}(n!)^2 = (n+1)\binom{n}{k}^2 (n!)^2$;

- $Q_{0,k,2n} = \binom{n-1}{k-1}\binom{n}{k}(n!)^2$;

- $Q_{0,k,2n+1} = \frac{(n+1)(n-k+1)}{k}\binom{n}{k-1}^2 (n!)^2$;

- $Q_{1,k,2n} = \binom{n-1}{k}\binom{n}{k}(n!)^2$;

- $Q_{1,k,2n+1} = \binom{n}{k}^2 n!(n+1)!$;

- $M_{k,2n} = \frac{n+1}{k+1}\binom{n-1}{k}\binom{n}{k}(n!)^2 = \binom{n-1}{k}\binom{n+1}{k+1}(n!)^2$;

- $M_{k,2n+1} = \frac{1}{n-k+1}\binom{n}{k}^2 ((n+1)!)^2 = \binom{n}{k}\binom{n+1}{k}n!(n+1)!$.

Distribution of descents according to parity can be viewed as distribution of consecutive occurrences of the following patterns. We use e, o, or $*$ as superscripts for a pattern's letters to require that in an occurrence of the pattern in a permutation, the corresponding letters must be even, odd or either. For example, the permutation 25314 has two occurrences of the pattern $\underline{2^*1^o}$ (they are 53 and 31, both of them are occurrences of the pattern $\underline{2^o1^o}$), one occurrence of the pattern $\underline{1^o2^e}$ (namely, 14), no occurrences of the pattern $\underline{1^o2^o}$, and no occurrences of the pattern $\underline{2^e1^*}$.

Using the notation above, we can state an alternative definition of the *Genocchi numbers*, defined in Subsection A.2.1:

Definition 9.1.2. The Genocchi number G_{2n} is the number of permutations $\sigma = \sigma_1\sigma_2\cdots\sigma_{2n+1}$ in $\mathcal{S}_{2n+1}$ that avoid simultaneously the patterns $\underline{1^e2^*}$ and $\underline{2^o1^*}$.

Kitaev and Remmel [525] generalized results in Theorem 9.1.1 by studying the problem of counting descents according to whether the first or the second element in a descent pair is equivalent to 0 mod k for $k \geq 2$. For any $k > 0$, let $k\mathbb{N} = \{0, k, 2k, 3k, \ldots\}$. Given set $X \subseteq \mathbb{N} = \{0, 1, \ldots\}$ and any $\sigma = \sigma_1\sigma_2\cdots\sigma_n \in \mathcal{S}_n$, we define the following:

- $\overleftarrow{Des}_X(\sigma) = \{i : \sigma_i > \sigma_{i+1} \ \& \ \sigma_i \in X\}$ and $\overleftarrow{des}_X(\sigma) = |\overleftarrow{Des}_X(\sigma)|$;

- $\overrightarrow{Des}_X(\sigma) = \{i : \sigma_i > \sigma_{i+1} \ \& \ \sigma_{i+1} \in X\}$ and $\overrightarrow{des}_X(\sigma) = |\overrightarrow{Des}_X(\sigma)|$;

- $A_n^{(k)}(x) = \sum_{\sigma \in \mathcal{S}_n} x^{\overleftarrow{des}_{k\mathbb{N}}(\sigma)} = \sum_{j=0}^{\lfloor \frac{n}{k} \rfloor} A_{j,n}^{(k)} x^j$;

- $B_n^{(k)}(x) = \sum_{\sigma \in \mathcal{S}_n} x^{\overrightarrow{des}_{k\mathbb{N}}(\sigma)} = \sum_{j=0}^{\lfloor \frac{n}{k} \rfloor} B_{j,n}^{(k)} x^j$;

- $B_n^{(k)}(x, z) = \sum_{\sigma \in \mathcal{S}_n} x^{\overrightarrow{des}_{k\mathbb{N}}(\sigma)} z^{\chi(\sigma_1 \in k\mathbb{N})} = \sum_{j=0}^{\lfloor \frac{n}{k} \rfloor} \sum_{i=0}^{1} B_{i,j,n}^{(k)} z^i x^j$.

Remark 9.1.3. Note that setting $k = 1$ gives us *usual* descents providing $A_n^{(1)}(x) = B_n^{(1)}(x) = A_n(x)$, whereas setting $k = 2$ gives $\overleftarrow{Des}_{\mathbb{E}}(\sigma)$ and $\overrightarrow{Des}_{\mathbb{E}}(\sigma)$.

In Section 3.6 we mentioned an application of such types of descents (occurrences of the pattern $\underline{21}$ subject to extra restrictions) in proving non-trivial identities. In the following theorem we record two selected enumerative results from [525].

Theorem 9.1.4. ([525]) *For all $0 \leq s \leq n$ and $0 \leq j \leq k - 1$,*

$$A_{n-s,kn+j}^{(k)} =$$
$$((k-1)n+j)!\left[\sum_{r=0}^{s}(-1)^{s-r}\binom{(k-1)n+j+r}{r}\binom{kn+j+1}{s-r}\prod_{i=1}^{n}(r+(k-1)i)\right].$$

For all $n \geq 0$, $k \geq 2$, and $0 \leq j \leq k - 1$,

$$B_{n-s,kn+j}^{(k)} =$$
$$((k-1)n+j)!\left[\sum_{r=0}^{s}(-1)^{s-r}\binom{(k-1)n+j+r}{r}\binom{kn+j+1}{s-r}\prod_{i=0}^{n-1}(r+j+(k-1)i)\right].$$

Kitaev et al. [518] continued in the same direction by fixing a set partition of $\mathbb{N}$, $(\mathbb{N}_1, \mathbb{N}_2, \ldots, \mathbb{N}_s)$ and studying the distribution of descents, levels, and ascents over the set of words over the alphabet $[k] = \{1, 2, \ldots, k\}$ according to whether the first letter of the descent, ascent, or level lies in $\mathbb{N}_i$. In particular, some of the results in [209] by Burstein and Mansour are refined and generalized in [518].

9.1.1   More on descents and ascents relative to equivalence classes mod k

Liese [561] studied the following variation of a pattern-matching problem. Suppose that $k \geq 2$ and we are given some sequence of distinct integers $\tau = \tau_1 \tau_2 \cdots \tau_j$. Then we say that a permutation $\sigma = \sigma_1 \sigma_2 \cdots \sigma_n \in \mathcal{S}_n$ has a τ-k-equivalence match at place i provided $\mathrm{red}(\sigma_i \sigma_{i+1} \cdots \sigma_{i+j-1}) = \mathrm{red}(\tau)$ and for all $s \in \{0, 1, \ldots, j-1\}$, $\sigma_{i+s} = \tau_{1+s}$ mod k. For example, if $\tau = 12$ and $\sigma = 51743682$, then σ has τ-matches (in this case, ascents) starting at positions 2, 5 and 6. However, if $k = 2$, then only the τ-match starting at position 5 is a τ-2-equivalence match.

More generally, if Υ is a set of sequences of distinct integers of length j, then we say that a permutation $\sigma = \sigma_1 \sigma_2 \cdots \sigma_n \in \mathcal{S}_n$ has a Υ-k-equivalence match at place i provided there is a $\tau \in \Upsilon$ such that $\mathrm{red}(\sigma_i \sigma_{i+1} \cdots \sigma_{i+j-1}) = \mathrm{red}(\tau)$ and for all $s \in \{0, 1, \ldots, j-1\}$, $\sigma_{i+s} = \tau_{1+s}$ mod k.

Let τ-k-$emch(\sigma)$ be the number of τ-k-equivalence matches in σ. Liese [561] studied the following polynomials

$$T_{\tau, k, n}(x) = \sum_{\sigma \in \mathcal{S}_n} x^{\tau\text{-}k\text{-}emch(\sigma)} = \sum_{s=0}^{n} T_{\tau, k, n}^s x^s;$$

$$U_{\tau, k, n}(x) = \sum_{\sigma \in \mathcal{S}_n} x^{\Upsilon\text{-}k\text{-}emch(\sigma)} = \sum_{s=0}^{n} U_{\Upsilon, k, n}^s x^s.$$

We provide here just one enumerative result from [561].

Theorem 9.1.5. *([561]) For all $y - k \leq j \leq y - 1$ and all $s \leq n$ such that $kn + j > 0$, we have*

$$U_{\Upsilon, k, kn+j} = ((k-1)n + j)! \left[\sum_{r=0}^{s} (-1)^{s-r} \binom{(k-1)n + r + j}{r} \binom{kn + j + 1}{s - r} \Gamma(r, j, n) \right],$$

where $\Gamma(r, j, n) = \prod_{i=0}^{n-1} ((k-1)n + r + j + 1 - \alpha - i(|\Upsilon| - 1))$.

9.1.2 k-descents and k-excedences in permutations

Liese and Remmel [564] considered the following problem. Given a permutation $\sigma = \sigma_1 \sigma_2 \cdots \sigma_n$ in $\mathcal{S}_n$, we say that σ has a k-*descent* at position i if $\sigma_i - \sigma_{i+1} = k$, and σ has a k-excedence at i if $\sigma_i - i = k$. Due to a bijection by Foata (see, e.g. [564] for the description), obtaining the distribution of k-descents in $\mathcal{S}_n$ is equivalent to obtaining the distribution of k-excedences in $\mathcal{S}_n$. These distributions are obtained in [564], where additionally some q-analogues are found.

Let $\mathrm{exc}_k(\sigma)$ denote the number of k-excedences in σ and

$$P_{n,k}(x) = \sum_{\sigma \in \mathcal{S}_n} x^{\mathrm{exc}_k(\sigma)} = \sum_{s=0}^{n-k} P_{n,k,s} x^s.$$

Examples of results obtained in [564] are as follows:

$$P_{n,k}(x) = \sum_{s \geq 0} x^s \sum_{i=s}^{n-k} (-1)^{i-s} (n-i)! \binom{i}{s} \binom{n-k}{i}$$

and thus

$$P_{n,k,s} = \sum_{i=s}^{n-k} (-1)^{i-s} (n-i)! \binom{i}{s} \binom{n-k}{i}.$$

Also, it is shown in [564] that

$$P_k(x,t) = \sum_{n \geq 0} \sum_{s=0}^{n} P_{n+k,k,s} \frac{x^s t^n}{n!} = \frac{k! e^{t(x-1)}}{(1-t)^{k+1}},$$

and

$$P(x,t,z) = \sum_{k \geq 0} \sum_{n \geq 0} \sum_{s=0}^{n} P_{n+k,k,s} \frac{x^s t^n z^k}{n! k!} = \frac{e^{t(x-1)}}{1-t-z}.$$

Three q-analogues for $P_{n,k,s}$ are provided in [564].

9.1.3 Descent pairs with prescribed tops and bottoms

Hall and Remmel [446] generalized results of Kitaev and Remmel [526, 525] by considering (X,Y)-*descents*. Given sets X and Y of positive integers and a permutation $\sigma = \sigma_1 \sigma_2 \cdots \sigma_n \in \mathcal{S}_n$, an (X,Y)-descent of σ is a descent pair $\sigma_i > \sigma_{i+1}$ whose "top" σ_i is in X and whose "bottom" σ_{i+1} is in Y. Two formulas are given in [446] for the number $P_{n,s}^{X,Y}$ of $\sigma \in \mathcal{S}_n$ with s (X,Y)-descents, which is shown to be a *hit number* of a certain *Ferrers board*. To state the formula we need the following definitions.

For any set $A \subseteq \mathbb{N}$, let $A_n = A \cap [n]$ and $A_n^c = (A^c)_n = [n] - A$, and for any j, $1 \leq j \leq n$, we define

$$\alpha_{A,n,j} = |A^c \cap \{j+1, j+2, \ldots, n\}| = |\{x \mid j < x \leq n \text{ and } x \notin A\}|, \text{ and}$$

$$\beta_{A,n,j} = |A^c \cap \{1, 2, \ldots, j-1\}| = |\{x \mid 1 \leq x < j \text{ and } x \notin A\}|.$$

Theorem 9.1.6. *([446])* $P_{n,s}^{X,Y}$ *is given by*

$$|X_n^c|! \sum_{r=0}^{s} (-1)^{s-r} \binom{|X_n^c| + r}{r} \binom{n+1}{s-r} \prod_{x \in X_n} (1 + r + \alpha_{X,n,x} + \beta_{Y,n,x}) =$$

$$|X_n^c|! \sum_{r=0}^{|X_n|-s} (-1)^{|X_n|-s-r} \binom{|X_n^c| + r}{r} \binom{n+1}{|X_n| - s - r} \prod_{x \in X_n} (r + \beta_{X,n,x} - \beta_{Y,n,x}).$$

For example, using Theorem 9.1.6, we have that $P_{6,2}^{X,Y} = 72$ if $X = \{2, 3, 4, 6, 7, 9\}$ and $Y = \{1, 4, 8\}$.

Theorem 9.1.6 was extended in [446] to words. Moreover, Hall and Remmel [446] studied a more general problem, namely the class of polynomials

$$P_n^{X,Y,Z}(x) = \sum_{s \geq 0} P_{n,s}^{X,Y,Z} x^s = \sum_{\sigma \in \mathcal{S}_n} x^{\mathrm{des}_{X,Y,Z}(\sigma)},$$

where for any subsets X, Y and Z of $\mathbb{N}$, and a permutation $\sigma = \sigma_1 \sigma_2 \cdots \sigma_n \in \mathcal{S}_n$,

$$\mathrm{des}_{X,Y,Z}(\sigma) = |\{i \mid \sigma_i > \sigma_{i+1}, \sigma_i \in X, \sigma_{i+1} \in Y, \sigma_i - \sigma_{i+1} \in Z\}|.$$

A particular result obtained in [446] with respect to the generalization is the following theorem.

Theorem 9.1.7. *(9.1.6)* $P_{k(r+m),s}^{k\mathbb{N}, k\mathbb{N}, \mathbb{N}_{\geq kr}}$ *is given by*

$$(kr + (k-1)m)! \sum_{r=0}^{s} (-1)^{s-r} \binom{kr + (k-1)m}{r} \binom{k(r+m) + 1}{s-r} (kr + (k-1)m + r)^m,$$

where $\mathbb{N}_{\geq kr} = \{kr, kr+1, kr+2, \ldots\}$.

Hall et al. [445] defined a statistic $\mathrm{stat}_{X,Y}(\sigma)$ on permutations σ and defined $P_{n,s}^{X,Y}(q)$ to be the sum of $q^{\mathrm{stat}_{X,Y}(\sigma)}$ over all $\sigma \in \mathcal{S}_n$ with s (X, Y)-descents, thus showing that there are natural q-analogues of the formulas in Theorem 9.1.6.

9.2 Place-difference-value patterns

This section is based on [528] by Kitaev and Remmel who offered a definition of a pattern whose particular cases are, for example, all of the patterns considered in the previous chapters. As usual, let $\mathbb{P}$ denote the set of positive integers and $k\mathbb{P}$ the set of all positive multiples of k.

Definition 9.2.1. A *place-difference-value pattern, PDVP*, is a quadruple $P = (p, X, Y, Z)$ where p is a permutation of length m, X is an $(m+1)$-tuple of non-empty, possibly infinite, sets of positive integers, Y is a set of triples $(s, t, Y_{s,t})$ where $0 \le s < t \le m+1$ and $Y_{s,t}$ is a non-empty, possibly infinite, set of positive integers, and Z is an m-tuple of non-empty, possibly infinite, sets of positive integers. A PDVP $P = (p_1 p_2 \cdots p_m, (X_0, X_1, \ldots, X_m), Y, (Z_1, \ldots, Z_m))$ occurs in a permutation $\pi = \pi_1 \pi_2 \cdots \pi_n$, if π has a subsequence $\pi_{i_1} \pi_{i_2} \cdots \pi_{i_m}$ with the following properties:

1. $\pi_{i_k} < \pi_{i_\ell}$ if and only if $p_k < p_\ell$ for $1 \le k < \ell \le m$;

2. $i_{k+1} - i_k \in X_k$ for $k = 0, 1, \ldots, m$, where we assume $i_0 = 0$ and $i_{m+1} = n + 1$;

3. for each $(s, t, Y_{s,t}) \in Y$, $|\pi_{i_s} - \pi_{i_t}| \in Y_{s,t}$ where we assume $\pi_{i_0} = \pi_0 = 0$ and $\pi_{i_{m+1}} = \pi_{n+1} = n + 1$; and

4. $\pi_{i_k} \in Z_k$ for $k = 1, 2, \ldots m$.

For example, as above, let $\mathbb{E}$ and $\mathbb{O}$ denote the set of even and odd numbers, respectively. Then the PDVP $(12, (\{1\}, \{3, 4\}, \{1, 2, 3\}), \{(1, 2, \mathbb{E})\}, (\mathbb{E}, \mathbb{P}))$ occurs in $\pi = 23154$ once as the subsequence 24. Indeed, each such occurrence must start at position 1 as required by the set X_0 and the second element of the sequence must occur either at position 4 or 5 as required by X_1. X_2 says the 4 must occur in one of the last three positions of the permutation. The condition that $Z_1 = \mathbb{E}$ says that the value in position 1 must be even. Finally the condition that $(1, 2, \mathbb{E}) \in Y$ rules out 25 as an occurrence of the pattern.

Classical patterns are PDVPs of the form $(p, (\mathbb{P}, \mathbb{P}, \ldots), \emptyset, (\mathbb{P}, \mathbb{P}, \ldots))$, whereas vincular patterns have the property that X_i is either $\mathbb{P}$ or $\{1\}$, $Y = \emptyset$, and $Z_i = \mathbb{P}$ for all i. Also, bivincular patterns have the property that each X_i is either $\mathbb{P}$ or $\{1\}$, $Z_i = \mathbb{P}$ for all i, and all the elements of Y are of the form $(i, j, \{1\})$. Similarly, the occurrences of the pattern $(21, (\mathbb{P}, \{1\}, \mathbb{P}), \emptyset, (X, Y))$ in a permutation π correspond to the (X, Y)-descents in π considered in Subsection 9.1.3 and the occurrences of the pattern $(21, (\mathbb{P}, \{1\}, \mathbb{P}), \{(1, 2, \{k\})\}, (\mathbb{P}, \mathbb{P}))$ in π correspond to the situation considered in Subsection 9.1.1. As is noted in [528], often there is more than one way to specify the same pattern.

Object in the literature	PDVP $P = (p, X, Y, Z)$
Classical patterns	$X_i = Z_j = \mathbb{P}$ for all i and j and $Y = \emptyset$
Vincular patterns	X_i is either $\mathbb{P}$ or $\{1\}$ for all i, $Z_j = \mathbb{P}$ for all j, and $Y = \emptyset$
Conditioning on parity of the elements in descent pairs as in [526]	$p = 21$ and $X_0 = \mathbb{P}$, $X_1 = \{1\}$, $X_2 = \mathbb{P}$, $Y = \emptyset$, and (Z_1, Z_2) equals $(\mathbb{E}, \mathbb{P})$, $(\mathbb{O}, \mathbb{P})$, $(\mathbb{P}, \mathbb{E})$, or $(\mathbb{P}, \mathbb{O})$
Patterns in [525]	Similar to the last patterns, except we allow Z_is of the form $k\mathbb{P}$ where $k \geq 3$
Patterns in [562]	$(21, (\mathbb{P}, \{1\}, \mathbb{P}), \{(1, 2, \{k\})\}, (\mathbb{P}, \mathbb{P}))$, where $k \geq 1$
Patterns in [446]	$(21, (\mathbb{P}, \{1\}, \mathbb{P}), \emptyset, (X, Y))$, X and Y are any fixed sets
Bivincular patterns	X_i is either $\mathbb{P}$ or $\{1\}$, the elements of Y are of the form $(i, j, \{1\})$, and $Z_i = \mathbb{P}$ for all i

Table 9.1: Objects studied in the literature using the language of place-difference-value patterns.

In Table 9.1, we list how several pattern conditions that have appeared in the literature can be expressed in terms of PDVPs.

Tauraso [763] found the number of permutations of size n avoiding simultaneously the PDVPs

$$(12, (\mathbb{P}, \{d\}, \mathbb{P}), \{(1, 2, \{d\})\}, (\mathbb{P}, \mathbb{P}))$$

and

$$(21, (\mathbb{P}, \{d\}, \mathbb{P}), \{(1, 2, \{d\})\}, (\mathbb{P}, \mathbb{P})),$$

where $2 \leq d \leq n - 1$ (see [726, A110128] for related objects).

One of the enumerative results by Kitaev and Remmel in [528] is a formula for $a_{n,k}$, the number of n-permutations avoiding simultaneously the patterns $\underline{231}$, $\underline{132}$ and $(12, (\mathbb{P}, \{k\}, \mathbb{P}), \{(1, 2, \{1\})\}, (\mathbb{P}, \mathbb{P}))$:

$$a_{n,k} = \begin{cases} F_n & \text{if } k = 1, \\ 2^{n-1} & \text{if } k \geq 2 \text{ and } n \leq k, \\ 3 \cdot 2^{n-3} & \text{if } k \geq 2 \text{ and } n \geq k + 1, \end{cases}$$

where F_n is the n-th Fibonacci number. The sequence of $a_{n,k}$ appears in [726, A042950]. A bijection is given in [528] between the permutations counted by $a_{n,k}$ and the set of ascents (occurrences of the VP $\underline{12}$) after n iterations of the morphism $1 \rightarrow 123$, $2 \rightarrow 13$, $3 \rightarrow 2$ (see Definition 8.5.2), starting with the letter 1. For example, for $k = 2$ and $n = 3$, there are three permutations avoiding the prohibitions, 123, 312, and 321, and there are three such ascents in 1231323.

Place-difference-value patterns in the case of words can be defined in a similar manner. Several results on this type of patterns are obtained in [528].

9.3 Pattern-avoidance in multisets

Albert et al. [14] considered *permutations of a multiset* $1^{a_1}2^{a_2}\cdots k^{a_k}$, which are sequences of length $a_1+a_2+\cdots+a_k$ containing a_i occurrences of i for each $1 \le i \le k$. We let $\mathcal{S}_{a_1,a_2,\ldots,a_k}(P)$ be the set of permutations of $1^{a_1}2^{a_2}\cdots k^{a_k}$ which avoid every pattern in a set P and $s_{a_1,a_2,\ldots,a_k}(P) = |\mathcal{S}_{a_1,a_2,\ldots,a_k}(P)|$. Albert et al. [14] provided a structural description of the permutations avoiding each possible set of classical 3-patterns, and in many cases a complete enumeration of such permutations according to the underlying multiset is given in [14]. We state here examples of enumerative results obtained in [14]:

$$s_{a_1,a_2,\ldots,a_k}(123,231) = \binom{a_1 + a_2 + \cdots + a_k}{a_1} + \sum_{2\le i<j\le k} a_i a_j;$$

$$s_{a_1,a_2,a_3}(123,321) = (a_2 + 1)\binom{a_1 + a_3}{a_1} \quad \text{and} \quad s_{a_1,a_2,a_3,a_4}(123,321) = 2\binom{a_1 + a_4}{a_1};$$

$$s_{a_1,a_2,\ldots,a_k}(132,231) = \binom{a_1 + a_2}{a_1} \prod_{t=3}^{k}(a_t + 1);$$

$$s_{a_1,a_2,\ldots,a_k}(123,132,231) = s_{a_1,a_2,\ldots,a_k}(132,213,231) = \binom{a_1 + a_2}{a_1} + \sum_{t=3}^{k} a_t;$$

$$s_{a_1,a_2,\ldots,a_k}(123,132,312) = \binom{a_1 + a_2 + \cdots + a_k}{a_k}.$$

Heubach and Mansour [459] enumerated multisets avoiding one or more word 3-patterns, in particular, obtaining the following results:

$$s_{a_1,a_2,\ldots,a_k}(111) = \begin{cases} \frac{(a_1+a_2+\cdots+a_k)!}{a_1!a_2!\cdots a_k!} & \text{if } a_i \le 2 \text{ for all } i, \\ 0 & \text{otherwise;} \end{cases}$$

$$s_{a_1,a_2,\ldots,a_k}(112) = s_{a_1,a_2,\ldots,a_k}(121) = \prod_{j=2}^{k}(a_j + a_{j+1} + \cdots + a_k + 1);$$

$$s_{a_1,a_2,\ldots,a_k}(221) = s_{a_1,a_2,\ldots,a_k}(212) = \prod_{j=1}^{k-1}(a_1 + a_2 + \cdots + a_j + 1).$$

The following theorem generalizes the fact that all classical 3-patterns are Wilf-equivalent on permutations.

Theorem 9.3.1. *([713]) Fix a multiset M. The number of permutations of M that avoid a classical pattern $p \in \mathcal{S}_3$ is independent of the choice of p.*

Myers [637] gave a bijective proof of Theorem 9.3.1 by generalizing the construction for ordinary permutations in [721] by Simion and Schmidt.

9.4 Patterns in finite approximations of sequences

In [508, 519] and [520], it is suggested to count occurrences of patterns in certain words (of increasing length), instead of the typical problem of avoiding/counting patterns in all permutations or all words. These words are chosen to be the set of all finite approximations of certain sequences and we discuss the topic in this section.

9.4.1 The Peano curve

The Peano (Hilbert) Curve was discovered in 1890 to construct a continuous mapping from the unit interval onto the unit square. This curve is an example of a *fractal space filling curve*. The Peano curve is constructed iteratively (see [519] for a proper explanation of the construction), and the first three iterations are shown in Figure 9.1.

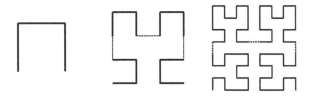

Figure 9.1: The first steps to construct the Peano curve.

A *Peano word* P_n is obtained by traveling along the Peano curve after the n-th iteration. P_n is over the alphabet $\Sigma = \{u, \bar{u}, r, \bar{r}\}$ where u stands for *up*, $\bar{u}$ stands for *down*, r stands for *right*, and $\bar{r}$ stands for *left*. For example, $P_1 = ur\bar{u}$ and $P_2 = r\bar{u}ur\bar{u}r\bar{u}ur\bar{u}$. To deal with patterns, the alphabet in question must be ordered, so we assume that $u < r < \bar{u} < \bar{r}$. The *Peano (infinite) word* $P = \lim_{n \to \infty} P_{2n+1}$.

Kitaev et al. [519] found the number of occurrences of several patterns in P_n. For example, the following theorems are proved in [519].

Theorem 9.4.1. *([519]) For any $n, k \in \mathbb{N} \backslash \{0\}$, the number of occurrences of the classical pattern $1^k = \underbrace{11 \cdots 1}_{k \ times}$ in P_n is given by*

$$\binom{4^{n-1} - 2^{n-1}}{k} + 2\binom{4^{n-1}}{k} + \binom{4^{n-1} + 2^{n-1} - 1}{k}.$$

Theorem 9.4.2. *([519]) The number of occurrences of ascents and descents in P_n is given by the following formulas:*

$$\underline{12}(P_{2k+1}) = \tfrac{2}{5}(4 \cdot 16^k + 1),$$
$$\underline{12}(P_{2k+2}) = \tfrac{2}{5}(16^{k+1} - 1),$$
$$\underline{21}(P_{2k+1}) = \tfrac{8}{5}(16^k - 1),$$
$$\underline{21}(P_{2k+2}) = \tfrac{2}{5}(16^{k+1} - 1).$$

Additionally, it was proved in [519] that the infinite word P cannot be generated by iterations of a morphism.

9.4.2 The Dragon curve sequence and the σ-sequence

The *Dragon curve sequence* was discovered by NASA physicist John E. Heighway and was described by Martin Gardner in his Scientific American column Mathematical Games in 1967. This sequence is also known as the *paperfolding sequence* which can be described as follows. Start with a rectangular piece of paper which we shall view from the edge. Fold the right half over the left half, with a sharp crease down the middle. Take the folded paper and fold again in the same way. Continue this folding process for a few more generations. After a number of folds, unfold the paper, and spread each fold to an angle of exactly 90 degrees. The resulting edge curve is our dragon presented in Figure 9.2.

This curve is a classic example of a recursively generated *fractal shape*. While traveling through the Dragon curve sequence, we can create a binary word indicating whether a turn to the right or a turn to the left was made. For Figure 9.2, this word would start like $\ell\ell r\ell\ell r r\ell\ell\ell r r\ell r \ldots$.

It turns out that the word corresponding to the Dragon curve sequence is equivalent to the σ-sequence w_σ defined below that was used by Evdokimov [372] in constructing certain chains of maximal length in the unit n-cube.

Figure 9.2: The Dragon curve sequence.

Any natural number n can be presented unambiguously as $n = 2^t(4s + \sigma)$, where $\sigma < 4$, and t is the greatest natural number such that 2^t divides n. If n runs through the natural numbers then σ runs through the sequence that we will call the *sequence of* σ, or the σ-*sequence*. We let w_σ denote that sequence. Obviously, w_σ consists of 1s and 3s. The initial letters of w_σ are $11311331113313\ldots$.

An equivalent definition of the σ-sequence is as follows:

$$C_1 = 1, \qquad\qquad D_1 = 3$$

$$C_{k+1} = C_k 1 D_k, \qquad\qquad D_{k+1} = C_k 3 D_k$$

$$k = 1, 2, \ldots$$

and $w_\sigma = \lim_{k \to \infty} C_k$. Even though there are several ways to define the σ-sequence, it cannot be defined by iteration of a morphism, as was shown in [506].

Kitaev [508] gave either explicit formulas or recurrence relations for the number of occurrences of certain classical and consecutive patterns in w_σ. Examples of these results are as follows, where $c_n(p)$ is the number of occurrences of a pattern p in C_n

and $1^k = \underbrace{11\cdots1}_{k \text{ times}}$:

$$c_n(1^k) = \tfrac{2^n-k}{2^{n-1}-k}\binom{2^{n-1}-1}{k};$$

$$c_n(12) = 2\cdot 4^{n-2} + (n-2)\cdot 2^{n-2}, \text{ for } n \geq 2;$$

$$c_n(21) = 2\cdot 4^{n-2} - n\cdot 2^{n-2}, \text{ for } n \geq 2;$$

$$c_n(\underline{12}) = 2^{n-2}, \text{ for } n \geq 2;$$

$$c_n(\underline{112}) = c_n(\underline{221}) + 1 = 3\cdot 2^{n-4}, \text{ for } n \geq 4.$$

Several enumerative results of general nature, in particular, dealing with consecutive patterns, are obtained in [508].

9.4.3 General study related to words generated by morphisms and patterns

Kitaev et al. [520] started a general study of counting occurrences of patterns in words generated by morphisms (see Definition 8.5.2). Let $f : A^* \to A^*$ be a morphism, where $A = \{a_1, a_2, \ldots, a_k\}$. We assume that $a_1 < a_2 < \cdots < a_k$ and, for $1 \leq i \leq k$, $f(a_i) = a_{i_1}a_{i_2}\cdots a_{i_{p_i}}$ with $p_i \geq 0$ ($p_i = 0$ if and only if $f(a_i) = \varepsilon$).

Definition 9.4.3. Let n be a non-negative integer. The *incidence matrix* of f^n is the $k \times k$ matrix

$$M(f^n) = (m_{n,i,j})_{1\leq i,j\leq k}$$

where $m_{n,i,j}$ is the number of occurrences of the letter a_i in the word $f^n(a_j)$, that is, $m_{n,i,j} = |f^n(a_j)|_{a_i}$. Also, we let $|f^n(a_j)|_p$ denote the number of occurrences of a pattern p in the word $f^n(a_j)$.

Proposition 9.4.4. *([36]) For every $n \in \mathbb{N}$, $M(f)^n = M(f^n)$.*

The following theorem gives the number of inversions and non-inversions (occurrences of the patterns 21 and 12, respectively) in words obtained by iterating a morphism f.

Proposition 9.4.5. *([520]) For each letter $a_\ell \in A$, let p_ℓ and q_ℓ be such that $f(a_\ell) = a_{\ell_1}a_{\ell_2}\cdots a_{\ell_{p_\ell}}$ and $f^n(a_\ell) = a_{\ell'_1}a_{\ell'_2}\cdots a_{\ell'_{q_\ell}}$. Then, for all $n \in \mathbb{N}$,*

$$|f^{n+1}(a_\ell)|_{12} = \sum_{1\leq i<j\leq p_\ell}\left(\sum_{r=1}^{k-1}\left(m_{n,r,\ell_i}\cdot\sum_{s=r+1}^{k}m_{n,s,\ell_j}\right)\right) + \sum_{t=1}^{k}|f^n(a_t)|_{12}\cdot m_{1,t,\ell}$$

$$= \sum_{1 \leq i < j \leq q_\ell} \left(\sum_{r=1}^{k-1} (m_{1,r,\ell'_i} \cdot \sum_{s=r+1}^{k} m_{1,s,\ell'_j}) \right) + \sum_{t=1}^{k} |f(a_t)|_{12} \cdot m_{n,t,\ell}$$

$$|f^{n+1}(a_\ell)|_{21} = \sum_{1 \leq i < j \leq p_\ell} \left(\sum_{r=2}^{k} (m_{n,r,\ell_i} \cdot \sum_{s=1}^{r-1} m_{n,s,\ell_j}) \right) + \sum_{t=1}^{k} |f^n(a_t)|_{21} \cdot m_{1,t,\ell}$$

$$= \sum_{1 \leq i < j \leq q_\ell} \left(\sum_{r=2}^{k} (m_{1,r,\ell'_i} \cdot \sum_{s=1}^{r-1} m_{1,s,\ell'_j}) \right) + \sum_{t=1}^{k} |f(a_t)|_{21} \cdot m_{n,t,\ell}.$$

Some applications of Proposition 9.4.5 are discussed in [520]. In particular, for the *Thue-Morse morphism* μ defined by $\mu(a_1) = a_1 a_2$ and $\mu(a_2) = a_2 a_1$ (this is essentially the morphism h in Subsection 8.5.1), one has

$$|\mu^n(a_1)|_{12} = |\mu^n(a_1)|_{21} = |\mu^n(a_2)|_{12} = |\mu^n(a_2)|_{21} = 2^{2n-3},$$

while for the *Fibonacci morphism* φ defined by $\varphi(a_1) = a_1 a_2$, $\varphi(a_2) = a_1$, one has

$$|\varphi^{n+2}(a_1)|_{21} = \sum_{p=0}^{n} F_p F_{n-p}^2,$$

$$|\varphi^{n+2}(a_1)|_{12} = \sum_{p=0}^{n} F_p F_{n-p}^2 + F_n + \begin{cases} 1 & \text{if } n \text{ is odd,} \\ -1 & \text{if } n \text{ is even,} \end{cases}$$

where F_n is the n-th Fibonacci number.

As particular cases of a more general result, in [520] the number of inversions and non-inversions was obtained for the *Prouhet morphisms* and the *Arshon morphisms* (see [520] for references) which we briefly discuss below. However, we note that [520] contains results on other morphisms, and general results with examples on occurrences of the consecutive patterns $\underline{12}$ and $\underline{21}$ for certain morphisms.

Prouhet morphisms

Let $k \geq 2$. The *Prouhet morphism* π_k is defined on A by

$$\pi_k(a_i) = a_i a_{i+1} \cdots a_k a_1 \cdots a_{i-1}, 1 \leq i \leq k.$$

Proposition 9.4.6. *([520]) For every i, $1 \leq i \leq k$, and for every $n \in \mathbb{N}$,*

$$|\pi_k^{n+1}(a_i)|_{12} = \frac{(k-1)k^n}{12} \left(3k^{n+1} + k - 2 \right),$$

$$|\pi_k^{n+1}(a_i)|_{21} = \frac{(k-1)k^n}{12} \left(3k^{n+1} - k + 2 \right).$$

Arshon morphisms

Let k be any even positive integer. The morphism β_k is defined, for every r, $1 \leq r \leq k/2$, by

$$
\begin{aligned}
a_{2r-1} &\mapsto a_{2r-1}a_{2r} \cdots a_{k-1}a_k a_1 a_2 \cdots a_{2r-3}a_{2r-2}, \\
a_{2r} &\mapsto a_{2r-1}a_{2r-2} \cdots a_2 a_1 a_k a_{k-1} \ldots a_{2r+1}a_{2r}.
\end{aligned}
$$

Proposition 9.4.7. *([520]) Let k be any even positive integer. For every i, $1 \leq i \leq k$, and for every $n \in \mathbb{N}$,*

$$
\begin{aligned}
|\beta_k^{n+1}(a_i)|_{12} &= \tfrac{k^{n-1}}{4}\left[(k-1)k^{n+2} + 2k\right], \\
|\beta_k^{n+1}(a_i)|_{21} &= \tfrac{k^{n-1}}{4}\left[(k-1)k^{n+2} - 2k\right].
\end{aligned}
$$

For example, for every i, $1 \leq i \leq k$, and for every $n \geq 1$,

$$
|\beta_6^{n+1}(a_i)|_{12} = 6^{n-1} \cdot (45 \cdot 6^n + 3), \quad |\beta_6^{n+1}(a_i)|_{21} = 6^{n-1} \cdot (45 \cdot 6^n - 3).
$$

9.5 Pattern-avoidance on partial permutations

Assume that $\diamond$ is a special symbol not belonging to an alphabet A. The symbol $\diamond$ is called a *hole*. A *partial word* over A is a word over the alphabet $A \cup \{\diamond\}$. In the study of partial words, the holes are usually treated as gaps that may be filled by an arbitrary letter of A. The *length* of a partial word is the number of its symbols, including the holes.

The study of partial words was initiated by Berstel and Boasson [112]. These words are used in comparing genes, and certain combinatorial aspects of partial words have been studied in the literature (see [258] for references).

Let V be a set of symbols not containing $\diamond$. A *partial permutation of V* is a partial word π such that each symbol of V appears in π exactly once, and all the remaining symbols of π are holes. Let P_n^k denote the set of all partial permutations of $[n-k] = \{1, 2, \ldots, n-k\}$ that have exactly k holes. For example,

$$
P_3^1 = \{12\diamond, 21\diamond, 1\diamond 2, 2\diamond 1, \diamond 12, \diamond 21\}.
$$

It is easy to see that $|P_n^k| = \binom{n}{k}(n-k)! = n!/k!$. Also, note that $P_n^0 = \mathcal{S}_n$. For a set $H \subseteq [n]$ of size k, we let P_n^H denote the set of partial permutations $\pi_1 \pi_2 \cdots \pi_n \in P_n^k$ such that $\pi_i = \diamond$ if and only if $i \in H$; $p_n^H = |P_n^H|$.

Claesson et al. [258] extended the notion of classical pattern-avoidance to the case of partial permutations as follows. Let $\pi \in P_n^k$ be a partial permutation and let $i_1 < i_2 < \cdots < i_{n-k}$ be the indices of the non-hole elements of π. A permutation $\sigma \in \mathcal{S}_n$ is an *extension* of π if

$$\mathrm{red}(\sigma_{i_1}\sigma_{i_2}\cdots\sigma_{i_{n-k}}) = \pi_{i_1}\pi_{i_2}\cdots\pi_{i_{n-k}}.$$

For example, the partial permutation $1\diamond2$ has three extensions, namely 213, 123 and 132. In general, the number of extensions of $\pi \in P_n^k$ is $\binom{n}{k}k! = n!/(n-k)!$.

We say that $\pi \in P_n^k$ *avoids* a *pattern* $p \in \mathcal{S}_\ell$ if each extension of π avoids p. For example, $\pi = 32\diamond154$ avoids 1234, but π does not avoid 123, since the permutation 325164 is an extension of π and it contains two occurrences of the pattern 123. Let $P_n^k(p)$ be the set of all the partial permutations in P_n^k that avoid p, and let $p_n^k(p) = |P_n^k(p)|$.

We can also define the *number of occurrences* of $p \in \mathcal{S}_\ell$ in $\pi \in P_n^k$ as the number of ℓ element subsequences in π that can be extended to a permutation whose reduced form is p. For example, the pattern 123 occurs 5 times in $3\diamond124$; its occurrences are $3\diamond4$, $\diamond12$, $\diamond14$, $\diamond24$ and 124. The notion of the number of occurrences was introduced in [258], but there have been no further results on it in the literature so far.

We say that two patterns p and q are *k-Wilf-equivalent* if $p_n^k(p) = p_n^k(q)$ for all n. Notice that 0-Wilf-equivalence coincides with the standard notion of Wilf-equivalence. We also say that two patterns p and q are *⋆-Wilf-equivalent* if p and q are k-Wilf-equivalent for all $k \geq 0$. Two patterns p and q are *strongly k-Wilf-equivalent* if $p_n^H(p) = p_n^H(q)$ for each n and for each k-element subset $H \subseteq [n]$. Finally, p and q are *strongly ⋆-Wilf-equivalent* if they are strongly k-Wilf-equivalent for all $k \geq 0$.

While strong k-Wilf-equivalence implies k-Wilf-equivalence, and strong ⋆-Wilf-equivalence implies ⋆-Wilf-equivalence, the converse implications are not true, as noted in [258]. Consider for example the patterns $p = 1342$ and $q = 2431$. A partial permutation avoids p if and only if its reverse avoids q, and thus p and q are ⋆-Wilf-equivalent. However, p and q are not strongly 1-Wilf-equivalent, and hence not strongly ⋆-Wilf-equivalent either. To see this, we fix $H = \{2\}$ and easily check that $s_5^H(p) = 13$ while $s_5^H(q) = 14$.

One of the main results in [258] states that a permutation pattern of the form $12\cdots\ell X$ is strongly ⋆-Wilf-equivalent to the pattern $\ell(\ell-1)\cdots1X$, where $X = x_{\ell+1}x_{\ell+2}\cdots x_m$ is any permutation of $\{\ell+1, \ell+2\ldots, m\}$. This result is a strengthening of Theorem 4.0.3 and it requires a different proof.

Another main result in [258] states that for any permutation X of the set $\{4, 5, \ldots, k\}$, the two patterns $312X$ and $231X$ are strongly ⋆-Wilf-equivalent. This is a refinement of Theorem 4.0.5, and it also requires a different proof.

Further, Claesson et al. [258] studied the k-Wilf-equivalence of patterns whose length is small in terms of k. It is not hard to see that all patterns of length ℓ are k-Wilf-equivalent whenever $\ell \leq k + 1$, because $p_n^k(p) = 0$ for every such p and every $n \geq \ell$. Thus, the shortest non-trivial patterns are of length $k + 2$. For these patterns, we show that k-Wilf-equivalence yields a new characterization of Baxter permutations (see Subsection 2.2.4): a pattern p of length $k + 2$ is a Baxter permutation if and only if $p_n^k(p) = \binom{n}{k}$. For any non-Baxter permutation q of length $k + 2$, $p_n^k(q)$ is strictly smaller than $\binom{n}{k}$ and is in fact a polynomial in n of degree at most $k - 1$.

Finally, Claesson et al. [258] obtained explicit formulas for $p_n^k(p)$ for every p of length at most four and every $k \geq 1$. These results for 4-patterns in the case $k = 1$ are recorded in the following theorem (a bijective proof is also given in [258] for the formula for $p_n^1(1234)$).

Theorem 9.5.1. *([258]) We have*

$$p_n^1(1234) = \binom{2n-2}{n-1};$$

$$p_n^1(1342) = \binom{2n-2}{n-1} - \binom{2n-2}{n-5};$$

$$p_n^1(2413) = \frac{2}{n+1}\binom{2n}{n} - 2^{n-1}.$$

9.6 Signed and colored permutations and patterns

We can regard the elements of the *hyperoctahedral group* B_n as *signed permutations* written as $\alpha = \alpha_1 \alpha_2 \cdots \alpha_n$ in which each of the letters $1, 2, \ldots, n$ appears, possibly barred. For example, $B_2 = \{12, \overline{1}2, 1\overline{2}, \overline{1}\,\overline{2}, 21, \overline{2}1, 2\overline{1}, \overline{2}\,\overline{1}\}$. Clearly, $|B_n| = 2^n n!$. The *barring operation* maps $\pi \in B_n$ to the signed permutation $\overline{\pi}$ which is obtained from π by changing the sign of all elements. Clearly, $\overline{\overline{\pi}} = \pi$.

More generally, let r be a nonnegative integer and let $\mathcal{CS}_n^r$ denote the set of permutations of $\{1, 2, \ldots, n\}$, written in one-line notation, in which each element has an associated *color* from $\{0, 1, \ldots, r\}$. The elements of $\mathcal{CS}_n^r$ are called *colored permutations*, and we write the colors of their entries as exponents, as in $2^0 1^4 4^0 3^3$. We can think of $\mathcal{CS}_n^r$ as the wreath product $\mathcal{CS}_n^r = \mathcal{S}_n \wr C_r$ where C_r is the cyclic group of order $r + 1$. When $r = 0$, we can omit the color of all elements of $\mathcal{CS}_n^r$ and obtain $\mathcal{S}_n$. When $r = 1$ we identify $\mathcal{CS}_n^r$ with the set B_n by omitting the color 0 and by replacing the color 1 with a bar.

Suppose that π and σ are colored permutations. A subsequence of π has *type* σ whenever it has all of the same pairwise comparisons as σ and each entry of the subsequence of π has the same color as the corresponding entry in σ. For example, the subsequence $3^2 2^2 8^0 4^3$ of $5^3 3^2 1^0 2^2 6^0 8^0 4^3 7^1$ has type $2^2 1^2 4^0 3^3$. We say that π avoids σ whenever π has no subsequence of type σ. For example, the colored permutation $2^1 10^4 0^5 2^3 1^8 0^7 0^6 2^9 1$ avoids $3^1 1^1 2^0$ and $1^1 3^2 2^2$, but it has $4^0 5^2 6^2$ as a subsequence so it does not avoid $1^0 2^2 3^2$; on the other hand, the signed permutation $3\bar{4}1\bar{2}5$ avoids $\overline{12}$. We let $\mathcal{CS}_n^r(P)$ denote the set of colored permutations in $\mathcal{CS}_n^r$ avoiding each colored permutation (referred to as a pattern) in P. We also let $cs_n^r(P) = |\mathcal{CS}_n^r(P)|$ and $cs_n^1(P) = |B_n(P)|$.

Simion [720] was the first one to consider pattern-avoidance in the context of signed permutations, which was studied further in [621] by Mansour and West.

Proposition 9.6.1. *([720]) If p is any signed pattern of length 2, then for each n,*

$$cs_n^1(p) = \sum_{k=0}^{n} k! \binom{n}{k}^2.$$

Proposition 9.6.2. *([720],[621])*

$$cs_n^1(12, \bar{2}1) = n! \left(1 + \sum_{i=1}^{n} \frac{1}{i} \sum_{j=0}^{i-1} \frac{1}{j!}\right);$$

$$cs_n^1(12, 21) = cs_n^1(12, 1\bar{2}) = cs_n^1(2\bar{1}, 1\bar{2}) = cs_n^1(2\bar{1}, \bar{1}2) = (n+1)!;$$

$$cs_n^1(1\bar{2}, \bar{1}2) = 2 \sum_{\ell=1}^{n} \sum_{i_1+i_2+\cdots+i_\ell=n,\ i_j\geq 1} \prod_{j=1}^{\ell} i_j!;$$

$$cs_n^1(12, \overline{12}) = cs_n^1(12, \overline{21}) = \binom{2n}{n}.$$

Formulas for the number of signed permutations avoiding more than two signed 2-patterns can be found in [621] by Mansour and West. Also, Dukes et al. [320] provided a complete classification of signed patterns of lengths three and four, also extending their studies to *involutive signed permutations*. The general results in [320] leading to the classification are the following two theorems. Moreover, signed involutions avoiding any subset of signed 2-patterns are enumerated in [319] by Dukes and Mansour.

Theorem 9.6.3. *([320]) For any pattern $\tau \in B_\ell$ and $k \geq \ell$, we have*

$$\tau(\ell+1)(\ell+2)\cdots k \sim \bar{\tau}(\ell+1)(\ell+2)\cdots k,$$

where the barring operation is defined at the beginning of this section. Moreover, the patterns are also Wilf-equivalent if we consider their avoidance by signed involutions.

	B_1	S_1	B_2	S_2	B_3	S_3	B_4	S_4	B_5	S_5
# symmetry classes	1	1	2	1	6	2	40	7	284	23
# Wilf classes	1	1	1	1	2	1	14	3	≥ 58 ≤ 137	16

Table 9.2: Number of symmetry and Wilf classes in B_k and S_k for $k \leq 5$ (avoidance in B_n and S_n, respectively).

Theorem 9.6.4. *([320]) For any pattern $\tau \in B_\ell$ and $k \geq \ell$, we have*

$$\tau k(k-1)\cdots(\ell+1) \sim \overline{\tau}k(k-1)\cdots(\ell+1).$$

Moreover, the patterns are also Wilf-equivalent if we consider their avoidance by signed involutions.

A comparison of the number of symmetry classes and Wilf classes in permutations and signed permutations is made in [320] and we record it in Table 9.2.

Mansour and Sun [613] studied *even signed permutations* avoiding signed 2-patterns. A signed permutation $\pi \in B_n$ is said to be *even* if the number of barred letters in π is an even number; π is said to be *odd* otherwise. All cases of even signed permutations avoiding any subset of signed 2-patterns are enumerated in [613]. We provide just one example of the enumerative results obtained in [613], where $d_n(1\overline{2})$ and $d'_n(1\overline{2})$ denote, respectively, the number of even and odd signed permutations avoiding the pattern $1\overline{2}$.

Theorem 9.6.5. *([613]) For any integer $n \geq 0$, we have*

$$d_n(1\overline{2}) = \frac{1}{2}\left(\sum_{j=0}^{n}\binom{n}{j}^2 j! + \frac{n!}{2^n}\binom{n}{\frac{n}{2}}\right) \text{ and } d'_n(1\overline{2}) = \frac{1}{2}\left(\sum_{j=0}^{n}\binom{n}{j}^2 j! - \frac{n!}{2^n}\binom{n}{\frac{n}{2}}\right),$$

where $\binom{a}{b}$ is assumed to be 0 whenever b or a is a non-integer number.

Proposition 9.6.1 was generalized by Mansour [586].

Theorem 9.6.6. *([586]) If p is any colored pattern of length 2, then for each n,*

$$cs_n^r(p) = \sum_{k=0}^{n} k!r^k\binom{n}{k}^2.$$

A number of other enumerative results are obtained in [586] on colored permutations avoiding pairs of colored 2-patterns. In particular, the following theorem was proved in [586].

Theorem 9.6.7. *([586]) For $n \geq 0$, $cs_n^r(T) = n!(n+r)r^{n-1}$ where*

- $T = \{1^0 2^0, 2^0 1^0\}$ *for $r \geq 0$;*

- $T = \{1^0 2^0, 1^0 2^1\}$ *for $r \geq 1$;*

- $T = \{1^0 2^1, 2^0 1^1\}$ *for $r \geq 1$;*

- $T = \{1^0 2^1, 2^1 1^0\}$ *for $r \geq 1$;*

- $T = \{1^0 2^1, 1^0 2^2\}$ *for $r \geq 2$.*

For a generalization of the results mentioned so far in this section see [590] by Mansour. Moreover, connections between colored permutations and Chebyshev polynomials of the second kind are discussed in [336] by Egge.

Recall from Table 2.2 that there are at least ten classes of restricted permutations counted by the large Schröder numbers, no two of which are trivially Wilf-equivalent. Egge [335] provided twelve such classes in the case of pattern-avoiding signed permutations. These classes are $B_n(T)$ where T is any of the following sets:

$$\{2\bar{1}, \overline{21}, \overline{3}12, 312\}, \quad \{\overline{21}, \bar{2}1, 1\overline{2}3, 312\}, \quad \{21, 2\bar{1}, \overline{321}, 31\bar{2}\},$$
$$\{\overline{21}, \bar{2}1, 1\bar{2}3, 123\}, \quad \{\overline{21}, \bar{2}1, 1\bar{2}3, 321\}, \quad \{21, 2\bar{1}, \overline{231}, 31\bar{2}\},$$
$$\{\overline{21}, \bar{2}1, 1\bar{2}3, 132\}, \quad \{21, 2\bar{1}, \overline{312}, 31\bar{2}\}, \quad \{21, 2\bar{1}, \overline{132}, 1\overline{32}\},$$
$$\{\overline{21}, \bar{2}1, 1\bar{2}3, 231\}, \quad \{21, 2\bar{1}, \overline{321}, \overline{321}\}, \quad \{21, \bar{2}1, \overline{321}, \overline{312}\}.$$

Kitaev et al. [523, 529] considered a modification of the notion of pattern-avoiding colored permutation discussed above. In what follows, it is convenient to think of colored permutations as pairs (σ, w) where $\sigma = \sigma_1 \sigma_2 \cdots \sigma_n \in \mathcal{S}_n$ and $w = w_1 w_2 \cdots w_n$ with $w_i \in \{0, 1, \ldots, r\}$ for $1 \leq i \leq n$.

Given a word w over the alphabet $\{0, 1, \ldots, r\}$, we modify the notion of reduced form by letting $\text{red}(w)$ denote the word obtained by replacing the i-th largest integer in w by $i - 1$. For example, if $w = 36356$ then $\text{red}(w) = 02012$. We now say that $(\tau, u) \in \mathcal{CS}_k^j$ *occurs* in $(\sigma, w) \in \mathcal{CS}_n^r$ if τ occurs in σ (in any specified sense: as a classical or a vincular pattern) and the reduced form of the word formed by the colors of the occurrence is exactly $\text{red}(u)$. That is, we drop the requirement for the corresponding signs to agree exactly like which we had above: they now need to agree as patterns. For this context, it is suitable to assume $u = \text{red}(u)$. For example, $(21543, 10010)$ contains an occurrence of $(12, 00)$ (for example, $2^1 4^1$), while $(21543, 11010)$ avoids $(12, 01)$.

Out of many enumerative results obtained in [523, 529], we state just the following theorem which was proved both algebraically and combinatorially.

Theorem 9.6.8. *([529]) The number of signed n-permutations simultaneously avoiding* $(12, 00)$ *and* $(12, 01)$, *that is,* $\mathcal{CS}_n^1((12, 00), (12, 01))$, *is given by the* $(n + 1)$-*th Catalan number* $C_{n+1} = \frac{1}{n+2}\binom{2n+2}{n+1}$.

We invite the interested reader to see [380] and [468] for more results which we do not discuss here.

9.7 Generalized factor order

Let $(P, \leq_P)$ be a partially ordered set and P^* the set of all words over P. Define the *generalized factor order* on P^* by letting $u \leq w$ if there is a factor w' of w having the same length as u such that $u \leq w'$, where the comparison of u and w' is done letterwise using the partial order in P. This factor w' is called an *embedding* of u into w. One obtains the (ordinary) *factor order* by insisting that $u = w'$ or, equivalently, by taking P to be an antichain.

Definition 9.7.1. A *language* is any $\mathcal{L} \subseteq P^*$. It has an associated generating function

$$f_{\mathcal{L}} = \sum_{w \in \mathcal{L}} w.$$

The language $\mathcal{L}$ is *regular* if $f_{\mathcal{L}}$ is rational.

Given any set P, a *nondeterministic finite automaton* or *NFA* over P is a directed graph Δ with vertices V and arcs $\vec{E}$ having the following properties.

1. The elements of V are called *states* and $|V|$ is finite.

2. There is a designated *initial state* α and a set Ω of *final states*.

3. Each arc of $\vec{E}$ is labeled with an element of P.

Given a (directed) path in Δ starting at α, we construct a word in P^* by concatenating the elements on the arcs on the path in the order in which they are encountered. The *language accepted by* Δ is the set of all such words which are associated with paths ending in a final state. It is a well-known theorem that, for $|P|$ finite, a language $\mathcal{L} \subseteq P^*$ is regular if and only if there is a NFA accepting $\mathcal{L}$ (see [513] for references).

Kitaev at el. [513] proved that for a given $u \in P^*$, the language $\mathcal{F}(u) = \{w \mid w \geq u\}$ is accepted by a finite state automaton. If P is finite then it follows that the generating function $F(u) = \sum_{w \geq u} w$ is rational. This is an analogue of a theorem of Björner and Sagan for *generalized subword order* (see [122]).

For the case where P is the set of positive integers $\mathbb{P}$ with the usual ordering, we define the weight of a word $w = w_1 w_2 \cdots w_n$ to be $wt(w) = x^{\sum_{i=1}^{n} w_i} t^n$ and we define the weight generating function $F(u; t, x) = \sum_{w \geq_P u} wt(w)$. Also, the related generating function $S(u; t, x) = \sum_{w \in \mathcal{S}(u)} wt(w)$ where $\mathcal{S}(u)$ is the set of all words w such that the only embedding of u into w is in the last $|u|$ letters.

Examples of enumerative results obtained in [513] are as follows:

$$S(123; t, x) = \frac{t^3 x^6}{(1-x)^2(1 - x - tx + tx^3 - t^2 x^4)};$$

$$S(213; t, x) = \frac{t^3 x^6 (1 + tx^3)}{(1-x)(1 - x + t^2 x^4)(1 - x - tx + tx^3 - t^2 x^4)}.$$

Definition 9.7.2. Two words u and v are Wilf-equivalent, denoted $u \sim v$, if and only if $F(u; t, x) = F(v; t, x)$.

A number of results on Wilf-equivalence are proved in [513] by Kitaev et al. Among them there are the following two theorems.

Theorem 9.7.3. *([513]) We have $u \sim v$ if and only if $S(u; t, x) = S(v; t, x)$.*

Theorem 9.7.4. *([513]) Let $x, y, z \in \{1, 2, \ldots, m\}^*$ and suppose that $n > m$. Then $xmynz \sim xnymz$.*

Langley et al. [552] gave an explicit formula for $S(u; t, x)$ if u factors into a weakly increasing word followed by a weakly decreasing word. This formula was used in [552] as an aid to classify Wilf-equivalence for all words of length 3. Many other results were obtained in [513] and [552], which we do not state in this book. However, we do mention the following two conjectures from these papers.

Conjecture 9.7.5. ([513]) If $u \sim v$, then v is a rearrangement of u.

Conjecture 9.7.6. ([552]) If $u \sim v$, then there is a weight preserving bijection $f : \mathbb{P} \to \mathbb{P}$ such that for all $w \in \mathbb{P}^*$, $f(w)$ is a rearrangement of w and $w \in \mathcal{F}(u)$ if and only if $f(w) \in \mathcal{F}(v)$.

The first conjecture above is called the *weak rearrangement conjecture* in [552].

9.8   Miscellaneous

In this chapter we have discussed several extensions of research on patterns in permutations and words, but there are still other relevant directions of interest, some of which are mentioned very briefly below.

9.8.1   Pattern-avoiding set partitions

A *set partition of size* n is a collection of disjoint blocks B_1, $B_2, \ldots, B_d$ whose union is the set $[n] = \{1, 2, \ldots, n\}$. We choose the ordering of the blocks to satisfy $\min B_1 < \min B_2 < \cdots < \min B_d$. Such a set partition is represented by a *canonical sequence* $\pi_1 \pi_2 \cdots \pi_n$, with $\pi_i = j$ if $i \in B_j$. A partition π *contains* a partition σ if the canonical sequence of π contains a subsequence that is order-isomorphic to the canonical sequence of σ.

Sagan [707] described equivalence classes of partitions of size at most 3, which was extended by Jelínek and Mansour [482] to determining all the equivalence classes of partitions of size at most 7. Goyt [426] studied multi-avoidance of partitions of a 3-element set.

The study of pattern-avoidance in set partitions was initiated by Klazar [532, 533, 534]. However, there is an alternative definition of an occurrence of a partition in another partition due to Callan [218], who used the notion of "flattening" a partition by erasing the dividers between its blocks when the blocks are ordered in a standard way — increasing entries in each block and blocks arranged in increasing order of their first entries.

For further information on this subject see [427, 428, 429, 609, 607].

9.8.2   Pattern-avoidance in matrices

Kitaev et al. [521] generalized the concept of pattern-avoidance from permutations and words to matrices (*numbered polyominos*), and considered binary matrices avoiding the smallest non-trivial patterns. The studies in [521] are motivated by a connection to so-called *bipartite Ramsey problems*. Research in this direction was continued in [507] by Kitaev, who considered simultaneous avoidance of two or more *right angled numbered polyomino patterns* on three cells. Many other results in this direction, some of which involve certain Wilf-equivalences, can be found in [732, 731] by Spiridonov. Finally, a variation on this theme in a less general setting can be found in [449] by Harmse and Remmel.

9.8.3   Pattern-avoidance in n-dimensional objects

Kitaev and Robbins [530] generalized the concept of a pattern occurrence in permutations, words or matrices to a pattern occurrence in n-*dimensional objects*, which are basically sets of $(n + 1)$-tuples. In the case $n = 3$, a possible interpretation of such patterns in terms of bipartite graphs is given in [530]. For so-called *zero-box*

Class	Enumeration
$\mathrm{Av}^c(123, 132)$	$n(2^{n-1} - 2)$
$\mathrm{Av}^c(123, 231)$	$n(n^2 - 3n + 4)/2$
$\mathrm{Av}^c(123, 321)$	0 for $n \geq 5$
$\mathrm{Av}^c(132, 213)$	$n(2^{n-1} - n + 1)$
$\mathrm{Av}^c(132, 231)$	$n2^{n-2}$
$\mathrm{Av}^c(132, 312)$	$n(2^{n-1} - n + 1)$
$\mathrm{Av}^c(3142, 2413)$	nS_{n-1}

Table 9.3: Enumeration of some cyclically closed pattern classes.

patterns, Kitaev and Robbins [530] studied *vanishing borders* related to bipartite Ramsey problems in the case of two dimensions. Also, the maximal number of 1s in binary objects avoiding (in two different senses) a zero-box pattern is considered in [530].

9.8.4 Cyclically closed pattern classes of permutations

Albert et al. [20] considered the notion of the *cyclic closure* of a permutation pattern class, which is defined as the set of all the cyclic rotations of its permutations. Examples of finitely based classes whose cyclic closure is also finitely based are given in [20], as well as an example where the cyclic closure is not finitely based, which is the class Av(265143). The following enumerative results are presented in [20] where $\mathrm{Av}^c(p)$ denotes the cyclic closure of $\mathrm{Av}(p)$, and C_n is the n-th Catalan number.

Theorem 9.8.1. *([20])* $\mathrm{Av}^c(231)$ *is enumerated by* $n\left(C_n - C_{n-1} - C_{n-2} - \cdots - C_1\right)$ *for* $n \geq 2$.

Theorem 9.8.2. *([20])* $\mathrm{Av}^c(321)$ *is enumerated by* $n\left(C_n - 2^n + \binom{n}{2} + 2\right)$ *for* $n \geq 4$.

More enumeration results from [20] are presented in Table 9.3 that, in particular, involves Av(3142, 2413), the class of separable permutations counted by the (large) Schröder numbers S_n and discussed in Subsection 2.2.5.

9.8.5 Pattern-matching in the cycle structure of permutations

Jones and Remmel [487] studied occurrences of patterns in the cycle structures of permutations. More precisely, suppose a permutation π is written in the cycle form where the i-th cycle is $C_i = (c_{0,i}, c_{1,i}, \ldots, c_{k_i-1,i})$, $c_{0,i}$ is the smallest element in C_i, k_i is the length of C_i and the cycles are arranged by increasing smallest elements. For a consecutive pattern $\tau = \tau_1 \tau_2 \ldots \tau_k$, we say that π has a *cycle τ-match (c-τ-match)* if there is an i such that $k_i \geq k$ and an r such that $\mathrm{red}(c_{r,i} c_{r+1,i} \cdots c_{r+k-1,i}) = \tau$, where we take indices of the form $r + s$ modulo k_i. Let c-τ-$\mathrm{mch}(\pi)$ be the number of cycle τ-matches in the permutation π. For example, if $\tau = \underline{213}$ and $\pi = (1, 10, 9)(2, 3)(4, 7, 5, 8, 6)$, then $91(10)$ is a cycle τ-match in the first cycle and 758 and 647 are cycle τ-matches in the third cycle so that c-τ-$\mathrm{mch}(\pi) = 3$. Similarly, one can define cycle τ-occurrences in π. Moreover, the notion of cycle matches and cycle occurrences can be extended to sets of patterns in the obvious way. One way to study cycle pattern-matching is to use the theory of exponential structures to reduce the problem down to studying pattern matching in n-cycles, which is discussed in [487]. See [487] for enumerative results in the direction of cycle pattern matching.

9.8.6 Equivalence classes of permutations avoiding a pattern

Given an equivalence relation on permutations, one can study the corresponding equivalence classes *all* of whose members avoid a given pattern p. Úlfarsson [771] studied four equivalence relations: *conjugacy, order isomorphism, Knuth-equivalence* and *toric equivalence*. Each of these relations, with respect to pattern-avoidance, produces known counting sequences. For example, in the case of toric equivalence, there is a class of permutations that are counted by the *Euler totient function* $\varphi(n)$ defined to be the number of positive integers less than or equal to n that are *coprime* to n; a subclass of these permutations is shown in [771] to be counted by the *number-of-divisors function*. An intriguing fact discussed in [771] is that the famous *Riemann Hypothesis* can be stated in terms of pattern-avoidance:

Conjecture 9.8.3. ([771]; equivalent to the Riemann Hypothesis) The following inequality holds for all $n \geq 5041$:

$$\sum_{\pi_1 \pi_2 \cdots \pi_n \in U} \pi_1 < e^\gamma \log \log n,$$

where γ is the *Euler-Mascheroni constant* and U is the union of the *toric equivalence classes* that avoid the bivincular pattern $\genfrac{}{}{0pt}{}{\overline{123}}{2\overline{1}3}$.

9.8.7 Extended pattern-avoidance

A 0-1 matrix is said to be *extendably τ-avoiding* if it can be the upper left corner of a τ-avoiding permutation matrix. Eriksson and Linusson [368] introduced this notion and proved that the number of extendably 321-avoiding rectangles are enumerated by the *ballot numbers*. Linusson [565] studied the other five 3-patterns. The main result in [565] is the following theorem.

Theorem 9.8.4. *([565]) The number of extendably τ-avoiding $r \times k$ rectangular matrices with d dots is*

$$\binom{r+k}{d} - \binom{r+k}{d-1}$$

for $\tau = 321, 312, 231$ and 132, and

$$\sum_{x=0}^{d} \binom{r}{x}\binom{k}{x} - \binom{r+k}{d-1}$$

for $\tau = 213$ and 123.

Another result in [565] is that the Simion-Schmidt-West bijection for permutations avoiding patterns 12τ and 21τ works also for extended pattern-avoidance. As an application, Linusson [565] used the results on extended pattern-avoidance to prove a sequence of refinements on the enumeration of permutations avoiding 3-patterns.

Appendix A

Useful notions and facts

In this appendix we introduce most of the statistics on permutations and words appearing in this book, as well as several well-known sequences of numbers and certain lattice paths. Also, we state in Table A.7 several classes of restricted permutations together with references and initial values of the corresponding sequences.

A.1 Statistics on permutations and words

Definition A.1.1. In a permutation $\pi = \pi_1\pi_2\ldots\pi_n$, π_i is a *left-to-right maximum* (resp., *left-to-right minimum*) if $\pi_i \geq \pi_j$ (resp., $\pi_i \leq \pi_j$) for $1 \leq j \leq i$. Also, π_i is a *right-to-left maximum* (resp., *right-to-left minimum*) if $\pi_i \geq \pi_j$ (resp., $\pi_i \leq \pi_j$) for $i \leq j \leq n$.

Example A.1.2. If $\pi = 3152647$ then the left-to-right maxima are 3, 5, 6 and 7, the left-to-right minima are 3 and 1, the right-to-left maximum is 7, and the right-to-left minima are 7, 4, 2 and 1.

In Table A.1 we define all permutation statistics appearing in the book. The list of statistics is essentially taken from [259]. Note that many of the statistics defined make sense for words, though not all. For example, the statistic head $.i$, the position of the smallest letter, is not defined for words as we may have several instances of the smallest letter. The expression $\pi\infty$ appearing in Table A.1 means the permutation beginning with π and ending with a letter that is larger that any letter in π.

Remark A.1.3. Six statistics in Table A.1 have complex names like comp $.r$ or last $.i$, which stands for respective compositions (see Definition 1.0.12 for meaning of the trivial bijections i and r). For example, to calculate the value of the statistic

Statistic	Description
asc	# of ascents (letters followed by a larger letter = occurrences of $1\underline{2}$)
comp	# of components (ways to factor $\pi = \sigma\tau$ so that each letter in non-empty σ is smaller than any letter in τ; the empty τ gives one such factorization)
comp $.r$	# of reverse components (ways to factor $\pi = \sigma\tau$ so that each letter in non-empty σ is larger than any letter in τ; the empty τ gives one factorization)
cyc	# of cycles
dasc	# of double ascents (# of occurrences of the pattern $\underline{123}$)
ddes	# of double descents (# of occurrences of the pattern $\underline{321}$)
des	# of descents (letters followed by a smaller letter = occurrences of $2\underline{1}$)
exc	# of excedances (positions i in $\pi = \pi_1\pi_2\ldots\pi_n$ such that $\pi_i > i$)
fp	# of fixed points (positions i in $\pi = \pi_1\pi_2\ldots\pi_n$ such that $\pi_i = i$)
head	the first (leftmost) letter
head $.i$	position of the smallest letter
inv	# of inversions (pairs $i < j$ such that $\pi_i > \pi_j$ in $\pi = \pi_1\pi_2\ldots\pi_n$)
last	the last (rightmost) letter
last $.i$	position of the largest letter
ldr	length of leftmost decreasing run (largest i such that $\pi_1 > \pi_2 > \cdots > \pi_i$)
lds	length of the longest decreasing subsequence
length	# of letters
lir	length of leftmost increasing run (largest i such that $\pi_1 < \pi_2 < \cdots < \pi_i$)
lir $.i$	$= zeil.c$, largest i such that $12\cdots i$ is a subsequence in π
lis	length of the longest increasing subsequence
lmax	# of left-to-right maxima
lmin	# of left-to-right minima
maj	sum of positions in which descents occur
peak	# of peaks (positions i in π such that $\pi_{i-1} < \pi_i > \pi_{i+1}$)
peak $.i$	$\#\{ i \mid i$ is to the right of both $i - 1$ and $i + 1\}$
slmax	largest i such that $\pi_1 \geq \pi_1$, $\pi_1 \geq \pi_2$, ..., $\pi_1 \geq \pi_i$ (# of letters to the left of second left-to-right maximum in $\pi\infty$)
rank	for $\pi = \pi_1\pi_2\ldots\pi_n$, largest k such that $\pi_i > k$ for all $i \leq k$ (see [361])
rdr	$= lir.r$, length of the rightmost decreasing run
rmax	# of right-to-left maxima
rmin	# of right-to-left minima
rir	$= ldr.r$, length of the rightmost increasing run
valley	# of valleys (positions i in π such that $\pi_{i-1} > \pi_i < \pi_{i+1}$)
valley $.i$	$\#\{ i \mid i$ is to the left of both $i - 1$ and $i + 1\}$
zeil	length of longest subsequence $n(n - 1)\cdots i$ in π of length n (see [815])

Table A.1: Definitions of permutation statistics appearing in the book.

For $\pi = 4523176$	For $\pi = 3265714$
asc$(\pi) = 3$, comp$(\pi) = 3$, des$(\pi) = 3$	asc$(\pi) = 3$, comp$(\pi) = 2$, des$(\pi) = 3$
comp $.r(\pi) = 1$, exc$(\pi) = 3$, fp$(\pi) = 0$	comp $.r(\pi) = 1$, exc$(\pi) = 4$, fp$(\pi) = 1$
head$(\pi) = 4$, head $.i(\pi) = 5$, last$(\pi) = 6$	head$(\pi) = 3$, head $.i(\pi) = 6$, last$(\pi) = 4$
last $.i(\pi) = 6$, ldr$(\pi) = 1$, lds$(\pi) = 3$	last $.i(\pi) = 5$, ldr$(\pi) = 2$, lds$(\pi) = 3$
lir$(\pi) = 2$, lir $.i(\pi) = 1$, slmax$(\pi) = 1$	lir$(\pi) = 1$, lir $.i(\pi) = 1$, slmax$(\pi) = 2$
lmax$(\pi) = 3$, lmin$(\pi) = 3$, peak$(\pi) = 3$	lmax$(\pi) = 3$, lmin$(\pi) = 3$, peak$(\pi) = 2$
peak $.i(\pi) = 1$, lis$(\pi) = 3$, rdr$(\pi) = 2$	peak $.i(\pi) = 2$, lis$(\pi) = 3$, rdr$(\pi) = 1$
rmax$(\pi) = 2$, rmin$(\pi) = 2$, cyc$(\pi) = 2$	rmax$(\pi) = 2$, rmin$(\pi) = 2$, cyc$(\pi) = 3$
rir$(\pi) = 1$, valley $.i(\pi) = 2$, zeil$(\pi) = 2$	rir$(\pi) = 2$, valley $.i(\pi) = 2$, zeil$(\pi) = 1$
dasc$(\pi) = 0$, ddes$(\pi) = 0$, maj$(\pi) = 12$	dasc$(\pi) = 0$, ddes$(\pi) = 0$, maj$(\pi) = 9$
inv$(\pi) = 9$, valley$(\pi) = 2$, rank$(\pi) = 2$	inv$(\pi) = 10$, valley$(\pi) = 3$, rank$(\pi) = 1$

Table A.2: Values of permutation statistics introduced in Table A.1 on the permutations 4523176 and 3265714 (length$(\pi) = 7$ in both cases).

comp $.r$ we can either follow the direct description of it in Table A.1, or we can take a permutation, apply to it the reverse operation r and calculate the value of the statistic comp on the obtained permutation. Similarly, a way to calculate last $.i$ is to take a permutation, apply to it the inverse operation i, and to calculate the value of the statistic last on the obtained permutation.

Table A.2 provides examples supporting the definitions in Table A.1.

Remark A.1.4. As noted, not all permutation statistics can be translated into statistics for words. Similarly, not all word statistics can be translated into permutation statistics. For example, the statistic lev, which is the number of levels, is defined on words as the number of times two equal letters stay next to each other (e.g., lev(3114244333) = 4), is of interest in words, but has the trivial value zero on permutations.

A.2 Numbers and lattice paths of interest

In this section we collect definitions and some basic facts on selected numbers appearing in this book. For more information on particular objects of interest discussed below, one should consult other sources, such as standard texts in combinatorics (e.g., [743, 745]).

A.2.1   Numbers involved

The *Fibonacci numbers* are named after Leonardo of Pisa (c. 1170 - c. 1250), who was known as Fibonacci. The Fibonacci numbers were known in ancient India, and there are many applications of these numbers. The n-th Fibonacci number is defined recursively as follows. $F_0 = 0$, $F_1 = 1$, and for $n \geq 2$, $F_n = F_{n-1} + F_{n-2}$. These numbers satisfy a countless number of identities. Letting

$$\varphi = \frac{1 + \sqrt{5}}{2} \approx 1.6180339887 \cdots$$

denote the *golden ratio*, the n-th Fibonacci number is given by

$$F_n = \frac{\varphi^n - (1 - \varphi)^n}{\sqrt{5}} = \frac{\varphi^n - (-1/\varphi)^n}{\sqrt{5}}.$$

Thus, asymptotically $F_n \sim \varphi^n/\sqrt{5}$. The generating function for the Fibonacci numbers is given by

$$F(x) = \sum_{n \geq 0} F_n x^n = \frac{x}{1 - x - x^2}.$$

The first few Fibonacci numbers, for $n = 0, 1, 2, \ldots$, are

$$0, 1, 1, 2, 3, 5, 8, 13, 21, 34, 55, 89, 144, \ldots.$$

The *Catalan numbers* are named after the Belgian mathematician Eugéne Charles Catalan (1814–1894) and they appear in many counting problems. Catalan defined these numbers in [235] in 1838. For $n \geq 0$, the n-th Catalan number is given by

$$C_n = \frac{1}{n + 1} \binom{2n}{n} = \frac{(2n)!}{(n + 1)!n!}.$$

A recursive way to define the Catalan numbers is as follows: $C_0 = 1$ and $C_{n+1} = \sum_{i=0}^{n} C_i C_{n-i}$ for $n \geq 0$, and asymptotically, the Catalan numbers grow as

$$C_n \sim \frac{4^n}{n^{3/2}\sqrt{\pi}}.$$

The generating function for the Catalan numbers is

$$C(x) = \sum_{n \geq 0} C_n x^n = \frac{1 - \sqrt{1 - 4x}}{2x}.$$

The first few Catalan numbers, for $n = 0, 1, 2, \ldots$, are

$$1, 1, 2, 5, 14, 42, 132, 429, 1430, 4862, 16796, 58786, 208012, 742900, \ldots.$$

We refer the reader to `www-math.mit.edu/~rstan/ec/catadd.pdf` for an impressive collection by Stanley of different structures counted by the Catalan numbers.

The *Bell numbers* are named after Eric Temple Bell (1883-1960). The n-th Bell number gives the number of *partitions* of a set with n elements. For example, there are five partitions for the 3-element set $\{a, b, c\}$: $\{\{a\}, \{b\}, \{c\}\}$, $\{\{a\}, \{b, c\}\}$, $\{\{b\}, \{a, c\}\}$, $\{\{c\}, \{a, b\}\}$, and $\{\{a, b, c\}\}$. The Bell numbers can be defined recursively as follows: $B_0 = B_1 = 1$ and

$$B_{n+1} = \sum_{k=0}^{n} \binom{n}{k} B_k.$$

The exponential generating function of the Bell numbers is

$$B(x) = \sum_{n \geq 0} \frac{B_n}{n!} x^n = e^{e^x - 1}.$$

The (ordinary) generating function of the Bell numbers can be written as the following continued fraction:

$$\sum_{n \geq 0} B_n x^n = \cfrac{1}{1 - 1 \cdot x - \cfrac{1 \cdot x^2}{1 - 2 \cdot x - \cfrac{2 \cdot x^2}{1 - 3 \cdot x - \cfrac{3 \cdot x^2}{\ddots}}}}.$$

Asymptotically, these numbers grow as

$$B_n \sim \frac{1}{\sqrt{n}} [\lambda(n)]^{n+\frac{1}{2}} e^{\lambda(n) - n - 1},$$

where $\lambda(n) = e^{W(n)} = \frac{n}{W(n)}$ and W is the *Lambert W function*. The first few Bell numbers, for $n = 0, 1, 2, \ldots$, are

$$1, 1, 2, 5, 15, 52, 203, 877, 4140, 21147, 115975, \ldots.$$

Stirling numbers of the first kind $s(n, k)$ are the coefficients in the following expansion

$$x(x - 1)(x - 2) \cdots (x - n + 1) = \sum_{k=0}^{n} s(n, k) x^k.$$

Unsigned Stirling numbers of the first kind

$$\begin{bmatrix} n \\ k \end{bmatrix} = |s(n, k)| = (-1)^{n-k} s(n, k)$$

count the number of permutations of n elements with k disjoint cycles. The unsigned
Stirling numbers arise as coefficients in the following expansion:

$$x(x+1)\cdots(x+n-1) = \sum_{k=0}^{n}\begin{bmatrix} n \\ k \end{bmatrix} x^k.$$

See Table A.3 for the values of $s(n,k)$ for $0 \le n \le 7$ and $0 \le k \le 7$. The following
recurrence relation is well-known:

$$\begin{bmatrix} n \\ k \end{bmatrix} = (n-1)\begin{bmatrix} n-1 \\ k \end{bmatrix} + \begin{bmatrix} n-1 \\ k-1 \end{bmatrix}$$

for $k > 0$, with the initial conditions $\begin{bmatrix} 0 \\ 0 \end{bmatrix} = 1$ and $\begin{bmatrix} n \\ 0 \end{bmatrix} = \begin{bmatrix} 0 \\ n \end{bmatrix} = 0.$

$n\backslash k$	0	1	2	3	4	5	6	7
0	1							
1	0	1						
2	0	-1	1					
3	0	2	-3	1				
4	0	-6	11	-6	1			
5	0	24	-50	35	-10	1		
6	0	-120	274	-225	85	-15	1	
7	0	720	-1764	1624	-735	175	-21	1

Table A.3: The initial values for the Stirling numbers of the first kind.

A *Stirling number of the second kind* $S(n,k)$ is the number of ways to partition
a set of n objects into k groups. For example, given $k = 2$ and the 3-element set
$\{a, b, c\}$ there are three such partitions $\{\{a\}, \{b, c\}\}$, $\{\{b\}, \{a, c\}\}$, and $\{\{c\}, \{a, b\}\}$.
The Stirling numbers of the second kind can be calculated using the following for-
mula:

$$S(n,k) = \frac{1}{k!}\sum_{j=0}^{k}(-1)^{k-j}\binom{k}{j}j^n.$$

See Table A.4 for the values of $S(n,k)$ for $0 \le n \le 7$ and $0 \le k \le 7$. It follows
from definitions that $B_n = \sum_{k=0}^{n} S(n,k)$ providing a well-known relation between
the Bell numbers and the Stirling numbers of the second kind.

The *Eulerian number* $A(n,m)$ (also denoted in some sources by $E(n,m)$ or
$\left\langle \begin{matrix} n \\ m \end{matrix} \right\rangle$) is the number of n-permutations with m ascents (elements followed by

$n\backslash k$	0	1	2	3	4	5	6	7
0	1							
1	0	1						
2	0	1	1					
3	0	1	3	1				
4	0	1	7	6	1			
5	0	1	15	25	10	1		
6	0	1	31	90	65	15	1	
7	0	1	63	301	350	140	21	1

Table A.4: The initial values for the Stirling numbers of the second kind.

larger elements). These numbers are coefficients of the *Eulerian polynomials*

$$A_n(x) = \sum_{m=0}^{n} A(n, m) x^{n-m}.$$

$A_n(x)$ appears as the numerator in an expression for the generating function of the sequence 1^n, 2^n, 3^n, ...:

$$\sum_{k=1}^{\infty} k^n x^k = \frac{\sum_{m=0}^{n} A(n, m) x^{m+1}}{(1 - x)^{n+1}}.$$

The Eulerian numbers can be calculated using the recursion

$$A(n, m) = (n - m)A(n - 1, m - 1) + (m + 1)A(n - 1, m)$$

with $A(0, 0) = 1$, or using a closed-form expression

$$A(n, m) = \sum_{k=0}^{m} (-1)^k \binom{n + 1}{k} (m + 1 - k)^n.$$

See Table A.5 for the values of $A(n, m)$ for $0 \leq n \leq 7$ and $0 \leq m \leq 7$.

Some of the properties of the Eulerian numbers, starting from two straightforward ones, and one more related to the *Bernoulli number* B_{n+1}, are as follows:

$$A(n, m) = A(n, n - m - 1);$$

$$\sum_{m=0}^{n-1} A(n, m) = n!;$$

$n \backslash m$	0	1	2	3	4	5	6	7
0	1							
1	1							
2	1	1						
3	1	4	1					
4	1	11	11	1				
5	1	26	66	26	1			
6	1	57	302	302	57	1		
7	1	120	1191	2416	1191	120	1	

Table A.5: The initial values for the Eulerian numbers.

$$\sum_{m=0}^{n-1} (-1)^m A(n,m) = \frac{2^{n+1}(2^{n+1} - 1)B_{n+1}}{n+1}.$$

The *Motzkin numbers* are named after Theodore Motzkin (1908–1970). There are at least 14 equivalent different ways to combinatorially define the Motzkin numbers. One way is to do it through the Motzkin paths defined in Subsection A.2.2 below. Another way is as follows. The n-th Motzkin number M_n is the number of different ways of drawing non-intersecting chords between n points on a circle. The generating function for the Motzkin numbers is

$$M(x) = \sum_{n \geq 0} M_n x^n = \frac{1 - x - \sqrt{1 - 2x - 3x^2}}{2x^2} = \cfrac{1}{1 - x - \cfrac{x^2}{1 - x - \cfrac{x^2}{1 - x - \cfrac{x^2}{1 - x - \ddots}}}}$$

and these numbers begin as follows:

$$1, 1, 2, 4, 9, 21, 51, 127, 323, 835, 2188, 5798, 15511, 41835, 113634, \ldots.$$

The *large Schröder numbers*, also called *Schröder numbers*, are defined by the following recurrence relation:

$$S_n = S_{n-1} + \sum_{k=0}^{n-1} S_k S_{n-1-k},$$

where $S_0 = 1$. These numbers have the following generating function:

$$\frac{1 - x - \sqrt{1 - 6x + x^2}}{2x}$$

and they begin, for $n = 0, 1, 2, \ldots$ as $1, 2, 6, 22, 90, 394, \ldots$. The *small Schröder numbers* have the following generating function

$$\frac{1 + x - \sqrt{1 - 6x + x^2}}{4x}$$

and they begin as $1, 1, 3, 11, 45, 197, \ldots$.

It was discovered by David Hough and noted by Stanley [740] that the small Schröder numbers were apparently known to Hipparchus over 2100 years ago. See Problem 6.39 in [745] for a list of 19 structures enumerated by the Schröder numbers, as well as for references for these numbers. We also refer to [717] and references therein, for discussions of several combinatorial objects counted by Schröder numbers.

For $n = 1, 2, 3, \ldots$ and $1 \leq k \leq n$, the *Narayana numbers* $N(n, k)$ are defined as follows:

$$N(n, k) = \frac{1}{n} \binom{n}{k} \binom{n}{k-1}.$$

These numbers are named after T. V. Narayana (1930–1987), a mathematician from India. It is known that $N(n, 1) + N(n, 2) + \cdots + N(n, n) = C_n$ linking the Narayana numbers to the Catalan numbers. The first eight rows of the *Narayana triangle* (presenting the non-zero numbers $N(n, k)$) are as in Table A.6.

$n \backslash k$	1	2	3	4	5	6	7	8
1	1							
2	1	1						
3	1	3	1					
4	1	6	6	1				
5	1	10	20	10	1			
6	1	15	50	50	15	1		
7	1	21	105	175	105	21	1	
8	1	28	196	490	490	196	28	1

Table A.6: The first eight rows of the Narayana triangle.

The *Padovan numbers* are named after Richard Padovan who attributed their discovery to Dutch architect Hans van der Lann in his 1994 essay *Dom. Hans van der Laan : Modern Primitive*. These numbers are defined recursively as follows: $P(0) = P(1) = P(2) = 1$ and $P(n) = P(n-2) + P(n-3)$. The generating function for the Padovan numbers is

$$P(x) = \sum_{n \geq 0} P(n) x^n = \frac{1 + x}{1 - x^2 - x^3}.$$

The initial values of the Padovan numbers are as follows:

$$1, 1, 1, 2, 2, 3, 4, 5, 7, 9, 12, 16, 21, 28, 37, 49, 65, 86, \ldots.$$

The n-th *harmonic number* is the sum of reciprocals of the first n natural numbers:

$$H_n = 1 + \frac{1}{2} + \frac{1}{3} + \cdots + \frac{1}{n}.$$

Harmonic numbers were studied in antiquity and are important, e.g., in various branches of number theory. The following integral representation of H_n was given by Euler:

$$H_n = \int_0^1 \frac{1 - x^n}{1 - x} dx.$$

The generating function for the harmonic numbers is

$$\sum_{n=1}^{\infty} H_n x^n = \frac{-\ln(1-x)}{1-x}.$$

Asymptotically, H_n is given by $\ln(n)$.

The study of *Genocchi numbers* goes back to Euler. The Genocchi numbers can be defined by the following generating function.

(A.1)
$$\frac{2t}{e^t + 1} = t + \sum_{n \geq 1} (-1)^n G_{2n} \frac{t^{2n}}{(2n)!}.$$

These numbers were studied intensively during the last three decades (see, e.g., [349] and references therein). Dumont [328] showed that the Genocchi number G_{2n} is the number of permutations $\sigma = \sigma_1 \sigma_2 \cdots \sigma_{2n+1}$ in S_{2n+1} such that

$$\sigma_i < \sigma_{i+1} \quad \text{if } \sigma_i \text{ is odd,}$$
$$\sigma_i > \sigma_{i+1} \quad \text{if } \sigma_i \text{ is even.}$$

The first few Genocchi numbers are $1, 1, 3, 17, 155, 2073, \ldots.$

A.2.2 Lattice paths involved

A *lattice path* for purposes of this book can be defined as a path in $\mathbb{R}^2$ all of whose coordinates are integers.

A *Dyck path of length $2n$* is a lattice path from $(0,0)$ to $(2n,0)$ with steps $u = (1,1)$ ("u" stands for "up") and $d = (1,-1)$ ("d" stands for "down") that never goes below the x-axis. It is easy to see that if D denotes the set of all Dyck paths (of arbitrary length) then one has the following relation for D:

(A.2)
$$D = 1 + udD + uDdD$$

where 1 stands for a single point $(0,0)$ which is a Dyck path by definition. In Figure A.1 one can find all Dyck paths of length 6.

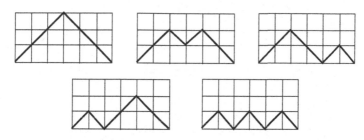

Figure A.1: All Dyck paths of length 6.

The number of Dyck paths is given by the Catalan numbers.

A *Motzkin path of length n* is a lattice path from $(0,0)$ to $(n,0)$ with steps $u = (1,1)$, $d - (1,-1)$ and $h = (1,0)$ ("h" stands for "horizontal") that never goes below the x-axis. It is easy to see that if M denotes the set of all Motzkin paths (of arbitrary length) then one has the following relation for M:

(A.3)
$$S = 1 + hM + uMdM$$

where 1 stands for a single point $(0,0)$ which is a Motzkin path by definition. In Figure A.2 one can find all Motzkin paths of length 4.

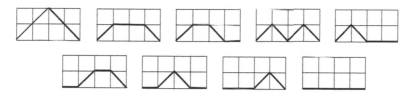

Figure A.2: All Motzkin paths of length 4.

The number of Motzkin paths is given by the Motzkin numbers.

A *Schröder path of length $2n$* is a lattice path from $(0,0)$ to $(2n,0)$ with steps $u = (1,1)$, $d = (1,-1)$ and $h = (1,0)$ that never goes below the x-axis and in which runs of h-steps must be of even length. It is easy to see that if S denotes the set of all Schröder paths (of arbitrary length) then one has the following relation for S:

(A.4) $$S = 1 + hhS + uSdS$$

where 1 stands for a single point (0,0) which is a Schröder path by definition. In Figure A.3 one can find all Schröder paths of length 4.

Figure A.3: All Schröder paths of length 4.

The number of Schröder paths is given by the large Schröder numbers. These paths were introduced by Bonin et al. [152] in 1993.

A.3   Pattern-avoiding permutations

In this book we meet many classes of pattern-avoiding permutations. Some of the classes have attracted more attention in the literature, other less. In Table A.7 we collect selected classes of pattern-avoiding permutations that have attracted considerable attention in the literature. The table provides most of the relevant references and initial values for corresponding sequences.

Permutations	Patterns	Initial values
Baxter [4, 98, 151] [167, 158, 272, 221, 257] [379, 250, 410, 326, 419] [436, 620, 581, 787]	2$\underline{41}$3, 3$\underline{14}$2	$1, 2, 6, 22, 92, 422, 2074, 10754, \ldots$
forest-like [160]	1324, 21$\overline{3}$54 = 1324, 2$\underline{14}$3	$1, 2, 6, 22, 89, 379, 1661, 7405, \ldots$
layered [140, 136] [132, 616, 661]	231, 312	$1, 2, 4, 8, 16, 32, 64, 128, \ldots$
Motzkin [359]	132, 1$\underline{23}$	Motzkin numbers $1, 2, 4, 9, 21, 51, 127, 323, \ldots$
deque-restricted [540, 801, 73] Schröder [73, 546, 56, 679] separable [27, 338] [155, 469, 546, 734, 798]	1324, 2314 4231, 4132 2413, 3142	large Schröder numbers $1, 2, 6, 22, 90, 394, 1806, 8558, \ldots$
simple [25, 48, 22, 183] [188, 46, 94, 186, 189] [187, 634, 714]	no	$1, 2, 0, 2, 6, 46, 338, 2926, \ldots$
smooth [706, 551, 807] [745, 130, 160, 444]	1324, 2143	$1, 2, 6, 22, 88, 366, 1552, 6652, \ldots$
vexillary [554, 742, 66, 764, 799]	2143	$1, 2, 6, 23, 103, 513, 2761, 15767, \ldots$
2-stack sortable [799] [815, 323, 423, 156] non-separable [324, 262]	2341, 3$\overline{5}$241 2413, 41$\overline{3}$52 = 2413, 3$\underline{14}$2	$1, 2, 6, 22, 91, 408, 1938, 9614, \ldots$

Table A.7: Some pattern-avoiding classes of permutations appearing in the book together with several references and initial values of the corresponding sequences for $1 \leq n \leq 8$. Note that one may need to apply trivial bijections to the prohibited patterns in order to get the patterns appearing under a particular reference.

Appendix B

Some algebraic background

In this appendix we provide just a very few basic definitions related to the book, referring to more substantial sources on algebra related to combinatorics like [422] by Goulden and Jackson and [743] by Stanley, if needed. We discuss below the ring of formal power series (related to generating functions — an important notion for us) and Chebyshev polynomials of the second kind which appear several times in the book.

B.1 The ring of formal power series

Definition B.1.1. A *ring* is a set R equipped with two binary operations $+$ from $R \times R$ to R and $\cdot$ from $R \times R$ to R, where $\times$ denotes the Cartesian product, called *addition* and *multiplication*, respectively. The set and two operations, $(R, +, \cdot)$, must satisfy the following *ring axioms*:

- $(R, +)$ is an *abelian group* under addition:

    1. **Closure under addition.** For all a, b in R, the result of the operation $a + b$ is also in R.

    2. **Associativity of addition.** For all a, b, c in R, we have $(a + b) + c = a + (b + c)$.

    3. **Existence of additive identity.** There exists an element 0 in R, such that for all elements a in R, we have $0 + a = a + 0 = a$.

    4. **Existence of additive inverse.** For each a in R, there exists an element b in R such that $a + b = b + a = 0$.

    5. **Commutativity of addition.** For all a, b in R, we have $a + b = b + a$.

401

- $(R, \cdot)$ is required to be a *monoid* under multiplication:

  1. **Closure under multiplication.** For all a, b in R, the result of the operation $a \cdot b$ is also in R.
  2. **Associativity of multiplication.** For all a, b, c in R, we have $(a \cdot b) \cdot c = a \cdot (b \cdot c)$.
  3. **Existence of multiplicative identity.** There exists an element 1 in R, such that for all elements a in R, we have $1 \cdot a = a \cdot 1 = a$.

- The distributive laws:

  1. For all a, b and c in R, we have $a \cdot (b + c) = (a \cdot b) + (a \cdot c)$.
  2. For all a, b and c in R, we have $(a + b) \cdot c = (a \cdot c) + (b \cdot c)$.

Definition B.1.2. A ring R is *commutative* if the multiplication operation is commutative, that is, if for all a, b in R, $a \cdot b = b \cdot a$.

In what follows, the term "formal" is used to indicate that convergence is not considered in the context, and thus formal power series need not represent functions on R.

Definition B.1.3. Let R be a ring. We define the set $R[[x]]$ of *formal power series* in the indeterminate x with coefficients from R to be all formal sums $\sum_{n \geq 0} a_n x^n$. Further, we define addition and multiplication on formal power series, respectively, as follows:

- $\left(\sum_{n \geq 0} a_n x^n \right) + \left(\sum_{n \geq 0} b_n x^n \right) = \sum_{n \geq 0} (a_n + b_n) x^n$;

- $\left(\sum_{n \geq 0} a_n x^n \right) \cdot \left(\sum_{n \geq 0} b_n x^n \right) = \sum_{n \geq 0} \left(\sum_{i+j=n} (a_i b_j) \right) x^n$.

It is easy to prove that if R is commutative then $R[[x]]$ is commutative. In what follows, we assume R is commutative.

If $f(x)$ and $g(x)$ are elements of $R[[x]]$ satisfying $f(x)g(x) = 1 = 1 + 0x + 0x^2 + \cdots$, then we write $g(x) = f(x)^{-1}$. Suppose $f(x) = \sum_{n \geq 0} a_n x^n$. Then it is easy to see that $f(x)^{-1}$ exists (in which case it is unique) if and only if $a_0 \neq 0$.

For $f(x)$ as above, one can define the *formal derivative* of $f(x)$ *with respect to* x as follows:

$$D_x f(x) = f'(x) = \frac{df}{dx} := \sum_{n \geq 0} (n + 1) a_{i+1} x^n.$$

The *product rule, chain rule*, and *Leibniz's theorem* hold for the formal derivative and can be proved by comparing coefficients.

One can also define a *formal integral*, which we do not do here.

B.2 Chebyshev polynomials of the second kind

Definition B.2.1. The *Chebyshev polynomials of the second kind* are defined by the following recurrence relation

$$U_{n+1}(x) = 2xU_n(x) - U_{n-1}(x)$$

with $U_0(x) = 1$ and $U_1(x) = 2x$. These polynomials can also be defined by the trigonometric identity:

$$U_n(\cos(\nu)) = \frac{\sin((n+1)\nu)}{\sin(\nu)}.$$

An example of a generating function for $U_n(x)$ is as follows:

$$\sum_{n \geq 0} U_n(x)t^n = \frac{1}{1 - 2tx + t^2}.$$

An example of an explicit formula for $U_n(x)$ is as follows:

$$U_n(x) = \frac{(x + \sqrt{x^2 - 1})^{n+1} - (x - \sqrt{x^2 - 1})^{n+1}}{2\sqrt{x^2 - 1}}.$$

The Chebyshev polynomials are named after Pafnuty Chebyshev.

Bibliography

[1] M. Abramson. A note on permutations with fixed pattern. *J. Combin. Theory Ser. A*, 19(2):237–239, 1975.

[2] M. Abramson. Permutations related to secant, tangent and Eulerian numbers. *Canad. Math. Bull.*, 22(3):281–291, 1979.

[3] M. Abramson and W. O. J. Moser. Permutations without rising or falling *w*-sequences. *Ann. Math. Statist.*, 38:1245–1254, 1967.

[4] E. Ackerman, G. Barequet, and R. Y. Pinter. A bijection between permutations and floorplans, and its applications. *Discrete Appl. Math.*, 154(12):1674–1684, 2006.

[5] R. M. Adin, F. Brenti, and Y. Roichman. Descent numbers and major indices for the hyperoctahedral group. *Adv. in Appl. Math.*, 27(2-3):210–224, 2001. Special issue in honor of Dominique Foata's 65th birthday (Philadelphia, PA, 2000).

[6] R. M. Adin, F. Brenti, and Y. Roichman. Descent representations and multivariate statistics. *Trans. Amer. Math. Soc.*, 357(8):3051–3082 (electronic), 2005.

[7] R. M. Adin, F. Brenti, and Y. Roichman. Equi-distribution over descent classes of the hyperoctahedral group. *J. Combin. Theory Ser. A*, 113(6):917–933, 2006.

[8] R. M. Adin and Y. Roichman. Descent functions and random Young tableaux. *Combin. Probab. Comput.*, 10(3):187–201, 2001.

[9] R. M. Adin and Y. Roichman. Shape avoiding permutations. *J. Combin. Theory Ser. A*, 97(1):162–176, 2002.

[10] R. M. Adin and Y. Roichman. Equidistribution and sign-balance on 321-avoiding permutations. *Sém. Lothar. Combin.*, 51:Art. B51d, 14 pp. (electronic), 2004/05.

[11] S. G. Akl. A new algorithm for generating derangements. *BIT*, 20:2–7, 1980.

[12] M. H. Albert. On the length of the longest subsequence avoiding an arbitrary pattern in a random permutation. *Random Structures Algorithms*, 31(2):227–238, 2007.

[13] M. H. Albert. Young classes of permutations. *arXiv:1008.4615v1 [math.CO]*, 2010.

[14] M. H. Albert, R. E. L. Aldred, M. D. Atkinson, C. Handley, and D. Holton. Permutations of a multiset avoiding permutations of length 3. *European J. Combin.*, 22(8):1021–1031, 2001.

[15] M. H. Albert, R. E. L. Aldred, M. D. Atkinson, and D. A. Holton. Algorithms for pattern involvement in permutations. In *Algorithms and Computation (Christchurch, 2001)*, volume 2223 of *Lecture Notes in Comput. Sci.*, pages 355–366. Springer, Berlin, 2001.

[16] M. H. Albert, R. E. L. Aldred, M. D. Atkinson, H. P. van Ditmarsch, B. D. Handley, C. C. Handley, and J. Opatrny. Longest subsequences in permutations. *Australas. J. Combin.*, 28:225–238, 2003.

[17] M. H. Albert, R. E. L. Aldred, M. D. Atkinson, H. P. van Ditmarsch, C. C. Handley, and D. A. Holton. Restricted permutations and queue jumping. *Discrete Math.*, 287(1-3):129–133, 2004.

[18] M. H. Albert, R. E. L. Aldred, M. D. Atkinson, H. P. van Ditmarsch, C. C. Handley, D. A. Holton, and D. J. McCaughan. Sorting classes. *Electron. J. Combin.*, 12:Research Paper 31, 25 pp. (electronic), 2005.

[19] M. H. Albert, R. E. L. Aldred, M. D. Atkinson, H. P. van Ditmarsch, C. C. Handley, D. A. Holton, and D. J. McCaughan. Compositions of pattern restricted sets of permutations. *Australas. J. Combin.*, 37:43–56, 2007.

[20] M. H. Albert, R. E. L. Aldred, M. D. Atkinson, H. P. van Ditmarsch, C. C. Handley, D. A. Holton, D. J. McCaughan, and C. W. Monteith. Cyclically closed pattern classes of permutations. *Australas. J. Combin.*, 38:87–100, 2007.

[21] M. H. Albert and M. D. Atkinson. Sorting with a forklift. *Electron. J. Combin.*, 9(2):Research paper 9, 23 pp. (electronic), 2002/03. Permutation patterns (Otago, 2003).

[22] M. H. Albert and M. D. Atkinson. Simple permutations and pattern restricted permutations. *Discrete Math.*, 300(1-3):1–15, 2005.

[23] M. H. Albert, M. D. Atkinson, and R. Brignall. Permutation classes of polynomial growth. *Ann. Comb.*, 11(3-4):249–264, 2007.

[24] M. H. Albert, M. D. Atkinson, C. C. Handley, D. A. Holton, and W. Stromquist. On packing densities of permutations. *Electron. J. Combin.*, 9(1):Research Paper 5, 20 pp. (electronic), 2002.

[25] M. H. Albert, M. D. Atkinson, and M. Klazar. The enumeration of simple permutations. *J. Integer Seq.*, 6(4):Article 03.4.4, 18 pp. (electronic), 2003.

[26] M. H. Albert, M. D. Atkinson, and N. Ruškuc. Regular closed sets of permutations. *Theoret. Comput. Sci.*, 306(1-3):85–100, 2003.

[27] M. H. Albert, M. D. Atkinson, and V. R. Vatter. Subclasses of the separable permutations. *arXiv:1007.1014v1 [math.CO]*, 2010.

[28] M. H. Albert, M. Coleman, R. Flynn, and I. Leader. Permutations containing many patterns. *Ann. Comb.*, 11(3-4):265–270, 2007.

[29] M. H. Albert, M. Elder, A. Rechnitzer, P. Westcott, and M. Zabrocki. On the Stanley-Wilf limit of 4231-avoiding permutations and a conjecture of Arratia. *Adv. in Appl. Math.*, 36(2):96–105, 2006.

[30] M. H. Albert, S. Linton, and N. Ruškuc. The insertion encoding of permutations. *Electron. J. Combin.*, 12:Research Paper 47, 31 pp. (electronic), 2005.

[31] M. H. Albert and S. A. Linton. Growing at a perfect speed. *Combin. Probab. Comput.*, 18(3):301–308, 2009.

[32] M. H. Albert and J. West. Universal cycles for permutation classes. *DMTCS proc. AK*, pages 39–50, 2009.

[33] D. Aldous and P. Diaconis. Longest increasing subsequences: from patience sorting to the Baik-Deift-Johansson theorem. *Bull. Amer. Math. Soc. (N.S.)*, 36(4):413–432, 1999.

[34] R. E. L. Aldred, M. D. Atkinson, and D. J. McCaughan. Avoiding consecutive patterns in permutations. *Adv. in Appl. Math.*, 45(3):449–461, 2010.

[35] R. E. L. Aldred, M. D. Atkinson, H. P. van Ditmarsch, C. C. Handley, D. A. Holton, and D. J. McCaughan. Permuting machines and priority queues. *Theoret. Comput. Sci.*, 349(3):309–317, 2005.

[36] J.-P. Allouche and J. Shallit. *Automatic sequences: theory, applications, generalizations*. Cambridge University Press, Cambridge, UK, 2003.

[37] N. Alon and E. Friedgut. On the number of permutations avoiding a given pattern. *J. Combin. Theory Ser. A*, 89(1):133–140, 2000.

[38] J. M. Amigó, S. Elizalde, and M. B. Kennel. Forbidden patterns and shift systems. *J. Combin. Theory Ser. A*, 115(3):485–504, 2008.

[39] J. M. Amigó, S. Elizalde, and M. B. Kennel. Pattern avoidance in dynamical systems. In *Sixth Conference on Discrete Mathematics and Computer Science (Spanish)*, pages 53–60. Univ. Lleida, Lleida, 2008.

[40] D. J. Anick. On the homology of associative algebras. *Trans. Amer. Math. Soc.*, 296(2):641–659, 1986.

[41] R. Arratia. On the Stanley-Wilf conjecture for the number of permutations avoiding a given pattern. *Electron. J. Combin.*, 6:Note, N1, 4 pp. (electronic), 1999.

[42] A. Asinowski, G. Barequet, M. Bousquet-Mélou, T. Mansour, and R. Pinter. Orders induced by segments in floorplan partitions and (2-14-3,3-41-2)-avoiding permutations. *arXiv:1011.1889v1 [math.CO]*, 2010.

[43] A. Asinowski and T. Mansour. Dyck paths with coloured ascents. *European J. Combin.*, 29(5):1262–1279, 2008.

[44] M. D. Atkinson. Generalized stack permutations. *Combin. Probab. Comput.*, 7(3):239–246, 1998.

[45] M. D. Atkinson. Permutations which are the union of an increasing and a decreasing subsequence. *Electron. J. Combin.*, 5:Research paper 6, 13 pp. (electronic), 1998.

[46] M. D. Atkinson. Restricted permutations. *Discrete Math.*, 195(1-3):27–38, 1999.

[47] M. D. Atkinson. Some equinumerous pattern-avoiding classes of permutations. *Discrete Math. Theor. Comput. Sci.*, 7(1):71–73 (electronic), 2005.

[48] M. D. Atkinson. On permutation pattern classes with two restrictions only. *Ann. Comb.*, 11(3-4):271–283, 2007.

[49] M. D. Atkinson and R. Beals. Priority queues and permutations. *SIAM J. Comput.*, 23(6):1225–1230, 1994.

[50] M. D. Atkinson and R. Beals. Permuting mechanisms and closed classes of permutations. In *Combinatorics, computation & logic '99 (Auckland)*, volume 21 of *Aust. Comput. Sci. Commun.*, pages 117–127. Springer, Singapore, 1999.

[51] M. D. Atkinson and R. Beals. Permutation involvement and groups. *Q. J. Math.*, 52(4):415–421, 2001.

[52] M. D. Atkinson, M. J. Livesey, and D. Tulley. Permutations generated by token passing in graphs. *Theoret. Comput. Sci.*, 178(1-2):103–118, 1997.

[53] M. D. Atkinson, M. M. Murphy, and N. Ruškuc. Partially well-ordered closed sets of permutations. *Order*, 19(2):101–113, 2002.

[54] M. D. Atkinson, M. M. Murphy, and N. Ruškuc. Sorting with two ordered stacks in series. *Theoret. Comput. Sci.*, 289(1):205–223, 2002.

[55] M. D. Atkinson, M. M. Murphy, and N. Ruškuc. Pattern avoidance classes and subpermutations. *Electron. J. Combin.*, 12:Research Paper 60, 18 pp. (electronic), 2005.

[56] M. D. Atkinson and J.-R. Sack. Pop-stacks in parallel. *Inform. Process. Lett.*, 70(2):63–67, 1999.

[57] M. D. Atkinson, B. E. Sagan, and V. R. Vatter. Counting (3+1) - avoiding permutations. *arXiv:1102.5568v1 [math.CO]*, 2011.

[58] M. D. Atkinson and T. Stitt. Restricted permutations and the wreath product. *Discrete Math.*, 259(1-3):19–36, 2002.

[59] M. D. Atkinson and M. Thiyagarajah. The permutational power of a priority queue. *BIT*, 33(1):2–6, 1993.

[60] S. Avgustinovich, A. Glen, B. V. Halldórsson, and S. Kitaev. On shortest crucial words avoiding abelian powers. *Discrete Appl. Math.*, 158(6):605–607, 2010.

[61] S. Avgustinovich and S. Kitaev. On uniquely k-determined permutations. *Discrete Math.*, 308(9):1500–1507, 2008.

[62] S. Avgustinovich, S. Kitaev, A. Pyatkin, and A. Valyuzhenich. On square-free permutations. *preprint*.

[63] D. Avis and M. Newborn. On pop-stacks in series. *Utilitas Math.*, 19:129–140, 1981.

[64] E. Babson and E. Steingrímsson. Generalized permutation patterns and a classification of the Mahonian statistics. *Sém. Lothar. Combin.*, 44:Art. B44b, 18 pp. (electronic), 2000.

[65] E. Babson and J. West. The permutations $123p_4 \cdots p_m$ and $321p_4 \cdots p_m$ are Wilf-equivalent. *Graphs Combin.*, 16(4):373–380, 2000.

[66] J. Backelin, J. West, and G. Xin. Wilf-equivalence for singleton classes. *Adv. in Appl. Math.*, 38(2):133–148, 2007.

[67] E. Bagno. Euler-Mahonian parameters on colored permutation groups. *Sém. Lothar. Combin.*, 51:Art. B51f, 16 pp. (electronic), 2004/05.

[68] E. Bagno, A. Butman, and D. Garber. Statistics on the multi-colored permutation groups. *Electron. J. Combin.*, 14(1):Research Paper 24, 16 pp. (electronic), 2007.

[69] E. Bagno and D. Garber. On the excedance number of colored permutation groups. *Sém. Lothar. Combin.*, 53:Art. B53f, 17 pp. (electronic), 2004/06.

[70] E. Bagno, D. Garber, and T. Mansour. Excedance number for involutions in complex reflection groups. *Sém. Lothar. Combin.*, 56:Art. B56d, 11 pp. (electronic), 2006/07.

[71] J. Baik, P. Deift, and K. Johansson. On the distribution of the length of the longest increasing subsequence of random permutations. *J. Amer. Math. Soc.*, 12:1119–1178, 1999.

[72] J. Baik and E. M. Rains. The asymptotics of monotone subsequences of involutions. *Duke Math. J.*, 109(2):205–281, 2001.

[73] J. Bandlow, E. S. Egge, and K. Killpatrick. A weight-preserving bijection between Schröder paths and Schröder permutations. *Ann. Comb.*, 6(3-4):235–248, 2002.

[74] J. Bandlow and K. Killpatrick. An area-to-inv bijection between Dyck paths and 312-avoiding permutations. *Electron. J. Combin.*, 8(1):Research Paper 40, 16 pp. (electronic), 2001.

[75] H. Barcelo, R. Maule, and S. Sundaram. On counting permutations by pairs of congruence classes of major index. *Electron. J. Combin.*, 9(1):Research Paper 21, 10 pp. (electronic), 2002.

[76] H. Barcelo, B. E. Sagan, and S. Sundaram. Counting permutations by congruence class of major index. *Adv. in Appl. Math.*, 39(2):269–281, 2007.

[77] E. Barcucci, A. Bernini, L. Ferrari, and M. Poneti. A distributive lattice structure connecting Dyck paths, noncrossing partitions and 312-avoiding permutations. *Order*, 22(4):311–328 (2006), 2005.

[78] E. Barcucci, A. Bernini, and M. Poneti. From Fibonacci to Catalan permutations. *Pure Math. Appl. (PU.M.A.)*, 17(1-2):1–17, 2006.

[79] E. Barcucci, A. Del Lungo, E. Pergola, and R. Pinzani. Directed animals, forests and permutations. *Discrete Math.*, 204(1-3):41–71, 1999.

[80] E. Barcucci, A. Del Lungo, E. Pergola, and R. Pinzani. ECO: a methodology for the enumeration of combinatorial objects. *J. Differ. Equations Appl.*, 5(4-5):435–490, 1999.

[81] E. Barcucci, A. Del Lungo, E. Pergola, and R. Pinzani. Some combinatorial interpretations of q-analogs of Schröder numbers. *Ann. Comb.*, 3(2-4):171–190, 1999. On combinatorics and statistical mechanics.

[82] E. Barcucci, A. Del Lungo, E. Pergola, and R. Pinzani. From Motzkin to Catalan permutations. *Discrete Math.*, 217(1-3):33–49, 2000. Formal power series and algebraic combinatorics (Vienna, 1997).

[83] E. Barcucci, A. Del Lungo, E. Pergola, and R. Pinzani. Permutations avoiding an increasing number of length-increasing forbidden subsequences. *Discrete Math. Theor. Comput. Sci.*, 4(1):31–44 (electronic), 2000.

[84] E. Barcucci, A. Del Lungo, E. Pergola, and R. Pinzani. Some permutations with forbidden subsequences and their inversion number. *Discrete Math.*, 234(1-3):1–15, 2001.

[85] J.-L. Baril. Gray code for permutations with a fixed number of cycles. *Discrete Math.*, 307(13):1559–1571, 2007.

[86] J.-L. Baril. Efficient generating algorithm for permutations with a fixed number of excedances. *Pure Math. Appl. (PU.M.A.)*, 19(2-3):61–69, 2008.

[87] J.-L. Baril. More restrictive Gray codes for some classes of pattern avoiding permutations. *Inform. Process. Lett.*, 109(14):799–804, 2009.

[88] J.-L. Baril and V. Vajnovszki. Gray code for derangements. *Discrete Appl. Math.*, 140(1-3):207–221, 2004.

[89] M. Barnabei, F. Bonetti, and M. Silimbani. The signed Eulerian numbers on involutions. *Pure Math. Appl. (PU.M.A.)*, 19(2-3):117–126, 2008.

[90] M. Barnabei, F. Bonetti, and M. Silimbani. The descent statistic on involutions is not log-concave. *European J. Combin.*, 30(1):11–16, 2009.

[91] M. Barnabei, F. Bonetti, and M. Silimbani. The Eulerian distribution on centrosymmetric involutions. *Discrete Math. Theor. Comput. Sci.*, 11(1):95–115, 2009.

[92] M. Barnabei, F. Bonetti, and M. Silimbani. The joint distribution of consecutive patterns and descents in permutations avoiding 3-1-2. *European J. Combin.*, 31(5):1360–1371, 2010.

[93] R. W. Barton. Packing densities of patterns. *Electron. J. Combin.*, 11:Research Paper 80, 16 pp. (electronic), 2004.

[94] F. Bassino, M. Bouvel, A. Pierrot, and D. Rossin. A polynomial algorithm for deciding the finiteness of simple permutations contained in permutation classes. *In preparation.*

[95] F. Bassino, M. Bouvel, A. Pierrot, and D. Rossin. Deciding the finiteness of the number of simple permutations contained in a wreath-closed class is polynomial. *arXiv:1002.3866v1 [cs.DS]*, 2010.

[96] B. Bauslaugh and F. Ruskey. Generating alternating permutations lexicographically. *BIT*, 30:17–26, 1990.

[97] A. M. Baxter and A. D. Jaggard. Some general results for even-Wilf-equivalence. *In preparation.*

[98] G. Baxter. On fixed points of the composite of commuting functions. *Proc. Amer. Math. Soc.*, 15:851–855, 1964.

[99] R. J. Baxter. Dichromatic polynomials and Potts models summed over rooted maps. *Ann. Comb.*, 5(1):17–36, 2001.

[100] B. I. Bayoumi, M. H. El-Zahar, and S. M. Khamis. Asymptotic enumeration of n-free partial orders. *Order*, 5(6):219–232, 1989.

[101] C. Bebeacua, T. Mansour, A. Postnikov, and S. Severini. On the X-rays of permutations. In *Proceedings of the Workshop on Discrete Tomography and its Applications*, volume 20 of *Electron. Notes Discrete Math.*, pages 193–203. Elsevier Sci. B. V., Amsterdam, 2005.

[102] A. Bernini. *Some properties of pattern avoiding permutations.* PhD thesis, Universita degli Studi di Firenze, 2006.

[103] A. Bernini, M. Bouvel, and L. Ferrari. Some statistics on permutations avoiding generalized patterns. *Pure Math. Appl. (PU.M.A.)*, 18(3-4):223–237, 2007.

[104] A. Bernini, I. Fanti, and E. Grazzini. An exhaustive generation algorithm for Catalan objects and others. *Pure Math. Appl. (PU.M.A.)*, 17(1-2):39–53, 2006.

[105] A. Bernini, L. Ferrari, and R. Pinzani. Enumerating permutations avoiding three Babson-Steingrímsson patterns. *Ann. Comb.*, 9(2):137–162, 2005.

[106] A. Bernini, L. Ferrari, and R. Pinzani. Enumeration of some classes of words avoiding two generalized patterns of length three. *J. Autom. Lang. Comb.*, 14(2), 2009.

[107] A. Bernini, L. Ferrari, and E. Steingrímsson. The Möbius function of the consecutive pattern poset. *arXiv:1103.0173v1 [math.CO]*, 2011.

[108] A. Bernini, E. Grazzini, E. Pergola, and R. Pinzani. A general exhaustive generation algorithm for Gray structures. *Acta Inform.*, 44(5):361 376, 2007.

[109] A. Bernini and E. Pergola. Enumerating permutations avoiding more than three Babson-Steingrímsson patterns. *J. Integer Seq.*, 10(6):Article 07.6.4, 21 pp. (electronic), 2007.

[110] D. Bernstein. MacMahon-type identities for signed even permutations. *Electron. J. Combin.*, 11(1):Research Paper 83, 18 pp. (electronic), 2004.

[111] D. Bernstein. Euler-Mahonian polynomials for $C_a \wr S_n$. *Adv. in Appl. Math.*, 37(2):198–208, 2006.

[112] J. Berstel and L. Boasson. Partial words and a theorem of Fine and Wilf. *Theoret. Comput. Sci.*, 218(1):135 – 141, 1999.

[113] R. Biagioli. Major and descent statistics for the even-signed permutation group. *Adv. in Appl. Math.*, 31(1):163–179, 2003.

[114] S. C. Billey. Pattern avoidance and rational smoothness of Schubert varieties. *Adv. Math.*, 139(1):141–156, 1998.

[115] S. C. Billey and T. Braden. Lower bounds for Kazhdan-Lusztig polynomials from patterns. *Transform. Groups*, 8(4):321–332, 2003.

[116] S. C. Billey, W. Jockusch, and R. P. Stanley. Some combinatorial properties of Schubert polynomials. *J. Algebraic Combin.*, 2(4):345–374, 1993.

[117] S. C. Billey and V. Lakshmibai. *Singular loci of Schubert varieties*, volume 182 of *Progress in Mathematics*. Birkhäuser, Boston, 2000.

[118] S. C. Billey and A. Postnikov. Smoothness of Schubert varieties via patterns in root subsystems. *Adv. in Appl. Math.*, 34(3):447–466, 2005.

[119] S. C. Billey and R. Vakil. Intersections of Schubert varieties and other permutation array schemes. In *Algorithms in algebraic geometry*, volume 146 of *IMA Vol. Math. Appl.*, pages 21–54. Springer, New York, 2008.

[120] S. C. Billey and G. S. Warrington. Kazhdan-Lusztig polynomials for 321-hexagon-avoiding permutations. *J. Algebraic Combin.*, 13(2):111–136, 2001.

[121] S. C. Billey and G. S. Warrington. Maximal singular loci of Schubert varieties in SL(n)/B. *Trans. Amer. Math. Soc.*, 355(10):3915–3945 (electronic), 2003.

[122] A. Björner and B. E. Sagan. Rationality of the Möbius function of a composition poset. *Theoret. Comput. Sci.*, 359(1-3):282–298, 2006.

[123] A. Björner and M. L. Wachs. Permutation statistics and linear extensions of posets. *J. Combin. Theory Ser. A*, 58(1):85–114, 1991.

[124] J. Bloom and D. Saracino. On bijections for pattern-avoiding permutations. *J. Combin. Theory Ser. A*, 116(8):1271–1284, 2009.

[125] E. D. Bolker and A. M. Gleason. Counting permutations. *J. Combin. Theory Ser. A*, 29(2):236–242, 1980.

[126] M. Bóna. *Exact and asymptotic enumeration of permutations with subsequence conditions*. PhD thesis, Massachusetts Institute of Technology, 1997.

[127] M. Bóna. Exact enumeration of 1342-avoiding permutations: a close link with labeled trees and planar maps. *J. Combin. Theory Ser. A*, 80(2):257–272, 1997.

[128] M. Bóna. The number of permutations with exactly r 132-subsequences is P-recursive in the size! *Adv. in Appl. Math.*, 18(4):510–522, 1997.

[129] M. Bóna. Permutations avoiding certain patterns: the case of length 4 and some generalizations. *Discrete Math.*, 175(1-3):55–67, 1997.

[130] M. Bóna. The permutation classes equinumerous to the smooth class. *Electron. J. Combin.*, 5:Research Paper 31, 12 pp. (electronic), 1998.

[131] M. Bóna. Permutations with one or two 132-subsequences. *Discrete Math.*, 181(1-3):267–274, 1998.

[132] M. Bóna. The solution of a conjecture of Stanley and Wilf for all layered patterns. *J. Combin. Theory Ser. A*, 85(1):96–104, 1999.

[133] M. Bóna. Corrigendum: "Symmetry and unimodality in *t*-stack sortable permutations". *J. Combin. Theory Ser. A*, 99(1):191–194, 2002.

[134] M. Bóna. A simplicial complex of 2-stack sortable permutations. *Adv. in Appl. Math.*, 29(4):499–508, 2002.

[135] M. Bóna. Symmetry and unimodality in *t*-stack sortable permutations. *J. Combin. Theory Ser. A*, 98(1):201–209, 2002.

[136] M. Bóna. A survey of stack-sorting disciplines. *Electron. J. Combin.*, 9(2):Article 1, 16 pp. (electronic), 2002/03. Permutation patterns (Otago, 2003).

[137] M. Bóna. *Combinatorics of permutations.* Discrete Mathematics and its Applications (Boca Raton). Chapman & Hall/CRC, Boca Raton, FL, 2004. With a foreword by Richard Stanley.

[138] M. Bóna. A simple proof for the exponential upper bound for some tenacious patterns. *Adv. in Appl. Math.*, 33(1):192–198, 2004.

[139] M. Bóna. A combinatorial proof of the log-concavity of a famous sequence counting permutations. *Electron. J. Combin.*, 11(2):Note 2, 4 pp. (electronic), 2004/06.

[140] M. Bóna. The limit of a Stanley-Wilf sequence is not always rational, and layered patterns beat monotone patterns. *J. Combin. Theory Ser. A*, 110(2):223–235, 2005.

[141] M. Bóna. *A walk through combinatorics: An introduction to enumeration and graph theory.* World Scientific, Hackensack, second edition, 2006. With a foreword by Richard Stanley.

[142] M. Bóna. New records in Stanley-Wilf limits. *European J. Combin.*, 28(1):75–85, 2007.

[143] M. Bóna. On three different notions of monotone subsequences. *arXiv:0711.4325v1 [math.CO]*, 2007.

[144] M. Bóna. Generalized descents and normality. *Electron. J. Combin.*, 15(1):Note 21, 8, 2008.

[145] M. Bóna. Where the monotone pattern (mostly) rules. *Discrete Math.*, 308(23):5782–5788, 2008.

[146] M. Bóna. Real zeros and normal distribution for statistics on Stirling permutations defined by Gessel and Stanley. *SIAM J. Discrete Math.*, 23(1):401–406, 2008/09.

[147] M. Bóna and R. Ehrenborg. A combinatorial proof of the log-concavity of the numbers of permutations with k runs. *J. Combin. Theory Ser. A*, 90(2):293–303, 2000.

[148] M. Bóna and R. Flynn. The average number of block interchanges needed to sort a permutation and a recent result of Stanley. *Inform. Process. Lett.*, 109(16):927–931, 2009.

[149] M. Bóna, A. McLennan, and D. White. Permutations with roots. *Random Structures Algorithms*, 17(2):157–167, 2000.

[150] M. Bóna, B. E. Sagan, and V. R. Vatter. Pattern frequency sequences and internal zeros. *Adv. in Appl. Math.*, 28(3-4):395–420, 2002. Special issue in memory of Rodica Simion.

[151] N. Bonichon, M. Bousquet-Mélou, and E. Fusy. Baxter permutations and plane bipolar orientations. *Electron. Notes in Discrete Math.*, 31:69–74, 2008. The International Conference on Topological and Geometric Graph Theory.

[152] J. Bonin, L. Shapiro, and R. Simion. Some q-analogues of the Schröder numbers arising from combinatorial statistics on lattice paths. *J. Stat. Planning and Inference*, 34:35–55, 1993.

[153] A. Borodin. Longest increasing subsequences of random colored permutations. *Electron. J. Combin.*, 6:Research Paper 13, 12 pp. (electronic), 1999.

[154] P. Bose, J. F. Buss, and A. Lubiw. Pattern matching for permutations. In *Algorithms and data structures (Montreal, PQ, 1993)*, volume 709 of *Lecture Notes in Comput. Sci.*, pages 200–209. Springer, Berlin, 1993.

[155] P. Bose, J. F. Buss, and A. Lubiw. Pattern matching for permutations. *Inform. Process. Lett.*, 65(5):277–283, 1998.

[156] M. Bousquet-Mélou. Multi-statistic enumeration of two-stack sortable permutations. *Electron. J. Combin.*, 5:Research Paper 21, 12 pp. (electronic), 1998.

[157] M. Bousquet-Mélou. Sorted and/or sortable permutations. *Discrete Math.*, 225(1-3):25–50, 2000. Formal power series and algebraic combinatorics (Toronto, ON, 1998).

[158] M. Bousquet-Mélou. Four classes of pattern-avoiding permutations under one roof: generating trees with two labels. *Electron. J. Combin.*, 9(2):Research paper 19, 31 pp. (electronic), 2002/03. Permutation patterns (Otago, 2003).

[159] M. Bousquet-Mélou. Counting permutations with no long monotone subsequence via generating trees and the kernel method. *arXiv:1006.0311v3 [math.CO]*, 2010.

[160] M. Bousquet-Mélou and S. Butler. Forest-like permutations. *Ann. Comb.*, 11(3-4):335–354, 2007.

[161] M. Bousquet-Mélou, A. Claesson, W. M. B. Dukes, and S. Kitaev. (2+2)-free posets, ascent sequences and pattern avoiding permutations. *J. Combin. Theory Ser. A*, 117(7):884–909, 2010.

[162] M. Bousquet-Mélou and E. Steingrímsson. Decreasing subsequences in permutations and Wilf equivalence for involutions. *J. Algebraic Combin.*, 22(4):383–409, 2005.

[163] M. Bouvel and L. Ferrari. On the enumeration of d-minimal permutations. *arXiv:1010.5963v1 [math.CO]*, 2010.

[164] M. Bouvel and E. Pergola. Posets and permutations in the duplication-loss model. *Pure Math. Appl. (PU.M.A.)*, 19(2-3):71–80, 2008.

[165] M. Bouvel and D. Rossin. The longest common pattern problem for two permutations. *Pure Math. Appl. (PU.M.A.)*, 17(1-2):55–69, 2006.

[166] M. Bouvel and D. Rossin. A variant of the tandem duplication – random loss model of genome rearrangement. *Theoret. Comput. Sci.*, 410(8-10):847–858, 2009.

[167] W. M. Boyce. Baxter permutations and functional composition. *Houston J. Math.*, 7(2):175–189, 1981.

[168] R. Brak, S. Corteel, J. Essam, R. Parviainen, and A. Rechnitzer. A combinatorial derivation of the PASEP stationary state. *Electron. J. Combin.*, 13(1):Research Paper 108, 23 pp. (electronic), 2006.

[169] P. Brändén. Actions on permutations and unimodality of descent polynomials. *European J. Combin.*, 29(2):514–531, 2008.

[170] P. Brändén, A. Claesson, and E. Steingrímsson. Catalan continued fractions and increasing subsequences in permutations. *Discrete Math.*, 258(1-3):275–287, 2002.

[171] P. Brändén and T. Mansour. Finite automata and pattern avoidance in words. *J. Combin. Theory, Series A*, 110(1):127–145, 2005.

[172] A. Brandstädt and D. Kratsch. On partitions of permutations into increasing and decreasing subsequences. *Elektron. Informationsverarb. Kybernet.*, 22:263–273, 1986.

[173] C. Brennan and A. Knopfmacher. The distribution of ascents of size d or more in samples of geometric random variables. In *2005 International Conference on Analysis of Algorithms*, Discrete Math. Theor. Comput. Sci. Proc., AD, pages 343–351 (electronic). Assoc. Discrete Math. Theor. Comput. Sci., Nancy, 2005.

[174] C. Brennan and A. Knopfmacher. The first and last ascents of size d or more in samples of geometric random variables. *Quaest. Math.*, 28(4):487–500, 2005.

[175] C. Brennan and A. Knopfmacher. The first ascent of size d or more in compositions. In *Fourth Colloquium on Mathematics and Computer Science Algorithms, Trees, Combinatorics and Probabilities*, Discrete Math. Theor. Comput. Sci. Proc., AG, pages 261–269. Assoc. Discrete Math. Theor. Comput. Sci., Nancy, 2006.

[176] C. Brennan and A. Knopfmacher. The distribution of ascents of size d or more in compositions. *Discrete Math. Theor. Comput. Sci.*, 11(1):1–10, 2009.

[177] C. Brennan, A. Knopfmacher, and S. Wagner. The distribution of ascents of size d or more in partitions of n. *Combin. Probab. Comput.*, 17(4):495–509, 2008.

[178] C. A. C. Brennan. Value and position of large weak left-to-right maxima for samples of geometrically distributed variables. *Discrete Math.*, 307(23):3016–3030, 2007.

[179] F. Brenti. Permutation enumeration symmetric functions, and unimodality. *Pacific J. Math.*, 157(1):1–28, 1993.

[180] F. Brenti. Combinatorial properties of the Kazhdan-Lusztig R-polynomials for S_n. *Adv. Math.*, 126(1):21–51, 1997.

[181] R. Brignall. Wreath products of permutation classes. *Electron. J. Combin.*, 14(1):Research Paper 46, 15 pp. (electronic), 2007.

[182] R. Brignall. Grid classes and partial well order. *arXiv:0906.3723v1 [math.CO]*, 2009.

[183] R. Brignall. A survey of simple permutations. *In Permutation Patterns*, vol. 376 of London Math. Soc. Lecture Note Series, pages 41–65, Cambridge University Press, Cambridge, 2010.

[184] R. Brignall, S. B. Ekhad, R. Smith, and V. R. Vatter. Almost avoiding permutations. *Discrete Math.*, 309(23-24):6626–6631, 2009.

[185] R. Brignall and N. Georgiou. Almost exceptional simple permutations. *In preparation.*

[186] R. Brignall, S. Huczynska, and V. R. Vatter. Decomposing simple permutations, with enumerative consequences. *Combinatorica*, 28(4):385–400, 2008.

[187] R. Brignall, S. Huczynska, and V. R. Vatter. Simple permutations and algebraic generating functions. *J. Combin. Theory Ser. A*, 115(3):423–441, 2008.

[188] R. Brignall, N. Ruškuc, and V. R. Vatter. Simple permutations: decidability and unavoidable substructures. *Theoret. Comput. Sci.*, 391(1-2):150–163, 2008.

[189] R. Brignall, N. Ruškuc, and V. R. Vatter. Simple extensions of combinatorial structures. *arXiv:0911.4378v2 [math.CO]*, 2010.

[190] W.G. Brown. Enumeration of non-separable planar maps. *Canad. J. Math.*, 15:526–545, 1963.

[191] W.G. Brown and W. T. Tutte. On the enumeration of rooted non-seperable planar maps. *Canad. J. Math.*, 16:572–577, 1964.

[192] S. Brunetti, A. Del Lungo, and F. Del Ristoro. An equipartition property for the distribution of multiset permutation inversions. *Adv. in Appl. Math.*, 27(1):41–50, 2001.

[193] E. M. Bullock. Improved bounds on the length of maximal abelian square-free words. *Electron. J. Combin.*, 11(1):Research Paper 17, 12 pp. (electronic), 2004.

[194] R. Bundschuh. Asymmetric exclusion process and extremal statistics of random sequences. *Phys. Rev. E*, 65, 2002.

[195] A. Burstein. *Enumeration of words with forbidden patterns*. PhD thesis, University of Pennsylvania, 1998.

[196] A. Burstein. Restricted Dumont permutations. *Ann. Comb.*, 9(3):269–280, 2005.

[197] A. Burstein. On some properties of permutation tableaux. *Ann. Comb.*, 11(3-4):355–368, 2007.

[198] A. Burstein, S. Elizalde, and T. Mansour. Restricted Dumont permutations, Dyck paths, and noncrossing partitions. *Discrete Math.*, 306(22):2851–2869, 2006.

[199] A. Burstein and N. Eriksen. Combinatorial properties of permutation tableaux. *DMTCS proc. AJ*, pages 625–640, 2008.

[200] A. Burstein and P. Hästö. Packing sets of patterns. *European J. Combin.*, 31(1):241–253, 2010.

[201] A. Burstein, P. Hästö, and T. Mansour. Packing patterns into words. *Electron. J. Combin.*, 9(2):Research paper 20, 13 pp. (electronic), 2002/03. Permutation patterns (Otago, 2003).

[202] A. Burstein, V. Jelínek, E. Jelínkova, and E. Steingrímsson. The Möbius function of separable and decomposable permutations. *arXiv:1102.1611v1 [math.CO]*, 2011.

[203] A. Burstein and S. Kitaev. On unavoidable sets of word patterns. *SIAM J. Discrete Math.*, 19(2):371–381 (electronic), 2005.

[204] A. Burstein and S. Kitaev. Partially ordered patterns and their combinatorial interpretations. *Pure Math. Appl. (PU.M.A.)*, 19(2-3):27–38, 2008.

[205] A. Burstein and I. Lankham. Combinatorics of patience sorting piles. *Sém. Lothar. Combin.*, 54A:Art. B54Ab, 19 pp. (electronic), 2005/07.

[206] A. Burstein and I. Lankham. A geometric form for the extended patience sorting algorithm. *Adv. in Appl. Math.*, 36(2):106–117, 2006.

[207] A. Burstein and I. Lankham. Restricted patience sorting and barred pattern avoidance. *arXiv:0512122v2 [math.CO]*, 2006.

[208] A. Burstein and T. Mansour. Words restricted by patterns with at most 2 distinct letters. *Electron. J. Combin.*, 9(2):Research paper 3, 14 pp. (electronic), 2002/03. Permutation patterns (Otago, 2003).

[209] A. Burstein and T. Mansour. Counting occurrences of some subword patterns. *Discrete Math. Theor. Comput. Sci.*, 6(1):1–11 (electronic), 2003.

[210] A. Burstein and T. Mansour. Words restricted by 3-letter generalized multi-permutation patterns. *Ann. Comb.*, 7(1):1–14, 2003.

[211] A. Burstein and H. S. Wilf. On cyclic strings without long constant blocks. *Fibonacci Quart.*, 35(3):240–247, 1997.

[212] F. Butler. Rook theory and cycle-counting permutation statistics. *Adv. in Appl. Math.*, 33(4):655–675, 2004.

[213] D. Callan. A recursive bijective approach to counting permutations containing 3-letter patterns. *arXiv:0211380v1 [math.CO]*, 2002.

[214] D. Callan. A Wilf equivalence related to two stack sortable permutations. *arXiv:0510211v1 [math.CO]*, 2005.

[215] D. Callan. A combinatorial interpretation of the eigensequence for composition. *J. Integer Seq.*, 9(1):Article 06.1.4, 12 pp. (electronic), 2006.

[216] D. Callan. Permutations avoiding a nonconsecutive instance of a 2- or 3-letter pattern. *arXiv:0610428v2 [math.CO]*, 2006.

[217] D. Callan. Bijections from Dyck paths to 321-avoiding permutations revisited. *arXiv:0711.2684v1 [math.CO]*, 2007.

[218] D. Callan. Pattern avoidance in "flattened" partitions. *Discrete Math.*, 309(12):4187 4191, 2009.

[219] D. Callan. A bijection to count (1-23-4)-avoiding permutations. *arXiv:1008.2375v1 [math.CO]*, 2010.

[220] P. J. Cameron. Homogeneous permutations. *Electron. J. Combin.*, 9(2):Research paper 2, 9 pp. (electronic), 2002/03. Permutation patterns (Otago, 2003).

[221] H. Canary. Aztec diamonds and Baxter permutations. *Electron. J. Combin.*, 17:Research Paper 105, 12 pp. (electronic), 2010.

[222] E. R. Canfield and H. S. Wilf. Counting permutations by their alternating runs. *J. Combin. Theory Ser. A*, 115(2):213–225, 2008.

[223] L. Carlitz. Sequences and inversions. *Duke Math. J.*, 37:193–198, 1970.

[224] L. Carlitz. Addendum: "Enumeration of up-down sequences" (Discrete Math. 4 (1973), 273–286). *Discrete Math.*, 5:291, 1973.

[225] L. Carlitz. Enumeration of up-down permutations by number of rises. *Pacific J. Math.*, 45:49–58, 1973.

[226] L. Carlitz. Enumeration of up-down sequences. *Discrete Math.*, 4:273–286, 1973.

[227] L. Carlitz. Permutations and sequences. *Advances in Math.*, 14:92–120, 1974.

[228] L. Carlitz. Permutations, sequences and special functions. *SIAM Rev.*, 17:298–322, 1975.

[229] L. Carlitz. Enumeration of permutations by sequences. *Fibonacci Quart.*, 16(3):259–268, 1978.

[230] L. Carlitz. The number of permutations with a given number of sequences. *Fibonacci Quart.*, 18(4):347–352, 1980.

[231] L. Carlitz. Enumeration of permutations by sequences. II. *Fibonacci Quart.*, 19(5):398–406, 465, 1981.

[232] L. Carlitz, D. P. Roselle, and R. A. Scoville. Permutations and sequences with repetitions by number of increases. *J. Combin. Theory*, 1:350–374, 1966.

[233] L. Carlitz, Richard Scoville, and Theresa Vaughan. Enumeration of permutations and sequences with restrictions. *Duke Math. J.*, 40:723–741, 1973.

[234] L. Carlitz, Richard Scoville, and Theresa Vaughan. Enumeration of pairs of permutations and sequences. *Bull. Amer. Math. Soc.*, 80:881–884, 1974.

[235] E. Catalan. Note sur une équation aux différences finies. *J. Math. Pures et Appliqués*, 3, 1838.

[236] M. Cerasoli. Enumeration of compositions with prescribed parts. *J. Math. Anal. Appl.*, 89(2):351–358, 1982.

[237] M.-S. Chang and F.-H. Wang. Efficient algorithms for the maximum weight clique and maximum weight independent set problems on permutation graphs. *Inform. Process. Lett.*, 43:293–295, 1992.

[238] R. Chapman and L. K. Williams. A conjecture of Stanley on alternating permutations. *Electron. J. Combin.*, 14(1):Note 16, 7 pp. (electronic), 2007.

[239] K. Chaudhuri, K. Chen, R. Mihaescu, and S. Rao. On the tandem duplication-random loss model of genome rearrangement. *SODA*, pages 564–570, 2006.

[240] D. Chebikin. Variations on descents and inversions in permutations. *Electron. J. Combin.*, 15(1):Research Paper 132, 34, 2008.

[241] D. Chebikin, R. Ehrenborg, P. Pylyavskyy, and M. Readdy. Cyclotomic factors of the descent set polynomial. *J. Combin. Theory Ser. A*, 116(2):247–264, 2009.

[242] W. Y. C. Chen, E. Y. P. Deng, and L. L. M. Yang. Riordan paths and derangements. *Discrete Math.*, 308(11):2222–2227, 2008.

[243] W. Y. C. Chen, Y.-P. Deng, and L. L. M. Yang. Motzkin paths and reduced decompositions for permutations with forbidden patterns. *Electron. J. Combin.*, 9(2):Research paper 15, 13 pp. (electronic), 2002/03. Permutation patterns (Otago, 2003).

[244] W. Y. C. Chen and L. H. Liu. Permutation tableaux and the dashed permutation pattern 32-1. *arXiv:1007.5019v1 [math.CO]*, 2010.

[245] Y. Cherniavsky and E. Bagno. Permutation representations on invertible matrices. *Linear Algebra Appl.*, 419(2-3):494–518, 2006.

[246] C.-O. Chow and W. C. Shiu. Counting simsun permutations by descents. *Ann. Combin.*, 2011. To appear.

[247] T. Chow and J. West. Forbidden subsequences and Chebyshev polynomials. *Discrete Math.*, 204(1-3):119–128, 1999.

[248] W.-C. Chuang, S.-P. Eu, T.-S. Fu, and Y.-J. Pan. On simsun and double simsun permutations and avoiding a pattern of length three.

[249] F. R. K. Chung, P. Diaconis, and R. Graham. Universal cycles for combinatorial structures. *Discrete Math.*, 110(1-3):43–59, 1992.

[250] F. R. K. Chung, R. L. Graham, V. E. Hoggatt, Jr., and M. Kleiman. The number of Baxter permutations. *J. Combin. Theory Ser. A*, 24(3):382–394, 1978.

[251] A. Claesson. Generalized pattern avoidance. *European J. Combin.*, 22(7):961–971, 2001.

[252] A. Claesson. *Permutation patterns, continued fractions, and a group determined by an ordered set*. PhD thesis, Göteborg University/Chalmers University of Technology, 2004.

[253] A. Claesson. Counting segmented permutations using bicoloured Dyck paths. *Electron. J. Combin.*, 12:Research Paper 39, 18 pp. (electronic), 2005.

[254] A. Claesson, W. M. B. Dukes, and S. Kitaev. A direct encoding of Stoimenow's matchings as ascent sequences. *arXiv:0910.1619v1 [math.CO]*, 2009.

[255] A. Claesson, W. M. B. Dukes, and M. Kubitzke. Partition and composition matrices. *arXiv:1006.1312v1 [math.CO]*, 2010.

[256] A. Claesson, W. M. B. Dukes, and E. Steingrímsson. Permutations sortable by n-4 passes through a stack. *Ann. Comb.*, 14:45–51, 2010.

[257] A. Claesson, V. Jelínek, E. Jelínková, and S. Kitaev. Pattern avoidance in partial permutations. *arXiv:1005.2216v1 [math.CO]*, 2010.

[258] A. Claesson, V. Jelínek, E. Jelínkova, and S. Kitaev. Pattern avoidance in partial permutations. *arXiv:1005.2216v1 [math.CO]*, 2010.

[259] A. Claesson and S. Kitaev. Classification of bijections between 321- and 132- avoiding permutations. *Sém. Lothar. Combin.*, 60:Art. B60d, 30, 2008.

[260] A. Claesson, S. Kitaev, and E. Steingrímsson. A hierarchy for a class of vincular permutation patterns and $\beta(0,1)$-trees. *In preparation*.

[261] A. Claesson, S. Kitaev, and E. Steingrímsson. An involution on trees. *In preparation*.

[262] A. Claesson, S. Kitaev, and E. Steingrímsson. Decompositions and statistics for $\beta(1,0)$-trees and nonseparable permutations. *Adv. in Appl. Math.*, 42(3):313–328, 2009.

[263] A. Claesson and S. Linusson. n! matchings, n! posets. *Discrete Math. Theor. Comput. Sci. Proc. AK*, pages 505–516, 2010. Proceedings of the 22nd International Conference on Formal Power Series & Algebraic Combinatorics, San Francisco State University, San Francisco, USA, August 2-6 (2010).

[264] A. Claesson and T. Mansour. Counting occurrences of a pattern of type $(1,2)$ or $(2,1)$ in permutations. *Adv. in Appl. Math.*, 29(2):293–310, 2002.

[265] A. Claesson and T. Mansour. Enumerating permutations avoiding a pair of Babson-Steingrímsson patterns. *Ars Combin.*, 77:17–31, 2005.

[266] E. Clark and R. Ehrenborg. Explicit expressions for the extremal excedance set statistics. *European J. Combin.*, 31(1):270–279, 2010.

[267] R. J. Clarke, E. Steingrímsson, and J. Zeng. New Euler-Mahonian permutation statistics. *Sém. Lothar. Combin.*, 35:Art. B35c, approx. 29 pp. (electronic), 1995.

[268] R. J. Clarke, E. Steingrímsson, and J. Zeng. The k-extensions of some new Mahonian statistics. *European J. Combin.*, 18(2):143–154, 1997.

[269] R. J. Clarke, E. Steingrímsson, and J. Zeng. New Euler-Mahonian statistics on permutations and words. *Adv. in Appl. Math.*, 18(3):237–270, 1997.

[270] M. Coleman. An answer to a question by Wilf on packing distinct patterns in a permutation. *Electron. J. Combin.*, 11(1):Note 8, 4 pp. (electronic), 2004.

[271] L. Comtet. *Advanced Combinatorics*. Reidel, Dordrecht, 1974.

[272] R. Cori, S. Dulucq, and G. Viennot. Shuffle of parenthesis systems and Baxter permutations. *J. Combin. Theory Ser. A*, 43(1):1–22, 1986.

[273] R. Cori and G. Schaeffer. Description trees and tutte formulas. *Theoret. Comput. Sci.*, 292:165–183, 1997.

[274] S. Corteel. A simple bijection between permutation tableaux and permutations. *arXiv:0609700v1 [math.CO]*, 2006.

[275] S. Corteel. Crossings and alignments of permutations. *Adv. in Appl. Math.*, 38(2):149 – 163, 2007.

[276] S. Corteel, I. M. Gessel, C. D. Savage, and H. S. Wilf. The joint distribution of descent and major index over restricted sets of permutations. *Ann. Comb.*, 11(3-4):375–386, 2007.

[277] S. Corteel and P. Hitczenko. Expected values of statistics on permutation tableaux. In *2007 Conference on Analysis of Algorithms, AofA 07*, Discrete Math. Theor. Comput. Sci. Proc., AH, pages 325–339. Assoc. Discrete Math. Theor. Comput. Sci., Nancy, 2007.

[278] S. Corteel, G. Louchard, and R. Pemantle. Common intervals in permutations. *Discrete Math. Theor. Comput. Sci.*, 8(1):189–214 (electronic), 2006.

[279] S. Corteel and P. Nadeau. Bijections for permutation tableaux. *European J. Combin.*, 30(1):295–310, 2009.

[280] S. Corteel, R. Stanley, D. Stanton, and L. K. Williams. Formulae for Askey-Wilson moments and enumeration of staircase tableaux. *arXiv:1007.5174v2 [math.CO]*, 2010.

[281] S. Corteel and L. K. Williams. A Markov chain on permutations which projects to the PASEP. *International Mathematics Research Notices*, 39:page article IDmm55, 2007.

[282] S. Corteel and L. K. Williams. Tableaux combinatorics for the asymmetric exclusion process. *Adv. in Appl. Math.*, 39(3):293–310, 2007.

[283] S. Corteel and L. K. Williams. Tableaux combinatorics for the asymmetric exclusion process and Askey-Wilson polynomials. *arXiv:0910.1858v1 [math.CO]*, 2009.

[284] A. Cortez. Singularités génériques et quasi-résolutions des variétés de Schubert pour le groupe linéaire. *Adv. Math.*, 178:396–445, 2003.

[285] L. L. Cristea and H. Prodinger. *q*-enumeration of up-down words by number of rises. *Fibonacci Quart.*, 46/47(2):126–134, 2008/09.

[286] A. Crites. Enumerating pattern avoidance for affine permutations. *Electron. J. Combin.*, 17(1):Research Paper 127, 13 pp. (electronic), 2010.

[287] L. J. Cummings and M. Mays. A one-sided Zimin construction. *Electron. J. Combin.*, 8(1):Research Paper 27, 9 pp. (electronic), 2001.

[288] D. Daly. Fibonacci numbers, reduced decompositions, and 321/3412 pattern classes. *Ann. Combin.*, 14:53–64, 2010. 10.1007/s00026-010-0051-8.

[289] J. A. Davis, R. C. Entringer, R. L. Graham, and G. J. Simmons. On permutations containing no long arithmetic progressions. *Acta Arith.*, 34(1):81–90, 1977/78.

[290] F. De Mari and M. A. Shayman. Generalized Eulerian numbers and the topology of the Hessenberg variety of a matrix. *Acta Appl. Math.*, 12(3):213–235, 1988.

[291] M. Denert. The genus zeta function of hereditary orders in central simple algebras over global fields. *Math. Comp.*, 54:449–465, 1990.

[292] E. Y. P. Deng, W. M. B. Dukes, T. Mansour, and S. Y. J. Wu. Symmetric Schröder paths and restricted involutions. *Discrete Math.*, 309(12):4108–4115, 2009.

[293] R. S. Deodhar and M. K. Srinivasan. A statistic on involutions. *J. Algebraic Combin.*, 13(2):187–198, 2001.

[294] V. Deodhar. Local Poincaré duality and nonsingularity of Schubert varieties. *Comm. Algebra*, 13:1379–1388, 1985.

[295] V. Deodhar. A combinatorial settting for questions in Kazhdan-Lusztig theory. *Geom. Dedicata*, 36(1):95–119, 1990.

[296] B. Derrida, E. Domany, and D. Mukamel. An exact solution of a one-dimensional asymmetric exclusion model with open boundaries. *J. Stat. Phys.*, 69:667–687, 1992.

[297] B. Derrida, M. Evans, V. Hakiml, and V. Pasquier. Exact solution of a 1D asymmetric exclusion model using a matrix formulation. *J. Phys. A: Math. Gen.*, 26:1493–1517, 1993.

[298] B. Derrida, J. Lebowitz, and E. Speer. Shock profiles for the partially asymmetric simple exclusion process. *J. Stat. Phys.*, 89(135):135–167, 1997.

[299] J. Désarménien and M. L. Wachs. Descent classes of permutations with a given number of fixed points. *J. Combin. Theory Ser. A*, 64(2):311–328, 1993.

[300] E. Deutsch. Dyck path enumeration. *Discrete Math.*, 204(1-3):167–202, 1999.

[301] E. Deutsch and S. Elizalde. Restricted simsun permutations. *arXiv:0912.1361v1 [math.CO]*, 2009.

[302] E. Deutsch, A. J. Hildebrand, and H. S. Wilf. Longest increasing subsequences in pattern-restricted permutations. *Electron. J. Combin.*, 9(2):Research paper 12, 8 pp. (electronic), 2002/03. Permutation patterns (Otago, 2003).

[303] E. Deutsch, S. Kitaev, and J. B. Remmel. Equidistribution of descents, adjacent pairs, and place-value pairs on permutations. *J. Integer Seq.*, 12(5):Article 09.5.1, 19, 2009.

[304] E. Deutsch, A. Robertson, and D. Saracino. Refined restricted involutions. *European J. Combin.*, 28(1):481–498, 2007.

[305] E. Deutsch and L. Shapiro. A survey of the Fine numbers. *Discrete Math.*, 241(1-3):241–265, 2001. Selected Papers in honor of Helge Tverberg.

[306] P. Diaconis, J. Fulman, and R. Guralnick. On fixed points of permutations. *J. Algebraic Combin.*, 28(1):189–218, 2008.

[307] P. Diaconis, M. McGrath, and J. Pitman. Riffle shuffles, cycles, and descents. *Combinatorica*, 15(1):11–29, 1995.

[308] F. Disanto, L. Ferrari, R. Pinzani, and S. Rinaldi. Catalan pairs: A relational-theoretic approach to Catalan numbers. *Adv. in Appl. Math.*, 45(4):505–517, 2010.

[309] R. W. Doran. The Gray code. *J.UCS*, 13(11):1573–1597, 2007.

[310] V. Dotsenko and A. Khoroshkin. Anick-type resolutions and consecutive pattern avoidance. *arXiv:1002.2761v1 [math.CO]*, 2010.

[311] T. A. Dowling. A class of geometric lattices based on finite groups. *J. Combin. Theory, Series B*, 14(1):61–86, 1973.

[312] A. Duane and J. B. Remmel. Minimal overlapping patterns. Permutation patterns (Dartmouth College, 2010).

[313] E. Duchi and D. Poulalhon. On square permutations. In *Fifth Colloquium on Mathematics and Computer Science*, Discrete Math. Theor. Comput. Sci. Proc., AI, pages 207–222. Assoc. Discrete Math. Theor. Comput. Sci., Nancy, 2008.

[314] E. Duchi and G. Schaeffer. A combinatorial approach to jumping particles. *J. Combin. Theory Ser. A*, 110:1–29, 2005.

[315] W. M. B. Dukes. Permutation statistics on involutions. *European J. Combin.*, 28(1):186–198, 2007.

[316] W. M. B. Dukes, M. F. Flanagan, T. Mansour, and V. Vajnovszki. Combinatorial Gray codes for classes of pattern avoiding permutations. *Theoret. Comput. Sci.*, 396(1-3):35–49, 2008.

[317] W. M. B. Dukes, V. Jelínek, T. Mansour, and A. Reifegerste. New equivalences for pattern avoiding involutions. *Proc. Amer. Math. Soc.*, 137(2):457–465, 2009.

[318] W. M. B. Dukes, S. Kitaev, J. Remmel, and E. Steingrímsson. Enumerating (2+2)-free posets by indistinguishable elements. *J. Combin.*, 2(1):139–163, 2011.

[319] W. M. B. Dukes and T. Mansour. Signed involutions avoiding 2-letter signed patterns. *Ann. Comb.*, 11(3-4):387–403, 2007.

[320] W. M. B. Dukes, T. Mansour, and A. Reifegerste. Wilf classification of three and four letter signed patterns. *Discrete Math.*, 308(15):3125–3133, 2008.

[321] W. M. B. Dukes and R. Parviainen. Ascent sequences and upper triangular matrices containing non-negative integers. *Electron. J. Combin.*, 17(1):Research paper 53, 16 pp. (electronic), 2010.

[322] W.M.B. Dukes and A. Reifegerste. The area above the Dyck path of a permutation. *Adv. in Appl. Math.*, 45(1):15–23, 2010.

[323] S. Dulucq, S. Gire, and O. Guibert. A combinatorial proof of J. West's conjecture. *Discrete Math.*, 187(1-3):71–96, 1998.

[324] S. Dulucq, S. Gire, and J. West. Permutations with forbidden subsequences and nonseparable planar maps. In *Proceedings of the 5th Conference on Formal Power Series and Algebraic Combinatorics (Florence, 1993)*, volume 153, pages 85–103, 1996.

[325] S. Dulucq and O. Guibert. Stack words, standard tableaux and Baxter permutations. In *Proceedings of the 6th Conference on Formal Power Series and Algebraic Combinatorics (New Brunswick, NJ, 1994)*, volume 157, pages 91–106, 1996.

[326] S. Dulucq and O. Guibert. Baxter permutations. In *Proceedings of the 7th Conference on Formal Power Series and Algebraic Combinatorics (Noisy-le-Grand, 1995)*, volume 180, pages 143–156, 1998.

[327] S. Dulucq and R. Simion. Combinatorial statistics on alternating permutations. *J. Algebraic Combin.*, 8(2):169–191, 1998.

[328] D. Dumont. Interprétations combinatoires des nombres de Genocchi. *Duke Math. J.*, 41:305–318, 1974.

[329] A. Dzhumadil'daev. MacMahon's theorem for a set of permutations with given descent indices and right-maximal records. *Electron. J. Combin.*, 17(1):Research Paper 34, 14 pp. (electronic), 2010.

[330] P. H. Edelman. On inversions and cycles in inversions. *European J. Combin.*, 8:269–279, 1987.

[331] P. H. Edelman, R. Simion, and D. White. Partition statistics on permutations. *Discrete Math.*, 99(1-3):63–68, 1992.

[332] S. Effler and F. Ruskey. A CAT algorithm for generating permutations with a fixed number of inversions. *Inform. Process. Lett.*, 86(2):107–112, 2003.

[333] E. S. Egge. Restricted permutations related to Fibonacci numbers and k-generalized Fibonacci numbers. *arXiv:0109219v1 [math.CO]*, 2001.

[334] E. S. Egge. Restricted 3412-avoiding involutions, continued fractions, and Chebyshev polynomials. *Adv. in Appl. Math.*, 33(3):451–475, 2004.

[335] E. S. Egge. Restricted signed permutations counted by the Schröder numbers. *Discrete Math.*, 306(6):552–563, 2006.

[336] E. S. Egge. Restricted colored permutations and Chebyshev polynomials. *Discrete Math.*, 307(14):1792–1800, 2007.

[337] E. S. Egge. Restricted symmetric permutations. *Ann. Comb.*, 11(3-4):405–434, 2007.

[338] E. S. Egge and T. Mansour. Permutations which avoid 1243 and 2143, continued fractions, and Chebyshev polynomials. *Electron. J. Combin.*, 9(2):Research paper 7, 35 pp. (electronic), 2002/03. Permutation patterns (Otago, 2003).

[339] E. S. Egge and T. Mansour. 132-avoiding two-stack sortable permutations, Fibonacci numbers, and Pell numbers. *Discrete Appl. Math.*, 143(1-3):72–83, 2004.

[340] E. S. Egge and T. Mansour. 231-avoiding involutions and Fibonacci numbers. *Australas. J. Combin.*, 30:75–84, 2004.

[341] E. S. Egge and T. Mansour. Restricted permutations, Fibonacci numbers, and k-generalized Fibonacci numbers. *Integers*, 5(1):A1, 12 pp. (electronic), 2005.

[342] E. S. Egge and T. Mansour. Bivariate generating functions for involutions restricted by 3412. *Adv. in Appl. Math.*, 36(2):118–137, 2006.

[343] R. Ehrenborg. The asymptotics of almost alternating permutations. *Adv. in Appl. Math.*, 28(3-4):421–437, 2002. Special issue in memory of Rodica Simion.

[344] R. Ehrenborg and Y. Farjoun. Asymptotics of the Euler number of bipartite graphs. *Adv. in Appl. Math.*, 24(44):155–167, 2010.

[345] R. Ehrenborg and J. Jung. Descent pattern avoidance. *In preparation.*

[346] R. Ehrenborg, S. Kitaev, and P. Perry. A spectral approach to consecutive pattern-avoiding permutations. *arXiv:1009.2119v1 [math.CO]*, 2010.

[347] R. Ehrenborg, M. Levin, and M. A. Readdy. A probabilistic approach to the descent statistic. *J. Combin. Theory Ser. A*, 98(1):150–162, 2002.

[348] R. Ehrenborg and E. Steingrímsson. The excedance set of a permutation. *Adv. in Appl. Math.*, 24(3):284–299, 2000.

[349] R. Ehrenborg and E. Steingrímsson. Yet another triangle for the Genocchi numbers. *European J. Combin.*, 21(5):593–600, 2000.

[350] M. J. Elder. Pattern avoiding permutations are context-sensitive. *arXiv:0412019v2 [math.CO]*, 2005.

[351] M. J. Elder. Permutations generated by a stack of depth 2 and an infinite stack in series. *Electron. J. Combin.*, 13(1):Research Paper 68, 12 pp. (electronic), 2006.

[352] M. J. Elder and V. R. Vatter. Problems and conjectures presented at the third international conference on permutation patterns, University of Florida, March 7–11, 2005. *arXiv:0505504v1 [math.CO]*, 2005.

[353] S. Elizalde. *Consecutive patterns and statistics on restricted permutations.* PhD thesis, Universitat Politècnica de Catalunya, 2004.

[354] S. Elizalde. Multiple pattern avoidance with respect to fixed points and excedances. *Electron. J. Combin.*, 11(1):Research Paper 51, 40 pp. (electronic), 2004.

[355] S. Elizalde. Asymptotic enumeration of permutations avoiding generalized patterns. *Adv. in Appl. Math.*, 36(2):138–155, 2006.

[356] S. Elizalde. Generating trees for permutations avoiding generalized patterns. *Ann. Comb.*, 11(3-4):435–458, 2007.

[357] S. Elizalde. The number of permutations realized by a shift. *SIAM J. Discrete Math.*, 23(2):765–786, 2009.

[358] S. Elizalde and E. Deutsch. A simple and unusual bijection for Dyck paths and its consequences. *Ann. Comb.*, 7(3):281–297, 2003.

[359] S. Elizalde and T. Mansour. Restricted Motzkin permutations, Motzkin paths, continued fractions and Chebyshev polynomials. *Discrete Math.*, 305(1-3):170–189, 2005.

[360] S. Elizalde and M. Noy. Consecutive patterns in permutations. *Adv. in Appl. Math.*, 30(1-2):110–125, 2003. Formal power series and algebraic combinatorics (Scottsdale, AZ, 2001).

[361] S. Elizalde and I. Pak. Bijections for refined restricted permutations. *J. Combin. Theory Ser. A*, 105(2):207–219, 2004.

[362] R. Entringer. Enumeration of permutations of $(1, \cdots, n)$ by number of maxima. *Duke Math. J.*, 36:575–579, 1969.

[363] P. Erdős and G. Szekeres. A combinatorial problem in geometry. *Compositio Math.*, 2:463–470, 1935.

[364] P. Erdős. Some unsolved problems. *Magyar Tud. Akad. Mat. Kutató Int. Közl.*, 6:221–254, 1961.

[365] N. Eriksen, R. Freij, and J. Wästlund. Enumeration of derangements with descents in prescribed positions. *Electron. J. Combin.*, 16(1):Research Paper 32, 19, 2009.

[366] H. Eriksson, K. Eriksson, S. Linusson, and J. Wästlund. Dense packing of patterns in a permutation. *Ann. Comb.*, 11(3-4):459–470, 2007.

[367] K. Eriksson. Eulerian numbers for many-signed permutations. In *Proceedings of the Seventh Nordic Combinatorial Conference (Turku, 1999)*, volume 15 of *TUCS Gen. Publ.*, pages 13–14. Turku Cent. Comput. Sci., Turku, 1999. Joint work with Niklas Eriksen and Henrik Eriksson.

[368] K. Eriksson and S. Linusson. The size of Fulton's essential set. *Electron. J. Combin.*, 2:Research Paper 6, approx. 18 pp. (electronic), 1995.

[369] K. Eriksson and S. Linusson. Combinatorics of Fulton's essential set. *Duke Math. J.*, 85(1):61–76, 1996.

[370] Ö. Eğecioğlu and J. B. Remmel. Brick tabloids and the connection matrices between bases of symmetric functions. *Discrete Appl. Math.*, 34(1-3):107–120, 1991.

[371] A. Evdokimov and S. Kitaev. Crucial words and the complexity of some extremal problems for sets of prohibited words. *J. Combin. Theory Ser. A*, 105(2):273–289, 2004.

[372] A. A. Evdokimov. The maximal length of a chain in the unit n-dimensional cube. *Mat. Zametki*, 6:309–319, 1969.

[373] S. Even and A. Itai. Queues, stacks, and graphs. In *Theory of machines and computations (Proc. Internat. Sympos., Technion, Haifa, 1971)*, pages 71–86. Academic Press, New York, 1971.

[374] H. L. M. Faliharimalala and J. Zeng. Derangements and Euler's difference table for $C_l \wr S_n$. *Electron. J. Combin.*, 15(1):Research paper 65, 22, 2008.

[375] J.-M. Fédou. Fonctions de Bessel, empilements et tresses. *In séries formelles et combinatoire algébrique, Publ. du LaCIM, Université a Montréal, Québec*, 11:189–202, 1992.

[376] J.-M. Fédou and D. Rawlings. More statistics on permutation pairs. *Electron. J. Combin.*, 1:Research Paper 11, approx. 17 pp. (electronic), 1994.

[377] J.-M. Fédou and D. Rawlings. Adjacencies in words. *Adv. in Appl. Math.*, 16(2):206–218, 1995.

[378] J.-M. Fédou and D. Rawlings. Statistics on pairs of permutations. *Discrete Math.*, 143(1-3):31–45, 1995.

[379] S. Felsner, E. Fusy, M. Noy, and D. Orden. Bijections for Baxter families and related objects. *arXiv:0803.1546v1 [math.CO]*, 2008.

[380] M. Fire. Statistics on signed permutations groups (extended abstract). *Formal Power Series and Algebraic Combinatorics, San Diego, California 2006*.

[381] G. Firro and T. Mansour. Restricted permutations and polygons. *The Third International Conference on Permutation Patterns, University of Florida, Gainesville, Florida, March 7–11*, 2005.

[382] G. Firro and T. Mansour. Three-letter-pattern-avoiding permutations and functional equations. *Electron. J. Combin.*, 13(1):Research Paper 51, 14 pp. (electronic), 2006.

[383] G. Firro and T. Mansour. Restricted k-ary words and functional equations. *Discrete Appl. Math.*, 157(4):602–616, 2009.

[384] G. Firro, T. Mansour, and M. C. Wilson. Longest alternating subsequences in pattern-restricted permutations. *Electron. J. Combin.*, 14(1):Research Paper 34, 17, 2007.

[385] P. C. Fishburn. Intransitive indifference with unequal indifference intervals. *J. Math. Psych.*, 7:144–149, 1970.

[386] P. Flajolet and R. Schott. Nonoverlapping partitions, continued fractions, Bessel functions and a divergent series. *European J. Combin.*, 11(5):421–432, 1990.

[387] P. Flajolet and R. Sedgewick. *Analytic combinatorics*. Cambridge University Press, Cambridge, 2009.

[388] D. Foata and G.-N. Han. Inverses of words. *Sém. Lothar. Combin.*, 39:Art. B39d, 8pp. (electronic), 1997.

[389] D. Foata and G.-N. Han. Transformations on words. *J. Algorithms*, 28(1):172–191, 1998.

[390] D. Foata and G.-N. Han. Signed words and permutations. II. The Euler-Mahonian polynomials. *Electron. J. Combin.*, 11(2):Research Paper 22, 18 pp. (electronic), 2004/06.

[391] D. Foata and G.-N. Han. Signed words and permutations. III. The MacMahon Verfahren. *Sém. Lothar. Combin.*, 54:Art. B54a, 20 pp. (electronic), 2005/07.

[392] D. Foata and G.-N. Han. Signed words and permutations. I. A fundamental transformation. *Proc. Amer. Math. Soc.*, 135(1):31–40 (electronic), 2007.

[393] D. Foata and G.-N. Han. Decreases and descents in words. *Sém. Lothar. Combin.*, 58:Art. B58a, 17, 2007/08.

[394] D. Foata and G.-N. Han. Signed words and permutations. IV. Fixed and pixed points. *Israel J. Math.*, 163:217–240, 2008.

[395] D. Foata and G.-N. Han. New permutation coding and equidistribution of set-valued statistics. *Theoret. Comput. Sci.*, 410(38-40):3743–3750, 2009.

[396] D. Foata and G.-N. Han. Signed words and permutations. V. A sextuple distribution. *Ramanujan J.*, 19(1):29–52, 2009.

[397] D. Foata and G.-N. Han. The q-tangent and q-secant numbers via basic Eulerian polynomials. *Proc. Amer. Math. Soc.*, 138(2):385–393, 2010.

[398] D. Foata and A. Randrianarivony. Two oiseau decompositions of permutations and their application to Eulerian calculus. *European J. Combin.*, 27(3):342–363, 2006.

[399] D. Foata and M.-P. Schützenberger. Nombres d'Euler et permutations alternantes. *A Survey of Combinatorial Theory*, pages 173–187, 1973.

[400] D. Foata and D. Zeilberger. Denert's permutation statistic is indeed Euler-Mahonian. *Stud. Appl. Math.*, 83(1):31–59, 1990.

[401] D. Foata and D. Zeilberger. Babson-Steingrímsson statistics are indeed Mahonian (and sometimes even Euler-Mahonian). *Adv. in Appl. Math.*, 27(2-3):390–404, 2001. Special issue in honor of Dominique Foata's 65th birthday.

[402] E. Fuller. *Generating functions for composition/word statistics.* PhD thesis, University of California, San Diego, 2009.

[403] E. Fuller and J. B. Remmel. Symmetric functions and generating functions for descents and major indices in compositions. *Ann. Comb.*, 14(1):103–121, 2010.

[404] J. Fulman. Applications of the Brauer complex: card shuffling, permutation statistics, and dynamical systems. *J. Algebra*, 243(1):96–122, 2001.

[405] J. Fulman. Applications of symmetric functions to cycle and increasing subsequence structure after shuffles. *J. Algebraic Combin.*, 16(2):165–194, 2002.

[406] J. Fulman. Stein's method and non-reversible Markov chains. *In Stein's Method: Expository Lectures and Applications*, Ser. 46:66–74, 2004.

[407] M. Fulmek. Enumeration of permutations containing a prescribed number of occurrences of a pattern of length three. *Adv. in Appl. Math.*, 30(4):607–632, 2003.

[408] W. Fulton. Flags, Schubert polynomials, degeneracy loci, and determinantal formulas. *Duke Math. J.*, 65(3):381–420, 1992.

[409] Z. Füredi and P. Hajnal. Davenport-Schinzel theory of matrices. *Discrete Math.*, 103(3):233–251, 1992.

[410] E. Fusi. Bijective counting of involutive Baxter permutations. *arXiv:1010.3850v1 [math.CO]*, 2010.

[411] J. Galovich and D. White. Recursive statistics on words. *Discrete Math.*, 157(1-3):169–191, 1996.

[412] A. M. Garsia and I. Gessel. Permutation statistics and partitions. *Adv. in Math.*, 31(3):288–305, 1979.

[413] A. M. Garsia and A. Goupil. Character polynomials, their q-analogs and the Kronecker product. *Electron. J. Combin.*, 16(2):Research Paper 19, 40 pp. (electronic), 2009.

[414] V. Gasharov and V. Reiner. Cohomology of smooth Schubert varieties in partial flag manifolds. *J. London Math. Soc. (2)*, 66(3):550–562, 2002.

[415] D. D. Gebhard and B. E. Sagan. Sinks in acyclic orientations of graphs. *J. Combin. Theory Ser. B*, 80(1):130–146, 2000.

[416] I. Gessel, J. Weinstein, and H. S. Wilf. Lattice walks in $\mathbf{Z}^d$ and permutations with no long ascending subsequences. *Electron. J. Combin.*, 5:Research Paper 2, 11 pp. (electronic), 1998.

[417] I. M. Gessel. Symmetric functions and P-recursiveness. *J. Combin. Theory Ser. A*, 53(2):257–285, 1990.

[418] I. M. Gessel and C. Reutenauer. Counting permutations with given cycle structure and descent set. *J. Combin. Theory Ser. A*, 64(2):189–215, 1993.

[419] S. Gire. *Arbres, permutations à motifs exclus et cartes planaires : quelques problèmes algorithmiques et combinatoires.* PhD thesis, Université Bordeaux, 1993.

[420] A. Glen, B. V. Halldórsson, and S. Kitaev. Crucial abelian k-power-free words. *Discrete Math. and Theoret. Comput. Sci.*, 12(5):83–96, 2010.

[421] I. P. Goulden and D. M. Jackson. An inversion theorem for cluster decompositions of sequences with distinguished subsequences. *J. London Math. Soc.*, 20(2):567–576, 1979.

[422] I. P. Goulden and D. M. Jackson. *Combinatorial enumeration*. Dover, Mineola, 2004. With a foreword by Gian-Carlo Rota, Reprint of the 1983 original.

[423] I. P. Goulden and J. West. Raney paths and a combinatorial relationship between rooted nonseparable planar maps and two-stack-sortable permutations. *J. Combin. Theory Ser. A*, 75(2):220–242, 1996.

[424] D. Gouyou-Beauchamps. Standard Young tableaux of height 4 and 5. *European J. Combin.*, 10(1):69–82, 1989.

[425] D. Gouyou-Beauchamps and X. G. Viennot. Equivalence of the two-dimensional directed animal problem to a one-dimensional path problem. *Adv. in Appl. Math.*, 9(3):334–357, 1988.

[426] A. M. Goyt. Avoidance of partitions of a three-element set. *Adv. in Appl. Math.*, 41(1):95–114, 2008.

[427] A. M. Goyt and D. Mathisen. Permutation statistics and q-Fibonacci numbers. *Electron. J. Combin.*, 16(1):Research Paper 101, 15, 2009.

[428] A. M. Goyt and L. K. Pudwell. Avoiding colored partitions of lengths two and three. *arXiv:1103.0239v1 [math.CO]*, 2011.

[429] A. M. Goyt and B. E. Sagan. Set partition statistics and q-Fibonacci numbers. *European J. Combin.*, 30(1):230–245, 2009.

[430] R. M. Green and J. Losonczy. Freely braided elements of Coxeter groups. *Ann. Comb.*, 6(3–4):337–348, 2002.

[431] C. Greene and T. Zaslavsky. On the interpretation of Whitney numbers through arrangements of hyperplanes, zonotopes, non-radon partitions, and orientations of graphs. *Transactions of the American Mathematical Society*, 280(1):97–126, 1983.

[432] R. Grimaldi and S. Heubach. Binary strings without odd runs of zeros. *Ars Combin.*, 75:241–255, 2005.

[433] H. H. Gudmundsson. Dyck paths, standard Young tableaux, and pattern avoiding permutations. *arXiv:0912.4747v1 [math.CO]*, 2009.

[434] O. Guibert. *Combinatoire des permutations à motifs exclus, en liaison avec mots, cartes planaires et tableaux de Young.* PhD thesis, LaBRI, Université Bordeaux 1, 1995.

[435] O. Guibert. Stack words, standard Young tableaux, permutations with forbidden subsequences and planar maps. *Discrete Math.*, 210(1-3):71–85, 2000. Formal power series and algebraic combinatorics (Minneapolis, MN, 1996).

[436] O. Guibert and S. Linusson. Doubly alternating Baxter permutations are Catalan. *Discrete Math.*, 217(1-3):157–166, 2000. Formal power series and algebraic combinatorics (Vienna, 1997).

[437] O. Guibert and T. Mansour. Restricted 132-involutions. *Sém. Lothar. Combin.*, 48:Art. B48a, 23 pp. (electronic), 2002.

[438] O. Guibert and T. Mansour. Some statistics on restricted 132 involutions. *Ann. Comb.*, 6(3-4):349–374, 2002.

[439] O. Guibert and E. Pergola. Enumeration of vexillary involutions which are equal to their mirror/complement. *Discrete Math.*, 224(1-3):281–287, 2000.

[440] O. Guibert, E. Pergola, and R. Pinzani. Vexillary involutions are enumerated by Motzkin numbers. *Ann. Comb.*, 5(2):153–174, 2001.

[441] İ. Ş. Güloğlu and C. Koç. Stack-sortable permutations and polynomials. *Turkish J. Math.*, 33(1):1–8, 2009.

[442] J. Haglund. q-rook polynomials and matrices over finite fields. *Adv. in Appl. Math*, 20:450–487, 1997.

[443] J. Haglund, N. Loehr, and J. B. Remmel. Statistics on wreath products, perfect matchings, and signed words. *European J. Combin.*, 26(6):835–868, 2005.

[444] M. Haiman. Smooth Schubert varieties. *preprint*, 1992.

[445] J. Hall, J. Liese, and J. B. Remmel. q-counting descent pairs with prescribed tops and bottoms. *Electron. J. Combin.*, 16(1):Research Paper 111, 25, 2009.

[446] J. T. Hall and J. B. Remmel. Counting descent pairs with prescribed tops and bottoms. *J. Combin. Theory Ser. A*, 115(5):693–725, 2008.

[447] G.-N. Han, A. Randrianarivony, and J. Zeng. Un autre q-analogue des nombres d'Euler. *Sém. Lothar. Combin.*, 42:Art. B42e, 22 pp. (electronic), 1999. The Andrews Festschrift (Maratea, 1998).

[448] M. Hardarson. Avoidance of partially ordered generalized patterns of the form k-σ-k. *arXiv:0805.1872v1 [math.CO]*, 2008.

[449] J. Harmse and J. B. Remmel. Patterns in column strict fillings of rectangular arrays. *arXiv:1103.0077v1 [math.CO]*, 2011.

[450] P. A. Hästö. On descents in standard Young tableaux. *Electron. J. Combin.*, 7:Research Paper 59, 13 pp. (electronic), 2000.

[451] P. A. Hästö. The packing density of other layered permutations. *Electronic J. Combin*, 9:Research Paper 1, 16 pp. (electronic), 2002.

[452] P. E. Haxell, J. J. McDonald, and S. K. Thomasson. Counting interval orders. *Order*, 4:269–272, 1987.

[453] G. A. Hedlund. Remarks on the work of Axel Thue on sequences. *Nordisk Mat. Tidskr.*, 15:148–150, 1967.

[454] I. N. Herstein. *Noncommutative rings*. Math. Assoc. Amer., Washington, DC, 1968.

[455] G. Hetyei. On the *cd*-variation polynomials of André and Simsun permutations. *Discrete Comput. Geom.*, 16(3):259–275, 1996.

[456] G. Hetyei and E. Reiner. Permutation trees and variation statistics. *European J. Combin.*, 19(7):847–866, 1998.

[457] S. Heubach, S. Kitaev, and T. Mansour. Avoidance of partially ordered patterns in compositions. *Pure Math. Appl. (PU.M.A.)*, 17(1-2):123–134, 2006.

[458] S. Heubach and T. Mansour. Counting rises, levels, and drops in compositions. *Integers*, 5(1):A11, 24 pp. (electronic), 2005.

[459] S. Heubach and T. Mansour. Avoiding patterns of length three in compositions and multiset permutations. *Adv. in Appl. Math.*, 36(2):156–174, 2006.

[460] S. Heubach and T. Mansour. Enumeration of 3-letter patterns in compositions. In *Combinatorial number theory*, pages 243–264. de Gruyter, Berlin, 2007.

[461] S. Heubach and T. Mansour. *Combinatorics of compositions and words*. Discrete Mathematics and its Applications (Boca Raton). CRC Press, Boca Raton, 2010.

[462] M. Hildebrand, B. E. Sagan, and V. R. Vatter. Bounding quantities related to the packing density of $1(l+1)l\cdots2$. *Adv. in Appl. Math.*, 33(3):633–653, 2004.

[463] F. Hivert, J.-C. Novelli, L. Tevlin, and J.-Y. Thibon. Permutation statistics related to a class of noncommutative symmetric functions and generalizations of the Genocchi numbers. *Selecta Math. (N.S.)*, 15(1):105–119, 2009.

[464] G. Hong. Catalan numbers in pattern-avoiding permutations. *MIT Undergraduate J. Mathematics*, 10:53–68, 2008.

[465] Q.-H. Hou and T. Mansour. Horse paths, restricted 132-avoiding permutations, continued fractions, and Chebyshev polynomials. *Discrete Appl. Math.*, 154(8):1183–1197, 2006.

[466] S. Huczynska and N. Ruškuc. Pattern classes of permutations via bijections between linearly ordered sets. *European J. Combin.*, 29(1):118–139, 2008.

[467] S. Huczynska and V. R. Vatter. Grid classes and the Fibonacci dichotomy for restricted permutations. *Electron. J. Combin.*, 13(1):Research Paper 54, 14 pp. (electronic), 2006.

[468] A. Hultman and K. Vorwerk. Pattern avoidance and Boolean elements in the Bruhat order on involutions. *J. Algebraic Combin.*, 30(1):87–102, 2009.

[469] L. Ibarra. Finding pattern matchings for permutations. *Inform. Process. Lett.*, 61(6):293–295, 1997.

[470] M. Ishikawa, A. Kasraoui, and J. Zeng. Euler-Mahonian statistics on ordered partitions and Steingrímsson's conjecture—a survey. In *Combinatorial representation theory and related topics*, RIMS Kôkyûroku Bessatsu, B8, pages 99–113. Res. Inst. Math. Sci. (RIMS), Kyoto, 2008.

[471] M. Ishikawa, A. Kasraoui, and J. Zeng. Euler-Mahonian statistics on ordered set partitions. *SIAM J. Discrete Math.*, 22(3):1105–1137, 2008.

[472] B. Jackson, B. Stevens, and G. Hurlbert. Research problems on Gray codes and universal cycles. *Discrete Math.*, 309:5341–5348, 2009.

[473] D. M. Jackson and I. P. Goulden. Algebraic methods for permutations with prescribed patterns. *Adv. in Math.*, 42(2):113–135, 1981.

[474] D. M. Jackson and R. C. Read. A note on permutations without runs of given length. *Aequationes Math.*, 17(2-3):336–343, 1978.

[475] D. M. Jackson and J. W. Reilly. Permutations with a prescribed number of p-runs. *Ars Combinatoria*, 1(1):297–305, 1976.

[476] B. Jacquard and G. Schaeffer. A bijective census of nonseparable planar maps. *J. Combin. Theory Ser. A*, 83(1):1–20, 1998.

[477] A. D. Jaggard. Prefix exchanging and pattern avoidance by involutions. *Electron. J. Combin.*, 9(2):Research paper 16, 24 pp. (electronic), 2002/03. Permutation patterns (Otago, 2003).

[478] A. D. Jaggard. Subsequence containment by involutions. *Electron. J. Combin.*, 12:Research Paper 14, 15 pp. (electronic), 2005.

[479] M. Jani and R. G. Rieper. Continued fractions and Catalan problems. *Electron. J. Combin.*, 7:Research Paper 45, 8 pp. (electronic), 2000.

[480] V. Jelínek. Dyck paths and pattern-avoiding matchings. *European J. Combin.*, 28(1):202–213, 2007.

[481] V. Jelínek. *Wilf-type classifications; Extremal and enumerative theory of ordered structures.* PhD thesis, Charles University in Prague, 2008.

[482] V. Jelínek and T. Mansour. On pattern-avoiding partitions. *Electron. J. Combin.*, 15(1):Research paper 39, 52, 2008.

[483] V. Jelínek and T. Mansour. Wilf-equivalence on k-ary words, compositions, and parking functions. *Electron. J. Combin.*, 16(1):Research Paper 58, 9, 2009.

[484] R. Johansson and S. Linusson. Pattern avoidance in alternating sign matrices. *Ann. Comb.*, 11(3-4):471–480, 2007.

[485] J. R. Johnson. Universal cycles for permutations. *Discrete Math.*, 309(17):5264–5270, 2009.

[486] S. M. Johnson. Generation of permutations by adjacent transposition. *Math. Comp.*, 17:282–285, 1963.

[487] M. E. Jones and J. B. Remmel. Pattern matching in the cycle structure of permutations. *arXiv:1102.3161v1 [math.CO]*, 2011.

[488] M. Josuat-Vergés. A q-enumeration of alternating permutations. *European J. Combin.*, 31(7):1892–1906, 2010.

[489] A. Juarna and V. Vajnovszki. Fast generation of Fibonacci permutations. *CDMTCS, Research Report Series, University of Aukland*, 2004.

[490] A. Juarna and V. Vajnovszki. Some generalizations of a Simion-Schmidt bijection. *Comput. J.*, 50:574–580, 2007.

[491] A. Juarna and V. Vajnovszki. Combinatorial isomorphism between Fibonacci classes. *J. Discrete Math. Sci. Cryptogr.*, 11(2):147–158, 2008.

[492] T. Kaiser and M. Klazar. On growth rates of closed permutation classes. *Electron. J. Combin.*, 9(2):Research paper 10, 20 pp. (electronic), 2002/03. Permutation patterns (Otago, 2003).

[493] A. Kasraoui. A classification of Mahonian maj-inv statistics. *Adv. in Appl. Math.*, 42(3):342–357, 2009.

[494] A. Kasraoui. Ascents and descents in 01-fillings of moon polyominoes. *European J. Combin.*, 31(1):87–105, 2010.

[495] A. Kasraoui and J. Zeng. Euler-Mahonian statistics on ordered set partitions. II. *J. Combin. Theory Ser. A*, 116(3):539–563, 2009.

[496] C. Kassel, A. Lascoux, and C. Reutenauer. The singular locus of a Schubert variety. *J. Algebra*, 269(1):74–108, 2003.

[497] D. Kazhdan and G. Lusztig. Representations of Coxeter groups and Hecke algebras. *Inv. Math.*, 53:165–184, 1979.

[498] A. E. Kézdy, H. S. Snevily, and C. Wang. Partitioning permutations into increasing and decreasing subsequences. *J. Combin. Theory Ser. A*, 73(2):353–359, 1996.

[499] S. M. Khamis. Height counting of unlabeled interval and n-free posets. *Discrete Math.*, 275:165–175, 2004.

[500] A. Khoroshkin and B. Shapiro. Using homological duality in consecutive pattern avoidance. *Electron. J. Combin.*, 18(2):Research Paper 9, 17 pp. (electronic), 2011.

[501] K. Killpatrick. A relationship between the major index for tableaux and the charge statistic for permutations. *Electron. J. Combin.*, 12:Research Paper 45, 9 pp. (electronic), 2005.

[502] K. Killpatrick. Some statistics for Fibonacci tableaux. *European J. Combin.*, 30(4):929–933, 2009.

[503] S. Kitaev. Generalized pattern avoidance with additional restrictions. *Sém. Lothar. Combin.*, 48:Art. B48e, 19 pp. (electronic), 2002.

[504] S. Kitaev. *Generalized patterns in words and permutations.* PhD thesis, Göteborg University/Chalmers University of Technology, 2003.

[505] S. Kitaev. Multi-avoidance of generalised patterns. *Discrete Math.*, 260(1-3):89–100, 2003.

[506] S. Kitaev. There are no iterated morphisms that define the Arshon sequence and the σ-sequence. *J. Autom. Lang. Comb.*, 8:43–50, January 2003.

[507] S. Kitaev. On multi-avoidance of right angled numbered polyomino patterns. *Integers*, 4:A21, 20 pp. (electronic), 2004.

[508] S. Kitaev. The sigma-sequence and occurrences of some patterns, subsequences and subwords. *Australas. J. Combin.*, 29:187–200, 2004.

[509] S. Kitaev. Partially ordered generalized patterns. *Discrete Math.*, 298(1-3):212–229, 2005.

[510] S. Kitaev. Segmental partially ordered generalized patterns. *Theoret. Comput. Sci.*, 349(3):420–428, 2005.

[511] S. Kitaev. Introduction to partially ordered patterns. *Discrete Appl. Math.*, 155(8):929–944, 2007.

[512] S. Kitaev. A survey on partially ordered patterns. *In Permutation Patterns*, vol. 376 of London Math. Soc. Lecture Note Series, pages 115–135, Cambridge University Press, Cambridge, 2010.

[513] S. Kitaev, J. Liese, J. B. Remmel, and B. E. Sagan. Rationality, irrationality, and Wilf equivalence in generalized factor order. *Electron. J. Combin.*, 16(2):Research Paper 22, 26 pp. (electronic), 2009.

[514] S. Kitaev and T. Mansour. Partially ordered generalized patterns and k-ary words. *Ann. Comb.*, 7(2):191–200, 2003.

[515] S. Kitaev and T. Mansour. A survey on certain pattern problems. *Preprint*, 2004.

[516] S. Kitaev and T. Mansour. On multi-avoidance of generalized patterns. *Ars Combin.*, 76:321–350, 2005.

[517] S. Kitaev and T. Mansour. Simultaneous avoidance of generalized patterns. *Ars Combin.*, 75:267–288, 2005.

[518] S. Kitaev, T. Mansour, and J. B. Remmel. Counting descents, rises, and levels, with prescribed first element, in words. *Discrete Math. Theor. Comput. Sci.*, 10(3):1–22, 2008.

[519] S. Kitaev, T. Mansour, and P. Séébold. Generating the Peano curve and counting occurrences of some patterns. *J. Autom. Lang. Comb.*, 9(4):439–455, 2004.

[520] S. Kitaev, T. Mansour, and P. Séébold. Counting ordered patterns in words generated by morphisms. *Integers*, 8:A03, 28, 2008.

[521] S. Kitaev, T. Mansour, and A. Vella. Pattern avoidance in matrices. *J. Integer Seq.*, 8(2):Article 05.2.2, 16 pp. (electronic), 2005.

[522] S. Kitaev, T. B. McAllister, and T. K. Petersen. Enumerating segmented patterns in compositions and encoding by restricted permutations. *Integers*, 6:A34, 16 pp. (electronic), 2006.

[523] S. Kitaev, A. Niedermaier, J. B. Remmel, and M. Riehl. New pattern matching conditions for wreath products of the cyclic groups with symmetric groups. *arXiv:0908.4076v1 [math.CO]*, 2009.

[524] S. Kitaev and A. Pyatkin. On avoidance of V- and Λ-patterns in permutations. *Ars Combinatoria*, 97:203–215, 2010.

[525] S. Kitaev and J. B. Remmel. Classifying descents according to equivalence mod k. *Electron. J. Combin.*, 13(1):Research Paper 64, 39 pp. (electronic), 2006.

[526] S. Kitaev and J. B. Remmel. Classifying descents according to parity. *Ann. Comb.*, 11(2):173–193, 2007.

[527] S. Kitaev and J. B. Remmel. Enumerating (2+2)-free posets by the number of minimal elements and other statistics. *Discrete Math. Theor. Comput. Sci. Proc. AK*, pages 689–700, 2010. Proceedings of the 22nd International Conference on Formal Power Series & Algebraic Combinatorics, San Francisco State University, San Francisco, USA, August 2-6 (2010).

[528] S. Kitaev and J. B. Remmel. Place-difference-value patterns: A generalization of generalized permutation and word patterns. *Integers*, 10:#A11, 129–154, 2010.

[529] S. Kitaev, J. B. Remmel, and M. Riehl. On a pattern avoidance condition for the wreath product of cyclic groups with symmetric groups. *arXiv:0910.3135v1 [math.CO]*, 2009.

[530] S. Kitaev and J. R. Robbins. On multi-dimensional patterns. *Pure Math. Appl. (PU.M.A.)*, 18(3-4):291–299, 2007.

[531] M. Klazar. A general upper bound in extremal theory of sequences. *Comment. Math. Univ. Carolin.*, 33(4):737–746, 1992.

[532] M. Klazar. On abab-free and abba-free set partitions. *European J. Combin.*, 17(1):53–68, 1996.

[533] M. Klazar. Counting pattern-free set partitions. I. A generalization of Stirling numbers of the second kind. *European J. Combin.*, 21(3):367–378, 2000.

[534] M. Klazar. Counting pattern-free set partitions. II. Noncrossing and other hypergraphs. *Electron. J. Combin.*, 7:Research Paper 34, 25 pp. (electronic), 2000.

[535] M. Klazar. The Füredi-Hajnal conjecture implies the Stanley-Wilf conjecture. In *Formal power series and algebraic combinatorics (Moscow, 2000)*, pages 250–255. Springer, Berlin, 2000.

[536] M. Klazar. On growth rates of permutations, set partitions, ordered graphs and other objects. *Electron. J. Combin.*, 15(1):Research Paper 75, 22, 2008.

[537] T. Kløve. Generating functions for the number of permutations with limited displacement. *Electron. J. Combin.*, 16(1):Research Paper 104, 11, 2009.

[538] D. E. Knuth. Permutations, matrices, and generalized Young tableaux. *Pacific J. Math.*, 34:709–727, 1970.

[539] D. E. Knuth. *The art of computer programming. Volume 3, Sorting and Searching*. Addison-Wesley, Reading, 1973. Addison-Wesley Series in Computer Science and Information Processing.

[540] D. E. Knuth. *The art of computer programming. Volume 1, Fundamental Algorithms*. Addison-Wesley, Reading, second edition, 1975. Addison-Wesley Series in Computer Science and Information Processing.

[541] R. Kolpakov. Efficient lower bounds on the number of repetition-free words. *J. Integer Seq.*, 10(3):Article 07.3.2, 16 pp. (electronic), 2007.

[542] M. Korn. Maximal abelian square-free words of short length. *J. Combin. Theory Ser. A*, 102(1):207–211, 2003.

[543] C. Krattenthaler. Permutations with restricted patterns and Dyck paths. *Adv. in Appl. Math.*, 27(2-3):510–530, 2001. Special issue in honor of Dominique Foata's 65th birthday (Philadelphia, PA, 2000).

[544] C. Krattenthaler. Growth diagrams, and increasing and decreasing chains in fillings of ferrers shapes. *Adv. in Appl. Math.*, 37(3):404–431, 2006. Special Issue in honor of Amitai Regev on his 65th Birthday.

[545] D. Kremer. Permutations with forbidden subsequences and a generalized Schröder number. *Discrete Math.*, 218(1-3):121–130, 2000.

[546] D. Kremer. Postscript: "Permutations with forbidden subsequences and a generalized Schröder number" [Discrete Math. 218 (2000), no. 1-3, 121–130; MR1754331 (2001a:05005)]. *Discrete Math.*, 270(1-3):333–334, 2003.

[547] D. Kremer and W. C. Shiu. Finite transition matrices for permutations avoiding pairs of length four patterns. *Discrete Math.*, 268(1-3):171–183, 2003.

[548] A. G. Kuznetsov, I. M. Pak, and A. E. Postnikov. Increasing trees and alternating permutations. *Uspekhi Mat. Nauk*, 49(6(300)):79–110, 1994.

[549] G. Labelle, P. Leroux, E. Pergola, and R. Pinzani. Stirling numbers interpolation using permutations with forbidden subsequences. *Discrete Math.*, 246(1-3):177–195, 2002. Formal power series and algebraic combinatorics (Barcelona, 1999).

[550] D. Lai. A comparative study on a permutation statistic. *Stat. Methodol.*, 2(3):155–167, 2005.

[551] V. Lakshmibai and B. Sandhya. Criterion for smoothness of Schubert varieties in Sl(n)/B. *Proc. Indian Acad. Sci. Math. Sci.*, 100(1):45–52, 1990.

[552] T. Langley, J. Liese, and J. B. Remmel. Generating functions for Wilf equivalence under generalized factor order. *arXiv:1005.4372v1 [math.CO]*, 2010.

[553] T. Langley and J. B. Remmel. Enumeration of m-tuples of permutations and a new class of power bases for the space of symmetric functions. *Adv. in Appl. Math.*, 36(1):30–66, 2006.

[554] A. Lascoux and M. P. Schützenberger. Polynomes de Schubert. *C. R. Acad. Sci. Paris, Série I*, 294:447–450, 1982.

[555] B. Lass. Orientations acycliques et le polynome chromatique. *European J. Combin.*, 22(8):1101–1123, 2001.

[556] I. Le. Wilf classes of pairs of permutations of length 4. *Electron. J. Combin.*, 12:Research Paper 25, 27 pp. (electronic), 2005.

[557] P. Levande. Two new interpretations of the fishburn numbers and their refined generating functions. *arXiv:1006.3013v1 [math.CO]*, 2010.

[558] J. B. Lewis. Alternating, pattern-avoiding permutations. *Electron. J. Combin.*, 16(1):Note 7, 8 pp. (electronic), 2009.

[559] J. B. Lewis. Generating trees and pattern avoidance in alternating permutations. *arXiv:1005.4046v1 [math.CO]*, 2010.

[560] J. B. Lewis. Pattern avoidance and RSK-like algorithms for alternating permutations and Young tableaux. *arXiv:0909.4966v2 [math.CO]*, 2010.

[561] J. Liese. Counting descents and ascents relative to equivalence classes mod k. *Ann. Comb.*, 11(3-4):481–506, 2007.

[562] J. Liese. *Counting patterns in permutations and words*. PhD thesis, University of California, San Diego, 2008.

[563] J. Liese and J. B. Remmel. Generating functions for permutations avoiding a consecutive pattern. *Ann. Comb.*, 14(1):123–141, 2010.

[564] J. Liese and J. B. Remmel. Q-analogues of the number of permutations with k-excedances. *Pure Math. Appl. (PU.M.A.)*, 21(2):285–320, 2010.

[565] S. Linusson. Extended pattern avoidance. *Discrete Math.*, 246(1-3):219–230, 2002. Formal power series and algebraic combinatorics (Barcelona, 1999).

[566] M. Lipson. Completion of the Wilf-classification of 3-5 pairs using generating trees. *Electron. J. Combin.*, 13(1):Research Paper 31, 19 pp. (electronic), 2006.

[567] N. A. Loehr. Conjectured statistics for the higher q, t-Catalan sequences. *Electron. J. Combin.*, 12:Research Paper 9, 54 pp. (electronic), 2005.

[568] N. A. Loehr. Permutation statistics and the q, t-Catalan sequence. *European J. Combin.*, 26(1):83–93, 2005.

[569] N. A. Loehr. The major index specialization of the q, t-Catalan. *Ars Combin.*, 83:145–160, 2007.

[570] B. F. Logan and L. A. Shepp. A variational problem for random Young tableaux. *Advances in Math.*, 26:206–222, 1977.

[571] M. Lothaire. *Combinatorics on words*. Cambridge Mathematical Library. Cambridge University Press, Cambridge, 1997. With a foreword by Roger Lyndon and a preface by Dominique Perrin, Corrected reprint of the 1983 original, with a new preface by Perrin.

[572] M. Lothaire. *Algebraic combinatorics on words*, volume 90 of *Encyclopedia of Mathematics and its Applications*. Cambridge University Press, Cambridge, 2002. A collective work by Jean Berstel, Dominique Perrin, Patrice Séébold, Julien Cassaigne, Aldo De Luca, Steffano Varricchio, Alain Lascoux, Bernard

Leclerc, Jean-Yves Thibon, Veronique Bruyère, Christiane Frougny, Filippo Mignosi, Antonio Restivo, Christophe Reutenauer, Dominique Foata, Guo-Niu Han, Jacques Désarménien, Volker Diekert, Tero Harju, Juhani Karhumaki and Wojciech Plandowski, With a preface by Berstel and Perrin.

[573] M. Lothaire. *Applied combinatorics on words*, volume 105 of *Encyclopedia of Mathematics and its Applications*. Cambridge University Press, Cambridge, 2005. A collective work by Jean Berstel, Dominique Perrin, Maxime Crochemore, Eric Laporte, Mehryar Mohri, Nadia Pisanti, Marie-France Sagot, Gesine Reinert, Sophie Schbath, Michael Waterman, Philippe Jacquet, Wojciech Szpankowski, Dominique Poulalhon, Gilles Schaeffer, Roman Kolpakov, Gregory Koucherov, Jean-Paul Allouche and Valérie Berthé, With a preface by Berstel and Perrin.

[574] G. Louchard and H. Prodinger. The number of inversions in permutations: a saddle point approach. *J. Integer Seq.*, 6(2):Article 03.2.8, 19 pp. (electronic), 2003.

[575] M. Lugo. The cycle structure of compositions of random involutions. *arXiv:0911.3604v1 [math.CO]*, 2009.

[576] M. Lugo. Profiles of permutations. *Electron. J. Combin.*, 16(1):Research Paper 99, 20, 2009.

[577] C. T. MacDonald, J. H. Gibbs, and A. C. Pipkin. Kinetics of biopolymerization on nucleic acid templates. *Biopolymers*, 6(1):1–25, 1968.

[578] P. A. MacMahon. *Combinatory analysis. Vol. I, II (bound in one volume)*. Dover Phoenix Editions. Dover, Mineola, 2004. Reprint of *An introduction to combinatory analysis* (1920) and *Combinatory analysis. Vol. I, II* (1915, 1916).

[579] C. L. Mallows. Problem 62-2, patience sorting. *SIAM Review*, 4:148–149, 1962.

[580] C. L. Mallows. Problem 62-2. *SIAM Review*, 5:375–376, 1963.

[581] C. L. Mallows. Baxter permutations rise again. *J. Combin. Theory Ser. A*, 27(3):394–396, 1979.

[582] P. Manara and P. C. Cippo. The fine structure of 321 avoiding involutions. *arXiv:1010.5919v1 [math.CO]*, 2010.

[583] L. Manivel. Generic singularities of Schubert varieties. *arXiv:0105239v1 [math.CO]*, 2001.

[584] L. Manivel. Le lieu singulier des variétés de Schubert. *Internat. Math. Res. Notices*, (16):849–871, 2001.

[585] T. Mansour. Permutations containing and avoiding certain patterns. In *Formal power series and algebraic combinatorics (Moscow, 2000)*, pages 704–708. Springer, Berlin, 2000.

[586] T. Mansour. Pattern avoidance in coloured permutations. *Sém. Lothar. Combin.*, 46:Art. B46g, 12 pp. (electronic), 2001/02.

[587] T. Mansour. Continued fractions and generalized patterns. *European J. Combin.*, 23(3):329–344, 2002.

[588] T. Mansour. Permutations avoiding a pattern from S_k and at least two patterns from S_3. *Ars Combin.*, 62:227–239, 2002.

[589] T. Mansour. Restricted 1-3-2 permutations and generalized patterns. *Ann. Comb.*, 6(1):65–76, 2002.

[590] T. Mansour. Coloured permutations containing and avoiding certain patterns. *Ann. Comb.*, 7(3):349–355, 2003.

[591] T. Mansour. Restricted 132-alternating permutations and Chebyshev polynomials. *Ann. Comb.*, 7(2):201–227, 2003.

[592] T. Mansour. 321-avoiding permutations and Chebyshev polynomials. In *Mathematics and computer science. III*, Trends Math., pages 37–38. Birkhäuser, Basel, 2004.

[593] T. Mansour. Continued fractions, statistics, and generalized patterns. *Ars Combin.*, 70:265–274, 2004.

[594] T. Mansour. Counting occurrences of 132 in an even permutation. *Int. J. Math. Math. Sci.*, (25-28):1329–1341, 2004.

[595] T. Mansour. On an open problem of Green and Losonczy: exact enumeration of freely braided permutations. *Discrete Math. Theor. Comput. Sci.*, (6):461–470, 2004.

[596] T. Mansour. Permutations containing a pattern exactly once and avoiding at least two patterns of three letters. *Ars Combin.*, 72:213–222, 2004.

[597] T. Mansour. Permutations restricted by patterns of type $(2, 1)$. *Ars Combin.*, 71:201–223, 2004.

[598] T. Mansour. Restricted 132-Dumont permutations. *Australas. J. Combin.*, 29:103–117, 2004.

[599] T. Mansour. Equidistribution and sign-balance on 132-avoiding permutations. *Sém. Lothar. Combin.*, 51:Art. B51e, 11 pp. (electronic), 2004/05.

[600] T. Mansour. The enumeration of permutations whose posets have a maximum element. *Adv. in Appl. Math.*, 37(4):434–442, 2006.

[601] T. Mansour. Restricted 132-avoiding k-ary words, Chebyshev polynomials, and continued fractions. *Adv. in Appl. Math.*, 36(2):175–193, 2006.

[602] T. Mansour. Restricted even permutations and Chebyshev polynomials. *Discrete Math.*, 306(12):1161–1176, 2006.

[603] T. Mansour. Longest alternating subsequences in pattern-restricted k-ary words. *Online J. Anal. Comb.*, (3):Art. 5, 10, 2008.

[604] T. Mansour. Longest alternating subsequences of k-ary words. *Discrete Appl. Math.*, (156):119–124, 2008.

[605] T. Mansour. Enumeration of words by the sum of differences between adjacent letters. *Discrete Math. Theor. Comput. Sci.*, 11(1):173–185, 2009.

[606] T. Mansour, E. Y. P. Deng, and R. R. X. Du. Dyck paths and restricted permutations. *Discrete Appl. Math.*, 154(11):1593–1605, 2006.

[607] T. Mansour and A. O. Munagi. Enumeration of partitions by long rises, levels, and descents. *J. Integer Seq.*, 12(1):Article 09.1.8, 17, 2009.

[608] T. Mansour and A. Robertson. Refined restricted permutations avoiding subsets of patterns of length three. *Ann. Comb.*, 6(3-4):407–418, 2002.

[609] T. Mansour, M. Shattuck, and S. H. F. Yan. Counting subwords in a partition of a set. *Electron. J. Combin.*, 17:Research Paper 19, 21 pp. (electronic), 2010.

[610] T. Mansour and C. Song. New permutation statistics: variation and a variant. *Discrete Appl. Math.*, 157(8):1974–1978, 2009.

[611] T. Mansour and Z. E. Stankova. 321-polygon-avoiding permutations and Chebyshev polynomials. *Electron. J. Combin.*, 9(2):Research paper 5, 16 pp. (electronic), 2002/03. Permutation patterns (Otago, 2003).

[612] T. Mansour and Y. Sun. Excedance numbers for permutations in complex reflection groups. *Sém. Lothar. Combin.*, 58:Art. B58b, 7, 2007/08.

[613] T. Mansour and Y. D. Sun. Even signed permutations avoiding 2-letter signed patterns. *J. Math. Res. Exposition*, 29(5):813–822, 2009.

[614] T. Mansour and A. Vainshtein. Avoiding maximal parabolic subgroups of S_k. *Discrete Math. Theor. Comput. Sci.*, 4(1):67–75 (electronic), 2000.

[615] T. Mansour and A. Vainshtein. Restricted permutations, continued fractions, and Chebyshev polynomials. *Electron. J. Combin.*, 7:Research Paper 17, 9 pp. (electronic), 2000.

[616] T. Mansour and A. Vainshtein. Layered restrictions and Chebyshev polynomials. *Ann. Comb.*, 5(3-4):451–458, 2001. Dedicated to the memory of Gian-Carlo Rota (Tianjin, 1999).

[617] T. Mansour and A. Vainshtein. Restricted 132-avoiding permutations. *Adv. in Appl. Math.*, 26(3):258–269, 2001.

[618] T. Mansour and A. Vainshtein. Restricted permutations and Chebyshev polynomials. *Sém. Lothar. Combin.*, 47:Article B47c, 17 pp. (electronic), 2001/02.

[619] T. Mansour and A. Vainshtein. Counting occurrences of 132 in a permutation. *Adv. in Appl. Math.*, 28(2):185–195, 2002.

[620] T. Mansour and V. Vajnovszki. Restricted 123-avoiding Baxter permutations and the Padovan numbers. *Discrete Appl. Math.*, 155(11):1430–1440, 2007.

[621] T. Mansour and J. West. Avoiding 2-letter signed patterns. *Sém. Lothar. Combin.*, 49:Art. B49a, 11 pp. (electronic), 2002/04.

[622] T. Mansour, S. H. F. Yan, and L. L. M. Yang. Counting occurrences of 231 in an involution. *Discrete Math.*, 306(6):564–572, 2006.

[623] A. Marcus and G. Tardos. Excluded permutation matrices and the Stanley-Wilf conjecture. *J. Combin. Theory Ser. A*, 107(1):153–160, 2004.

[624] D. Marinov and R. Radoičić. Counting 1324-avoiding permutations. *Electron. J. Combin.*, 9(2):Research paper 13, 9 pp. (electronic), 2002/03. Permutation patterns (Otago, 2003).

[625] B. D. McKay, J. Morse, and H. S. Wilf. The distributions of the entries of Young tableaux. *J. Combin. Theory Ser. A*, 97(1):117–128, 2002.

[626] A. Mendes. *Bulding generating functions brick by brick*. PhD thesis, University of California, San Diego, 2004.

[627] A. Mendes and J. B. Remmel. *Symmetric Functions and Generating Functions for Permutations and Words*. In preparation.

[628] A. Mendes and J. B. Remmel. Generating functions for statistics on $C_k \wr S_n$. *Sém. Lothar. Combin.*, 54A:Art. B54At, 40 pp. (electronic), 2005/07.

[629] A. Mendes and J. B. Remmel. Permutations and words counted by consecutive patterns. *Adv. in Appl. Math.*, 37(4):443–480, 2006.

[630] A. Mendes and J. B. Remmel. Descents, inversions, and major indices in permutation groups. *Discrete Math.*, 308(12):2509–2524, 2008.

[631] A. Miller. Asymptotic bounds for permutations containing many different patterns. *J. Combin. Theory Ser. A*, 116(1):92–108, 2009.

[632] E. Munarini and N. Zagaglia Salvi. Binary strings without zigzags. *Sém. Lothar. Combin.*, 49:Art. B49h, 15 pp. (electronic), 2002/04.

[633] E. Munarini and N. Zagaglia Salvi. Circular binary strings without zigzags. *Integers*, 3:A19, 17 pp. (electronic), 2003.

[634] M. M. Murphy. *Restricted permutations, antichains, atomic classes, and stack sorting*. PhD thesis, University of St Andrews, 2002.

[635] M. M. Murphy and V. R. Vatter. Profile classes and partial well-order for permutations. *Electron. J. Combin.*, 9(2):Research paper 17, 30 pp. (electronic), 2002/03. Permutation patterns (Otago, 2003).

[636] A. N. Myers. Counting permutations by their rigid patterns. *J. Combin. Theory Ser. A*, 99(2):345–357, 2002.

[637] A. N. Myers. Pattern avoidance in multiset permutations: bijective proof. *Ann. Comb.*, 11(3-4):507–517, 2007.

[638] A. N. Myers. Forbidden substrings on weighted alphabets. *Australas. J. Combin.*, 45:59–65, 2009.

[639] A. N. Myers and H. S. Wilf. Left-to-right maxima in words and multiset permutations. *Israel J. Math.*, 166:167–183, 2008.

[640] J. S. Myers. The minimum number of monotone subsequences. *Electron. J. Combin.*, 9(2):Research paper 4, 17 pp. (electronic), 2002/03. Permutation patterns (Otago, 2003).

[641] B. Nakamura. Computational approaches to consecutive pattern avoidance in permutations. *arXiv:1102.2480v1 [math.CO]*, 2011.

[642] A. Niedermaier. *Statistics on wreath products.* PhD thesis, University of California, San Diego, 2009.

[643] J. Noonan. The number of permutations containing exactly one increasing subsequence of length three. *Discrete Math.*, 152(1-3):307–313, 1996.

[644] J. Noonan and D. Zeilberger. The enumeration of permutations with a prescribed number of "forbidden" patterns. *Adv. in Appl. Math.*, 17(4):381–407, 1996.

[645] J. Noonan and D. Zeilberger. The Goulden-Jackson cluster method: extensions, applications and implementations. *J. Diff. Equations Appl.*, 5(4–5):355–377, 1999.

[646] A. M. Odlyzko and E. M. Rains. On longest increasing subsequences in random permutations. In *Analysis, geometry, number theory: the mathematics of Leon Ehrenpreis (Philadelphia, PA, 1998)*, volume 251 of *Contemp. Math.*, pages 439–451. Amer. Math. Soc., Providence, RI, 2000.

[647] E. Ouchterlony. *On Young tableau involutions and patterns in permutations.* PhD thesis, Linköpings Universitet, 2005.

[648] I. Pak and A. Redlich. Long cycles in *abc*-permutations. *Funct. Anal. Other Math.*, 2(1):87–92, 2008.

[649] G. Panova. Bijective enumeration of permutations starting with a longest increasing subsequence. *arXiv:0905.2013v2 [math.CO]*, 2009.

[650] R. Parviainen. Lattice path enumeration of permutations with k occurrences of the pattern 2-13. *J. Integer Seq.*, 9(3):Article 06.3.2, 8 pp. (electronic), 2006.

[651] R. Parviainen. Permutations, cycles and the pattern 2-13. *Electron. J. Combin.*, 13(1):Research Paper 111, 13 pp. (electronic), 2006.

[652] R. Parviainen. Cycles and patterns in permutations. *arXiv:0610616v3 [math.CO]*, 2007.

[653] T. K. Petersen. The sorting index. *arXiv:1007.1207v1 [math.CO]*, 2010.

[654] F. Petursson. A bijection between a class of directed rooted animals and certain permutations. *In preparation.*

[655] M. Poneti and V. Vajnovszki. Generating restricted classes of involutions, Bell and Stirling permutations. *European J. Combin.*, 31(2):553–564, 2010.

[656] A. Postnikov. Webs in totally positive Grassmann cells. *In preparation*.

[657] M. R. Pournaki. On the number of even permutations with roots. *Australas. J. Combin.*, 45:37–42, 2009.

[658] V. R. Pratt. Computing permutations with double-ended queues. Parallel stacks and parallel queues. In *Fifth Annual ACM Symposium on Theory of Computing (Austin, Tex., 1973)*, pages 268–277. Assoc. Comput. Mach., New York, 1973.

[659] C. B. Presutti. Determining lower bounds for packing densities of non-layered patterns using weighted templates. *Electron. J. Combin.*, 15:Research paper 50, 10 pp. (electronic), 2008.

[660] C. B. Presutti and W. Stromquist. Packing rates of measures and a conjecture for the packing density of 2413. *Permutation Patterns*, vol. 376 of London Math. Soc. Lecture Note Series, pages 287–316, Cambridge University Press, Cambridge, 2010.

[661] A. Price. *Packing densities of layered patterns*. PhD thesis, University of Pennsylvania, 1997.

[662] L. K. Pudwell. *Enumeration schemes for pattern-avoiding words and permutations*. PhD thesis, State University of New Jersey, 2008.

[663] L. K. Pudwell. Enumeration schemes for words avoiding patterns with repeated letters. *Integers*, 8:A40, 19, 2008.

[664] E. M. Rains. Increasing subsequences and the classical groups. *Electron. J. Combin.*, 5:Research Paper 12, 9 pp. (electronic), 1998.

[665] D. Rawlings. The r-major index. *J. Combin. Theory Ser. A*, 31(2):175–183, 1981.

[666] D. Rawlings. Enumeration of permutations by descents, idescents, imajor index, and basic components. *J. Combin. Theory Ser. A*, 36(1):1–14, 1984.

[667] D. Rawlings. Multicolored permutations, sequences, and tableaux. *Discrete Math.*, 83(1):63–79, 1990.

[668] D. Rawlings. Bernoulli trials and permutation statistics. *Internat. J. Math. Math. Sci.*, 15(2):291–311, 1992.

[669] D. Rawlings. A binary tree decomposition space of permutation statistics. *J. Combin. Theory Ser. A*, 59(1):111–124, 1992.

[670] D. Rawlings. Restricted words by adjacencies. *Discrete Math.*, 220(1-3):183–200, 2000.

[671] D. Rawlings. The q-exponential generating function for permutations by consecutive patterns and inversions. *J. Combin. Theory Ser. A*, 114(1):184–193, 2007.

[672] D. Rawlings and M. Tiefenbruck. Consecutive patterns: from permutations to column-convex polyominoes and back. *Electron. J. Combin.*, 17:Research Paper 62, 33 pp. (electronic), 2010.

[673] N. Ray and J. West. Posets of matrices and permutations with forbidden subsequences. *Ann. Comb.*, 7(1):55–88, 2003.

[674] A. Regev. Asymptotic values for degrees associated with strips of Young diagrams. *Adv. in Math.*, 41(2):115–136, 1981.

[675] A. Regev. Asymptotics of the number of k-words with an l-descent. *Electron. J. Combin.*, 5:Research Paper 15, 4pp. (electronic), 1998.

[676] A. Regev and Y. Roichman. Permutation statistics on the alternating group. *Adv. in Appl. Math.*, 33(4):676–709, 2004.

[677] A. Regev and Y. Roichman. Generalized statistics on S_n and pattern avoidance. *European J. Combin.*, 26(1):29–57, 2005.

[678] A. Reifegerste. A generalization of Simion-Schmidt's bijection for restricted permutations. *Electron. J. Combin.*, 9(2):Research paper 14, 9 pp. (electronic), 2002/03. Permutation patterns (Otago, 2003).

[679] A. Reifegerste. On the diagram of Schröder permutations. *Electron. J. Combin.*, 9(2):Research paper 8, 23 pp. (electronic), 2002/03. Permutation patterns (Otago, 2003).

[680] A. Reifegerste. On the diagram of 132-avoiding permutations. *European J. Combin.*, 24(6):759–776, 2003.

[681] A. Reifegerste. Permutation sign under the Robinson-Schensted correspondence. *Ann. Comb.*, 8(1):103–112, 2004.

[682] A. Reifegerste. Refined sign-balance on 321-avoiding permutations. *European J. Combin.*, 26(6):1009–1018, 2005.

[683] V. Reiner. Signed permutation statistics. *European J. Combin.*, 14(6):553–567, 1993.

[684] V. Reiner. Signed permutation statistics and cycle type. *European J. Combin.*, 14(6):569–579, 1993.

[685] V. Reiner. Upper binomial posets and signed permutation statistics. *European J. Combin.*, 14(6):581–588, 1993.

[686] V. Reiner. The distribution of descents and length in a Coxeter group. *Electronic J. Combin.*, 2:25, 1995.

[687] V. Reiner. Note on the expected number of Yang-Baxter moves applicable to reduced decompositions. *European J. Combin.*, 26(6):1019–1021, 2005.

[688] J. B. Remmel. Bijective proofs of formulae for the number of standard Young tableaux. *Linear and Multilinear Algebra*, 11(1):45–100, 1982.

[689] J. B. Remmel and M. Riehl. Generating functions for permutations which contain a given descent set. *Electron. J. Combin.*, 17:Research Paper 27, 33pp. (electronic), 2010.

[690] J. B. Remmel and R. Whitney. A bijective proof of the hook formula for the number of column strict tableaux with bounded entries. *European J. Combin.*, 4(1):45–63, 1983.

[691] D. Richards. Ballot sequences and restricted permutations. *Ars Combin.*, 25:83–86, 1988.

[692] A. Riehl. *Ribbon Schur functions and permutation patterns.* PhD thesis, University of California, San Diego, 2008.

[693] R. Rieper and M. Zeleke. Valleyless sequences. *Congressus Numerantium*, 145:33–42, 2000.

[694] A. Robertson. Permutations containing and avoiding 123 and 132 patterns. *Discrete Math. Theor. Comput. Sci.*, 3(4):151–154 (electronic), 1999.

[695] A. Robertson. Permutations restricted by two distinct patterns of length three. *Adv. in Appl. Math.*, 27(2-3):548–561, 2001. Special issue in honor of Dominique Foata's 65th birthday.

[696] A. Robertson. Restricted permutations from Catalan to Fine and back. *Sém. Lothar. Combin.*, 50:Art. B50g, 13 pp. (electronic), 2003/04.

[697] A. Robertson, D. Saracino, and D. Zeilberger. Refined restricted permutations. *Ann. Comb.*, 6(3-4):427–444, 2002.

[698] A. Robertson, H. S. Wilf, and D. Zeilberger. Permutation patterns and contin-
       ued fractions. *Electron. J. Combin.*, 6:Research Paper 38, 6 pp. (electronic),
       1999.

[699] O. Rodriguez. Note sur les inversions, ou dérangements produits dans les
       permutations. *J. de Math.*, 4:236–240, 1839.

[700] D. Roelants van Baronaigien. Constant time generation of involutions. *Congr.
       Numer.*, 90:87–96, 1992.

[701] D. G. Rogers. Ascending sequences in permutations. *Discrete Math.*, 22(1):35–
       40, 1978.

[702] M. Ronco. Shuffle bialgebras. *arXiv:0703437v3 [math.CO]*, 2008.

[703] P. Rosenstiehl and R. E. Tarjan. Gauss codes, planar Hamiltonian graphs,
       and stack-sortable permutations. *J. Algorithms*, 5(3):375–390, 1984.

[704] D. Rotem. On a correspondence between binary trees and a certain type of
       permutation. *Information Processing Lett.*, 4(3):58–61, 1975/76.

[705] D. Rotem. Stack sortable permutations. *Discrete Math.*, 33(2):185–196, 1981.

[706] K. M. Ryan. On Schubert varieties in the flag manifold of Sl(n,$\mathbb{C}$). *Math.
       Ann.,*, 276(2):205–224, 1987.

[707] B. E. Sagan. Pattern avoidance in set partitions. *arXiv:0604292v1 [math.CO]*,
       2008.

[708] B. E. Sagan and V. R. Vatter. The Möbius function of a composition poset.
       J. Algebraic Combin., 24(2):117–136, 2006.

[709] A. Sapounakis, I. Tasoulas, and P. Tsikouras. Counting strings in Dyck paths.
       Discrete Math., 307(23):2909–2924, 2007.

[710] D. Saracino. On two bijections from $s_n(321)$ to $s_n(132)$. *arXiv:1008.4557v1
       [math.CO]*, 2010.

[711] T. Sasamoto. One-dimensional partially asymmetric simple exclusion process
       with open boundaries: Orthogonal polynomials approach. *J. Phys.*, 32:7109–
       7131, 1999.

[712] C. D. Savage. A survey of combinatorial Gray codes. *SIAM Rev.*, 39(4):605–
       629, 1997.

[713] C. D. Savage and H. S. Wilf. Pattern avoidance in compositions and multiset permutations. *Adv. in Appl. Math.*, 36(2):194–201, 2006.

[714] J. H. Schmerl and W. T. Trotter. Critically indecomposable partially ordered sets, graphs, tournaments and other binary relational structures. *Discrete Math.*, 113(1-3):191–205, 1993.

[715] M. Schreckenberg and D. Wolf. *Traffic and Granular Flow'97*, 1998. Springer, Singapore.

[716] L. W. Shapiro and A. B. Stephens. Bootstrap percolation, the Schröder numbers, and the n-kings problem. *SIAM J. Discrete Math.*, 4:275–280, 1991.

[717] L. W. Shapiro and R. A. Sulanke. Bijections for the Schröder numbers. *Math. Mag.*, 73(5):369–376, 2000.

[718] A. Sharma. Enumerating permutations that avoid three term arithmetic progressions. *Electron. J. Combin.*, 16(1):Research Paper 63, 15, 2009.

[719] R. Simion. Partially ordered sets associated with permutations. *European J. Combin.*, 10(4):375–391, 1989.

[720] R. Simion. Combinatorial statistics on type-B analogues of noncrossing partitions and restricted permutations. *Electron. J. Combin.*, 7:Research Paper 9, 27 pp. (electronic), 2000.

[721] R. Simion and F. W. Schmidt. Restricted permutations. *European J. Combin.*, 6(4):383–406, 1985.

[722] R. Simion and D. Stanton. Octabasic Laguerre polynomials and permutation statistics. *J. Comput. Appl. Math.*, 68(1-2):297–329, 1996.

[723] M. Skandera. An application of Dumont's statistic. In *Formal power series and algebraic combinatorics (Moscow, 2000)*, pages 743–753. Springer, Berlin, 2000.

[724] M. Skandera. Dumont's statistic on words. *Electron. J. Combin.*, 8(1):Research Paper 11, 19 pp. (electronic), 2001.

[725] M. Skandera. An Eulerian partner for inversions. *Sém. Lothar. Combin.*, 46:Art. B46d, 19 pp. (electronic), 2001/02.

[726] N. J. A. Sloane. The on-line encyclopedia of integer sequences. *Notices Amer. Math. Soc.*, 50(8):912–915, 2003. Available on-line at http://www.research.att.com/~njas/sequences/index.html.

[727] R. Smith. Comparing algorithms for sorting with t stacks in series. *Ann. Comb.*, 8(1):113–121, 2004.

[728] R. Smith. *Combinatorial algorithms involving pattern containing and avoiding permutations*. PhD thesis, University of Florida, 2005.

[729] R. Smith and V. R. Vatter. The enumeration of permutations sortable by pop stacks in parallel. *Inform. Process. Lett.*, 109(12):626–629, 2009.

[730] D. A. Spielman and M. Bóna. An infinite antichain of permutations. *Electron. J. Combin.*, 7:Note 2, 4 pp. (electronic), 2000.

[731] A. Spiridonov. Pattern-avoidance in binary fillings of grid shapes (short version). *DMTCS proc. AJ*, pages 677–690, 2008.

[732] A. Spiridonov. *Pattern-avoidance in binary fillings of grid shapes*. PhD thesis, Massachusetts Institute of Technology, 2009.

[733] F. Spitzer. Interaction of markov processes. *Adv. Math.*, 5:246–290, 1970.

[734] Z. E. Stankova. Forbidden subsequences. *Discrete Math.*, 132(1-3):291–316, 1994.

[735] Z. E. Stankova. Classification of forbidden subsequences of length 4. *European J. Combin.*, 17(5):501–517, 1996.

[736] Z. E. Stankova. Shape-Wilf-ordering on permutations of length 3. *Electron. J. Combin.*, 14:Research Paper 56, 44, 2007.

[737] Z. E. Stankova and J. West. A new class of Wilf-equivalent permutations. *J. Algebraic Combin.*, 15(3):271–290, 2002.

[738] Z. E. Stankova and J. West. Explicit enumeration of 321-hexagon-avoiding permutations. *Discrete Math.*, 280(1-3):165–189, 2004.

[739] R. P. Stanley. Binomial posets, Möbius inversion, and permutation enumeration. *J. Combin. Theory Ser. A*, 20(3):336 – 356, 1976.

[740] R. P. Stanley. Hipparchus, Plutarch, Schröder, and Hough. *Amer. Math. Monthly*, 104:344–350, 1979.

[741] R. P. Stanley. Differentiably finite power series. *European J. Combin.*, 1:175–188, 1980.

[742] R. P. Stanley. On the number of reduced decompositions of elements of Coxeter groups. *European J. Combin.*, 5:359–372, 1984.

[743] R. P. Stanley. *Enumerative combinatorics. Vol. I.* The Wadsworth & Brooks/Cole Mathematics Series. Wadsworth & Brooks/Cole Advanced Books & Software, Monterey, 1986. With a foreword by Gian-Carlo Rota.

[744] R. P. Stanley. Log-concave and unimodal sequences in algebra, combinatorics and geometry. *Ann. NY Acad. Sci.*, 576:500–535, 1989. Graph Theory and Its Applications: East and West.

[745] R. P. Stanley. *Enumerative combinatorics. Vol. 2*, volume 62 of *Cambridge Studies in Advanced Mathematics*. Cambridge University Press, Cambridge, 1999. With a foreword by Gian-Carlo Rota and appendix 1 by Sergey Fomin.

[746] R. P. Stanley. The descent set and connectivity set of a permutation. *J. Integer Seq.*, 8(3):Article 05.3.8, 9 pp. (electronic), 2005.

[747] R. P. Stanley. Longest alternating subsequences of permutations. *Michigan Math. J.*, 57:675–687, 2008. Special volume in honor of Melvin Hochster.

[748] R. P. Stanley. Catalan addendum to enumerative combinatorics. *Available at www-math.mit.edu/~rstan/ec/catadd.pdf*, 2009.

[749] R. P. Stanley. A survey of alternating permutations. *arXiv:0912.4240v1 [math.CO]*, 2009.

[750] E. Steingrímsson. Permutation statistics of indexed permutations. *European J. Combin.*, 15(2):187–205, 1994.

[751] E. Steingrímsson. Generalized permutation patterns – a short survey. *Permutation Patterns*, vol. 376 of London Math. Soc. Lecture Note Series, pages 137–152, Cambridge University Press, Cambridge, 2010.

[752] E. Steingrímsson and B. E. Tenner. The Möbius function of the permutation pattern poset. *arXiv:0902.4011v3 [math.CO]*, 2010.

[753] E. Steingrímsson and L. K. Williams. Permutation tableaux and permutation patterns. *J. Combin. Theory Ser. A*, 114(2):211–234, 2007.

[754] A. Stoimenow. Enumeration of chord diagrams and an upper bound for Vassiliev invariants. *J. Knot Theory Ramifications*, 7(1):93–114, 1998.

[755] V. Strehl. Inversions in 2-ordered permutations—a bijective counting. *Bayreuth. Math. Schr.*, (28):127–138, 1989.

[756] V. Strehl. Comment on: "Permutations which are the union of an increasing and a decreasing subsequence" [Electron. J. Combin. **5** (1998), no. 1, Research Paper 6, 13 pp. (electronic); MR1490467 (98k:05001)] by M. D. Atkinson. *Electron. J. Combin.*, 5:Research Paper 6, comment 1, 1 HTML document; approx. 3 pp. (electronic), 1998.

[757] W. Stromquist. Packing layered posets into posets. *Preprint*, 1993.

[758] S. Sundaram. The homology of partitions with an even number of blocks. *Journal of Algebraic Combinatorics*, 4:69–92, 1995. 10.1023/A:1022437708487.

[759] S. Tanimoto. A study of Eulerian numbers by means of an operator on permutations. *European J. Combin.*, 24(1):33–43, 2003.

[760] S. Tanimoto. On the numbers of orbits of permutations under an operator related to Eulerian numbers. *Ann. Comb.*, 8(2):239–250, 2004.

[761] S. Tanimoto. A study of Eulerian numbers for permutations in the alternating group. *Integers*, 6:A31, 12 pp. (electronic), 2006.

[762] R. Tarjan. Sorting using networks of queues and stacks. *J. Assoc. Comput. Mach.*, 19:341–346, 1972.

[763] R. Tauraso. The dinner table problem: the rectangular case. *arXiv:0507293v1*, 2005.

[764] B. E. Tenner. Reduced decompositions and permutation patterns. *J. Algebraic Combin.*, 24(3):263–284, 2006.

[765] B. E. Tenner. Pattern avoidance and the Bruhat order. *J. Combin. Theory Ser. A*, 114(5):888–905, 2007.

[766] A. Thue. Über unendliche Zeichenreihen. *Norske Vid. Selsk. Skr. I Math-Nat. Kl.*, 7:1–22, 1906.

[767] J. Treadway and D. Rawlings. Bernoulli trials and Mahonian statistics: a tale of two q's. *Math. Mag.*, 67(5):345–354, 1994.

[768] H. F. Trotter. Algorithm 115: PERM. *Communications of the ACM*, 5(8):434–435, 1962.

[769] W. T. Tutte. A census of planar maps. *Canad. J. Math.*, 67(15):249–271, 1963.

[770] W. T. Tutte. *Graph Theory*, volume 21 of *Encyclopedia of mathematics and its applications*. Addison-Wesley, Reading, 1984. With a foreword by C. St. J. A. Nash-Williams.

[771] H. Úlfarsson. Equivalence classes of permutations avoiding a pattern. *arXiv:1005.5419v1 [math.CO]*, 2010.

[772] H. Úlfarsson. A unification of permutation patterns related to Schubert varieties. *arXiv:1002.4361v1 [math.CO]*, 2010.

[773] H. Úlfarsson and A. Woo. Which Schubert varieties are local complete intersections? *In preparation*.

[774] W. Unger. On the *k*-colouring of circle graphs. *Proc. 5th Annual Symposium on Theoretical Aspects of Computer Science, 1988, Lecture Notes in Computer Sci., Vol. 294*, pages 61–72, Springer, Berlin, 1988.

[775] W. Unger. The complexity of colouring circle graphs. *Proc. 9th Annual Symposium on Theoretical Aspects of Computer Science, 1992, Lecture Notes in Computer Sci., Vol. 577*, pages 389–400, Springer, Berlin, 1992.

[776] J. H. van Lint and R. M. Wilson. *A Course in Combinatorics*. Cambridge University Press, Cambridge, 2001.

[777] V. R. Vatter. Permutations avoiding two patterns of length three. *Electron. J. Combin.*, 9(2):Research paper 6, 19 pp. (electronic), 2002/03. Permutation patterns (Otago, 2003).

[778] V. R. Vatter. *The enumeration and structure of permutation classes*. PhD thesis, Rutgers University, 2005.

[779] V. R. Vatter. Finitely labeled generating trees and restricted permutations. *J. Symbolic Comput.*, 41(5):559–572, 2006.

[780] V. R. Vatter. Small permutation classes. *arXiv:0712.4006v2 [math.CO]*, 2007.

[781] V. R. Vatter. Enumeration schemes for restricted permutations. *Combin. Probab. Comput.*, 17(1):137–159, 2008.

[782] V. R. Vatter. Finding regular insertion encodings for permutation classes. *arXiv:0911.2683v1 [math.CO]*, 2009.

[783] V. R. Vatter. Permutation classes of every growth rate above 2.48188. *Mathematika*, 56:182–192, 2010.

[784] V. R. Vatter and S. Waton. On partial well-order for monotone grid classes of permutations. *Order*, 2010. DOI: 10.1007/s11083-010-9165-1.

[785] A. Vella. Pattern avoidance in permutations: linear and cyclic orders. *Electron. J. Combin.*, 9(2):Research paper 18, 43 pp. (electronic), 2002/03. Permutation patterns (Otago, 2003).

[786] A. M. Vershik and S. V. Kerov. Asymptotics of the Plancherel measure of the symmetric group and the limiting form of Young tableaux. *Soviet Math. Dokl.*, 18:527–531, 1977.

[787] X. G. Viennot. A bijective proof for the number of Baxter permutations. *Sém. Lothar. Combin., Le Klebach*, 1981.

[788] X. G. Viennot. A survey of polyominoes enumeration. *SFCA Proceedings, Montréal*, pages 399–420, 1992.

[789] X. G. Viennot. Catalan tableaux and the asymmetric exclusion process. *Proceedings of the conference Formal Power Series and Algebraic Combinatorics*, 2007.

[790] M. L. Wachs. The major index polynomial for conjugacy classes of permutations. *Discrete Math.*, 91(3):283–293, 1991.

[791] T. Walsh. Gray codes for involutions. *J. Combin. Math. Combin. Comput.*, 36:95–118, 2001.

[792] R. Warlimont. Permutations avoiding consecutive patterns. *Ann. Univ. Sci. Budapest. Sect. Comput.*, 22:373–393, 2003.

[793] R. Warlimont. Permutations avoiding consecutive patterns. II. *Arch. Math. (Basel)*, 84(6):496–502, 2005.

[794] D. Warren. Optimal packing behavior of some 2-block patterns. *Ann. Comb.*, 8(3):355–367, 2004.

[795] D. Warren. Packing densities of more 2-block patterns. *Adv. in Appl. Math.*, 36(2):202–211, 2006.

[796] D. I. Warren and E. Seneta. Peaks and Eulerian numbers in a random sequence. *J. Appl. Probab.*, 33(1):101–114, 1996.

[797] S. D. Waton. *On permutation classes defined by token passing networks, gridding matrices and pictures: three flavours of involvement.* PhD thesis, University of St Andrews, 2007.

[798] F. Wei. The weak Bruhat order and separable permutations. *arXiv:1009.5740v1 [math.CO]*, 2010.

[799] J. West. *Permutations with forbidden subsequences; and, stack-sortable permutations*. PhD thesis, Massachusetts Institute of Technology, 1990.

[800] J. West. Sorting twice through a stack. *Theoret. Comput. Sci.*, 117(1-2):303–313, 1993. Conference on Formal Power Series and Algebraic Combinatorics (Bordeaux, 1991).

[801] J. West. Generating trees and the Catalan and Schröder numbers. *Discrete Math.*, 146(1–3):247–262, 1995.

[802] J. West. Generating trees and forbidden subsequences. *Discrete Math.*, 157(1–3):363–374, 1996.

[803] H. Widom. On the limiting distribution for the length of the longest alternating sequence in a random permutation. *Electron. J. Combin.*, 13(1):Research Paper 25, 7, 2006.

[804] H. S. Wilf. Ascending subsequences of permutations and the shapes of Young tableaux. *J. Combin. Theory Ser. A*, 60:155–157, 1992.

[805] H. S. Wilf. The patterns of permutations. *Discrete Math.*, 257(2-3):575–583, 2002. Kleitman and combinatorics: a celebration (Cambridge, MA, 1999).

[806] L. Williams. Enumeration of totally positive Grassmann cells. *Advances in Math.*, 190:319–342, 2005.

[807] J. Wolper. A combinatorial approach to the singularities of Schubert varieties. *Adv. Math.*, 76:184–193, 1989.

[808] A. Woo. Permutations with Kazhdan-Lusztig polynomial $P_{id,w}(q) = 1 + q^h$. *Electron. J. Combin.*, 16(2, Special volume in honor of Anders Bjorner):Research Paper 10, 32, 2009. With an appendix by Sara Billey and Jonathan Weed.

[809] A. Woo and A. Yong. When is a Schubert variety Gorenstein? *Adv. in Math.*, 207(1):205–220, 2006.

[810] A. Woo and A. Yong. Governing singularities of Schubert varieties. *J. Algebra*, 320(2):495–520, 2008.

[811] S. H. F. Yan. On a conjecture about enumerating $(2 + 2)$-free posets. *arXiv:1006.1226v1 [math.CO]*, 2010.

[812] D. Zagier. Vassiliev invariants and a strange identity related to the Dedekind eta-function. *Topology*, 40:945–960, 2001.

[813] C. Zara. Parking functions, stack-sortable permutations, and spaces of paths in the Johnson graph. *Electron. J. Combin.*, 9(2):Research paper 11, 11 pp. (electronic), 2002/03. Permutation patterns (Otago, 2003).

[814] D. Zeilberger. Holonomic systems for special functions. *J. Comput. and Appl. Math.*, 32:321–368, 1990.

[815] D. Zeilberger. A proof of Julian West's conjecture that the number of two-stack-sortable permutations of length n is $2(3n)!/((n+1)!(2n+1)!)$. *Discrete Math.*, 102(1):85–93, 1992.

[816] D. Zeilberger. Enumeration schemes, and more importantly, their automatic generation. *Ann. Combin.*, 2:185–195, 1998.

Index